A Collection of Articles in Honor of Phillip Griffith

Edited by

Luchezar L. Avramov
Sankar P. Dutta
Hans-Bjørn Foxby

Department of Mathematics
University of Illinois at Urbana-Champaign

Originally published as Volume 51 (2007), Issue 1, of the Illinois Journal of Mathematics (ISSN 0019-2082).

Cover design by Tori Corkery.

Photos courtesy of Hiram Paley. The photo on the back cover depicts Altgeld Hall, the historic building that houses the Mathematics Department of the University of Illinois at Urbana-Champaign and which has been Phil Griffith's "home" for most of his professional life.

Printed by Cushing-Malloy, Ann Arbor, Michigan.

ISBN 0-9746986-2-8

Illinois Journal of Mathematics
Department of Mathematics
University of Illinois at Urbana-Champaign
Urbana, IL 61801, USA
`http://www.math.uiuc.edu/ijm`

Table of Contents

Foreword

Phil Griffith came to the University of Illinois at Urbana-Champaign in 1970, just two years after receiving his Ph.D. degree. On September 16–18, 2005, the University hosted a *Commutative Algebra Conference honoring the Contributions of Phillip Griffith*, organized by Andrew Kustin, Matthew Miller, Anurag Singh and Sandra Spiroff, with financial support from the National Security Agency, the University of Illinois, and the University of South Carolina. Fifteen invited lectures were given, several of them by former students, covering many aspects of commutative algebra on which Phil's work has exerted a significant influence.

Two announcements were made at the end of the meeting. One was Phil's decision to retire. The other was the decision of the *Illinois Journal of Mathematics* to mark that occasion, Phil Griffith's 35 years of service to the University of Illinois, and his contributions to mathematics with the publication of a special issue. Invitations to contribute were extended to the speakers at the meeting, its organizers, and to a number of experts whose work connects with Phil's own research. The result is the present volume.

We want to thank the authors for accepting our invitations, the referees for responding to our requests, the Managing Editor, A. J. Hildebrand, for support and valuable advice, and the Editorial Assistant, Debbie Broadrick, for expert help throughout the process.

Luchezar L. Avramov, Lincoln, Nebraska
Sankar P. Dutta, Urbana, Illinois
Hans-Bjørn Foxby, Copenhagen, Denmark

August, 2007

Illinois Journal of Mathematics
Volume 51, Number 1, Spring 2007, Pages vii–ix
S 0019-2082

MATHEMATICAL BIOGRAPHY OF PHIL GRIFFITH

By the late 1950s and throughout the 1960s the way in which problems in algebra were formulated and solved was greatly influenced by methods of homological algebra. Cartan and Eilenberg's seminal text, Homological Algebra, was published in 1956. No doubt for that reason Phil Griffith's early forays into mathematical research centered on problems that had a hint of homological flavor and for the most part could be translated into that language. His PhD advisor, Paul Hill, was a master in transfinite methods that he mostly applied to questions arising in the theory of abelian groups. The combination of view points proved to be lucky for the research fortunes of Griffith during the summer of 1967 in which he solved the Baer splitting problem for abelian groups (an abelian group G was called a Baer group provided any extension of a torsion group by G must be split exact; the problem had remained open for 30 years). Other results in the context of abelian groups were obtained by Griffith during this early period, e.g., he was first to construct, for prescribed $n > 0$, non-free abelian groups in which each subgroup of cardinality less than "aleph n" is free (a problem from Fuch's volumes on abelian groups). Perhaps it is fair to say that Griffith's main research success came from attempts to solve problems rather than to formulate and develop complex theoretical foundation. In 1970 he wrote a monograph, Infinite Abelian Group Theory, that was published by University of Chicago Press.

In 1968–1970 Griffith held a postdoctoral position at University of Chicago. Under the tutelage of Irving Kaplansky he developed a keen interest in the work of M. Auslander and H. Bass. While at Chicago, Griffith was fortunate to be in the company of D. Eisenbud, E.G. Evans and J.C. Robson. The research accomplished by Griffith during his Chicago experience was mainly focused on questions surrounding finite dimensional algebras and Artin rings. He collaborated with Eisenbud and Robson on these topics. However his interest in commutative algebra had begun in earnest, and he would later begin a ten year collaboration with E.G. Evans in that subject.

In 1970 Griffith became a member of the University of Illinois faculty, and in 1971 he was awarded an Alfred P. Sloan Foundation fellowship. In 1972 he was a visiting faculty member at Aarhus Universitet (Denmark). It was here while listening to lectures of many algebraic geometers (especially the French algebraic geometers) that Griffith's research became directed towards the newly developed area one might call "homological commutative algebra".

He collaborated with R. Fossum and H.-B. Foxby, the outgrowth of which was a joint article published in IHES on the structure of minimal injective resolutions (also with I. Reiten). The "homological era" in commutative algebra really began to flourish during this period as a result of the immense influence and success of sheaf theoretic tools developed by Serre and Grothendieck in tackling problems in algebraic geometry. In 1968 C. Peskine and L. Szpiro wrote their famous ground-breaking joint work published in IHES based on the methods of Serre and Grothendieck together with the "method of the Frobenius" inspired by E. Kunz. In this article they solved several of the "homological conjectures" in important cases, and they established relationships between many of these conjectures. By 1973 a second major breakthrough was to take place; namely M. Hochster would describe a procedure to construct a class of infinitely generated modules known as "maximal Cohen-Macaulay modules". These modules would have many applications by way of a general notion of "intersection theory" in settling a great number of the homological conjectures, and would later play a major role in the Evans-Griffith collaboration on the theory of syzygies. Hochster's construction (valid for equicharacteristic local rings) was based in part on his uncanny ability to formulate and solve "large" systems of equations for theoretical purposes.

Around 1975 Griffith teamed up with E.G. Evans to work on a "syzygy problem" about which Evans had been ruminating. The problem had been first stated in an unpublished PhD thesis by P. Hackman (Stockholm, Sweden) as a problem about "exterior powers and homology". The problem comes down to showing that a kth syzygy of finite projective dimension has rank at least k—or else it is free. In 1980 Evans and Griffith solved the problem for equicharacteristic local rings (later they discovered a proof in the mixed characteristic case for standard graded rings). A key point in their solution was to establish that order ideals of minimal generators for such non-free syzygy modules have height at least k. In turn the order ideal result rested upon proving an important fact about the length of certain finite free complexes that has come to be known as the "Improved New Intersection Theorem" (explicitly stated and named by Hochster). Although Hochster and Dutta (and later Hochster and Huneke) gave alternate proofs of the syzygy theorem, Evans and Griffith used Hochster's maximal Cohen-Macaulay modules as a basic tool for their arguments. Their original proof appeared in the Annals of Mathematics in 1981. In 1985 they co-authored a monograph, Syzygies, published by the London Mathematical Society. In all, Evans and Griffith collaborated on 15 articles related to the syzygy theorem. One of the pleasant byproducts of this theorem was the number of diverse applications that resulted, ranging from non-vanishing of cohomology of vector bundles to Cohen-Macaulay properties of three-generated ideals.

In 1983–1984 Fossum and Griffith were co-organizers for an NSF sponsored "Special Year" in commutative algebra. It should be mentioned that

Griffith benefited from year long visits at Illinois by renowned commutative algebraists, M. Auslander, L. Avramov, W. Bruns, H.-B. Foxby, C. Huneke, R.Y. Sharp and A. Singh. His weekly tea/coffee meetings with S. Dutta also have served as a time for amusement and inspiration.

Griffith, a native of Danville, Illinois, graduated from Northern Michigan University with a BS degree in 1963. After two years and receiving an MS degree at University of Missouri, he graduated from University of Houston with a PhD degree in 1968. He was married to his wife Judith in July 1960. They have two daughters, Keri and Lesli, and three grandchildren.

During his thirty-five years as a regular faculty member Griffith served on virtually every committee in the department. He served as research advisor for ten PhD students and as Director of Graduate Studies 2000-05.

Illinois Journal of Mathematics
Volume 51, Number 1, Spring 2007, Pages 1–20
S 0019-2082

CONSTRUCTING MODULES WITH PRESCRIBED COHOMOLOGICAL SUPPORT

LUCHEZAR L. AVRAMOV AND SRIKANTH B. IYENGAR

To Phil Griffith, algebraist and friend.

ABSTRACT. A cohomological support, $\operatorname{Supp}^*_{\mathcal{A}}(M)$, is defined for finitely generated modules M over a left noetherian ring R, with respect to a ring $\mathcal{A}$ of central cohomology operations on the derived category of R-modules. It is proved that if the $\mathcal{A}$-module $\operatorname{Ext}^*_R(M,M)$ is noetherian and $\operatorname{Ext}^*_R(M,R) = 0$ for $i \gg 0$, then every closed subset of $\operatorname{Supp}^*_{\mathcal{A}}(M)$ is the support of some finitely generated R-module. This theorem specializes to known realizability results for varieties of modules over group algebras, over local complete intersections, and over finite dimensional algebras over a field. The theorem is also used to produce large families of finitely generated modules of finite projective dimension over commutative local noetherian rings.

Introduction

Quillen introduced methods from algebraic geometry to the study of cohomology rings of finite groups in a seminal paper, [21]. His ideas and techniques have led to the appearance of a number of highly developed theories, which provide insight into the structure of an algebraic object through some geometric 'variety' attached to it. Use of such geometric invariants has been crucial to progress on a number of difficult problems.

Variety theories share certain formal properties needed in applications. Some of them guarantee that homologically similar modules, such as all syzygy modules of a given module, have the same variety. Modules with distinct varieties are therefore expected to exhibit quantifiable differences in homological behavior. For this reason, a description of all the varieties produced by a given theory is a useful tool for classifying homological patterns.

The prototype theory applies to all finite dimensional representations of a finite group; see [8] for a detailed exposition. It has been extended to

Received xxx; received in final form July 30, 2007.
2000 *Mathematics Subject Classification.* 13D03, 13D05, 13H10, 16E40, 20C05, 20J06.
Research partly supported by NSF grants DMS 0201904 (L.L.A), DMS 0602498 (S.I.).

representations of finite dimensional cocommutative Hopf algebras, [13], [25]. Parallel theories have been constructed for finitely generated modules over finite dimensional self-injective algebras, [12], [22], and over local complete intersection rings, [1], [2]. Historically, in each concrete case the proofs of the formal properties of a theory and of the relevant realizability theorem have involved delicate arguments specific to that context.

We are interested in modules over a fixed associative ring R.

The vehicle for passing from algebra to geometry is provided by a choice of commutative graded ring $\mathcal{A}$ of central cohomology operations on the derived category of R. In the examples above there are natural candidates for $\mathcal{A}$: the even cohomology ring of a group (or a Hopf algebra); the even subalgebra of the Hochschild cohomology of an associative algebras; the polynomial ring of Gulliksen operators over a complete intersection. However, other choices are possible and sometimes are desirable.

For each pair (M,N) of R-modules the graded group $\operatorname{Ext}^*_R(M,N)$ has a natural structure of graded $\mathcal{A}$-module. The set

$$\operatorname{Supp}^*_{\mathcal{A}}(M,N) = \{\mathfrak{p} \in \operatorname{Proj}\mathcal{A} \mid \operatorname{Ext}^*_R(M,N)_{\mathfrak{p}} \neq 0\},$$

where $\operatorname{Proj}\mathcal{A}$ is the space of all essential homogeneous prime ideals in $\mathcal{A}$ with the Zariski topology, is called the *cohomological support* of (M,N). The cohomological support of M is the set $\operatorname{Supp}^*_{\mathcal{A}}(M,M)$.

The principal contribution of this work is a method for constructing modules with prescribed cohomological supports. Part of our main result reads:

THEOREM 1. *Let R be a noetherian ring and let M and N be finite R-modules, such that the graded $\mathcal{A}$-module $\operatorname{Ext}^*_R(M,N)$ is noetherian.*

If $\operatorname{Ext}^i_R(M,R)=0$ holds for all $i \gg 0$, then for every closed subset X of $\operatorname{Supp}^*_{\mathcal{A}}(M,N)$ there exist finite R-modules M_X and N_X such that

$$\operatorname{Supp}^*_{\mathcal{A}}(M_X,N) = X = \operatorname{Supp}^*_{\mathcal{A}}(M,N_X)\,.$$

Moreover, when $N=M$ one can choose $N_X=M_X$.

Suitable specializations of Theorem 1 yield several known realizibility results: See Section 5 for Hopf algebras and Section 6 for associative algebras. Their earlier proofs were modeled on Carlson's Tensor Product Theorem [11] for varieties over group algebras; they rely heavily on the nature of R (in the first case) or on that of $\mathcal{A}$ (in the second).

Theorem 1 is proved in Section 4, based on work in Sections 1 and 3. Our argument requires few structural restrictions on R and none on $\mathcal{A}$ itself. The crucial input is the noetherian property of $\operatorname{Ext}^*_R(M,N)$ as a module over $\mathcal{A}$.

Another application of Theorem 1 goes into a completely different direction:

THEOREM 2. *Let $(Q,\mathfrak{q},k)$ be a commutative noetherian local ring and $\boldsymbol{f}$ a Q-regular sequence of length c contained in $\mathfrak{q}^2$.*

For $R = Q/Q\boldsymbol{f}$ and $\bar{k}$ an algebraic closure of k there exists a map

$$V\colon \left\{\begin{array}{c} \textit{isomorphism classes } [M] \\ \textit{of finite } R\textit{-modules with} \\ \operatorname{proj\,dim}_Q M < \infty \end{array}\right\} \longrightarrow \left\{\begin{array}{c} \textit{closed algebraic} \\ \textit{sets } \ X \subseteq \mathbb{P}^{c-1}_{\bar{k}} \\ \textit{defined over } k \end{array}\right\}$$

with the following properties:

(1) *V is surjective.*
(2) *$V([M]) = \varnothing$ if and only if* $\operatorname{proj\,dim}_R M < \infty$.
(3) *$V([M]) = V([\Omega^R_n(M)])$ for every syzygy module $\Omega^R_n(M)$.*
(4) *$V([M]) = V([M/\boldsymbol{x}M])$ for every M-regular sequence $\boldsymbol{x}$ in R.*

This result is surprising. Indeed, it exhibits large families of modules of finite projective dimension over any ring Q with $\operatorname{depth} Q \geq 2$, contrary to a commonly held perception that finite projective dimension is 'rare' over singular commutative rings. Furthermore, the remaining statements ascertain that modules mapping to distinct closed cones in $\bar{k}^c$ cannot be linked by any sequence of standard operations known to preserve finite projective dimension.

In Section 7 we prove Theorem 2, and deduce from it a recent theorem on the existence of cohomological varieties for modules over complete intersection local rings. For the latter we establish a descent result of independent interest.

In this paper varieties of modules are discussed in the broader context of varieties of complexes. The resulting marginal technical complications are easily offset by a gain in flexibility: We first realize a given set as the cohomological support of a bounded complex by using constructions whose effect is easy to track. To show that this set is also the support of a module we use 'syzygy complexes', a notion introduced and discussed in Section 1.

This paper is part of an ongoing study of cohomological supports of modules over general associative rings. In [5] we focus on proving existence of variety theories with desirable properties under a small set of conditions on a ring, its module(s), and a ring of central cohomological operators. The properties that have to be established are clarified in [9] by Benson, Iyengar, and Krause, who investigate a notion of support for triangulated categories equipped with an action by a central ring of operators. On the other hand, the methods of this paper can be adapted to prove realizability results in that context. Of particular interest is the case of certain monoidal categories, where work of Suarez-Alvarez, [24], provides natural candidates for rings of operators.

1. Syzygy complexes

In this section we recall a few basic concepts of DG homological algebra, following [4], and extend the notion of syzygy from modules to complexes.

Let R be an associative ring and $\mathsf{D}(R)$ the full derived category of left R-modules. We write $\simeq$ to indicate a quasi-isomorphism of complexes; these are the isomorphisms in $\mathsf{D}(R)$. The symbol $\cong$ is reserved for isomorphisms of complexes, and $\equiv$ is used to denote homotopy equivalences. Given a complex M of $\mathsf{D}(R)$, we write $\mathsf{Thick}_R(M)$ for its *thick closure*, that is to say, the intersection of the thick subcategories of $\mathsf{D}(R)$ containing M.

1.1. *Semiprojective complexes.* A complex P of R-modules is called *semiprojective* if $\operatorname{Hom}_R(P,-)$ preserves surjective quasi-isomorphisms; equivalently, if P is a complex of projective modules and $\operatorname{Hom}_R(P,-)$ preserves quasi-isomorphisms. The following properties are used in the proofs below.

Every quasi-isomorphism of semiprojective complexes is a homotopy equivalence. Every surjective quasi-isomorphism to a semiprojective complex has a left inverse. Every semiprojective complex C with $\mathrm{H}(C)=0$ is equal to $\mathsf{cone}(\mathrm{id}^B)$ for some complex B of projective modules with zero differential.

LEMMA 1.2. *If $\pi\colon P\to Q$ is a quasi-isomorphism of semiprojective complexes of R-modules and n is an integer, then there is a homotopy equivalence*

$$P_{\geqslant n}\oplus\Sigma^n Q'\equiv Q_{\geqslant n}\oplus\Sigma^n P',$$

where P' and Q' are projective R-modules.

Proof. Assume first that π is surjective. It then has a left inverse, hence one gets $P\cong Q\oplus E$ with $E=\operatorname{Ker}(\pi)$. This implies that E is semiprojective with $\mathrm{H}(E)=0$, and hence $E=\mathsf{cone}(\mathrm{id}^F)$ for some complex F of projective R-modules with $\partial^F=0$. Hence one gets a quasi-isomorphism

$$P_{\geqslant n}\cong Q_{\geqslant n}\oplus\mathsf{cone}(\mathrm{id}^{F_{\geqslant n}})\oplus\Sigma^n F_{n-1}\,.$$

The canonical map $P_{\geqslant n}\to Q_{\geqslant n}\oplus\Sigma^n F_{n-1}$ is thus a homotopy equivalence, as $\mathsf{cone}(\mathrm{id}^{F_{\geqslant n}})$ is homotopy equivalent to 0. This settles the surjective case.

In general, π factors as $P\to\widetilde{P}\xrightarrow{\psi}Q$, where $\widetilde{P}$ is equal to $P\oplus\Sigma^{-1}\mathsf{cone}(\mathrm{id}^Q)$ and ψ is the sum of π and the canonical surjection $\Sigma^{-1}\mathsf{cone}(\mathrm{id}^Q)\to Q$. Thus, ψ is a surjective quasi-isomorphism of semi-projective complexes. So is the canonical map $\widetilde{P}\to P$. The already settled case yields homotopy equivalences

$$P_{\geqslant n}\oplus\Sigma^n Q'\leftarrow\widetilde{P}\to Q_{\geqslant n}\oplus\Sigma^n P'\,.$$

for appropriate projective modules P' and Q'. □

1.3. *Syzygy complexes.* Let M be a complex of R-modules.

A *semiprojective resolution* of M is a quasi-isomorphism $P\to M$ from a semiprojective complex P. Every complex M has one, and it is unique up to homotopy equivalence. Thus, the preceding result may be viewed as a homotopical version of Shanuel's Lemma. Based on it, we introduce a homotopical version of the notion of syzygy module.

For each $n \in \mathbb{Z}$ let $\Omega_n^R(M)$ stand for any complex $\Sigma^{-n}(P_{\geqslant n})$, where P is a semiprojective resolution of M, and call it an nth *syzygy complex* of M over R. Its dependence on the choice of P is made precise by the preceding lemma.

Any complex P of projective modules with $P_i = 0$ for $i \ll 0$ is semiprojective. Thus, when M is an R-module and P is its projective resolution the complex $\Sigma^{-n}(P_{\geqslant n})$ is isomorphic in $\mathsf{D}(R)$ to an nth syzygy module of M.

The next lemma expands upon the last observation.

LEMMA 1.4. *If M is a complex of R-modules, $s = \sup\{i \mid \mathrm{H}_i(M) \neq 0\}$, and n is an integer with $n \geq s$, then $\Omega_n^R(M)$ is quasi-isomorphic to $\mathrm{H}_0(\Omega_n^R(M))$.*

Proof. Let $P \to N$ be a semiprojective resolution with $\Omega_n^R(M) = \Sigma^{-n}(P_{\geqslant n})$. For $i \geq n+1$ one has isomorphisms $\mathrm{H}_i(P_{\geqslant n}) \cong \mathrm{H}_i(P) \cong \mathrm{H}_i(M) = 0$, the first one of which comes from the exact sequence of complexes

(1.4.1) $$0 \to P_{<n} \to P \to P_{\geqslant n} \to 0\,. \qquad \square$$

1.5. *Cohomology.* Let M be a complex of R-modules and $P \to M$ a semiprojective resolution. For every complex N and each $i \in \mathbb{Z}$ the abelian group

$$\mathrm{Ext}_R^i(M,N) = \mathrm{H}_{-i}(\mathrm{Hom}_R(P,N)) = \mathrm{H}^i(\mathrm{Hom}_R(P,N))$$

is independent of the choice of resolution P; see 1.1. It is a module over R^{c}, the center of the ring R. For modules M, N this is the usual gadget; see 1.3.

Over noetherian rings syzygy modules inherit finiteness properties of the original module. We show that syzygy complexes behave similarly.

LEMMA 1.6. *If R is a noetherian ring and M is a complex with $\mathrm{H}(M)$ a finite R-module, then one can find a syzygy complex $\Omega_n^R(M)$ in $\mathsf{Thick}_R(M \oplus R)$.*

Furthermore, for every complex $C \in \mathsf{Thick}_R(M \oplus R)$ the following hold.

1. *The R-module $\mathrm{H}(C)$ is noetherian.*
2. $\mathrm{Ext}_R^{\gg 0}(M,N) = 0$ *for a bounded complex N implies* $\mathrm{Ext}_R^{\gg 0}(C,N) = 0$.
3. $\mathrm{Ext}_R^{\gg 0}(M,R) = 0$ *implies* $\mathrm{Ext}_R^{\gg 0}(C,F) = 0$ *for every projective R-module F.*

Proof. Under the hypotheses on R and M, one can choose a semiprojective resolution $P \simeq M$ with each P_i finite and $P_i = 0$ for $i \ll 0$. It follows that $P_{<n}$ is in $\mathsf{Thick}_R(R)$, so the exact sequence (1.4.1) yields $P_{\geqslant n} \in \mathsf{Thick}_R(M \oplus R)$.

The complexes L with $\mathrm{H}(L)$ finite form a thick subcategory of $\mathsf{D}(R)$. As it contains M and R, it contains $\mathsf{Thick}_R(M \oplus R)$ as well. This proves (1). A similar argument settles (2). For M and F as in (3) there is an isomorphism

$$\mathrm{Ext}_R^*(M,F) \cong \mathrm{Ext}_R^*(M,R) \otimes_R F\,,$$

which one can get by using the resolution P above. Thus, $\mathrm{Ext}_R^{\gg 0}(M,R) = 0$ implies $\mathrm{Ext}_R^{\gg 0}(M,F) = 0$. Now (2) yields $\mathrm{Ext}_R^{\gg 0}(C,F) = 0$, as desired. $\square$

2. Graded rings

Here we describe notation and terminology for dealing with graded objects.

2.1. *Graded modules.* Let $\mathcal{A}$ be a commutative ring that is *non-negatively graded*: $\mathcal{A} = \bigoplus_{i\in\mathbb{Z}} \mathcal{A}^i$ with $\mathcal{A}^i\mathcal{A}^j \subseteq \mathcal{A}^{i+j}$ and $\mathcal{A}^i = 0$ for $i < 0$.

Modules over $\mathcal{A}$ are $\mathbb{Z}$-graded: $\mathcal{M} = \bigoplus_{j\in\mathbb{Z}} \mathcal{M}^j$ with $\mathcal{A}^i\mathcal{M}^j \subseteq \mathcal{M}^{i+j}$. For such an $\mathcal{M}$ *finite* means finitely generated, *eventually noetherian* means $\mathcal{M}^{\geqslant j}$ is noetherian for $j \gg 0$, and *eventually zero* means $\mathcal{M}^{\geqslant j} = 0$ for $j \gg 0$.

The *annihilator* of $\mathcal{M}$ is the set $\operatorname{ann}_{\mathcal{A}} \mathcal{M} = \{a \in \mathcal{A} \mid a\mathcal{M} = 0\}$. It is a homogeneous ideal in $\mathcal{A}$, so $\mathcal{A}/\operatorname{ann}_{\mathcal{A}} \mathcal{M}$ is a graded ring and $\mathcal{M}$ is a graded module over it. When $\mathcal{M}$ is noetherian so is $\mathcal{A}/\operatorname{ann}_{\mathcal{A}} \mathcal{M}$, so modulo $\operatorname{ann}_{\mathcal{A}} \mathcal{M}$ every ideal in $\mathcal{A}$ is generated by finitely many homogeneous elements.

2.2. *Supports.* Let $\operatorname{Spec} \mathcal{A}$ be the space of prime ideals of $\mathcal{A}$, with the Zariski topology. For an $\mathcal{A}$-module $\mathcal{M}$, set

$$\begin{aligned}
\operatorname{Supp}_{\mathcal{A}} \mathcal{M} &= \{\mathfrak{p} \in \operatorname{Spec} \mathcal{A} \mid \mathcal{M}_{\mathfrak{p}} \neq 0\}\,;\\
\operatorname{Proj} \mathcal{A} &= \{\mathfrak{p} \in \operatorname{Spec} \mathcal{A} \mid \mathfrak{p} \text{ homogeneous and } \mathfrak{p} \not\supseteq \mathcal{A}^{\geqslant 1}\}\,;\\
\operatorname{Supp}^{+}_{\mathcal{A}} \mathcal{M} &= \operatorname{Supp}_{\mathcal{A}} \mathcal{M} \cap \operatorname{Proj} \mathcal{A}\,.
\end{aligned}$$

The following properties of graded $\mathcal{A}$-modules $\mathcal{L}, \mathcal{M}$, and $\mathcal{N}$ follow from the definition of support and the exactness of localization.

(1) If $\mathcal{L} \xrightarrow{\iota} \mathcal{M} \xrightarrow{\varepsilon} \mathcal{N}$ is an exact sequence, then

$$\operatorname{Supp}^{+}_{\mathcal{A}} \mathcal{M} \subseteq \operatorname{Supp}^{+}_{\mathcal{A}} \mathcal{L} \cup \operatorname{Supp}^{+}_{\mathcal{A}} \mathcal{N}\,;$$

equality holds when ι is injective and ε is surjective.

(2) For each $i \in \mathbb{Z}$, one has $\operatorname{Supp}^{+}_{\mathcal{A}} (\mathcal{M}^{\geqslant i}) = \operatorname{Supp}^{+}_{\mathcal{A}} \mathcal{M}$.

(3) If some $\mathcal{M}^{\geqslant n}$ is finite, say, if $\mathcal{M}$ is eventually noetherian, then

$$\operatorname{Supp}^{+}_{\mathcal{A}} \mathcal{M} = \{\mathfrak{p} \in \operatorname{Proj} \mathcal{A} \mid \mathfrak{p} \supseteq \operatorname{ann}_{\mathcal{A}}(\mathcal{M}^{\geqslant i})\}$$

holds for every $i \geq n$; thus, $\operatorname{Supp}^{+}_{\mathcal{A}} \mathcal{M}$ is a closed subset of $\operatorname{Proj} \mathcal{A}$.

(4) If the $\mathcal{A}$-modules $\mathcal{M}$ and $\mathcal{N}$ are finite, then

$$\operatorname{Supp}^{+}_{\mathcal{A}}(\mathcal{M} \otimes_{\mathcal{A}} \mathcal{N}) = \operatorname{Supp}^{+}_{\mathcal{A}} \mathcal{M} \cap \operatorname{Supp}^{+}_{\mathcal{A}} \mathcal{N}\,.$$

(5) If $\mathcal{M}$ is eventually zero, then $\operatorname{Supp}^{+}_{\mathcal{A}} \mathcal{M} = \varnothing$. The converse holds when $\mathcal{M}$ is eventually noetherian over $\mathcal{A}$.

In some cases, supports have a natural geometric interpretation.

2.3. *Varieties.* Let k be a field and $\bar{k}$ an algebraic closure of k. Assume that the graded ring $\mathcal{A}$ has $\mathcal{A}^0 = k$ and is generated over k by finitely many homogeneous elements of positive degree. For each graded $\mathcal{A}$-module $\mathcal{M}$ set

$$V_{\mathcal{A}}(\mathcal{M}) = \big(\operatorname{Supp}_{\bar{\mathcal{A}}} (\mathcal{M} \otimes_k \bar{k}) \cap \operatorname{Max} \bar{\mathcal{A}}\big) \cup \{\bar{\mathcal{A}}^{\geqslant 1}\},$$

where $\bar{\mathcal{A}}$ denotes the ring $\mathcal{A} \otimes_k \bar{k}$ and $\operatorname{Max} \bar{\mathcal{A}}$ the set of its maximal ideals.

Let $\mathcal{M}$ be a finite graded $\mathcal{A}$-module. The subset $V_{\mathcal{A}}(\mathcal{M})$ of $\operatorname{Max}\bar{\mathcal{A}}$ then is closed in the Zariski topology; it is also k-rational and conical, in the sense that it can be defined by homogeneous elements in $\mathcal{A}$. The Nullstellensatz implies that each one of the sets $V_{\mathcal{A}}(\mathcal{M})$ and $\operatorname{Supp}^+_{\mathcal{A}} \mathcal{M}$ determines the other.

The graded rings and modules of interest in this paper are generated by cohomological constructions, which we recall below.

2.4. *Products in cohomology.* Let M and N be complexes of R-modules, and let $P \to M$ and $Q \to N$ be semiprojective resolutions. For each $i \in \mathbb{Z}$ one has

$$\mathrm{H}_{-i}(\operatorname{Hom}_R(P,Q)) = \mathrm{H}^i(\operatorname{Hom}_R(P,Q)) \cong \mathrm{H}^i(\operatorname{Hom}_R(P,N)) = \operatorname{Ext}^i_R(M,N)$$

in view of properties discussed in 1.1 and 1.5. We set

$$\operatorname{Ext}^*_R(M,N) = \bigoplus_{i\in\mathbb{Z}} \operatorname{Ext}^i_R(M,N)\,.$$

This is a graded module over R^{c}, the center of the ring R.

Composition of homomorphisms turns $\operatorname{Hom}_R(Q,Q)$ and $\operatorname{Hom}_R(P,P)$ into DG algebras over the center R^{c} of R, and $\operatorname{Hom}_R(P,Q)$ into a left DG module over the first and a right DG module over the second. The actions are compatible, so $\operatorname{Ext}^*_R(N,N)$ and $\operatorname{Ext}^*_R(M,M)$ become graded R^{c}-algebras and $\operatorname{Ext}^*_R(M,N)$ a left-right graded bimodule over them.

These structures do not depend on choices of resolutions.

3. Cohomological supports

In this section R denotes an associative ring.

3.1. *Cohomology operations.* A *ring of central cohomology operations* is a commutative graded ring $\mathcal{A}$ equipped with a homomorphism of graded rings

$$\zeta_M \colon \mathcal{A} \longrightarrow \operatorname{Ext}^*_R(M,M)$$

for each $M \in \mathsf{D}(R)$, such that for all $N \in \mathsf{D}(R)$ and $\xi \in \operatorname{Ext}^*_R(M,N)$ one has

$$\xi \cdot \zeta_M(a) = \zeta_N(a) \cdot \xi \quad \text{for every} \quad a \in \mathcal{A}\,. \tag{3.1.1}$$

For $N = M$ this formula implies that $\zeta_M(\mathcal{A})$ is in the center of $\operatorname{Ext}^*_R(M,M)$.

We assume that $\mathcal{A}$ is non-negatively graded and that $\mathcal{A}^i = 0$ for i odd or $2\mathcal{A} = 0$; this hypothesis covers existing examples and avoids sign trouble.

3.2. *Scalars.* Using the standard identifications of rings

$$\operatorname{Ext}^*_R(R,R) = \operatorname{Hom}_R(R,R) = R^{\circ}\,,$$

where R° denotes the opposite ring of R, one sees from (3.1.1) that the homomorphism of rings $\zeta_R \colon \mathcal{A} \to R^{\circ}$ maps every element $a \in \mathcal{A}^0$ to the center of R°. We identify the centers of R° and R. Formula (3.1.1) then shows that the action of a on $\operatorname{Ext}^*_R(M,N)$ coincides with the maps induced by left multiplication with $\zeta_R(a)$ on M or on N.

For the next definition we use the notion of support introduced in 2.2.

3.3. *Cohomological supports.* Let $\mathcal{A}$ be a graded ring of central cohomology operations, as above. For each pair (M,N) of complexes we call the subset

$$\operatorname{Supp}^*_{\mathcal{A}}(M,N) = \operatorname{Supp}^+_{\mathcal{A}}\left(\operatorname{Ext}^*_R(M,N)\right) \subseteq \operatorname{Proj}\mathcal{A}$$

the *cohomological support* of (M,N). The cohomological support of M is

$$\operatorname{Supp}^*_{\mathcal{A}}(M) = \operatorname{Supp}^*_{\mathcal{A}}(M,M)\,.$$

The theorem below is the main result of this section.

THEOREM 3.4. *Let M and N be complexes of R-modules.*

If the graded $\mathcal{A}$-module $\operatorname{Ext}^*_R(M,N)$ is noetherian, then for every closed subset X of $\operatorname{Supp}^*_{\mathcal{A}}(M,N)$ there exist complexes M_X in $\mathsf{Thick}_R(M)$ and N_X in $\mathsf{Thick}_R(N)$, such that the following equalities hold:

$$X = \operatorname{Supp}^*_{\mathcal{A}}(M_X,N) = \operatorname{Supp}^*_{\mathcal{A}}(M_X,N_X) = \operatorname{Supp}^*_{\mathcal{A}}(M,N_X)\,.$$

Moreover, when $N = M$ one can take $N_X = M_X$.

The proof appears at the end of the section. Some of the preparatory material is used repeatedly throughout the paper.

Let d be an integer. The dth *shift* of a complex M is the complex ΣM with $(\Sigma^d M)_n = M_{n-d}$ for all n and $\partial^{\Sigma^d M} = (-1)^d\partial^M$. The dth *twist* of a graded $\mathcal{A}$-module $\mathcal{M}$ is the graded module $\mathcal{M}(d)$ with $\mathcal{M}(d)^j = \mathcal{M}^{d+j}$ for all j.

3.5. *Functoriality.* Let M, M', M'' and N, N', N'' be complexes of R-modules.

There exist canonical isomorphisms of graded $\mathcal{A}$-modules:

$$\operatorname{Ext}^*_R(\Sigma M,N)(1) \cong \operatorname{Ext}^*_R(M,N) \cong \operatorname{Ext}^*_R(M,\Sigma N)(-1)\,; \tag{3.5.1}$$

$$\begin{aligned} \operatorname{Ext}^*_R(M' \oplus M'',N) &\cong \operatorname{Ext}^*_R(M',N) \oplus \operatorname{Ext}^*_R(M'',N)\,; \\ \operatorname{Ext}^*_R(M,N' \oplus N'') &\cong \operatorname{Ext}^*_R(M,N') \oplus \operatorname{Ext}^*_R(M,N'')\,. \end{aligned} \tag{3.5.2}$$

Indeed, basic properties of the functor $\operatorname{Hom}_{\mathsf{D}(R)}(-,-)$ show that for a fixed N (respectively, M) the canonical isomorphisms of graded R^{c}-modules are linear for the action of $\operatorname{Ext}^*_R(N,N)$ on the left (respectively, of $\operatorname{Ext}^*_R(M,M)$ on the right). They are $\mathcal{A}$-linear because of the centrality of $\mathcal{A}$; see (3.1.1).

Similarly, exact triangles $M' \to M \to M'' \to$ and $N' \to N \to N'' \to$ in $\mathsf{D}(R)$ induce exact sequences of graded $\mathcal{A}$-modules

$$\begin{gathered}\operatorname{Ext}^*_R(M'',N) \longrightarrow \operatorname{Ext}^*_R(M,N) \longrightarrow \operatorname{Ext}^*_R(M',N) \to \\ \operatorname{Ext}^*_R(M'',N)(1) \to \operatorname{Ext}^*_R(M,N)(1) \\ \operatorname{Ext}^*_R(M,N') \longrightarrow \operatorname{Ext}^*_R(M,N) \longrightarrow \operatorname{Ext}^*_R(M,N'') \to \\ \operatorname{Ext}^*_R(M,N')(1) \to \operatorname{Ext}^*_R(M,N)(1)\end{gathered} \tag{3.5.3}$$

Putting together the remarks in 2.2 and 3.5, one gets:

LEMMA 3.6. *In the notation of* 3.5 *the following statements hold.*

$$\operatorname{Supp}^*_{\mathcal{A}}(\Sigma M,N) = \operatorname{Supp}^*_{\mathcal{A}}(M,N) = \operatorname{Supp}^*_{\mathcal{A}}(M,\Sigma N)\,. \tag{3.6.1}$$

$$\begin{gathered}\operatorname{Supp}^*_{\mathcal{A}}(M'\oplus M'',N) = \operatorname{Supp}^*_{\mathcal{A}}(M',N) \cup \operatorname{Supp}^*_{\mathcal{A}}(M'',N)\,. \\ \operatorname{Supp}^*_{\mathcal{A}}(M,N'\oplus N'') = \operatorname{Supp}^*_{\mathcal{A}}(M,N') \cup \operatorname{Supp}^*_{\mathcal{A}}(M,N'')\,.\end{gathered} \tag{3.6.2}$$

$$\begin{gathered}\operatorname{Supp}^*_{\mathcal{A}}(M,N) \subseteq \operatorname{Supp}^*_{\mathcal{A}}(M',N) \cup \operatorname{Supp}^*_{\mathcal{A}}(M'',N)\,. \\ \operatorname{Supp}^*_{\mathcal{A}}(M,N) \subseteq \operatorname{Supp}^*_{\mathcal{A}}(M,N') \cup \operatorname{Supp}^*_{\mathcal{A}}(M,N'')\,.\end{gathered} \tag{3.6.3}$$

If $\operatorname{Ext}^*_R(M,N)$ *is eventually zero, then* $\operatorname{Supp}^*_{\mathcal{A}}(M,N) = \varnothing$. *The converse holds when* $\operatorname{Ext}^*_R(M,N)$ *is eventually noetherian over* $\mathcal{A}$. □

The exact sequences (3.5.3) imply the following statement:

LEMMA 3.7. *Let* M *be a complex of* R*-modules.*

The full subcategory of $\mathsf{D}(R)$ *consisting of complexes* L *with* $\operatorname{Ext}^*_R(M,L)$ *(respectively,* $\operatorname{Ext}^*_R(L,M)$*) eventually noetherian over* $\mathcal{A}$ *is thick.* □

3.8. *Mapping cone.* Let M be a complex of R-modules.

For each $\varphi \in \mathcal{A}^d$ the morphism $\zeta_M(\varphi)\colon M \to \Sigma^d M$ defines an exact triangle

$$M \xrightarrow{\zeta_M(\varphi)} \Sigma^d M \longrightarrow M/\!\!/\varphi \longrightarrow\,, \tag{3.8.1}$$

which is unique up to isomorphism.

Let N be a complex of R-modules and set $\mathcal{M} = \operatorname{Ext}^*_R(M,N)$. By (3.5.3) and (3.5.1), the triangle above yields an exact sequence of graded $\mathcal{A}$-modules

$$\mathcal{M}(-d-1) \longrightarrow \mathcal{M}(-1) \longrightarrow \operatorname{Ext}^*_R(M/\!\!/\varphi,N) \longrightarrow \mathcal{M}(-d) \longrightarrow \mathcal{M}\,. \tag{3.8.2}$$

The maps at both ends are given by multiplication with φ, so from (3.8.2) one can extract an exact sequence of graded $\mathcal{A}$-modules

$$0 \longrightarrow (\mathcal{M}/\mathcal{M}\varphi)(-1) \longrightarrow \operatorname{Ext}^*_R(M/\!\!/\varphi,N) \longrightarrow (0 :_{\mathcal{M}} \varphi)(-d) \longrightarrow 0\,. \tag{3.8.3}$$

Let $\boldsymbol{\varphi} = \varphi_1,\dots,\varphi_n$ be homogeneous elements in $\mathcal{A}$. Set $\boldsymbol{\varphi}' = \varphi_1,\dots,\varphi_{n-1}$. A complex $(M/\!\!/\boldsymbol{\varphi}')/\!\!/\varphi_n$ is defined uniquely up to isomorphism in $\mathsf{D}(R)$; we let $M/\!\!/\boldsymbol{\varphi}$ denote any such complex. Iterated references to (3.8.1) yield

$$M/\!\!/\boldsymbol{\varphi} \in \mathsf{Thick}_R(M)\,. \tag{3.8.4}$$

EXAMPLE 3.9. If $\varphi_1, \dots, \varphi_n$ are in $\mathcal{A}^0$, then $\zeta_M(\varphi_i)$ is the homothety $M \to M$ defined by the central element $z_i = \zeta_R(\varphi_i) \in R$; see 3.2. Thus, in $\mathsf{D}(R)$ one has $M/\!\!/\boldsymbol{\varphi} \simeq M \otimes_{R^c} K(\boldsymbol{z})$, where $K(\boldsymbol{z})$ is the Koszul complex on $\boldsymbol{z} = z_1, \dots, z_n$.

PROPOSITION 3.10. *Let M, N be complexes of R-modules and $\boldsymbol{\varphi} = \varphi_1, \dots, \varphi_n$ a sequence of homogeneous elements in $\mathcal{A}$.*

If the $\mathcal{A}$-module $\operatorname{Ext}^*_R(M,N)$ is eventually noetherian, then so are the $\mathcal{A}$-modules $\operatorname{Ext}^*_R(M/\!\!/\boldsymbol{\varphi}, N)$, $\operatorname{Ext}^*_R(M, N/\!\!/\boldsymbol{\varphi})$, and $\operatorname{Ext}^*_R(M/\!\!/\boldsymbol{\varphi}, N/\!\!/\boldsymbol{\varphi})$, and

$$\begin{aligned}\operatorname{Supp}^*_{\mathcal{A}}(M/\!\!/\boldsymbol{\varphi}, N) = \operatorname{Supp}^*_{\mathcal{A}}(M/\!\!/\boldsymbol{\varphi}, N/\!\!/\boldsymbol{\varphi}) &= \operatorname{Supp}^*_{\mathcal{A}}(M, N/\!\!/\boldsymbol{\varphi}) \\ &= \operatorname{Supp}^*_{\mathcal{A}}(M,N) \cap \operatorname{Supp}^+_{\mathcal{A}}(\mathcal{A}/\mathcal{A}\boldsymbol{\varphi})\,.\end{aligned}$$

Proof. It suffices to treat the case when $\boldsymbol{\varphi}$ has a single element, φ. From the exact sequence (3.8.3) one sees that $\operatorname{Ext}^*_R(M/\!\!/\varphi, N)$ is eventually noetherian.

Set $\mathcal{M} = \operatorname{Ext}^*_R(M,N)$. The inclusion below holds because $(0 :_{\mathcal{M}} \varphi)$ is a submodule of $\mathcal{M}$ and a module over $\mathcal{A}/\mathcal{A}\varphi$; the equality comes from 2.2(5):

$$\operatorname{Supp}^+_{\mathcal{A}}(0 :_{\mathcal{M}} \varphi) \subseteq \operatorname{Supp}^+_{\mathcal{A}} \mathcal{M} \cap \operatorname{Supp}^+_{\mathcal{A}}(\mathcal{A}/\mathcal{A}\varphi)\,;$$

$$\operatorname{Supp}^+_{\mathcal{A}}\left(\mathcal{M}/\mathcal{M}\varphi\right) = \operatorname{Supp}^+_{\mathcal{A}} \mathcal{M} \cap \operatorname{Supp}^+_{\mathcal{A}}(\mathcal{A}/\mathcal{A}\varphi)\,.$$

The exact sequence (3.8.3) and 2.2(2) now imply an equality

$$\operatorname{Supp}^*_{\mathcal{A}}(M/\!\!/\varphi, N) = \operatorname{Supp}^+_{\mathcal{A}} \mathcal{M} \cap \operatorname{Supp}^+_{\mathcal{A}}(\mathcal{A}/\mathcal{A}\varphi)\,.$$

By a similar argument, $\operatorname{Ext}^*_R(M, N/\!\!/\varphi)$ is eventually noetherian and one has

$$\operatorname{Supp}^*_{\mathcal{A}}(M, N/\!\!/\varphi) = \operatorname{Supp}^+_{\mathcal{A}} \mathcal{M} \cap \operatorname{Supp}^+_{\mathcal{A}}(\mathcal{A}/\mathcal{A}\varphi)\,.$$

The remaining equality is a formal consequence of those already available. □

Proof of Theorem 3.4. Set $\mathcal{M} = \operatorname{Ext}^*_R(M,N)$ and $\mathcal{I} = \operatorname{ann}_{\mathcal{A}} \mathcal{M}$.

From 2.2(3) and 2.2(4) we get

$$\operatorname{Supp}^*_{\mathcal{A}}(M,N) = \operatorname{Supp}^+_{\mathcal{A}} \mathcal{M} = \operatorname{Supp}^+_{\mathcal{A}}(\mathcal{A}/\mathcal{I}).$$

As $\mathcal{M}$ is noetherian the closed subset X of $\operatorname{Supp}^+_{\mathcal{A}}(\mathcal{A}/\mathcal{I})$ has the form

$$X = \operatorname{Supp}^+_{\mathcal{A}}(\mathcal{A}/\mathcal{I}) \cap \operatorname{Supp}^+_{\mathcal{A}}(\mathcal{A}/\mathcal{A}\boldsymbol{\varphi}),$$

where $\boldsymbol{\varphi}$ is a finite set of homogeneous elements of $\mathcal{A}$; see 2.1. Thus, one gets

$$X = \operatorname{Supp}^*_{\mathcal{A}}(M,N) \cap \operatorname{Supp}^+_{\mathcal{A}}(\mathcal{A}/\mathcal{A}\boldsymbol{\varphi})\,.$$

Choose complexes M_X and N_X representing $M/\!\!/\boldsymbol{\varphi}$ and $N/\!\!/\boldsymbol{\varphi}$, respectively. One has $M_X \in \mathsf{Thick}_R(M)$ and $N_X \in \mathsf{Thick}_R(N)$; see (3.8.4). Also, one gets

$$X = \operatorname{Supp}^*_{\mathcal{A}}(M_X, N) = \operatorname{Supp}^*_{\mathcal{A}}(M_X, N_X) = \operatorname{Supp}^*_{\mathcal{A}}(M, N_X)$$

from Proposition 3.10. Clearly, when $M = N$ one may choose $N_X = M_X$. □

4. Realizability by modules

In this section R is an associative ring and $\mathcal{A}$ is a ring of central cohomology operations on $\mathsf{D}(R)$; see 3.1. The principal result here is a partial enhancement of Theorem 3.4. It contains Theorem 1 from the introduction.

Existence Theorem 4.1. *Let R be a left noetherian ring.*

When M and N are complexes of R-modules with $\mathrm{H}(M)$ *and* $\mathrm{H}(N)$ *finite, and X is a closed subset of* $\operatorname{Supp}^*_{\mathcal{A}}(M,N)$ *the following hold.*

(1) *If* $\operatorname{Ext}^*_R(M,N)$ *is eventually noetherian over $\mathcal{A}$ (and X is irreducible), then there exists a finite (and indecomposable) R-module M_X with*

$$X = \operatorname{Supp}^*_{\mathcal{A}}(M_X,N) \quad \textit{and} \quad M_X \in \mathsf{Thick}_R(M\oplus R)\,.$$

(2) *If, furthermore,* $\operatorname{Ext}^*_R(M,R)$ *is eventually zero (and X is irreducible), then there exists a finite (and indecomposable) R-module N_X with*

$$X = \operatorname{Supp}^*_{\mathcal{A}}(M,N_X) \quad \textit{and} \quad N_X \in \mathsf{Thick}_R(N\oplus R)\,.$$

When $N = M$ one may choose $N_X = M_X$.

Proof. Using Theorem 3.4, choose complexes C in $\mathsf{Thick}_R(M)$ and D in $\mathsf{Thick}_R(N)$, satisfying $\operatorname{Supp}^*_{\mathcal{A}}(C,N) = X = \operatorname{Supp}^*_{\mathcal{A}}(M,D)$.

By Lemma 1.6(1), the R-modules $\mathrm{H}(C)$ and $\mathrm{H}(D)$ are noetherian, so one has $\mathrm{H}_{\geqslant n}(C) = 0 = \mathrm{H}_{\geqslant n}(D)$ for some n. Lemma 1.6 provides syzygy complexes $\Omega^R_n(C)$ in $\mathsf{Thick}_R(M\oplus R)$ and $\Omega^R_n(D)$ in $\mathsf{Thick}_R(N\oplus R)$. Another application of Lemma 1.6(1) shows that the following R-modules are finite:

$$M_X = \mathrm{H}_0(\Omega^R_n(C)) \quad \text{and} \quad N_X = \mathrm{H}_0(\Omega^R_n(D))\,.$$

Lemma 1.4 yields $\Omega^R_n(C) \simeq M_X$ and $\Omega^R_n(D) \simeq N_X$.

(1) comes from the equalities below, the second one given by Lemma 4.2(1):

$$X = \operatorname{Supp}^*_{\mathcal{A}}(C,N) = \operatorname{Supp}^*_{\mathcal{A}}(\Omega^R_n(C),N) = \operatorname{Supp}^*_{\mathcal{A}}(M_X,N)\,.$$

(2) As $\operatorname{Ext}^*_R(M,R)$ is eventually zero, so is $\operatorname{Ext}^*_R(M,F)$; see Lemma 1.6(3). Thus, referring to Lemma 4.2(2) for the second equality, one obtains

$$X = \operatorname{Supp}^*_{\mathcal{A}}(M,D) = \operatorname{Supp}^*_{\mathcal{A}}(M,\Omega^R_n(D)) = \operatorname{Supp}^*_{\mathcal{A}}(M,N_X)\,.$$

When $N = M$ one can choose $D = C$ by Theorem 3.4, and hence get

$$N_X = M_X \simeq \Omega^R_n(C) \in \mathsf{Thick}_R(M\oplus R)\,.$$

Lemma 1.6(3) now shows that $\operatorname{Ext}^*_R(M_X,F)$ is eventually zero when F is free. Thus, the already established assertion of the theorem apply to M_X and give

$$X = \operatorname{Supp}^*_{\mathcal{A}}(M_X,M_X)\,.$$

It remains to establish the additional property when X is irreducible. Being a noetherian module, M_X is a finite direct sum of indecomposables. It follows from (3.6.2) that one can replace M_X with such a summand, without changing $\operatorname{Supp}^*_{\mathcal{A}}(M_X,N)$. A similar argument works for N_X. □

The following general property of syzygy complexes was used above.

LEMMA 4.2. *Let M, N be complexes of R-modules with bounded homology. For every integer n the following hold.*

(1) *There is an equality* $\mathrm{Supp}^*_{\mathcal{A}}(M,N) = \mathrm{Supp}^*_{\mathcal{A}}(\Omega^R_n(M),N)$.

(2) *If* $\mathrm{Ext}^*_R(M,F)$ *is eventually zero for every free module F, then also*

$$\mathrm{Supp}^*_{\mathcal{A}}(M,N) = \mathrm{Supp}^*_{\mathcal{A}}(M,\Omega^R_n(N)).$$

Proof. (2) Replacing N with a semiprojective resolution, we may assume that each N_j is projective and $N_j = 0$ for all $j \ll 0$. An elementary argument using the formulas in 3.5 shows that the complexes C of R-modules, for which $\mathrm{Ext}^*_R(M,C)$ is eventually zero, form a thick subcategory of $\mathsf{D}(R)$. It contains the free modules by hypothesis, and hence it contains all bounded complexes of projective modules. Therefore, $\mathrm{Ext}^{\geqslant i}_R(M,N_{<n}) = 0$ holds for all $i \gg 0$.

The inclusion $N_{<n} \subseteq N$ gives rise to an exact triangle $N_{<n} \to N \to N_{\geq n} \to$ in $\mathsf{D}(R)$. Again by (3.5.3), it induces an exact sequence of graded $\mathcal{A}$-modules

$$\mathrm{Ext}^*_R(M,N_{<n}) \longrightarrow \mathrm{Ext}^*_R(M,N) \longrightarrow \mathrm{Ext}^*_R(M,N_{\geq n}) \to$$

$$\mathrm{Ext}^*_R(M,N_{<n})(1) \to \mathrm{Ext}^*_R(M,N)(1)$$

In view of the preceding discussion it yields $\mathrm{Ext}^{\geqslant i}_R(M,N) \cong \mathrm{Ext}^{\geqslant i}_R(M,N_{\geqslant n})$ for all $i \gg 0$. On the other hand, one has $N_{\geqslant n} = \Sigma^n\Omega^R_n(N)$ because N is semiprojective. The desired equality now follows from 2.2(2).

(1) This follows from a similar, and simpler, argument. □

5. Bialgebras and Hopf algebras

In this section k denotes a field. We recall some notions concerning bialgebras and Hopf algebras, referring to [19] for details.

A *bialgebra* over k is a k-algebra R with structure map $\eta\colon k \to R$ and product $\mu\colon R\otimes_k R \to k$, equipped with homomorphisms of rings $\varepsilon\colon R \to k$, the *augmentation* and $\Delta\colon R \to R\otimes_k R$, the *co-product*, satisfying equalities

$$\varepsilon\eta = \mathrm{id}^k, \qquad (\Delta\otimes \mathrm{id}^R)\Delta = \Delta(\Delta\otimes\mathrm{id}^R),$$

$$\mu(\mathrm{id}^R \otimes \eta\varepsilon)\Delta = \mathrm{id}^R = \mu(\eta\varepsilon\otimes \mathrm{id}^R)\Delta.$$

Given R-modules M, N over a bialgebra R, the natural $R\otimes_k R$-module structure on $M\otimes_k N$ restricts along Δ to produce a canonical R-module structure:

$$r\cdot(m\otimes n) = \sum_{i=1}^{n}(r'_i m\otimes_k r''_i n) \quad \text{when} \quad \Delta(r) = \sum_{i=1}^{n}(r'_i\otimes_k r''_i).$$

This extends to tensor products of complexes of R-modules. Let M be such a complex. The canonical isomorphisms below are easily seen to be R-linear:

(5.0.1) $$k\otimes_k M \cong M \quad\text{and}\quad M\otimes_k k\cong M.$$

5.1. *Cohomology operations.* Let R be a bialgebra over k, and view k as an R-module via the augmentation ε. The ring $\operatorname{Ext}_R^*(k,k)$ has $\operatorname{Ext}_R^0(k,k) = k$ and is *graded-commutative*: for all $\alpha \in \operatorname{Ext}_R^i(k,k)$ and $\beta \in \operatorname{Ext}_R^j(k,k)$ one has

$$\alpha \cdot \beta = (-1)^{ij} \beta \cdot \alpha\,;$$

see [17, (VIII.4.7), (VIII.4.3)] or [16, (5.5)]. Thus, every graded subring

$$\text{(a)} \qquad \mathcal{A} \subseteq \operatorname{Ext}_R^{\bullet}(k,k) = \begin{cases} \bigoplus_{i\geqslant 0} \operatorname{Ext}_R^{2i}(k,k) & \text{if } \operatorname{char}(k) \neq 2\,; \\ \bigoplus_{i\geqslant 0} \operatorname{Ext}_R^{i}(k,k) & \text{if } \operatorname{char}(k) = 2\,. \end{cases}$$

is commutative. The functor $- \otimes_k M$ preserves quasi-isomorphisms of complexes of R-modules, so it induces a functor $- \otimes_k M \colon \mathsf{D}(R) \to \mathsf{D}(R)$. In view of the isomorphism $k \otimes_k M \cong M$; see (5.0.1), for each M one gets a map

$$\text{(b)} \qquad \zeta_M \colon \operatorname{Ext}_R^{\bullet}(k,k) \to \operatorname{Ext}_R^*(M,M)\,.$$

It is readily verified to be a central homomorphism of graded k-algebras.

The results in Section 3 apply to any algebra $\mathcal{A}$ as above. More comprehensive information is available for special classes of bialgebras.

A *Hopf algebra* is a bialgebra R with a k-linear map $\sigma \colon R \to R$, the *antipode*, satisfying $\varepsilon\sigma = \varepsilon$ and $\mu(1 \otimes \sigma)\Delta = \mu(\sigma \otimes 1)\Delta$. Quantum groups offer prime examples. A Hopf algebra is *cocommutative* if $\tau\Delta = \Delta$ holds, where $\tau(r \otimes s) = s \otimes r$. For instance, for a group G the k-linear maps defined by

$$\varepsilon(g) = 1\,, \quad \Delta(g) = g \otimes g\,, \quad \text{and} \quad \sigma(g) = g^{-1} \quad \text{for} \quad g \in G$$

turn the group algebra kG into a cocommutative Hopf algebra. Other classical examples are universal enveloping algebras of Lie algebras and restricted universal enveloping algebras of p-Lie algebras, where $p = \operatorname{char}(k) > 0$.

5.2. *Finiteness.* Let R be a Hopf algebra such that $\operatorname{rank}_k R$ finite.

If R is cocommutative, then $\operatorname{Ext}_R^{\bullet}(k,k)$ is finitely generated as a k-algebra, and $\operatorname{Ext}_R^*(M,N)$ is a finite $\operatorname{Ext}_R^{\bullet}(k,k)$-module for all R-modules M, N of finite k-rank: This is a celebrated theorem of Friedlander and Suslin [13, (1.5.2)], which extends earlier results for group algebras (Evens, Golod, Venkov) and for restricted Lie algebras (Friedlander and Parshall).

It is not known whether cohomology has similar finiteness properties when R is not cocommutative; for positive solutions in interesting classes of such Hopf algebras see Pevtsova and Witherspoon [20], and the bibliography there.

5.3. *Cohomological varieties.* Let R be a Hopf algebra with $\operatorname{rank}_k R$ finite, set $\mathcal{A} = \operatorname{Ext}_R^{\bullet}(k,k)$ (see 5.1(a)), and let $\bar{k}$ be an algebraic closure of k.

For a complex M with $\operatorname{Ext}_R^*(M,M)$ eventually noetherian over $\mathcal{A}$ define (with notation as in 2.3) the *cohomological variety* of M to be the subset

$$V_R^*(M) = V_{\mathcal{A}}(\operatorname{Ext}_R^*(M,M)) \subseteq \operatorname{Max}(\mathcal{A} \otimes_k \bar{k})\,.$$

EXISTENCE THEOREM 5.4. *Let R be a Hopf algebra over k, such that $\operatorname{rank}_k R$ is finite; set $\mathcal{A} = \operatorname{Ext}^{\bullet}_R(k,k)$. Let M be a complex with $\mathrm{H}(M)$ finite over R.*

If $\operatorname{Ext}^*_R(M,M)$ is eventually noetherian over $\mathcal{A}$, then for each closed conical k-rational subset X of $V^*_{\mathcal{A}}(M)$ there is a finite R-module M_X, such that

$$X = V^*_{\mathcal{A}}(M_X) \quad \text{and} \quad M_X \in \mathsf{Thick}_R(M \oplus R)\,.$$

Proof. Hopf algebras of finite rank are self-injective (see [19, (2.1.3)(4)]), so one has $\operatorname{Ext}^{\geqslant 1}_R(-,R) = 0$. It remains to invoke Theorem 4.1 and refer to 2.3. □

5.5. *Applications.* In view of 5.2, for $M = k$ the theorem specializes to results of Carlson [11], Suslin, Friedlander, and Bendel [25, (7.5)], Pevtsova and Witherspoon [20, (4.5)], among others. It is clear that there are also versions dealing with supports of pairs of modules, and with complexes.

6. Associative algebras

Here k is a field and R is a k-algebra. Let R° denote the opposite algebra of R, set $R^{\mathsf{e}} = R \otimes_k R^{\circ}$, and turn R into a left R^{e}-module by $(r \otimes r')\cdot s = rsr'$.

6.1. *Cohomology operations.* The *Hochschild cohomology* of R is the k-algebra

$$\mathrm{H}^*(R\,|\,k) = \bigoplus_{i\geqslant 0} \operatorname{Ext}^i_{R^{\mathsf{e}}}(R,R)\,.$$

Gerstenhaber [14, Cor. 1] proved that it is graded-commutative, so any subring

(a) $$\mathcal{A} \subseteq \mathrm{H}^{\bullet}(R\,|\,k) = \bigoplus_{i\geqslant 0} \mathrm{H}^{2i}(R\,|\,k)$$

is commutative. The map $r \mapsto 1 \otimes r$ is a homomorphism of rings $R^{\circ} \to R^{\mathsf{e}}$. It turns each complex of R^{e}-modules into one of right R-modules. Thus, $- \otimes_R M$ is an additive functor from complexes of R^{e}-modules to complexes of R-modules, where R acts on the target via the homomorphism of rings $R \to R^{\mathsf{e}}$ given by $r \mapsto r \otimes 1$. It induces an exact functor $- \otimes^{\mathbf{L}}_R M \colon \mathsf{D}(R^{\mathsf{e}}) \to \mathsf{D}(R)$ of derived categories, which produces a homomorphism of graded rings

$$\operatorname{Ext}^*_{R^{\mathsf{e}}}(R,R) \to \operatorname{Ext}^*_R(R \otimes^{\mathbf{L}}_R M, R \otimes^{\mathbf{L}}_R M)\,.$$

The isomorphism $R \otimes^{\mathbf{L}}_R M \simeq M$ now yields a natural homomorphism

(b) $$\zeta_M \colon \mathrm{H}^{\bullet}(R\,|\,k) \to \operatorname{Ext}^*_R(M,M)$$

of graded rings. These maps satisfy condition (3.1.1); see [23, (10.1)].

The results in Sections 3 and 4 apply to any algebra $\mathcal{A}$ as above. Once again, we focus on a special case to relate them to available literature.

6.2. *Finiteness.* Let R be a k-algebra with $\operatorname{rank}_k R$ finite, J the Jacobson radical of R, and set $K = R/J$. It is rarely the case that the $\mathrm{H}^*(R|k)$-module $\operatorname{Ext}^*_R(M,K)$ is noetherian for every finite R-module M; see [12, §1]. Examples when this property holds include the Hopf algebras in (5.2), exterior algebras, and commutative complete intersections rings; see (7.1).

6.3. *Cohomological varieties.* For R as in (6.2), let $\mathcal{A}$ be a subring of $\mathrm{H}^\bullet(R|k)$ with $\mathcal{A}^0 = k$ (see 6.1), and let $\bar{k}$ be an algebraic closure of k.

For a complex of R-modules M, such that the $\mathcal{A}$-module $\operatorname{Ext}^*_R(M,K)$ is eventually noetherian, define the *cohomological variety* of M to be the subset

$$V^*_{\mathcal{A}}(M) = V_{\mathcal{A}}(\operatorname{Ext}^*_R(M,K)) \subseteq \operatorname{Max}(\mathcal{A} \otimes_k \bar{k})\,.$$

As $\mathcal{A}$ acts on $\operatorname{Ext}^*_R(M,K)$ through $\operatorname{Ext}^*_R(K,K)$, one has $V^*_{\mathcal{A}}(M) \subseteq V^*_{\mathcal{A}}(K)$.

EXISTENCE THEOREM 6.4. *Let R and $\mathcal{A}$ be as in 6.3, and let M be a complex of R-modules with $\mathrm{H}(M)$ finite over R.*

If $\operatorname{Ext}^*_R(M,K)$ is eventually noetherian over $\mathcal{A}$, then for each closed conical k-rational subset X of $V^*_{\mathcal{A}}(M)$ there is a finite R-module M_X, such that

$$X = V^*_{\mathcal{A}}(M_X) \quad \text{and} \quad M_X \in \mathsf{Thick}_R(M \oplus R)\,.$$

Proof. The R-module R admits a finite filtration with subquotients isomorphic to direct summands of K. Thus, when the $\mathcal{A}$-module $\operatorname{Ext}^*_R(M,K)$ is noetherian, so is $\operatorname{Ext}^*_R(M,R)$. On it $\mathcal{A}$ acts through $\operatorname{Ext}^*_R(R,R) = R^{\varsigma}$; see 3.2. This means $\operatorname{Ext}^{\gg 0}_R(M,R) = 0$, so we may use Theorem 4.1, then 2.3. □

6.5. *Application.* When the $\mathcal{A}$-module $\operatorname{Ext}^*_R(K,K)$ is noetherian, Theorem 6.4 with $M = K$ yields a result of Erdmann *et al.*; see [12, (3.4)].

7. Commutative local rings

We say that $(R,\mathfrak{m},k)$ is a *local ring* if R is a commutative noetherian ring with unique maximal ideal $\mathfrak{m}$ and $k = R/\mathfrak{m}$.

An *embedded deformation* of codimension c of R is a surjective homomorphism $\varkappa\colon Q \to R$ of rings with $(Q,\mathfrak{q},k)$ a local ring and $\operatorname{Ker}(\varkappa)$ an ideal generated by a Q-regular sequence in $\mathfrak{q}^2$, of length c.

7.1. *Cohomology operations.* Let $(R,\mathfrak{m},k)$ be a local ring with an embedded deformation $\varkappa\colon Q \to R$ of codimension c. Set

(a) $$\mathcal{A} = R[\chi_1,\dots,\chi_c]$$

where $\chi_1,\dots,\chi_c$ are indeterminates of degree 2. For each $M \in \mathsf{D}(R)$ Avramov and Sun [7, (2.7), p. 700] construct a natural homomorphism of graded rings

(b) $$\zeta_M\colon \mathcal{A} \to \operatorname{Ext}^*_R(M,M)$$

satisfying condition (3.1.1); when M and N are R-modules the resulting structure of graded $\mathcal{A}$-module on $\operatorname{Ext}_R^*(M,N)$ coincides with that defined by Gulliksen [15]. For complexes M and N, the $\mathcal{A}$-module $\operatorname{Ext}_R^*(M,N)$ is finite if and only if the Q-module $\operatorname{Ext}_Q^*(M,N)$ is finite; see [7, (5.1)] and [3, (4.2)].

The action of $\mathcal{A}$ on $\operatorname{Ext}_R^*(M,k)$ factors through the graded ring

$$\mathcal{R} = \mathcal{A} \otimes_R k = k[\chi_1, \dots, \chi_c]. \tag{c}$$

Recall that a complex of Q-modules is said to be *perfect* if it is isomorphic, in $\mathsf{D}(Q)$, to a bounded complex of finite free Q-modules.

LEMMA 7.2. *Let Q, R, and $\mathcal{R}$ be the rings in 7.1.*
The following conditions are equivalent for each complex M of R-modules:
(i) *M is perfect over Q.*
(ii) *$\mathrm{H}(M)$ is finite over R and $\operatorname{Ext}_R^*(M,k)$ is finite over $\mathcal{R}$.*

Proof. (i) $\implies$ (ii): As M is perfect over Q, the Q-module $\operatorname{Ext}_Q^*(M,k)$ is finite, and hence the $\mathcal{R}$-module $\operatorname{Ext}_R^*(M,k)$ is finite; see 7.1.

(ii) $\implies$ (i): It follows from 7.1 that $\operatorname{Ext}_Q^*(M,k)$ is finite over Q, and hence is eventually zero. Since $\mathrm{H}(M)$ is finite over Q, the complex M admits a semiprojective resolution F with each F_i finite, $F_i = 0$ for $i \ll 0$, and $\partial(F) \subseteq \mathfrak{m}F$; see, for example, [4]. This yields an isomorphism $\operatorname{Ext}_Q^i(M,k) \cong \operatorname{Hom}_Q(F_i,k)$. Thus $\operatorname{Ext}_Q^i(M,k) = 0$ for $i \geq n$ implies $F_i = 0$ for $i \geq n$, so F is a perfect complex of Q-modules that is quasi-isomorphic to M. □

7.3. *Cohomological varieties.* Let $(R,\mathfrak{m},k)$ be a local ring with an embedded deformation $\varkappa$ of codimension c, as in 7.1, and $\bar{k}$ an algebraic closure of k. Let M be a complex of R-modules with $\operatorname{Ext}_R^*(M,k)$ noetherian over $\mathcal{R}$. The *cohomological variety* of M is the subset $V_\varkappa^*(M)$ of $\bar{k}^c$ defined by the formula

$$V_\varkappa^*(M) = V_{\mathcal{R}}(\operatorname{Ext}_R^*(M,k)) \subseteq \operatorname{Max}(\mathcal{R} \otimes_k \bar{k}) = \bar{k}^c,$$

where the second equality comes from Hilbert's Nullstellensatz. When M is a module and $\boldsymbol{f}$ is a Q-regular sequence that generates $\operatorname{Ker}(\varkappa)$, the construction above yields the cone $V_R^*(\boldsymbol{f};M)$ defined in [1].

A Q-module is a perfect complex in $\mathsf{D}(Q)$ if and only if $\operatorname{proj\,dim}_Q M$ is finite. Thus, Theorem 2 from the introduction is obtained from the next result by replacing affine cones by their projectivizations.

EXISTENCE THEOREM 7.4. *Let $(Q,\mathfrak{q},k)$ be a local ring, $\boldsymbol{f} \subset \mathfrak{q}^2$ a Q-regular sequence of length c, and $\varkappa\colon Q \to Q/Q\boldsymbol{f} = R$ the canonical surjection.*

The assignment $M \mapsto V_\varkappa^*(M)$, which maps complexes in $\mathsf{D}(R)$ that are perfect over Q to closed k-rational cones in $\bar{k}^c$, has the following properties:
(1) *It is surjective, even when restricted to modules.*
(2) *$V_\varkappa^*(M) = \{0\}$ if and only if M is perfect over R.*

(3) $V^*_\varkappa(M) = V^*_\varkappa(\Omega^R_n(M))$ *for every syzygy complex* $\Omega^R_n(M)$.
(4) $V^*_\varkappa(M) = V^*_\varkappa(M/\boldsymbol{x}M)$ *for a module* M *and an* M*-regular sequence* $\boldsymbol{x}$.

We import some material for the proof of part (1) of the theorem.

7.5. Under the hypotheses of the theorem Avramov, Gasharov, and Peeva [3, (3.3), (3.11), (6.2)] give a nonzero finite R-module G with

$$\operatorname{Ext}^*_R(G,k) \cong \mathcal{R} \otimes_k \operatorname{Ext}^*_Q(G,k)\,,$$

which also satisfies the conditions $\operatorname{proj\,dim}_Q G < \infty$ and $\operatorname{Ext}^{\geqslant 1}_R(G,R) = 0$.

Proof of Theorem 7.4. (1) One has $V_{\mathcal{R}}(\mathcal{R}) = \bar{k}^c$ by the Nullstellensatz. In view of 2.3 it thus suffices to find a module G with $\operatorname{Supp}^*_{\mathcal{R}}(G,k) = \operatorname{Spec}\mathcal{R}$, and to show that every closed subset X of $\operatorname{Supp}^*_{\mathcal{R}}(G,k)$ is realizable by a module M_X with $\operatorname{proj\,dim}_Q M_X < \infty$.

The R-module G from 7.5 has the necessary property, as $\operatorname{Ext}^*_R(G,k)$ is a nonzero graded free $\mathcal{R}$-module. Theorem 4.1 yields a module M_X with the desired cohomological support and is in $\mathsf{Thick}_R(G\oplus R)$. Since $G\oplus R$ is perfect over Q, the last condition implies that so does M_X.

(2) Evidently $V^*_\varkappa(M) = \{0\}$ if and only if $\operatorname{Supp}_{\mathcal{R}} \operatorname{Ext}^*_R(M,k) = \varnothing$. As the $\mathcal{R}$-module $\operatorname{Ext}^*_R(M,k)$ is finite, this is equivalent to $\operatorname{Ext}^{\gg 0}_R(M,k) = 0$. Lemma 7.2, applied with $Q = R$, yields the desired equivalence.

(3) This follows from Lemma 4.2.

(4) As noted in Example 3.9, the complex $M/\!\!/\boldsymbol{x}$ is quasi-isomorphic to the Koszul complex on $\boldsymbol{x}$, and hence to the R-module $M/\boldsymbol{x}M$. Thus, Proposition 3.10 implies $V^*_\varkappa(M/\boldsymbol{x}M) = V^*_\varkappa(M)$; see 2.3. □

In Theorem 7.4 the hypothesis that R has an embedded deformation can be weakened in a useful way. The main property is (1), so we focus on it.

7.6. *Completions.* The $\mathfrak{m}$-adic completion of $(R,\mathfrak{m},k)$ is a local ring, $(\widehat{R},\widehat{\mathfrak{m}},k)$. The maps $R \to \widehat{R}$ and $M \to \widehat{R}\otimes_R M = \widehat{M}$ induce isomorphisms

$$\text{(d)}\qquad \operatorname{Ext}^*_{\widehat{R}}(k,k) \longrightarrow \operatorname{Ext}^*_R(k,k) \quad\text{and}\quad \operatorname{Ext}^*_{\widehat{R}}(\widehat{M},k) \longrightarrow \operatorname{Ext}^*_R(M,k)\,,$$

the first one of graded k-algebras, the second of graded modules, equivariant over the first. Thus, when $\widehat{R}$ has an embedded deformation $\varkappa$ of codimension c the ring $\mathcal{R}$ from 7.1 acts on $\operatorname{Ext}^*_R(M,k)$ for each $M \in \mathsf{D}(R)$. As in 7.3, when $\operatorname{Ext}^*_R(M,k)$ is noetherian over $\mathcal{R}$ we define a *cohomological variety* by:

$$V^*_\varkappa(M) = V_{\mathcal{R}}(\operatorname{Ext}^*_R(M,k)) \subseteq \bar{k}^c\,.$$

Observe that if $\varkappa$ is an embedded deformation of R, then $\widehat{\varkappa}$, completion with respect to the maximal ideal of Q, is an embedded deformation of $\widehat{R}$, so (d) above yields $V^*_\varkappa(M) = V^*_{\widehat{\varkappa}}(\widehat{M})$.

The following descent result is of independent interest.

THEOREM 7.7. *Let $(R,\mathfrak{m},k)$ be a local ring, $Q \to R$ an embedded deformation, and L a complex of $\widehat{R}$-modules.*

If L is perfect over Q, then there exists a finite R-module M with

$$V^*_\varkappa(M) = V^*_\varkappa(L) \quad \textit{and} \quad \operatorname{proj\,dim}_Q \widehat{M} < \infty\,.$$

Proof. By 2.3, it suffices to prove $\operatorname{Supp}^+_{\mathcal{A}} \operatorname{Ext}^*_{\widehat{R}}(M,k) = \operatorname{Supp}^+_{\mathcal{A}} \operatorname{Ext}^*_{\widehat{R}}(L,k)$.

Choose a set of generators of the ideal $\mathfrak{m}$ and let $\boldsymbol{x}$ be its image under the composition $R \to \widehat{R} \to \mathcal{A}$. The equality $\boldsymbol{x} \operatorname{Ext}^*_{\widehat{R}}(L,k) = 0$ implies an inclusion

$$\operatorname{Supp}^+_{\mathcal{A}} \operatorname{Ext}^*_{\widehat{R}}(L,k) \subseteq \operatorname{Supp}^+_{\mathcal{A}} (\mathcal{A}/(\boldsymbol{x})\mathcal{A})\,.$$

This yields the second equality below; the first one holds by Lemma 3.7:

$$\begin{aligned}\operatorname{Supp}^+_{\mathcal{A}} \operatorname{Ext}^*_{\widehat{R}}(L/\!\!/\boldsymbol{x},k) &= \operatorname{Supp}^+_{\mathcal{A}} \operatorname{Ext}^*_{\widehat{R}}(L,k) \cap \operatorname{Supp}^+_{\mathcal{A}} (\mathcal{A}/(\boldsymbol{x})\mathcal{A}) \\ &= \operatorname{Supp}^+_{\mathcal{A}} \operatorname{Ext}^*_{\widehat{R}}(L,k)\,.\end{aligned}$$

The complex $L/\!\!/\boldsymbol{x}$ is quasi-isomorphic to the Koszul complex on $\boldsymbol{x}$ with coefficients in L; see 3.9. Thus, $\mathrm{H}(L/\!\!/\boldsymbol{x})$ has finite length over $\widehat{R}$, hence also over R. Let $F \to L/\!\!/\boldsymbol{x}$ be a semi-projective resolution over R with F a finite free complex. One then has quasi-isomorphisms

$$\widehat{R} \otimes_R F \simeq \widehat{R} \otimes_R (L/\!\!/\boldsymbol{x}) \simeq L/\!\!/\boldsymbol{x}$$

due to the flatness of $\widehat{R}$ over R and, for the second one, also to the finiteness of the length of $\mathrm{H}(L/\!\!/\boldsymbol{x})$ over R. Fix n so that $\mathrm{H}_{\geqslant n}(L/\!\!/\boldsymbol{x}) = 0$ holds and set $M = \mathrm{H}_n(F_{\geqslant n})$. One then has quasi-isomorphisms of complexes of $\widehat{R}$-modules

$$\widehat{M} \cong \widehat{R} \otimes_R M \simeq \widehat{R} \otimes_R (F_{\geqslant n})\,.$$

They imply $\widehat{M} \cong \Omega^{\widehat{R}}_i(L/\!\!/\boldsymbol{x})$, hence the first equality below:

$$\begin{aligned}\operatorname{Supp}^+_{\mathcal{A}} \operatorname{Ext}^*_{\widehat{R}}(\widehat{M},k) &= \operatorname{Supp}^+_{\mathcal{A}} \operatorname{Ext}^*_{\widehat{R}}(\Omega^{\widehat{R}}_i(L/\!\!/\boldsymbol{x}),k) \\ &= \operatorname{Supp}^+_{\mathcal{A}} \operatorname{Ext}^*_{\widehat{R}}(L/\!\!/\boldsymbol{x},k)\,.\end{aligned}$$

Lemma 1.4 gives the second one. It remains to note that since L and $\widehat{R}$ are both perfect over Q, so is any complex in $\mathsf{Thick}_{\widehat{R}}(L \oplus \widehat{R})$. Thus, $L/\!\!/\boldsymbol{x}$ is perfect over Q, by (3.8.4), and hence so is $\widehat{M} = \Omega^{\widehat{R}}_i(L/\!\!/\boldsymbol{x})$, by Lemma 1.4. □

EXISTENCE THEOREM 7.8. *Let $(R,\mathfrak{m},k)$ be a local ring.*

If $\varkappa\colon Q \to \widehat{R}$ is an embedded deformation of codimension c, then for each closed k-rational cone $X \subseteq \bar{k}^c$ there exists a finite R-module M with

$$X = V^*_\varkappa(M) \quad \textit{and} \quad \operatorname{proj\,dim}_Q \widehat{M} < \infty\,.$$

Proof. Theorem 7.4(1) provides a finite $\widehat{R}$-module L with $X = V^*_\varkappa(L)$, and of finite projective dimension over Q. Now apply Theorem 7.7. □

7.9. *Application.* A local ring R is *complete intersection* if $\widehat{R}$ admits an embedded deformation $\varkappa\colon Q \to R$ where Q is a regular local ring; see [18, §29]. For such an R Theorem 7.8 specializes to a result proved by Bergh [10, (2.3)], who uses Tate cohomology, and by Avramov and Jorgensen [6], who establish an existence theorem for cohomology modules by using equivalences of triangulated categories and Koszul duality.

REFERENCES

[1] L. L. Avramov, *Modules of finite virtual projective dimension*, Invent. Math. **96** (1989), 71–101. MR 981738 (90g:13027)

[2] L. L. Avramov and R.-O. Buchweitz, *Support varieties and cohomology over complete intersections*, Invent. Math. **142** (2000), 285–318. MR 1794064 (2001j:13017)

[3] L. L. Avramov, V. N. Gasharov, and I. V. Peeva, *Complete intersection dimension*, Inst. Hautes Études Sci. Publ. Math. (1997), 67–114 (1998). MR 1608565 (99c:13033)

[4] L. L. Avramov, H.-B. Foxby, and S. Halperin, *Differential graded homological algebra*, in preparation.

[5] L. L. Avramov and S. Iyengar, *Cohomologically noetherian rings and modules*, in preparation.

[6] L. L. Avramov and D. A. Jorgensen, *Reverse homological algebra over some local rings*, in preparation.

[7] L. L. Avramov and L.-C. Sun, *Cohomology operators defined by a deformation*, J. Algebra **204** (1998), 684–710. MR 1624432 (2000e:13021)

[8] D. J. Benson, *Representations and cohomology. II*, second ed., Cambridge Studies in Advanced Mathematics, vol. 31, Cambridge University Press, Cambridge, 1998. MR 1634407 (99f:20001b)

[9] D. J. Benson, S. Iyengar, and H. Krause, *Local cohomology and support for triangulated categories*, preprint, `arXiv:math.KT/0702610`

[10] P. Bergh, *On support varieties of modules over complete intersections*, Proc. Amer. Math. Soc., to appear.

[11] J. F. Carlson, *The variety of an indecomposable module is connected*, Invent. Math. **77** (1984), 291–299. MR 752822 (86b:20009)

[12] K. Erdmann, M. Holloway, R. Taillefer, N. Snashall, and Ø. Solberg, *Support varieties for selfinjective algebras*, K-Theory **33** (2004), 67–87. MR 2199789 (2007f:16014)

[13] E. M. Friedlander and A. Suslin, *Cohomology of finite group schemes over a field*, Invent. Math. **127** (1997), 209–270. MR 1427618 (98h:14055a)

[14] M. Gerstenhaber, *The cohomology structure of an associative ring*, Ann. of Math. (2) **78** (1963), 267–288. MR 0161898 (28 #5102)

[15] T. H. Gulliksen, *A change of ring theorem with applications to Poincaré series and intersection multiplicity*, Math. Scand. **34** (1974), 167–183. MR 0364232 (51 #487)

[16] S. Iyengar, *Modules and cohomology over group algebras: one commutative algebraist's perspective*, Trends in commutative algebra, Math. Sci. Res. Inst. Publ., vol. 51, Cambridge Univ. Press, Cambridge, 2004, pp. 51–86. MR 2132648 (2005m:13001)

[17] S. MacLane, *Homology*, first ed., Springer-Verlag, Berlin, 1967, Grundlehren der mathematischen Wissenschaften, Vol. 114. MR 0349792 (50 #2285)

[18] H. Matsumura, *Commutative ring theory*, Cambridge Studies in Advanced Mathematics, vol. 8, Cambridge University Press, Cambridge, 1986. MR 879273 (88h:13001)

[19] S. Montgomery, *Hopf algebras and their actions on rings*, CBMS Regional Conference Series in Mathematics, vol. 82, Amer. Math. Soc., Providence, RI, 1993. MR 1243637 (94i:16019)

[20] J. Pevtsova and S. Witherspoon, *Varieties for modules of quantum elementary abelian groups*, preprint, `arXiv:math.QA/0603409`.

[21] D. Quillen, *The spectrum of an equivariant cohomology ring. I, II*, Ann. of Math. (2) **94** (1971), 549–572; ibid. **94** (1971), 573–602. MR 0298694 (45 #7743)

[22] N. Snashall and Ø. Solberg, *Support varieties and Hochschild cohomology rings*, Proc. London Math. Soc. (3) **88** (2004), 705–732. MR 2044054 (2005a:16014)

[23] Ø. Solberg, *Support varieties for modules and complexes*, Trends in representation theory of algebras and related topics, Contemp. Math., vol. 406, Amer. Math. Soc., Providence, RI, 2006, pp. 239–270. MR 2258047 (2007f:16018)

[24] M. Suarez-Alvarez, *The Hilton-Heckmann argument for the anti-commutativity of cup products*, Proc. Amer. Math. Soc. **132** (2004), 2241–2246. MR 2052399 (2005a:18016)

[25] A. Suslin, E. M. Friedlander, and C. P. Bendel, *Support varieties for infinitesimal group schemes*, J. Amer. Math. Soc. **10** (1997), 729–759. MR 1443547 (98h:14055c)

L. L. Avramov, Department of Mathematics, University of Nebraska, Lincoln, NE 68588, U.S.A.

E-mail address: `avramov@math.unl.edu`

S. B. Iyengar, Department of Mathematics, University of Nebraska, Lincoln, NE 68588, U.S.A.

E-mail address: `iyengar@math.unl.edu`

Illinois Journal of Mathematics
Volume 51, Number 1, Spring 2007, Pages 21–27
S 0019-2082

APOLARITY AND COVARIANT FORMS

JOSEPH P. BRENNAN, JAYDEEP V. CHIPALKATTI, AND ROBERT M. FOSSUM

Dedicated to Phillip Griffith on the occasion of his retirement

ABSTRACT. This paper examines the existence of apolar covariants of order greater than or equal to n. These are shown to exist for all $n > 2$.

1. Introduction

This paper examines the concept of the apolarity of binary forms. A form g of order m is apolar to the form f of order $n \leq m$ if g is in the kernel of the linear mapping from forms of order m to forms of order $m - n$ defined by the n-th transvectant (Überschiebung) of the form f. The transvectant of two covariant forms (forms invariant under the action of the special linear group of 2×2-matrices) is again a form that is a covariant. This construction is central to the determination by Gordan [9] of the finite generation of the rings of covariant binary forms and to new approaches to finite generation [3].

The vanishing of the transvectant and more generally the value of the transvectant of two binary n-ics has been a subject of renewed interest in recent years. As can be seen from Theorem 3.1, it is closely connected with the Waring problem for forms [4], [6], [7], [11], [12], [13]. The question of the existence of apolar covariant forms is one that has not been adequately addressed by previous work on the topic. Examples of apolar covariant forms that were explicitly known in the classical literature are limited. A form of odd degree is apolar to itself [8, §207]. A form of odd order $(2n - 1)$ is apolar to its canonizant which is of order n [8, §206–207]. Lastly the binary form of order 4 is apolar to its covariant of order 6 [10, §89].

This note focuses on forms of order $m \geq n$ that are apolar to the standard form of order n. We call such forms *major apolar covariants* to distinguish them from the covariants of smaller order. These arose from considerations of the construction of covariants of the binary n-ic from those of the binary $(n - 1)$-ic (see [2]).

Received February 19, 2007; received in final form August 22, 2007.
2000 *Mathematics Subject Classification*. Primary 15A72. Secondary 13A50, 14L24.
The third author was supported by NSF CCF-TF-0514955.

2. Definitions

The transvectant has played an important role in classical invariant theory [5], [8], [14] at least since the middle of the 19th century. It is related to the Poisson bracket.

In this paper $\Bbbk$ is a field of characteristic zero. Suppose that A is an algebra over the field $\Bbbk$. A form in $A[x, y]$ is a homogeneous polynomial in $A[x, y]$.

DEFINITION 2.1. The Poisson bracket of two forms $f, g \in A[x, y]$ is denoted by $P(f, g)$ and is defined by

$$P(f,g) = \frac{\partial f}{\partial x}\frac{\partial g}{\partial y} - \frac{\partial f}{\partial y}\frac{\partial g}{\partial x}.$$

Additional notations used for this bracket are $T(f, g), (f, g), [f, g]$, among others. Interest will be focused on the iterations of the operator. These are denoted variously as $P^{(n)}(f, g), T^{(n)}(f, g), (f, g)_n, [f, g]_n$.

For fixed f the bracket is a derivation in the second variable and it is linear in each variable.

There is an important expansion that will be needed in the computations that follow.

PROPOSITION 2.2. *Suppose f, g are forms in $A[x, y]$. Then*

$$P^{(r)}(f,g) = \sum_{j=0}^{r} (-1)^j \binom{r}{j} \frac{\partial^r f}{\partial x^{r-j}\partial y^j} \frac{\partial^r g}{\partial x^j \partial y^{r-j}}$$

This can be demonstrated by induction. The expansion is found in many places—for example in [14].

In particular, it follows that

PROPOSITION 2.3.

$$P^{(r)}(f,g) = (-1)^r P^{(r)}(g,f).$$

If f is a form of degree n and g is a form of degree m, then $P^{(r)}(f, g)$ is a form of degree $n + m - 2r$. In the literature on invariant theory, this form is called the rth *transvectant* of the forms f, g.

The results to be found below will be focused on some very specific expressions of representations of the additive group.

DEFINITION 2.4. The *standard form* (of degree n) in the polynomial ring $\Bbbk[a_0, a_1, \ldots, a_n, x, y]$ is the form

$$f_n(a_0, \ldots, a_n, x, y) = \sum_{i=0}^{n} \binom{n}{i} a_i x^{n-i} y^i. \tag{2.1}$$

Suppose

$$g(x,y) = B_0x^m + \binom{m}{1}B_1x^{m-1}y + \cdots + \binom{m}{j}B_jx^{m-j}y^j + \cdots + B_my^m$$

with coefficients $B_i \in \Bbbk[a_0, \ldots, a_n]$ and where $m \geq n$. Then

$$\begin{aligned} P^{(n)}(f_n, g) &= \frac{n!m!}{(m-n)!}\sum_{j=0}^{n}\sum_{s=0}^{m-n}(-1)^j\binom{n}{j}\binom{m-n}{s}a_jB_{s+n-j}x^{m-n-s}y^s \\ &= \frac{n!m!}{(m-n)!}\sum_{s=0}^{m-n}\binom{m-n}{s}\left(\sum_{j=0}^{n}(-1)^j\binom{n}{j}a_jB_{n+s-j}\right)x^{m-n-s}y^s. \end{aligned}$$

The n-th transvectant of the standard form f_n with

$$g = (x - by)^m$$

can be computed using Proposition 2.2. Because

$$\frac{\partial^n(x-by)^m}{\partial x^j \partial y^{n-j}} = \frac{m!}{(m-n)!}(-1)^{n-j}b^{n-j}(x-by)^{m-n}$$

one obtains using Proposition 2.2

$$\begin{aligned} P^{(n)}(f_n, (x-by)^m) &= (-1)^n\frac{n!m!}{(m-n)!}\left(\sum_{j=0}^{n}\binom{n}{j}a_jb^{n-j}\right)(x-by)^{m-n} \\ &= (-1)^n\frac{n!m!}{(m-n!)}f_n(a_0, a_1, \ldots, a_n, b, 1)(x-by)^{m-n}. \end{aligned}$$

In particular, if b is a root of f_n, then the transvectant

$$P^{(n)}(f_n, (x-by)^m) = 0. \tag{2.2}$$

DEFINITION 2.5. A form $g \in \Bbbk[a_0, \ldots, a_n, x, y]$ of degree m is *apolar* to the form f of degree n (with $n \leq m$) if

$$P^{(n)}(f, g) = 0.$$

DEFINITION 2.6. A covariant form g of degree $m \geq n$ that is apolar to the standard form f_n will be called a *major apolar covariant form.*

In particular, the canonizant of a form of odd order is not to be considered as an major apolar covariant form in this sense.

3. Apolarity and sums of powers

The key result that characterizes the forms apolar to the form f is given by the following theorem whose proof has its foundation in the proof of [10].

THEOREM 3.1. *Let $f = \sum_{i=1}^{n} \binom{n}{i} a_i x^{n-i} y^i$ be a form of degree n with roots $(\xi_1 : 1), \ldots, (\xi_n : 1)$ in a splitting field $\Bbbk[\xi_1, \ldots, \xi_n]$ of the form f. A form g of degree m greater than n with coefficients in $k[\xi_1, \ldots, \xi_n]$ is apolar to f (i.e., the transvectant $(f, g)_n = 0$) if and only if there exist $c_i \in k[\xi_1, \ldots, \xi_n]$ such that*

$$g = \sum_{i=1}^{n} c_i (x - \xi_i y)^m.$$

Proof. First consider the forms $(x - \xi_i y)^m$. By (2.2) these forms are all apolar to the standard form f_n. It suffices to show that these form a basis of the vector space of apolar forms of degree m.

From the fullness of rank of the matrix

$$\begin{pmatrix} a_0 & -\binom{n}{1}a_1 & \cdots & (-1)^n a_n & 0 & 0 & 0 \\ 0 & a_0 & -\binom{n}{1}a_1 & \cdots & (-1)^n a_n & 0 & 0 \\ \vdots & \vdots & \ddots & \vdots & \vdots & \ddots & \vdots \\ 0 & 0 & \cdots & a_0 & -\binom{n}{1}a_1 & \cdots & (-1)^n a_n \end{pmatrix},$$

the dimension of apolar forms of degree m is exactly n so it will suffice to show that the forms $(x - \xi_i y)^m$ are linearly independent. This follows immediately from the distinctness of the roots of f and the consequent non-vanishing of the associated Vandermonde determinant. □

4. Theorems and examples

Although the condition of being both a covariant form and an apolar form may seem rare, there are several families of examples, both classical and new, that should be noted. The first example follows immediately from Proposition 2.3.

PROPOSITION 4.1. *If n is odd, then the standard form is an apolar covariant form.*

Simple computations show that for the standard cubic there is no other apolar covariant form. Considering the quintic, there are *two* independent forms that are apolar: the form f_5 and the Hessian (which is up to a scalar $P^{(2)}(f_5, f_5)$).

PROPOSITION 4.2. *The Hessian of the standard form f_5 is a major apolar covariant of f_5.*

Proof. The Hessian of the standard quintic is the covariant of degree 2 and order 6. The coefficient of x^6 of the Hessian is the $\mathbb{G}_a$-invariant $a_0 a_2 - a_1^2$. Consider the 5-th transvectant of the standard quintic and the Hessian. If this transvectant is a non-zero form, it is a form of degree 3 and order 1. But the quintic has no covariant forms of degree smaller than 5 of order 1, as can be seen in the table found in [10, §116]. □

This does however not lead to a general family of major apolar covariants, as is the consequence of the following result which was the starting place for our investigation of major apolar covariants.

PROPOSITION 4.3 ([2]). *The Hessian $P^{(2)}(f_n, f_n)$ is not an apolar covariant for any form of order $n > 5$.*

Proof. It suffices to show that $E_n = P^{(n)}(f_n, P^{(2)}(f_n, f_n)) \neq 0$ for the special case that $f_n = x^n + xy^{n-1}$. By a direct computation using Proposition 2.2, one obtains

$$E_n = -2\,(n!)^2(n-1)^2\left[\binom{2n-4}{n-4} + (-1)^n(n-3)(n-2)\right] y^{n-4}.$$

It is thus clear that $E_n \neq 0$ for n even and greater than 4.

For odd $n \geq 5$, let

$$\alpha_n = \binom{2n-4}{n-4}, \qquad \beta_n = (n-3)(n-2).$$

For $n = 5$, either direct computation or the use of Proposition 4.2 gives the ratio $\alpha_5/\beta_5 = 1$.

Observe that

$$\frac{\alpha_{n+2}/\beta_{n+2}}{\alpha_n/\beta_n} = 16\frac{\left(n-\frac{1}{2}\right)}{n+1}\frac{\left(n-\frac{3}{2}\right)}{n+2} \geq 16 \cdot \frac{3}{4} \cdot \frac{1}{2} = 6.$$

It follows that if n is odd and greater than 5, then $\alpha_n/\beta_n > 1$. Hence $E_n \neq 0$ for $n > 5$. □

This will be the subject of further discussion in [2]. These were the only major apolar covariants of forms of odd order that were noticed classically. That there are others is the content of the following proposition.

PROPOSITION 4.4. *If $n \geq 7$ is odd, then the covariant $\chi_n = P^{(2)}(f_n, P^{(n-3)}(f_n, f_n))$ is a non-zero major apolar covariant of order $n+2$ and degree three.*

Proof. The transvectant $P^{(n)}(\chi, f_n)$ is nominally a covariant of degree four and order 2 of the binary n-ic. It is required to show that this is zero. This is done by showing that the dimension of the vector space of covariants of degree four and order 2 the binary n-ic is zero. By Hermite reciprocity [8, §131] this is equivalent to showing that the dimension of the vector space of covariants of degree n and order 2 of the binary quartic is zero. The generators of the ring of covariants of the binary quartic have orders zero, four, and six [10, §89]. Therefore the binary quartic has no covariants of order two.

It is left as an exercise to the reader to confirm that $\chi_n \neq 0$ for $n \geq 7$ by evaluating the special case where $f_n = x^n + x^2y^{n-2}$. □

The situation is even less well known for standard forms for even degree. Classically the only major apolar covariant that appears to have been (explicitly) known is the sextic covariant of the quartic [10, §89 p. 94]. This form is (up to sign) denoted by t in [10] and by G in [8]. This covariant is the Jacobian (first transvectant) of the standard form f_4 with the Hessian. The form has as the coefficient of x^6 the $\mathbb{G}_a$-invariant $a_0^2a_3 - 3a_0a_1a_2 + 2a_1^3$.

For every even $n = 2k > 2$, there is a covariant of degree 3 and order $n+2$ whose leading coefficient is of the form $a_0^2a_{n-1} + \cdots$ [8, §165(bis)]. For the standard sextic form, this covariant is called $C_{3,8}$ in [8, §244] or the Jacobian of f_6 and the fourth transvectant of f_6 with itself in [10, §134].

PROPOSITION 4.5. *For n even and greater than 2, the covariant of degree 3 and order $n+2$ of the standard form f_n associated with the $\mathbb{G}_a$-invariants arising from the minimal degree protomorphs described in* [8, §165(bis)] *is a major apolar covariant.*

Proof. This covariant ξ_n is the one obtained as $\psi = P^{(1)}(P^{(n-2)}(f_n, f_n), f_n)$. The transvectant $P^{(n)}(\xi_n, f_n)$ is by construction a covariant of degree 4 and order 2 of the binary n-ic. By the same reasoning as in the proof of Proposition 4.4 the transvectant vanishes. □

There are additional apolar covariants that are not of the type described in Proposition 4.5 as can be seen from the following result.

PROPOSITION 4.6. *The binary octavic has a major apolar covariant of degree four and order ten.*

Proof. From either the description of the covariants of the octavic in [1] or from the Cayley-Sylvester formula given in [15, Corollary 4.2.8], there are the following facts. The dimension of the vector space $V_{8;4,10}$ of covariants of the binary octavic of degree four and order ten is two. The dimension of the vector space $V_{8;5,2}$ of the binary octavic of covariants of degree five and order two is one.

The 8-th transvectant with the standard form of order eight defines a transformation from $V_{8;4,10}$ to $V_{8;5,2}$. This mapping therefore has a non-trivial kernel and hence there is a major apolar covariant of degree four and order ten. □

REFERENCES

[1] L. Bedratyuk, *On complete system of covariants for the binary form of degree 8*, preprint, 2006, arXiv:math.AG/0612113.

[2] J. P. Brennan and R. M. Fossum, *Apolarity and differential algebra II*, in preparation.

[3] A. Brini, F. Regonati, and A. Teolis, *Combinatorics, transvectants and superalgebras. An elementary constructive approach to Hilbert's finiteness theorem*, Adv. in Appl. Math. **37** (2006), 287–308. MR 2261176

[4] J. Chipalkatti, *Apolar schemes of algebraic forms*, Canad. J. Math. **58** (2006), 476–491. MR 2223453 (2007i:14052)

[5] J. Dixmier, *Transvectants des formes binaires et représentations des groupes binaires polyhédraux*, Portugal. Math. **49** (1992), 349–396. MR 1201295 (94a:11053)

[6] R. Ehrenborg, *On apolarity and generic canonical forms*, J. Algebra **213** (1999), 167–194. MR 1674676 (2000a:15050)

[7] R. Ehrenborg and G.-C. Rota, *Apolarity and canonical forms for homogeneous polynomials*, European J. Combin. **14** (1993), 157–181. MR 1215329 (94e:15062)

[8] E. B. Elliott, *An introduction to the algebra of quantics*, Chelsea Publishing Company, 1964.

[9] P. A. Gordan, *Beweis, dass jede Covariante und Invariante einer binären Form eine ganze Function mit numerischen Coefficienten einer endlichen Anzahl solcher Formen ist*, J. Reine Angew. Math. **69** (1868), 323–354.

[10] J. H. Grace and A. Young, *The algebra of invariants*, Chelsea Publishing Company, 1965.

[11] J. P. S. Kung, *Canonical forms of binary forms: variations on a theme of Sylvester*, Invariant theory and tableaux (Minneapolis, MN, 1988), IMA Vol. Math. Appl., vol. 19, Springer, New York, 1990, pp. 46–58. MR 1035488 (91b:11046)

[12] A. Lascoux, *Forme canonique d'une forme binaire*, Invariant theory, Lecture Notes in Math., vol. 1278, Springer, Berlin, 1987, pp. 44–51. MR 924164 (89a:12005)

[13] R. Oldenburger, *Binary forms*, Proc. Nat. Acad. Sci. U.S.A. **26** (1940), 497—499. MR 0002834 (2,119c)

[14] P. J. Olver, *Classical invariant theory*, London Mathematical Society Student Texts, vol. 44, Cambridge University Press, Cambridge, 1999. MR 1694364 (2001g:13009)

[15] B. Sturmfels, *Algorithms in invariant theory*, Texts and Monographs in Symbolic Computation, Springer-Verlag, Vienna, 1993. MR 1255980 (94m:13004)

Joseph P. Brennan, Department of Mathematics, University of Central Florida, Orlando, FL 32816-1364, USA

E-mail address: jpbrenna@mail.ucf.edu

Jaydeep V. Chipalkatti, Department of Mathematics, University of Manitoba, Winnipeg, MB R3T 2N2, Canada

E-mail address: chipalka@cc.umanitoba.ca

Robert M. Fossum, Department of Mathematics and Beckman Institute, University of Illinois, Urbana, IL 61801, USA

E-mail address: rmfossum@uiuc.edu

Illinois Journal of Mathematics
Volume 51, Number 1, Spring 2007, Pages 29–39
S 0019-2082

ASYMPTOTIC GROWTH OF POWERS OF IDEALS

CĂTĂLIN CIUPERCĂ, FLORIAN ENESCU, AND SANDRA SPIROFF

To Phil Griffith

ABSTRACT. Let A be a locally analytically unramified local ring and $J_1, \ldots, J_k, I$ ideals such that $J_i \subseteq \sqrt{I}$ for all i, the ideal I is not nilpotent, and $\bigcap_k I^k = (0)$. Let $C = C(J_1, \ldots, J_k; I) \subseteq \mathbb{R}^{k+1}$ be the cone generated by $\{(m_1, \ldots, m_k, n) \in \mathbb{N}^{k+1} \mid J_1^{m_1} \ldots J_k^{m_k} \subseteq I^n\}$. We prove that the topological closure of C is a rational polyhedral cone. This generalizes results by Samuel, Nagata, and Rees.

Introduction

In this note we continue the study of the asymptotic properties of powers of ideals initiated by Samuel in [8]. Let A be a commutative noetherian ring with identity and I, J ideals in A with $J \subseteq \sqrt{I}$. Also, assume that the ideal I is not nilpotent and $\bigcap_k I^k = (0)$. Then for each positive integer m one can define $v_I(J, m)$ to be the largest integer n such that $J^m \subseteq I^n$. Similarly, $w_J(I, n)$ is defined to be the smallest integer m such that $J^m \subseteq I^n$. Under the above assumptions, Samuel proved that the sequences $\{v_I(J, m)/m\}_m$ and $\{w_J(I, n)/n\}_n$ have limits $l_I(J)$ and $L_J(I)$, respectively, and $l_I(J)L_J(I) = 1$ [8, Theorem 1]. It is also observed that these limits are actually the supremum and infimum of the respective sequences. One of the questions raised in Samuel's paper is whether $l_I(J)$ is always rational. This has been positively answered by Nagata [4] and Rees [5]. The approach used by Rees is described in the next section of this paper.

We consider the following generalization of the problem described above. Let $J_1, \ldots, J_k, I$ be ideals in a locally analytically unramified ring A such that $J_i \subseteq \sqrt{I}$ for all i, I is not nilpotent, and $\bigcap_k I^k = (0)$, and let $C = C(J_1, \ldots, J_k; I) \subseteq \mathbb{R}^{k+1}$ be the cone generated by $\{(m_1, \ldots, m_k, n) \in \mathbb{N}^{k+1} \mid J_1^{m_1} \ldots J_k^{m_k} \subseteq I^n\}$. We prove that the topological closure of C is a rational

Received June 26, 2006; received in final form October 26, 2006.
2000 *Mathematics Subject Classification.* Primary 13A15. Secondary 13A18.
The second author gratefully acknowledges partial financial support from the National Science Foundation, CCF-0515010 and Georgia State University, Research Initiation Grant.

polyhedral cone; i.e., a polyhedral cone bounded by hyperplanes whose equations have rational coefficients. Note that the case $k = 1$ follows from the results proved by Samuel, Nagata, and Rees; the cone C is the intersection of the half-planes given by $n \geq 0$ and $n \leq l_I(J)m_1$. In Section 3 we look at the periodicity of the rate of change of the sequence $\{v_I(J,m)\}_m$, more precisely, the periodicity of the sequence $\{v_I(J,m+1) - v_I(J,m)\}_m$. The last part of the paper describes a method of computing the limits studied by Samuel in the case of monomial ideals.

1. The Rees valuations of an ideal

In this section we give a brief description of the Rees valuations associated to an ideal.

For a noetherian ring A that is not necessarily an integral domain, a discrete valuation on A is defined as follows.

DEFINITION 1.1. Let A be a noetherian ring. We say that $v : A \to \mathbb{Z} \cup \{\infty\}$ is a discrete valuation on A if $\{x \in A \mid v(x) = \infty\}$ is a prime ideal P, v factors through $A \to A/P \to \mathbb{Z} \cup \{\infty\}$, and the induced function on A/P is a rank one discrete valuation on A/P. If I is an ideal in A, then we denote $v(I) := \min\{v(x) \mid x \in I\}$.

If R is a noetherian ring, we denote by $\overline{R}$ the integral closure of R in its total quotient ring $Q(R)$.

DEFINITION 1.2. Let I be an ideal in a noetherian ring A. An element $x \in A$ is said to be integral over I if x satisfies an equation $x^n + a_1x^{n-1} + \cdots + a_n = 0$ with $a_i \in I^i$. The set of all elements in A that are integral over I is an ideal $\overline{I}$, and the ideal I is called integrally closed if $I = \overline{I}$. If all the powers I^n are integrally closed, then I is said to be normal.

Given an ideal I in a noetherian ring A, for each $x \in A$ let $v_I(x) = \sup\{n \in \mathbb{N} \mid x \in I^n\}$. Rees [5] proved that for each $x \in A$ one can define

$$\overline{v}_I(x) = \lim_{k\to\infty} \frac{v_I(x^k)}{k},$$

and for each integer n one has $\overline{v}_I(x) \geq n$ if and only if $x \in \overline{I^n}$. Moreover, there exist discrete valuations $v_1, \ldots, v_h$ on A in the sense defined above, and positive integers $e_1, \ldots, e_h$ such that, for each $x \in A$,

$$\overline{v}_I(x) = \min\left\{\frac{v_i(x)}{e_i} \mid i = 1, \ldots, h\right\}. \tag{1.1}$$

We briefly describe a construction of the Rees valuations $v_1, \ldots, v_h$. Let $\mathfrak{p}_1, \ldots, \mathfrak{p}_g$ be the minimal prime ideals $\mathfrak{p}$ in A such that $\mathfrak{p} + I \neq A$, and let $\mathcal{R}_i(I)$ be the Rees ring $(A/\mathfrak{p}_i)[It, t^{-1}]$. Denote by $W_{i1}, \ldots, W_{ih_i}$ the rank one discrete valuation rings obtained by localizing the rings $\overline{\mathcal{R}_i(I)}$ at the minimal primes over $t^{-1}\overline{\mathcal{R}_i(I)}$, let w_{ij} $(i = 1, \dots, g,\ 1 \le j \le h_i)$ be the corresponding discrete valuations, and let $V_{ij} = W_{ij} \cap Q(A/\mathfrak{p}_i)$ $(i = 1, \dots, g)$. Then define $v_{ij}(x) := w_{ij}(x + \mathfrak{p}_i)$ and $e_{ij} := w_{ij}(t^{-1})(= v_{ij}(I))$ for all i, and for simplicity, renumber them as $e_1, \dots, e_h$ and $v_1, \dots, v_h$, respectively.

Rees [5] proved that $v_1, \dots, v_h$ are valuations satisfying (1.1). We refer the reader to the original article [5] for more details on this construction.

REMARK 1.3. With the notation established above, for every positive integer n we have

$$\overline{I^n} = \bigcap_{i=1}^{h} I^n V_i \cap R.$$

In particular, we have the following.

REMARK 1.4. If K, L are ideals in A, $v_1, \dots, v_h$ are the Rees valuations of L, and $v_i(K) \ge v_i(L)$ for all $i = 1, \dots, h$, then $\overline{K} \subseteq \overline{L}$.

The rationality of $l_I(J)$ can now be obtained as consequence of the results of Rees. Indeed, by [8, Theorem 2], if $J = (a_1, \dots a_s)$, then $l_I(J) = \min\{l_I(a_i) \mid i = 1, \dots s\}$, and for each i we have $l_I(a_i) = \overline{v}_I(a_i)$, which is rational.

Finally, recall the following definition.

DEFINITION 1.5. A local noetherian ring $(A, \mathfrak{m})$ is analytically unramified if its $\mathfrak{m}$-adic completion $\hat{A}$ is reduced.

Rees [6] proved that for every ideal I in an analytically unramified ring there exists an integer k such that for all $n \ge 0$, $\overline{I^{n+k}} \subseteq I^n$.

2. The cone structure

Throughout this section A is a locally analytically unramified ring and I and $\underline{J} = J_1, \dots, J_k$ are ideals in A such that $J_i \subseteq \sqrt{I}$ for all i. Let $C = C(J_1, \dots, J_k; I) \subseteq \mathbb{R}^{k+1}$ denote the cone generated by $\{(m_1, \dots, m_k, n) \in \mathbb{N}^{k+1} \mid J_1^{m_1} \dots J_k^{m_k} \subseteq I^n\}$. Also, for $(m_1, \dots, m_k) \in \mathbb{N}^k$, let $v_I(\underline{J}, m_1, \dots, m_k)$ denote the largest nonnegative integer n such that $J_1^{m_1} \dots J_k^{m_k} \subseteq I^n$.

For each Rees valuation v_j of I, denote $\alpha_{ij} = v_j(J_i)/e_j$ for all i, j, where $e_j = v_j(I)$. Then we consider

$$D_j = \left\{ (m_1, \dots, m_k) \in \mathbb{R}^k_{\ge 0} \,\middle|\, \sum_{s=1}^{k} m_s \alpha_{sj} \le \sum_{s=1}^{k} m_s \alpha_{sl} \text{ for all } l \ne j \right\},$$

and we say that a Rees valuation v_j is relevant if $D_j \ne \{0\}$. After a renumbering, assume that $v_1, v_2, \dots, v_r$ $(r \le h)$ are the relevant Rees valuations.

Note that each D_j is an intersection of half-spaces (hence a polyhedral cone), $\bigcup_{j=1}^r D_j = \mathbb{R}^k_{\geq 0}$, and two cones D_i, D_j $(i \neq j)$ either intersect along one common face or have only the origin in common. Let

$$E_j = \left\{ (m_1, \dots, m_k, n) \in \mathbb{R}^{k+1}_+ \,\middle|\, (m_1, \dots, m_k) \in D_j \text{ and } n < \sum_{s=1}^k m_s \alpha_{sj} \right\}$$

and

$$\overline{E}_j = \left\{ (m_1, \dots, m_k, n) \in \mathbb{R}^{k+1}_+ \,\middle|\, (m_1, \dots, m_k) \in D_j \text{ and } n \leq \sum_{s=1}^k m_s \alpha_{sj} \right\}.$$

THEOREM 2.1. *Let A be a locally analytically unramified ring. Then for each $j = 1, \dots, r$ we have*

$$E_j \cap \mathbb{Q}^{k+1} \subseteq C \cap (D_j \times \mathbb{R}_{\geq 0}) \subseteq \overline{E}_j.$$

Proof. Let $(m_1, \dots, m_k, n) \in C \cap (D_j \times \mathbb{R}_{\geq 0})$. Then there exists $t \in \mathbb{R}$ such that $tm_1, \dots, tm_k$ are positive integers and

$$J_1^{tm_1} \dots J_k^{tm_k} \subseteq I^{tn}.$$

Hence, for each Rees valuation v_j of I we obtain

$$tm_1 v_j(J_1) + \dots + tm_k v_j(J_k) \geq tn v_j(I),$$

or equivalently,

$$n \leq \sum_{s=1}^k m_s \alpha_{sj}.$$

For the other inclusion, first observe that it is enough to prove that $E_j \cap \mathbb{Z}^{k+1} \subseteq C \cap (D_j \times \mathbb{R}_{\geq 0})$. Indeed, if $E_j \cap \mathbb{Z}^{k+1} \subseteq C \cap (D_j \times \mathbb{R}_{\geq 0})$, then for each $\alpha \in E_j \cap \mathbb{Q}^{k+1}$ there exists a positive integer L such that $\alpha L \in E_j \cap \mathbb{Z}^{k+1} \subseteq C \cap (D_j \times \mathbb{R}_{\geq 0})$. This implies that $\alpha \in (1/L)\big(C \cap (D_j \times \mathbb{R}_{\geq 0})\big) = C \cap (D_j \times \mathbb{R}_{\geq 0})$

Let $(m_1, \dots, m_k, n) \in E_j \cap \mathbb{Z}^{k+1}$. Set $\alpha = \sum_{s=1}^k m_s \alpha_{sj}$. Since the ring A is analytically unramified, there exists an integer N such that $\overline{I^t} \subseteq I^{t-N}$ for all t. (The convention is that $I^n = A$ for $n \leq 0$.) Let g be the integer part of α. For any Rees valuation v_i of A we then get

$$v_i(I^g) = g e_i \leq \alpha e_i \leq \left(\sum_{s=1}^k m_s \alpha_{si} \right) e_i = v_i(J_1^{m_1} \dots J_k^{m_k}),$$

and hence, by Remark 1.4,

$$J_1^{m_1} \dots J_k^{m_k} \subseteq \overline{I^g} \subseteq I^{g-N}.$$

This implies that

$$v_I(\underline{J}, m_1, \dots, m_k) \geq g - N > \alpha - 1 - N. \tag{2.1}$$

Since $n < \alpha$, we can find $\delta > 0$ such that $n < \alpha - \delta$. Choose l such that $l\delta > N + 1$ and $lm_1, \dots, lm_k, ln$ are integers. By (2.1), we obtain $v_I(\underline{J}, lm_1, \dots, lm_k) > l\alpha - N - 1$, and by the choice of l, we also have $nl < l\alpha - N - 1$. Then $nl < v_I(\underline{J}, lm_1, \dots, lm_k)$, which implies that $J_1^{lm_1} \dots J_k^{lm_k} \subseteq I^{ln}$; i.e., $(m_1, \dots, m_k, n) \in C$. □

COROLLARY 2.2. *The topological closure of C is a rational polyhedral cone.*

Proof. From the previous theorem it follows that the topological closure of $C \cap (D_j \times \mathbb{R}_{\geq 0})$ is $\overline{E}_j$, and hence the topological closure of C is the polyhedral cone bounded by the hyperplanes $n = \sum_{s=1}^{k} m_s \alpha_{sj}$ $(j = 1, \dots, r)$ and the coordinate hyperplanes. □

A detailed example of Corollary 2.2 is given below in Example 2.5.

COROLLARY 2.3. *Let $a_1, a_2, \dots, a_k$ be real numbers. The limit*

$$\lim_{m_1, \dots, m_k \to \infty} \frac{v_I(\underline{J}, m_1, \dots, m_k)}{a_1 m_1 + \cdots + a_k m_k} \tag{2.2}$$

exists if and only if there exists a rational number l such that $la_s = \alpha_{s1} = \alpha_{s2} = \cdots = \alpha_{sr}$ for all $s = 1, \dots, k$. In this case the limit is equal to l.

Proof. Since the polyhedral cones D_j form a partition of $\mathbb{R}^k_{\geq 0}$, the limit (2.2) exists and is equal to l if and only if for each j we have

$$\lim_{\substack{m_1, \dots, m_k \to \infty \\ (m_1, \dots, m_k) \in D_j}} \frac{v_I(\underline{J}, m_1, \dots, m_k)}{a_1 m_1 + \cdots + a_k m_k} = l. \tag{2.3}$$

On the other hand, (2.3) holds if and only if $la_s = \alpha_{sj}$ for all $s = 1, \dots, k$. Indeed, this limit exists and is equal to l if and only if over D_j the topological closure of C is bounded by the hyperplane $n = la_1 m_1 + \cdots + la_k m_k$, which therefore should coincide with the hyperplane $n = \sum_{s=1}^{k} m_s \alpha_{sj}$.

In conclusion, the limit (2.2) exists and is equal to l if and only if all the hyperplanes $n = \sum_{s=1}^{k} m_s \alpha_{sj}$ $(j = 1, \dots, r)$ coincide with $n = la_1 m_1 + \cdots + la_k m_k$, or equivalently, $la_s = \alpha_{s1} = \alpha_{s2} = \cdots = \alpha_{sr}$ for all $s = 1, \dots, k$. □

COROLLARY 2.4. *Assume that the ideal I has only one Rees valuation. Then the limit*

$$\lim_{m_1, \dots, m_k \to \infty} \frac{v_I(\underline{J}, m_1, \dots, m_k)}{a_1 m_1 + \cdots + a_k m_k}$$

exists if and only if $l_I(J_1)/a_1 = \cdots = l_I(J_k)/a_k$.

Proof. This is a particular case of the previous Corollary. □

EXAMPLE 2.5. Let $A = \mathbb{R}[[X,Y,Z]]/(XY^2 - Z^9)$ and $I = (x,y,z)A$ be as in [3, Example 3.1]. Then $\mathcal{R}(I) = A[It, t^{-1}]$, $\mathcal{R}(I)/t^{-1}\mathcal{R}(I) \cong \mathbb{R}[xt, yt, zt]/(xt)(yt)^2$, and there are two Rees valuations v_1 and v_2, corresponding to the minimal primes $\mathfrak{p}_1 = (xt, t^{-1})$ and $\mathfrak{p}_2 = (yt, t^{-1})$, over $t^{-1}\mathcal{R}(I)$. As shown in [3, Example 3.1], we have $v_1(x) = 7, v_1(y) = v_1(z) = 1$ and $v_2(x) = v_2(z) = 1, v_2(y) = 4$. Thus $v_1(I) = \min\{v_1(x), v_1(y), v_1(z)\} = 1$. Likewise $v_2(I) = 1$. Set $J_1 = (x, z^2)$ and $J_2 = (y^2, z^3)$. Then $v_1(J_1) = 2, v_2(J_1) = 1$, and $v_1(J_2) = 2, v_2(J_2) = 3$. Therefore, $E_1 = \{(m_1, m_2, n) | n \leq 2m_1 + 2m_2\}$ and $E_1 = \{(m_1, m_2, n) | n \leq m_1 + 3m_2\}$. The boundary planes of E_1 and E_2 in $\mathbb{R}^3$ are $z = 2x + 2y$ and $z = x + 3y$, respectively. Thus, according to the results of Corollary 2.2, the topological closure of the cone generated by $\{(m_1, m_2, n) | J_1^{m_1} J_2^{m_2} \subseteq I^n\}$ is as pictured below.

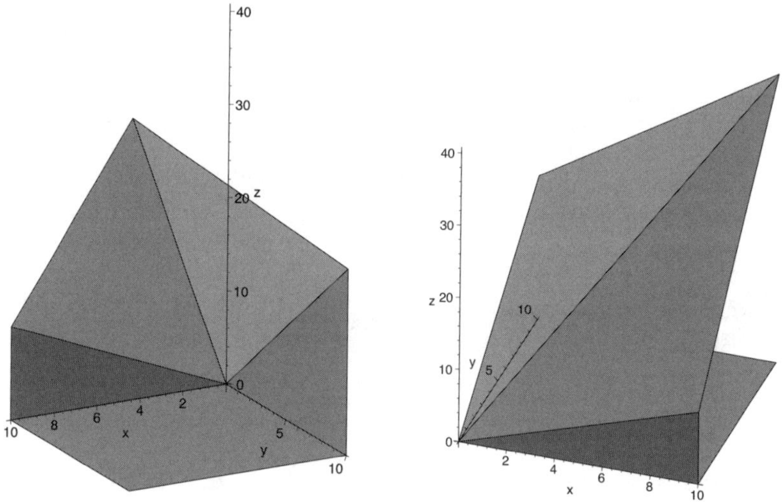

FIGURE 1. View from the front and rotated 90° counter-clockwise around the z-axis.

EXAMPLE 2.6. Let $A = k[[X,Y]]$, with k a field, and $I = (x^3, x^2y, y^2)$. As shown in [7], I has only one associated Rees valuation. Let $J_1 = (x^3y^7)$, $J_2 = (x^4y^6)$, and $J_3 = (x^5y^2)$. Using the methods in Section 4, we can compute $l_I(J_1) = 9/2, l_I(J_2) = 13/3$, and $l_I(J_3) = 8/3$. Then by Corollary 2.4, the limit

$$\lim_{m_1,m_2,m_3\to\infty} \frac{v_I(J_1, J_2, J_3, m_1, m_2, m_3)}{27m_1 + 26m_2 + 16m_3}$$

exists and equals 1/6 since

$$\frac{l_I(J_1)}{27} = \frac{l_I(J_2)}{26} = \frac{l_I(J_3)}{16} = \frac{1}{6}.$$

3. Periodic increase

In this section we take a closer look at the sequence $\{v_I(J,m)\}_m$. To simplify the notation we will simply write $v(m)$ instead of $v_I(J,m)$.

We address the question of whether this sequence increases eventually in a periodic way; that is, whether or not there exists a positive integer t such that $v(m+t) - v(m+t-1) = v(m) - v(m-1)$ for $m \gg 0$, or equivalently, $v(m+t) - v(m) = \textit{constant}$, for $m \gg 0$. Our work is partly motivated by [4, Theorem 8], where Nagata proves that the deviation $v(m) - l_I(J)m$ is bounded. In particular, this implies that there exists a positive constant C such that $0 \leq v(m+t) - v(m) - v(t) < C$ for all m, t.

We begin by defining noetherian filtrations.

DEFINITION 3.1. A family of ideals $\mathcal{F} = \{F_m\}_{m\geq 0}$ in a noetherian ring A is called a filtration if $F_0 = A$, $F_{m+1} \subseteq F_m$, and $F_m F_n \subseteq F_{m+n}$ for all $m, n \geq 0$. We say that the filtration $\{F_m\}_{m\geq 0}$ is noetherian if the associated graded ring $\oplus_{m\geq 0} F_m$ is noetherian. Equivalently, the filtration $\mathcal{F}$ is noetherian if and only if there exists t such that $F_{m+t} = F_m F_t$ for all $m \geq t$ ([1, 4.5.12]).

PROPOSITION 3.2. *Let I, J be ideals in a noetherian local ring A such that $J \subseteq \sqrt{I}$, the ideals I, J are not nilpotent, and $\bigcap_k I^k = (0)$. Assume that J is principal and the ring $\mathcal{B} = \oplus_{m,n} J^m \cap I^n$ is noetherian. Then there exists a positive integer t such that $v(m+t) = v(m) + v(t)$ for all $m \geq t$.*

Proof. In the ring $\oplus_{n\geq 0} I^n$ consider the filtration $\{F_m\}$ with $F_m = \oplus_{n\geq 0} J^m \cap I^n$. Since $\mathcal{B} = \oplus_{m\geq 0} F_m$ is noetherian, there exists a positive integer t such that $F_{m+t} = F_m F_t$ for all $m \geq t$. We will prove that this implies $v(m+t) = v(m) + v(t)$ for all $m \geq t$. First note that the inequality $v(m+t) \geq v(m) + v(t)$ always holds. By contradiction, assume that $v(m+t) > v(m) + v(t)$ for some $m \geq t$. This implies that the component of degree $v(m) + v(t) + 1$ in F_{m+t} is J^{m+t}, and since $F_{m+t} = F_m F_t$ we then obtain

$$J^{m+t} = J^t(J^m \cap I^{v(m)+1}) + J^m(J^t \cap I^{v(t)+1}).$$

Let $J = (z)$. Then we have

$$(z)^{m+t} = z^{m+t}(I^{v(m)+1} : z^m) + z^{m+t}(I^{v(t)+1} : z^t).$$

From the definition of $v(-)$, both $(I^{v(m)+1} : z^m)$ and $(I^{v(t)+1} : z^t)$ are contained in the maximal ideal, and by the Nakayama Lemma, we must have z nilpotent, contradicting our assumptions. □

REMARK 3.3. It is not always true that the ring $\mathcal{B}$ is noetherian. For such an example see [2, Lemma 5.6].

Note that there are a few other natural conditions that ensure the periodic increase of the sequence $\{v(m)\}_m$. We comment on these below.

REMARK 3.4. If the ring $\mathcal{G}(I) = \oplus_{n\geq 0} I^n/I^{n+1}$ is reduced, then we have $v(m) = mv(1)$ for all m. In particular, the sequence $v(m+1) - v(m)$ is constant. Indeed, let $x \in J \setminus I^{v(1)+1}$. The image of x in $I^{v(1)}/I^{v(1)+1} \subseteq \mathcal{G}(I)$ is nonzero, and since $\mathcal{G}(I)$ is reduced, so is the image of x^m in $I^{mv(1)}/I^{mv(1)+1}$. This implies that $J^m \not\subseteq I^{mv(1)+1}$, and hence $v(m) \leq mv(1)$.

The point of view formulated in the above remark can be refined to include the case when J is not necessarily principal, but it comes at the expense of strengthening the hypotheses.

REMARK 3.5. Assume that I is normal and $J = (a_1, \ldots, a_s)$. Then for every m we have $v_I(J,m) = \min\{v_I((a_j),m) \mid j = 1,\ldots,s\}$. Indeed, if $n := \min\{v_I((a_j),m) \mid j = 1,\ldots,s\}$, then $a_j^m \in I^n$ for all $j = 1,\ldots,s$. This implies that $J^m \subseteq \overline{J^m} = \overline{(a_1^m,\ldots,a_s^m)} \subseteq \overline{I^n} = I^n$, so $v_I(J,m) \geq n$. On the other hand, if $v_I(J,m) > n$, we have $J^m \subseteq I^{v_I((a_j),m)+1}$ for some j and hence $a_j^m \in I^{v_I((a_j),m)+1}$, a contradiction. If I is normal and all the rings $\oplus_{m,n}(a_j^m) \cap I^n$ are noetherian ($j = 1,\ldots,s$), by Proposition 3.2 we obtain that there exists t_j such that $v_I((a_j),m+t_j) = v_I((a_j),m) + v_I((a_j),t_j)$ for $m \geq t_j$. If we have $t_1 = t_2 = \cdots = t_s = t$ (i.e., the sequences $v_I((a_j),m)$ increase eventually in a periodic way with the same period), then we have $v_I(J,m+t) = v_I(J,m) + v_I(J,t)$ for $m \geq t$. Indeed, by the above observation, $v_I(J,m+t) = v_I((a_j),m+t_j)$ for some j, and hence $v_I(J,m+t) = v_I((a_j),m) + v_I((a_j),t) \leq v_I(J,m) + v_I(J,t)$. The other inequality always holds.

Note that in the situation described in Remark 3.4, when the associated graded ring $\mathcal{G}(I) = \oplus_{n\geq 0} I^n/I^{n+1}$ is reduced (which implies that I is normal), we have $t_1 = t_2 = \cdots = t_s = 1$.

Our final observation introduces a bigraded ring associated to the ideals J and I that can be used in examining the periodicity of the rate of change of the sequence $\{v(m)\}_m$.

REMARK 3.6. Let $\mathcal{C}$ be the ring $\oplus_{m\geq 0, n\geq 0} F_{m,n}$, with $F_{m,n} = J^m \cap I^n / J^m \cap I^{n+1}$ and multiplication defined naturally such that $F_{m,n}F_{m',n'} \subseteq F_{m+m',n+n'}$. Let $F_m = \oplus_{n\geq 0} F_{m,n}$. Note that F_m is a filtration on $\mathcal{G}(I) = \oplus_{n\geq 0} I^n/I^{n+1}$ and $F_{m,n} = 0$ for $n < v(m)$, while $F_{m,v(m)} \neq 0$ for all m. As in the above remark, one can check that $v(m+t) = v(m) + v(t)$ is equivalent to $F_{m,v(m)}F_{t,v(t)} \neq 0$.

So, if there exists t such that $F_{t,v(t)}$ contains a nonzerodivisor on $\mathcal{C}$, then $v(m+t) = v(m) + v(t)$ for all m. However, note that $\mathcal{C}$ a domain implies that $F_0 = \mathcal{G}(I)$, the associated graded ring of I, is a domain as well, and then Remark 3.4 applies.

4. Computations

In this section we describe a method of determining $L_J(I) = \inf\{m/n \mid J^m \subseteq I^n\}$ (and $l_I(J) = 1/L_J(I)$) for two monomial ideals I and J in a polynomial ring $k[x_1, \ldots, x_r]$ over a field k. Whenever $J = (a_1, \ldots, a_s)$, one has $L_J(I) = \max\{L_{(a_j)}(I) \mid j = 1, \ldots, s\}$ ([8, Theorem 2]), so we may assume that J is a principal ideal. Let $I = (x_1^{b_{i1}} x_2^{b_{i2}} \ldots x_r^{b_{ir}} \mid i = 1, \ldots, t)$ and $J = (x_1^{c_1} x_2^{c_2} \ldots x_r^{c_r})$.

First observe that $J^m \subseteq I^n$ if and only if there exist nonnegative integers $y_1, \ldots, y_t$ with $y_1 + \cdots + y_t = n$ such that

$$\sum_{i=1}^{t} b_{ij} y_i \leq c_j m \quad \text{for all} \quad j = 1, \ldots, r. \tag{4.1}$$

Set $B_{ij} = (1/c_j) b_{ij}$, $z_i = y_i/(y_1 + \cdots + y_t) = y_i/n$ and $z = (z_1, \ldots, z_t) \in \mathbb{Q}^t$.

So $J^m \subseteq I^n$ if and only if there exist $z_i = y_i/n$ with $y_1 + \cdots + y_t = n$ such that

$$\frac{m}{n} \geq \frac{1}{nc_j} \sum_{i=1}^{t} b_{ij} y_i = \sum_{i=1}^{t} B_{ij} z_i \text{ for all } j = 1, \ldots, r. \tag{4.2}$$

Consider the function $\alpha : \mathbb{R}^t \to \mathbb{R}$, $\alpha(z) = \max_{1 \leq j \leq r}\{\sum_{i=1}^{t} B_{ij} z_i\}$ and the subsets of the rationals $\Lambda_1 = \{m/n \mid J^m \subseteq I^n\}$ and $\Lambda_2 = \{\alpha(z) \mid z_1, \ldots, z_t \in \mathbb{Q}_{\geq 0}, z_1 + \cdots + z_t = 1\}$. We will prove that

$$\inf \Lambda_1 = \inf \Lambda_2 \tag{4.3}$$

The inequality $\geq$ follows from (4.2). For the other inequality, we will show that $\Lambda_2 \subseteq \Lambda_1$. Let $\alpha(z) \in \Lambda_2$ with $z_i = p_i/q$ ($1 \leq i \leq t$, $p_1 + \cdots + p_t = q$, and p_i, q nonnegative integers). The coefficients B_{ij} are rationals, so after clearing the denominators we obtain $\alpha(z) = h/lq$ for some nonnegative integers h, l. By (4.2), since $z_i = lp_i/lq$ for all i, we have $h/lq \in \Lambda_1$, which finishes the proof of (4.3).

Note that

$$\inf \Lambda_2 = \inf \{\alpha(z) \mid z_1, \ldots, z_t \in \mathbb{R}_{\geq 0}, z_1 + \cdots + z_t = 1\},$$

so we need to minimize the function

$$\alpha(z) = \max \left\{ \sum_{i=1}^{t} B_{ij} z_i \,\middle|\, j = 1, \ldots, r \right\}$$

subject to the constraints

$$z_1, \ldots, z_t \geq 0 \quad \text{and} \quad z_1 + \cdots + z_t = 1.$$

Let

$$\Delta_k = \left\{ z \in \mathbb{R}_{\geq 0}^t \,\middle|\, \sum_{i=1}^{t} B_{ik} z_i \geq \sum_{i=1}^{t} B_{ij} z_i \text{ for all } j \neq k \right\}.$$

Clearly $\Delta_1 \cup \cdots \cup \Delta_r = \mathbb{R}^t_{\geq 0}$, so it is enough to minimize the function α on each Δ_k.

In conclusion, for each $k = 1, \dots, r$, the problem reduces to minimizing the objective function

$$\alpha(z) = \sum_{i=1}^{t} B_{ik} z_i$$

subject to the constraints

$$z_1, \dots, z_t \geq 0, \quad z_1 + \cdots + z_t = 1$$

and

$$\sum_{i=1}^{t} B_{ik} z_i \geq \sum_{i=1}^{t} B_{ij} z_i \quad \text{for all} \quad j \neq k.$$

This is a classical problem linear programming problem which can be algorithmically solved using the simplex method.

REMARK 4.1. In general, the limits $l_I(J)$ and $L_j(I)$ need not be reached by an element of the sequences $\{v_I(J, m)/m\}_m$ and $\{w_J(I, n)/n\}_n$, respectively. However, in the monomial case, as the procedure described above shows, there exists a pair (m, n) with $J^m \subseteq I^n$ and $L_J(I) = n/m$.

EXAMPLE 4.2. Let $A = k[x, y]$ and $I = (x^3, x^2y, y^2)$, $J = (x^3y^7)$. In this case, $b_{11} = 3, b_{12} = 0, b_{21} = 2, b_{22} = 1, b_{31} = 0, b_{32} = 2$, $c_1 = 3, c_2 = 7$ and $B_{11} = 3/3 = 1, B_{12} = 0/7 = 0, B_{21} = 2/3, B_{22} = 1/7, B_{31} = 0, B_{32} = 2/7$. Then

$$\Delta_1 = \left\{(z_1, z_2, z_3) \in \mathbb{R}^3_{\geq 0} \mid z_1 + (2/3)z_2 \geq (1/7)z_2 + (2/7)z_3\right\}$$

and

$$\Delta_2 = \left\{(z_1, z_2, z_3) \in \mathbb{R}^3_{\geq 0} \mid (1/7)z_2 + (2/7)z_3 \geq z_1 + (2/3)z_2\right\}.$$

By using a computer algebra system that has the simplex method implemented, one can obtain that the minimum on each of the sets Δ_1 and Δ_2 is $2/9$, and hence $L_J(I) = 2/9$.

In fact, the minimum can occur only at the intersection of various regions Δ_k (in our case on $\Delta_1 \cap \Delta_2$), for there are no critical points in the interior of Δ_k.

ACKNOWLEDGEMENT. The authors would like to thank Robert Lazarsfeld for a talk which inspired them to consider the problem treated in the article. They also thank Mel Hochster for pointing out to them the example mentioned in Remark 3.3.

REFERENCES

[1] W. Bruns and J. Herzog, *Cohen-Macaulay rings*, Cambridge Studies in Advanced Mathematics, vol. 39, Cambridge University Press, Cambridge, 1993. MR 1251956 (95h:13020)

[2] J. B. Fields, *Lengths of Tors determined by killing powers of ideals in a local ring*, J. Algebra **247** (2002), 104–133. MR 1873386 (2003a:13019)

[3] R. Hübl and I. Swanson, *Discrete valuations centered on local domains*, J. Pure Appl. Algebra **161** (2001), 145–166. MR 1834082 (2002f:13006)

[4] M. Nagata, *Note on a paper of Samuel concerning asymptotic properties of ideals*, Mem. Coll. Sci. Univ. Kyoto. Ser. A. Math. **30** (1957), 165–175. MR 0089836 (19,727c)

[5] D. Rees, *Valuations associated with ideals. II*, J. London Math. Soc. **31** (1956), 221–228. MR 0078971 (18,8b)

[6] ———, *A note on analytically unramified local rings*, J. London Math. Soc. **36** (1961), 24–28. MR 0126465 (23 #A3761)

[7] J. D. Sally, *One-fibered ideals*, Commutative algebra (Berkeley, CA, 1987), Math. Sci. Res. Inst. Publ., vol. 15, Springer, New York, 1989, pp. 437–442. MR 1015533 (90h:13003)

[8] P. Samuel, *Some asymptotic properties of powers of ideals*, Ann. of Math. (2) **56** (1952), 11–21. MR 0049166 (14,128c)

CĂTĂLIN CIUPERCĂ, DEPARTMENT OF MATHEMATICS, NORTH DAKOTA STATE UNIVERSITY, FARGO, ND 58105, USA

E-mail address: catalin.ciuperca@ndsu.edu

FLORIAN ENESCU, DEPARTMENT OF MATHEMATICS AND STATISTICS, GEORGIA STATE UNIVERSITY, ATLANTA, GA 30303, USA

E-mail address: fenescu@gsu.edu

SANDRA SPIROFF, DEPARTMENT OF MATHEMATICS, SEATTLE UNIVERSITY, SEATTLE, WA 98122, USA

E-mail address: spiroffs@seattleu.edu

Illinois Journal of Mathematics
Volume 51, Number 1, Spring 2007, Pages 41–56
S 0019-2082

FACTORIZATIONS OF BIRATIONAL EXTENSIONS OF LOCAL RINGS

STEVEN DALE CUTKOSKY AND HEMA SRINIVASAN

In honor of Phillip Griffith

ABSTRACT. We give a proof of local strong factorization of a birational, monomial extension of regular local rings along a valuation of rank 1 and maximal rational rank. Our proof uses methods from linear algebra, and is in the spirit of Christensen's proof of this result in dimension 3. This has also been proven by Karu using toric geometry.

1. Introduction

Suppose that R and S are regular local rings such that S dominates R ($R \subset S$ and the maximal ideal m_S of S contracts to the maximal ideal m_R of R).

$R \to S$ is monomial if there exist regular parameters $x_1, \ldots, x_m$ in R, $y_1, \ldots, y_n$ in S, an $m \times n$ matrix $A = (a_{ij})$ of rank m whose entries are nonnegative integers and units $\delta_i \in S$ such that

$$x_i = \prod_{j=1}^{n} y_j^{a_{ij}} \delta_i$$

for $1 \leq i \leq m$.

Suppose that $P \subset R$ is a regular prime (R/P is a regular local ring) and $0 \neq f \in P$. The regular local ring $R_1 = R[\frac{P}{f}]_m$, where m is a maximal ideal of $R[\frac{P}{f}]$ containing m_R, is called a monoidal transform of R.

Suppose that V is a valuation ring of the quotient field of S which dominates S (and thus dominates R). Then given a regular prime P of R (or of S) there exists a unique monoidal transform R_1 of R (or S_1 of S) obtained from P such that V dominates R_1 (or V dominates S_1).

Received January 19, 2006; received in final form November 14, 2006.
2000 *Mathematics Subject Classification*. 13B10, 13A18, 14E05.
Research of Steven Dale Cutkosky partially supported by NSF.

The local monomialization theorem of [C2] and [C4] shows that given an extension $R \to S \subset V$ as above such that R, S are essentially of finite type over a field k of characteristic zero, there exists a commutative diagram

$$\begin{array}{ccc} R_1 & \to & S_1 \quad \subset V \\ \uparrow & & \uparrow \\ R & \to & S \end{array}$$

such that the vertical arrows are products of monoidal transforms and $R_1 \to S_1$ is monomial.

Suppose that we further have that $R \to S$ is birational (the induced homomorphism of quotient fields is an isomorphism). If $R \to S$ is monomial and birational, then we can find regular parameters $\overline{y}_1, \ldots, \overline{y}_n$ in S such that

$$x_i = \prod_{j=1}^{n} \overline{y}_j^{a_{ij}}$$

for $1 \leq i \leq n$ (since $B = A^{-1}$ has integral coefficients).

We may now state Abhyankar's local factorization conjecture (page 237 of [Ab]). Suppose that $R \to S$ is a birational extension of regular local rings of dimension $n \geq 3$ and V is a valuation ring of the quotient field of S such that V dominates R. The conjecture is that there exists a commutative diagram

$$(1.1) \qquad \begin{array}{ccccc} & & T & \subset & V \\ & \nearrow & & \nwarrow & \\ R & & \to & & S \end{array}$$

where the northeast and northwest arrows are products of monoidal transforms.

It is proven in [Z] and [Ab1] that there is a direct factorization of $R \to S$ by monoidal transforms if $n = 2$. However, examples of the failure of a direct factorization of $R \to S$ by monoidal transforms are given in [Sa] and [Sh] when $n \geq 3$.

The local factorization theorem is proven when $n = 3$ (and R is essentially of finite type over a field of characteristic 0) in [C1, Theorem A].

In [C2, Theorem 1.9] it is proven that the local monomialization theorem ([C2, Theorem 1.1]) and "strong factorization" of birational toric morphisms of nonsingular toric varieties implies the local factorization theorem in all dimensions (in characteristic zero).

There are two published proofs of "strong factorization" of birational toric morphisms, [Mo] and [AMR]. They have both been found to have errors (as explained in the correction [AMR1] to [AMR]).

Suppose that R is essentially of finite type over a field. In [C2], a strong version of local monomialization is used to reduce the proof of local factorization to the following problem, which is essentially in linear algebra.

We assume that $R \to S$ is monomial, with respect to regular parameters $x_1, \dots, x_n$ in R and $y_1, \dots, y_n$ in S, the value group of V is contained in $\mathbf{R}$, and if ν is a valuation of the quotient field of S whose valuation ring is V, then

$$\tau_1 = \nu(y_1), \dots, \tau_n = \nu(y_n)$$

are rationally independent real numbers.

In this special case, we can assume that $R = k[x_1, \dots, x_n]_{(x_1,\dots,x_n)}$ and $S = k[y_1, \dots, y_n]_{(y_1,\dots,y_n)}$, where k is a field. We have expressions $x_i = \prod_{j=1}^n y_j^{a_{ij}}$ for $1 \le i \le n$. If

$$f = \sum \alpha_{i_1,\dots,i_n} y_1^{i_1} \cdots y_n^{i_n} \in k[y_1, \dots, y_n],$$

we have $\nu(f) = \min\{i_1\tau_1 + \cdots + i_n\tau_n \mid \alpha_{i_1,\dots,i_n} \neq 0\}$. We will call the local factorization conjecture in this special case the "monomial problem".

When $n = 3$, the monomial problem is solved by Christensen [Ch]. In [C2, Theorem 1.6], the first author extends this to prove a weaker form of the monomial problem for all n. By combining this with the local monomialization theorem of [C2], it was proved in [C2] that a birational extension $R \to S$ can be factored by $n - 2$ triangles of monoidal transforms.

Recently, there has been a proof by Karu [K] of this monomial problem, using toric geometry.

In this paper, we give a self-contained proof of the monomial problem. We solve the problem in the spirit of Christensen's original theorem in dimension 3. In particular, the problem can be stated completely in the language of linear algebra, and we prove it using linear algebra. As a result, we give an explicit algorithm for the solution of the monomial problem. This theorem (Theorem 2.1) is proven in Section 2 of this paper. The solution to the monomial problem is given in Theorem 3.1.

We show in Theorem 3.3 of Section 3 of this paper how the local monomialization theorem, [C2, Theorem 1.1] and Theorem 2.1 of this paper prove the local factorization conjecture. This provides a complete proof to Theorem 1.9 of [C2].

A monoidal transform affects the coefficient matrix A as a column addition. The valuation can be understood as a column vector $\vec{v}$ of positive rational numbers. To preserve the property that the valuation ring dominates the monodial transform of the local ring, we allow only those column operations on A that keep both A and $A^{-1}\vec{v}$ positive. We construct an algorithm here for finding a sequence of permissible column additions and interchanges to be followed by a sequence of permissible subtractions that results in the identity matrix.

2. Matrix factorization

Suppose that $A = (a_{ij})$ is an $n \times n$ matrix with coefficients which are nonnegative integers and $\mathrm{Det}(A) = \pm 1$. Further suppose that $\vec{v} = (v_1, \ldots, v_n)^t$ is a $n \times 1$ column vector with coefficients which are positive rationally independent real numbers, and $\vec{w} = (w_1, \ldots, w_n)^t = A^{-1}\vec{v}$ is a vector with positive coefficients (which are necessarily rationally independent). $(A, \vec{v}, \vec{w})$ satisfying these conditions will be called an n-dimensional triple.

The column addition C_{ij} of A which adds the j-th column of A to the i-th column is called *permissible* for $(A, \vec{v}, \vec{w})$ if $w_j - w_i > 0$. The triple $(A, \vec{v}, \vec{w})$ is transformed under the permissible column addition C_{ij} to the triple $(A(1), \vec{v}(1), \vec{w}(1))$, where $A(1) = (a(1)_{ij})$ is obtained from A by adding the j-th column of A to the i-th column, $\vec{v}(1) = \vec{v}$ and $\vec{w}(1) = (w(1)_1, \ldots, w(1)_n)^t = A(1)^{-1}\vec{v}(1)$. $\vec{w}(1)$ is obtained from $\vec{w}$ by subtracting the i-th coefficient w_i from the j-th coefficient w_j of $\vec{w}$.

The row subtraction R_{ji} of A which subtracts the i-th row of A from the j-th row is called *permissible* for $(A, \vec{v}, \vec{w})$ if $a_{jk} \geq a_{ik}$ for $1 \leq k \leq n$. The triple $(A, \vec{v}, \vec{w})$ is transformed under the permissible row subtraction R_{ji} to the triple $(A(1), \vec{v}(1), \vec{w}(1))$, where $A(1) = (a(1)_{ij})$ is obtained from A by subtracting the i-th row of A from the j-th row, $\vec{v}(1)$ is obtained from $\vec{v}$ by subtracting the i-th coefficient v_i from the j-th coefficient v_j and $\vec{w}(1) = (w(1)_1, \ldots, w(1)_n)^t = A(1)^{-1}\vec{v}(1)$. We have that $\vec{w}(1) = \vec{w}$.

The row interchange T_{ij} of A interchanges the i-th and j-th rows of A. T_{ij} transforms the triple $(A, \vec{v}, \vec{w})$ into the triple $(A(1), \vec{v}(1), \vec{w}(1))$, where $A(1)$ is obtained from A by interchanging the i-th and j-th row, $\vec{w}(1) = \vec{w}$ and $\vec{v}(1)$ is obtained from $\vec{v}$ by interchanging the i-th and j-th row of $\vec{v}$.

In this section, we prove the following theorem:

THEOREM 2.1. *Suppose that $A = (a_{ij})$ is an $n \times n$ matrix with coefficents which are nonnegative integers and* $\mathrm{Det}(A) = \pm 1$. *Further suppose that $\vec{v} = (v_1, \ldots, v_n)^t$ is a $n \times 1$ vector with coefficients which are positive rationally independent real numbers. Then there exists a sequence of permissible column additions and row interchanges*

$$(A, \vec{v}, \vec{w}) \to (A(1), \vec{v}, \vec{w}(1)) \to \cdots \to (A(s), \vec{v}, \vec{w}(s))$$

followed by a sequence of permissible row subtractions

$$(A(s), \vec{v}, \vec{w}(s)) \to (A(s+1), \vec{v}(s+1), \vec{w}(s)) \to \cdots \to (A(t), \vec{v}(t), \vec{w}(t))$$

such that $A(t)$ is the $n \times n$ identity matrix.

We will denote the inverse of a matrix A by $B = (b_{ij}) = A^{-1}$. If a permissible column addition C_{ij} is performed by adding the j-th column of A to the i-th column, with a resulting transformation of triples $(A, \vec{v}, \vec{w}) \to (A(1), \vec{v}, \vec{w}(1))$, then $B(1) = (b(1)_{ij}) = A(1)^{-1}$ is obtained from $B = A^{-1}$ by subtracting the i-th row of B from the j-th row, since $C_{ij}^{-1} = R_{ji}$ and

$$B(1) = A(1)^{-1} = (AC_{ij})^{-1} = C_{ij}^{-1}A^{-1} = R_{ji}A^{-1}.$$

Similarly, if a permissible row subtraction R_{ji} is performed by subtracting the i-th row of A from the j-th row, with a resulting transformation of triples $(A, \vec{v}, \vec{w}) \to (A(1), \vec{v}(1), \vec{w})$, then $B(1) = (b(1)_{ij}) = A(1)^{-1}$ is obtained from $B = A^{-1}$ by adding the j-th column of B to the i-th column.

If a permissible row interchange T_{ij} is performed, then $B(1) = A(1)^{-1}$ is obtained from B by interchanging the i-th and j-th column.

Given a triple $(A, \vec{v}, \vec{w})$, we define $\beta = \max_k\{|b_{k1}|\}$. We will write $A = (C_1, \ldots, C_n)$.

To simplify notation, we will denote the inverse of a matrix $A(t)$ by $B(t) = (b_{ij}(t))$, and define $\beta(t) = \max_k\{|b_{k1}(t)|\}$. We will denote $A(t) = (C_1(t), \ldots, C_n(t))$.

REMARK 2.2. Fix i and j. Either C_{ji} is permissible or C_{ij} is permissible (but not both). If C_{ij} is permissible, then after performing C_{ij} a finite number of times, C_{ji} becomes permissible. This is because C_{ij} decreases w_j by a positive integral multiple of w_i.

DEFINITION 2.3. A permissible C_{ij} is allowable for the triple $(A, \vec{v}, \vec{w})$ if b_{i1} and b_{j1} are both non-zero and have the same sign.

DEFINITION 2.4. A permissible C_{ij} is *-allowable for the triple $(A, \vec{v}, \vec{w})$ if either $b_{i1}b_{j1} = 0$, or C_{ij} is allowable.

REMARK 2.5. (1) If we perform a *-allowable C_{ij} on the triple $(A, \vec{v}, \vec{w})$ to get $(A(1), \vec{v}, \vec{w}(1))$, then $b_{j1}(1) = b_{j1} - b_{i1}$, $b_{k1}(1) = b_{k1}$ if $k \neq j$ and thus

$$\beta(1) = \max_k\{|b_{k1}(1)|\} \leq \max_k\{|b_{k1}|\} = \beta.$$

(2) Suppose that we fix i and j. Then after a finite sequence consisting of allowable C_{ij} and C_{ji}, both C_{ij} and C_{ji} are not allowable. If at least one of b_{i1}, b_{j1} is nonzero, then after a finite sequence consisting of *-allowable C_{ij} and C_{ji}, both C_{ij} and C_{ji} are not *-allowable.

Proof of (2). If b_{i1} and b_{j1} are nonzero of the same sign, and we perform C_{ij} (or C_{ji}) to obtain the new triple $(A(1), \vec{v}, \vec{w}(1))$, and $b_{i1}(1)$, $b_{j1}(1)$ have the same sign, we then obtain that $(|b_{i1}|, |b_{j1}|) > (|b_{i1}(1)|, |b_{j1}(1)|)$ in the Lex order on $\mathbf{Z}^2$.

Suppose that $b_{i1} \neq 0$ and $b_{j1} = 0$. If C_{ij} is *-allowable, then after performing C_{ij}, we obtain that both C_{ij} and C_{ji} are not *-allowable. If C_{ji} is *-allowable, and we perform C_{ji}, then $b_{i1}(1) = b_{i1}$, $b_{j1}(1) = 0$. By Remark 2.2, we can only perform C_{ji} a finite number of consecutive times. $\square$

LEMMA 2.6. *There exists a sequence of allowable column additions*

$$(A,\vec{v},\vec{w}) \to (A(1),\vec{v},\vec{w}(1)) \to \cdots \to (A(t),\vec{v},\vec{w}(t))$$

such that at most two entries of the first column of $B(t)$ are nonzero.

The proof of this lemma is immediate from [C2, Theorem 6.3].

LEMMA 2.7. *There exists a sequence of *-allowable column additions*

$$(A,\vec{v},\vec{w}) \to (A(1),\vec{v},\vec{w}(1)) \to \cdots \to (A(s),\vec{v},\vec{w}(s))$$

such that there are indices i and j with $b_{i1}(s) = 1$, $b_{j1}(s) = -1$ and $b_{l1} = 0$ if $l \neq i$ and $l \neq j$.

Proof. By Lemma 2.6, there exists a sequence of allowable column additions $(A,\vec{v},\vec{w}) \to (A(t_1),\vec{v},\vec{w}(t_1))$ such that at most two entries of the first column of $B(t_1)$ are nonzero. Without loss of generality, we may assume that $b_{k1} = 0$ if $k \neq 1$ or 2.

First assume that one of b_{11} or b_{21} is zero. We may suppose that $b_{21} = 0$. Then since $\mathrm{Det}(B) = \pm 1$, we have that $b_{11} = \pm 1$. As in Remark 2.2, we can (if necessary) perform the permissible column addition C_{21} a finite number of times so that the column addition C_{12} is permissible. We can then perform C_{12} to get a matrix which satisfies the conclusions of the lemma in this case.

Now assume that both b_{11} and b_{21} are nonzero. Since $b_{11}C_1 + b_{21}C_2 = e_1$, where $e_1 = (1, 0, \ldots, 0)^t$, it follows that b_{11} and b_{21} have opposite signs. Recall that $\beta = \max\{|b_{11}|, |b_{21}|\}$. If $\beta = 1$, then we have obtained the conclusions of the theorem.

Assume that $\beta > 1$. We will show that we can construct a sequence of column additions in the first 3 columns which are *-allowable

$$(A,\vec{v},\vec{w}) \to (A(1),\vec{v},\vec{w}(1)) \to \cdots \to (A(s_1),\vec{v},\vec{w}(s_1)) \tag{2.1}$$

such that $\beta(s_1) < \beta$.

Once we have established the existence of the sequence (2.1), we can apply Lemma 2.6 to construct a sequence of allowable column additions

$$(A(s_1),\vec{v},\vec{w}(s_1)) \to (A(s_1+1),\vec{v},\vec{w}(s_1+1)) \to \cdots \to (A(s_2),\vec{v},\vec{w}(s_2)) \tag{2.2}$$

such that at most two of the entries in the first column of $B(s_2)$ are nonzero, and $\beta(s_2) \leq \beta(s_1) < \beta$. We can thus alternate sequences (2.1) and (2.2) to eventually obtain the conclusions of the theorem.

It remains to prove that we can construct a sequence (2.1).

Since $\mathrm{Det}(B) = \pm 1$, and $\beta > 1$, we must have that the maximum β is obtained by only one of $|b_{11}|$ and $|b_{21}|$. Without loss of generality, we may assume that

$$\beta = |b_{11}| > |b_{21}|.$$

We now perform a finite sequence of *-allowable column additions C_{32}, followed by a *-allowable column addition C_{23} to obtain a sequence of transformations of triples

$$(A, \vec{v}, \vec{w}) \to \cdots \to (A(t_1), \vec{v}, \vec{w}(t_1)),$$

where the first column of $B(t_1)$ is

$$(b_{11}(t_1), b_{21}(t_2), \ldots, b_{n1}(t_1))^t = (b_{11}, b_{21}, -b_{21}, 0, \ldots, 0)^t,$$

with $\beta(t_1) = |b_{11}(t_1)| = |b_{11}| = \beta$, and either C_{13} or C_{31} is allowable.

If C_{31} is allowable on $(A(t_1), \vec{v}, \vec{w}(t_1))$, we perform it to get

$$|b_{11}(t_1 + 1)| = |b_{11}(t_1) - b_{31}(t_1)| = |b_{11} + b_{21}| < \beta(t_1) = \beta$$

and we stop.

If not, we have that C_{13} is allowable and after that, $b_{11}(t_1 + 1) = b_{11}$ and

$$b_{31}(t_1 + 1) = b_{31}(t_1) - b_{11}(t_1) = -b_{21} - b_{11}$$

have opposite signs. Further, $\beta(t_1+1) = \beta(t_1)$ and $w_3(t_1+1) = w_3(t_1) - w_1 \leq w_3 - w_1$. Now, C_{32} or C_{23} must be allowable.

Now we perform a finite sequence of *-allowable column additions C_{32}, and *-allowable column additions C_{23}, to obtain a sequence of transformations of triples

$$(A(t_1 + 1), \vec{v}, \vec{w}(t_1 + 1)) \to \cdots \to (A(t_2), \vec{v}, \vec{w}(t_2)),$$

where $\beta(t_2) = \beta(t_1 + 1) = \beta$,

$$\max\{| b_{21}(t_2) |, | b_{31}(t_2) |\} <| b_{11}(t_2) |= \beta(t_2) = \beta,$$

and $b_{21}(t_2)$ and $b_{31}(t_2)$ have opposite signs. One of C_{13}, C_{31}, C_{12} or C_{21} must now be allowable.

Performing an allowable C_{31} or C_{21} decreases β and we stop. If not, we perform C_{13} or C_{12} to get $\beta(t_2+1) = \beta(t_2)$ and none of the four C_{13}, C_{31}, C_{12} and C_{21} are allowable.

Further, $w_2(t_2 + 1)$ or $w_3(t_2 + 1)$ is reduced by w_1, so that,

$$w_2(t_2 + 1) + w_3(t_2 + 1) = w_2(t_2) + w_3(t_2) - w_1 \leq w_2 + w_3 - 2w_1.$$

Now either C_{32} or C_{32} becomes allowable and we repeat this process. Since we can perform a C_{13} or a C_{12} at most $[(w_2 + w_3)/w_1]$ times, we must achieve a reduction in β after a finite number of steps. □

LEMMA 2.8. *Let $(A, \vec{v}, \vec{w})$ be a triple such that $A = (C_1, \ldots, C_n)$ satisfies the relation*

$$C_k = C_1 - e_1$$

for some k, where $e_1 = (1, 0, \ldots, 0)^t$. Let A_{11} be the matrix obtained from A be deleting the first row and column. Then

$$\mathrm{Det}(A_{11}) = \mathrm{Det}(A) = \pm 1.$$

Let $\tilde{v} = (v_2, \ldots, v_n)^t$ *and* $\tilde{w} = (\tilde{w}_2, \ldots, \tilde{w}_n) = A_{11}^{-1}\tilde{v}$. *Then*

$$\tilde{w}_j = w_j \text{ for } j \neq k$$

and

$$\tilde{w}_k = w_1 + w_k.$$

Proof. Set $\lambda = \text{Det}(A) = \pm 1$. Subtracting the k-th column of A from the first column, we see that $\text{Det}(A_{11}) = \lambda$. We thus have that $B = A^{-1} = \lambda\,\text{adj}(A)$, and $A_{11}^{-1} = \lambda\,\text{adj}(A_{11})$. Let

$$A_{11}^{-1} = \lambda\,\text{adj}(A_{11}) = \begin{pmatrix} x_{22} & x_{23} & \cdots & x_{2n} \\ x_{32} & x_{33} & \cdots & x_{3n} \\ & \vdots & & \\ x_{n2} & x_{n3} & \cdots & x_{nn} \end{pmatrix}.$$

Since $C_k = C_1 - e_1$ and $\text{adj}(A) = \lambda A^{-1}$, the first column of $\text{adj}(A)$ is

$$(\lambda, 0, \ldots, 0, -\lambda, 0, \ldots, 0)^t,$$

where $-\lambda$ occurs in the k-th row.

We will compute the entry λb_{ij} in the i-th row and j-th column of $\text{adj}(A)$. Let A_{ji} be the matrix obtained from A by deleting the j-th row and i-th column.

First suppose that $i \neq 1$, $i \neq k$ and $j > 1$. Subtracting the k-th column of A_{ji} (the $(k-1)$-st column of A_{ji} if $i < k$) from the first column, and expanding the determinant along the first column, we get that the (i,j)-th entry of $\text{adj}(A)$ is

$$\begin{aligned}(-1)^{i+j}\,\text{Det}(A_{ji}) &= (-1)^{i+j}\,\text{Det}[(A_{ji})_{11}] \\ &= (-1)^{i+j}\,\text{Det}[(A_{11})_{j-1,i-1}] \\ &= \lambda x_{i,j}.\end{aligned}$$

For $j > 1$, set

$$t_j = \lambda(-1)^{1+j}\,\text{Det}(A_{j1}).$$

We have that the $(1,j)$-th entry of $\text{adj}(A)$ is λt_j. We expand

$$
\begin{aligned}
(-1)^{1+j}\operatorname{Det}(A_{j1}) &= (-1)^{1+j}\operatorname{Det}\begin{pmatrix} a_{12} & \cdots & a_{11}-1 & \cdots \\ & \vdots & & \\ a_{n2} & \cdots & a_{n1} & \cdots \end{pmatrix} \\
&= (-1)^{1+j}\operatorname{Det}\begin{pmatrix} a_{12} & \cdots & a_{11} & \cdots \\ & \vdots & & \\ a_{n2} & \cdots & a_{n1} & \cdots \end{pmatrix} \\
&\qquad + (-1)^{1+j+1+k-2}\operatorname{Det}[(A_{11})_{j-1,k-1}] \\
&= (-1)^{1+j+k-2}\operatorname{Det}(A_{jk}) + (-1)^{j+k}\operatorname{Det}[(A_{11})_{j-1,k-1}] \\
&= -(-1)^{j+k}\operatorname{Det}(A_{jk}) + \lambda x_{kj}.
\end{aligned}
$$

We see that, for $j > 1$, the (k, j)-th entry of $\operatorname{adj}(A)$ is $\lambda(x_{k,j} - t_j)$.

In conclusion,

$$
B = A^{-1} = \lambda \operatorname{adj}(A) = \begin{pmatrix} 1 & t_2 & \cdots & t_n \\ 0 & x_{22} & \cdots & x_{2n} \\ & \vdots & & \\ -1 & x_{k2} - t_2 & \cdots & x_{kn} - t_n \\ & \vdots & & \\ 0 & x_{n2} & \cdots & x_{nn} \end{pmatrix}.
$$

Now we see that

$$
\begin{aligned}
\tilde{w}_i &= (x_{i2}, \ldots, x_{in})(v_2, \ldots, v_n)^t \\
&= (0, x_{i2}, \ldots, x_{in})(v_1, \ldots, v_n)^t \\
&= w_i
\end{aligned}
$$

if $i \neq k$, and

$$
\begin{aligned}
\tilde{w}_k &= \sum_{j=2}^{n} x_{kj} v_j \\
&= \left[-v_1 + \sum_{j=2}^{n} (x_{kj} - t_j) v_j \right] + \left[v_1 + \sum_{j=2}^{n} t_j v_j \right] \\
&= w_k + w_1
\end{aligned}
$$

□

Now we prove Theorem 2.1.

A quadruple $(A, \vec{v}, \vec{w}, k)$ is a triple $(A, \vec{v}, \vec{w})$ and a number k with $1 < k \leq n$ such that if $A = (C_1, \ldots, C_n)$, then $C_k = C_1 - e_1$. To a quadruple $(A, \vec{v}, \vec{w}, k)$ we associate an $(n-1)$-dimensional triple $(\tilde{A} = A_{11}, \tilde{v}, \tilde{w})$ with the notation of Lemma 2.8.

A permissible transformation for the quadruple $(A,\vec{v},\vec{w},k)$ is a series of permissible column additions and row interchanges which transform the triple $(A,\vec{v},\vec{w})$ to a triple $(A(1),\vec{v},\vec{w}(1))$ and a number j with $1<j\le n$, such that $(A(1),\vec{v},\vec{w}(1),j)$ is a quadruple.

By Lemma 2.7, and since $B=A^{-1}$, there exists a sequence of permissible column additions, possibly followed by some row interchanges T_{ij}, $(A,\vec{v},\vec{w})\to(A(1),\vec{v},\vec{w}(1))$, and a number $k(1)$ such that $(A(1),\vec{v},\vec{w}(1),k(1))$ is a quadruple. Without loss of generality, we may assume that there exists a number k such that $(A,\vec{v},\vec{w},k)$ is a quadruple.

If $n=3$, then after expanding the determinant of $\tilde{A}$, we see that after possibly performing the row interchange T_{23}, there is a sequence of permissible row subtractions R_{ji} which transform $\tilde{A}$ into the identity matrix. If $n>3$, we assume by induction that there exists a sequence of permissible column additions and row interchanges

$$(\tilde{A},\tilde{v},\tilde{w})\to(\tilde{A}(1),\tilde{v},\tilde{w}(1))\to\cdots\to(\tilde{A}(s),\tilde{v},\tilde{w}(s)) \tag{2.3}$$

followed by a sequence of permissible row subtractions

$$(\tilde{A}(s),\tilde{v},\tilde{w}(s))\to(\tilde{A}(s+1),\tilde{v}(s+1),\tilde{w}(s))\to\cdots\to(\tilde{A}(l),\tilde{v}(l),\tilde{w}(s)) \tag{2.4}$$

such that $\tilde{A}(l)$ is the $(n-1)\times(n-1)$ identity matrix.

We will first construct a sequence of permissible transformations of quadruples

$$(A,\vec{v},\vec{w},k)\to(A(1),\vec{v},\vec{w}(1),k(1))\to\cdots\to(A(s),\vec{v},\vec{w}(s),k(s)) \tag{2.5}$$

such that for $1\le t\le s$, we have $A(t)_{11}=\tilde{A}(t)$ and $\tilde{w}(t)=A(t)_{11}^{-1}(v_2,\ldots,v_n)^t$.

Suppose that we have constructed (2.5) out to $(A(t),\vec{v},\vec{w}(t),k(t))$, and $t<s$. We will construct $(A(t+1),\vec{v},\vec{w}(t+1),k(t+1))$.

First suppose that $\tilde{A}(t+1)$ is obtained from $\tilde{A}(t)$ by interchanging the i-th and j-th row. Let the triple $(A(t+1),\vec{v}(t+1),\vec{w}(t+1))$ be obtained from the triple $(A(t),\vec{v},\vec{w}(t))$ by performing the row interchange T_{ij}. Then the row interchange T_{ij} determines a permissible transformation of $(A(t),\vec{v},\vec{w}(t),k(t))$ to $(A(t+1),\vec{v},\vec{w}(t+1),k(t))$, such that $A(t+1)_{11}=\tilde{A}(t+1)$.

Suppose that $\tilde{A}(t+1)$ is obtained from $\tilde{A}(t)$ by adding the j-th column of $\tilde{A}(t)$ to the i-th column. We necessarily have that $\tilde{w}_j(t)>\tilde{w}_i(t)$. Set $k=k(t)$.

If $i\ne k$ and $j\ne k$, then we have (by Lemma 2.8) that $\vec{w}_j(t)>\vec{w}_i(t)$, and thus the column addition C_{ij} determines a permissible transformation of $(A(t),\vec{v},\vec{w}(t),k(t))$ to $(A(t{+}1),\vec{v},\vec{w}(t{+}1),k(t))$, such that $A(t{+}1)_{11}=\tilde{A}(t{+}1)$.

Suppose that $i=k$. Then $\tilde{w}_j(t)>\tilde{w}_k(t)$. Since $\tilde{w}_k(t)=w_1(t)+w_k(t)$ and $\tilde{w}_j(t)=w_j(t)$ (by Lemma 2.8), we can construct a permissible transformation of quadruples $(A(t),\vec{v},\vec{w}(t),k(t))\to(A(t+1),\vec{v},\vec{w}(t+1),k(t))$ by first performing the permissible column addition C_{kj} followed by the permissible column addition C_{1j}. We have that $A(t+1)_{11}=\tilde{A}(t+1)$.

Suppose that $j=k$. Then $\tilde{w}_k(t)>\tilde{w}_i(t)$.

If $w_1(t) > w_i(t)$, then we define a permissible transformation of quadruples

$$(A(t), \vec{v}, \vec{w}(t), k(t)) \to (A(t+1), \vec{v}, \vec{w}(t+1), k(t))$$

by performing the permissible column addition C_{i1}.

Suppose that $w_1(t) < w_i(t)$. If $(\overline{A}, \vec{v}, \overline{w})$ is the triple obtained from $(A(t), \vec{v}, \vec{w}(t))$ by C_{1i}, then we have that the i-th coefficient of $\overline{w}$ is $\overline{w}_i = w_i(t) - w_1(t)$. Since $\tilde{w}_k(t) > \tilde{w}_i(t)$, we must have that $w_1(t) + w_k(t) > w_i(t)$, which implies that $\overline{w}_k(t) > \overline{w}_i(t)$. Thus we can construct a permissible transformation of quadruples

$$(A(t), \vec{v}, \vec{w}(t), k(t)) \to (A(t+1), \vec{v}, \vec{w}(t+1), k(t+1) = i)$$

by first performing the permissible column addition C_{1i} followed by the permissible column addition C_{ik}. We have that $A(t+1)_{11} = \tilde{A}(t+1)$.

We can thus inductively construct the sequence (2.5). Let $k = k(s)$. Since $C_k(s) = C_1(s) - e_1$, where $C_k(s)$, $C_1(s)$ are the k-th and first columns of $A(s)$, the sequence of permissible row subtractions of (2.4) gives a sequence of permissible row subtractions $(A(s), \vec{v}, \vec{w}(s)) \to (A(l), \vec{v}(l), \vec{w}(s))$ such that $A(l)$ is a matrix, where $A(l)_{11}$ is an identity matrix, $a_{1k}(l) = a_{11}(tl) - 1$, $a_{k1}(l) = 1$, and $a_{i1}(l) = 0$ if $i \neq 1$ and $i \neq k$.

Now we perform the successive permissible row subtractions on $A(l)$ of subtracting $a_{11}(l) - 1$ times the k-th row from the first row, subtracting $a_{1i}(l)$ times the i-th row from the first row for all $i \neq k$, and finally subtracting the first row from the k-th row, to transform $A(l)$ into the identity matrix. This completes the proof of Theorem 2.1.

We say that a column subtraction is permissible on $(A, \vec{v}, \vec{w})$ if it leaves the entries of A nonnegative. If we subtract the i-th column from the j-th column, then $\vec{v}$ is unchanged, but the coefficient w_i of $\vec{w}$ is added to w_j. So as a corollary to Theorem 2.1, or by a simple modification of the proof of Theorem 2.1, we obtain:

THEOREM 2.9. *Suppose that $(A, \vec{v}, \vec{w})$ is a triple. Then there exists a sequence of permissible column additions and interchanges, followed by a sequence of permissible column subtractions, that transforms A to the identity matrix.*

3. Local factorization of birational extensions

Suppose that R is a regular local ring with quotient field K, and ν is a valuation of K, with valuation ring V, such that V dominates R ($R \subset V$ and the maximal ideal of V intersects R in its maximal ideal). A monoidal transform of R along ν is a regular local ring $R(1)$ such that $R(1) = R[\frac{P}{f}]_m$, where P is a regular prime of R, $f \in P$ is such that $\nu(f) = \min\{\nu(g) \mid g \in P\}$, and $m = \{g \in R[\frac{P}{f}] \mid \nu(g) > 0\}$. We have that V dominates $R(1)$ and $R(1)$ dominates R.

THEOREM 3.1. *Suppose that k is a field, $k[x_1,\ldots,x_n]$, $k[y_1,\ldots,y_n]$ are polynomial rings and there exists a matrix (a_{ij}) of nonnegative integers satisfying*

$$x_i = \prod_{j=1}^{n} y_j^{a_{ij}}$$

for $1 \le i \le n$ with $\mathrm{Det}(a_{ij}) = \pm 1$. Let $R = k[x_1,\ldots,x_n]_{(x_1,\ldots,x_n)}$ and $S = k[y_1,\ldots,y_n]_{(y_1,\ldots,y_n)}$. Suppose that ν is a rank 1 valuation of $k(y_1,\ldots,y_n)$ with valuation ring V which dominates S, such that $\nu(y_1),\ldots,\nu(y_n)$ are rationally independent. Then there exists a commutative diagram

$$\begin{array}{ccccc} & & T & & \\ & \nearrow & & \nwarrow & \\ R & & \rightarrow & & S \end{array} \tag{3.1}$$

such that T is a regular local ring dominated by V, and the northeast and northwest arrows are products of monoidal transforms along ν.

Proof. Let $A = (a_{ij})$, $\vec{v} = (\nu(x_1),\ldots,\nu(x_n))^t$ and $\vec{w} = (\nu(y_1),\ldots,\nu(y_n))$. By Theorem 2.1, there exists a sequence of permissible column additions and row interchanges

$$(A,\vec{v},\vec{w}) \rightarrow (A(1),\vec{v},\vec{w}(1)) \rightarrow \cdots \rightarrow (A(s),\vec{v},\vec{w}(s))$$

followed by a sequence of permissible row subtractions

$$(A(s),\vec{v},\vec{w}(s)) \rightarrow (A(s+1),\vec{v}(s+1),\vec{w}(s)) \rightarrow \cdots \rightarrow (A(t),\vec{v}(t),\vec{w}(t))$$

such that $A(t)$ is the $n \times n$ identity matrix.

We will construct a diagram (3.1), in which the northwest arrow is a product of monoidal transforms along ν,

$$S \rightarrow S(1) \rightarrow \cdots \rightarrow S(s) = T, \tag{3.2}$$

and the northeast arrow is a product of monoidal transforms along ν

$$R \rightarrow R(1) \rightarrow \cdots \rightarrow R(t-s) = T. \tag{3.3}$$

We inductively construct (3.2), with a system of regular parameters $(y_1(l),\ldots,y_n(l))$ in each $S(l)$, so that $x_i = \prod_j y_j(l)^{a_{ij}(l)}$ for $1 \le i \le n$, $\vec{v}(l) = (\nu(x_1),\ldots,\nu(x_n))^t = \vec{v}$ and $\vec{w}(l) = (\nu(y_1(l)),\ldots,\nu(y_n(l)))^t$ for $1 \le l \le s$.

Suppose that $A(l+1)$ is obtained from $A(l)$ by the row interchange T_{ij}. We define $S(l+1)$ to be $S(l)$, and we interchange the regular parameters x_i and x_j of R.

Suppose that $A(l+1)$ is obtained from $A(l)$ by the permissible column addition C_{ij}. We define $S(l+1)$ to be the local ring of the blow up of the prime ideal $(y_i(l), y_j(l))$ which is dominated by V. Since $\nu(y_j(l)) > \nu(y_i(l))$, we have that

$$S(l+1) = S(l)\left[\frac{y_j(l)}{y_i(l)}\right]_{(y_1(l+1),\ldots,y_n(l+1))},$$

where

$$y_k(l+1) = \begin{cases} y_k(l) & \text{if } k \neq j \\ \dfrac{y_j(l)}{y_i(l)} & \text{if } k = j \end{cases}$$

are regular parameters in $S(l+1)$.

We now inductively construct (3.3), with a system of regular parameters $(x_1(l), \ldots, x_n(l))$ in each $R(l)$, so that $x_i(l) = \prod_j y_j(s)^{a_{ij}(l+s)}$ for $1 \leq i \leq n$, $\vec{v}(l+s) = (\nu(x_1(l)), \ldots, \nu(x_n(l)))^t$ and $\vec{w}(l+s) = (\nu(y_1(s)), \ldots, \nu(y_n(s)))^t = \vec{w}(s)$ for $1 \leq l \leq t-s$.

Suppose that $A(l+1+s)$ is obtained from $A(l+s)$ by the permissible row subtraction R_{ji}. We define $R(l+1)$ to be the local ring of the blow up of the prime ideal $(x_i(l), x_j(l))$ which is dominated by V. Since $x_i(l)$ divides $x_j(l)$ in T, we have that

$$R(l+1) = R(l)\left[\frac{x_j(l)}{(x_i(l)}\right]_{(x_1(l+1),\ldots,x_n(l+1))},$$

where

$$x_k(l+1) = \begin{cases} x_k(l) & \text{if } k \neq j \\ \dfrac{x_j(l)}{x_i(l)} & \text{if } k = j \end{cases}$$

are regular parameters in $R(l+1)$.

Since $A(t) = Id$, we have that $x_i(t-s) = y_i(s)$ for $1 \leq i \leq n$, and thus T satisfies the conclusions of the theorem. □

REMARK 3.2. If $S \to S(1)$ is a monoidal transform of a regular local ring S, then S is called an inverse monoidal transform of $S(1)$ (Chapter 6 of [C2]). With the notation of Theorem 3.1, a permissible column subtraction of $A = (a_{ij})$ induces an inverse monoidal transform $R \to S(1) \to S$ of S. We can use Theorem 2.9, instead of Theorem 2.1, to prove Theorem 3.1. Theorem 2.9 proves the equivalent statement that a diagram (3.1) can be constructed, where the northwest arrow is factored by a sequence of inverse monoidal transforms from T to S.

THEOREM 3.3. *Suppose that $R \subset S$ are regular local rings, essentially of finite type over a field k of characteristic zero, with a common quotient field K, such that S dominates R. Let V be a valuation ring of K which dominates S. Then there exists a regular local ring T, with quotient field K, such that T*

dominates S, V dominates T, and the inclusions $R \to T$ and $S \to T$ can be factored by sequences of monoidal transforms

$$\begin{array}{ccccc} & & V & & \\ & & \uparrow & & \\ & & T & & \\ & \nearrow & & \nwarrow & \\ R & & \to & & S. \end{array} \tag{3.4}$$

Proof. Let $r = \operatorname{rank}(V)$. We can perform monoidal transforms on R and S so that the assumptions of [C2, Theorem 5.5] hold. By [C2, Theorem 5.5], there exists a commutative diagram of regular local rings

$$\begin{array}{ccc} R' & \to & S' \\ \uparrow & & \uparrow \\ R & \to & S \end{array}$$

such that R', S' have respective regular parameters $(z_1, \ldots, z_n)$, $(w_1, \ldots, w_n)$ satisfying the conclusions of [C2, Theorem 5.5]. In particular, there exists a matrix a_{ij} such that $z_i = \prod_j w_j^{a_{ij}}$ for $1 \le i \le n$, where $A = (a_{ij})$ has the block form

$$A = \begin{pmatrix} G_1 & & & & & & 0 \\ & Id & & & & & \\ & & G_2 & & & & \\ & & & Id & & & \\ 0 & & & & \cdots & & \\ & & & & & Id & \\ & & & & & & G_r \end{pmatrix}.$$

Here, $G_i = (g_{jk}(i))$ is an $s_i \times s_i$ matrix of determinant ± 1, so that we have

$$z_{t_1+\cdots+t_{i-1}+1} = w_{t_1+\cdots+t_{i-1}+1}^{g_{11}(i)} \cdots w_{t_1+\cdots+t_{i-1}+s_i}^{g_{1s_i}(i)}$$

$$\vdots$$

$$z_{t_1+\cdots+t_{i-1}+s_i} = w_{t_1+\cdots+t_{i-1}+1}^{g_{s_i1}(i)} \cdots w_{t_1+\cdots+t_{i-1}+s_i}^{g_{s_is_i}(i)}$$

for $1 \le i \le r$. We further have that $\nu(z_{t_1+\cdots+t_{i-1}+1}), \cdots, \nu(z_{t_1+\cdots+t_{i-1}+s_i})$ are rationally independent, and if $V_i = V \cap k(z_{t_1+\cdots+t_{i-1}+1}, \cdots, z_{t_1+\cdots+t_{i-1}+s_i})$, then V_i has rank 1 (and rational rank s_i).

Let

$$\overline{R}_i = k\left[z_{t_1+\cdots+t_{i-1}+1}, \ldots, z_{t_1+\cdots+t_{i-1}+s_i}\right]_{(z_{t_1+\cdots+t_{i-1}+1}, \ldots, z_{t_1+\cdots+t_{i-1}+s_i})},$$

$$\overline{S}_i = k\left[w_{t_1+\cdots+t_{i-1}+1}, \ldots, w_{t_1+\cdots+t_{i-1}+s_i}\right]_{(w_{t_1+\cdots+t_{i-1}+1}, \ldots, w_{t_1+\cdots+t_{i-1}+s_i})}.$$

By Theorem 3.1, there exists a regular local ring $\overline{T}_i$ which is dominated by V_i and a commutative diagram

$$\begin{array}{ccc} & V_i & \\ & \uparrow & \\ & \overline{T}_i & \\ \nearrow & & \nwarrow \\ \overline{R}_i & \rightarrow & \overline{S}_i \end{array}$$

such that the northeast and northwest arrows are products of monoidal transforms. By performing the corresponding sequences of monoidal transforms on R' and S' for $1 \leq i \leq r$, we obtain the conclusions of the theorem. □

References

[Ab1] S. Abhyankar, *On the valuations centered in a local domain*, Amer. J. Math. **78** (1956), 321–348. MR 0082477 (18,556b)

[Ab] ———, *Algebraic geometry for scientists and engineers*, Mathematical Surveys and Monographs, vol. 35, American Mathematical Society, Providence, RI, 1990. MR 1075991 (92a:14001)

[AMR] D. Abramovich, K. Matsuki, and S. Rashid, *A note on the factorization theorem of toric birational maps after Morelli and its toroidal extension*, Tohoku Math. J. (2) **51** (1999), 489–537. MR 1725624 (2000i:14073)

[AMR1] ———, *Correction: "A note on the factorization theorem of toric birational maps after Morelli and its toroidal extension"*, Tohoku Math. J. (2) **52** (2000), 629–631. MR 1793939 (2001j:14069)

[Ch] C. Christensen, *Strong domination/weak factorization of three-dimensional regular local rings*, J. Indian Math. Soc. (N.S.) **45** (1981), 21–47 (1984). MR 828858 (88a:14010a)

[C1] S. D. Cutkosky, *Local factorization of birational maps*, Adv. Math. **132** (1997), 167–315. MR 1491444 (99c:14018)

[C2] ———, *Local monomialization and factorization of morphisms*, Astérisque **260** (1999). MR 1734239 (2001c:14027)

[C3] ———, *Monomialization of morphisms from 3-folds to surfaces*, Lecture Notes in Mathematics, vol. 1786, Springer-Verlag, Berlin, 2002. MR 1927974 (2004a:14014)

[C4] ———, *Local monomialization of transcendental extensions*, Ann. Inst. Fourier (Grenoble) **55** (2005), 1517–1586. MR 2172273

[C5] ———, *Toroidalization of birational morphism of projective 3-folds*, Mem. Amer. Math. Soc., to appear.

[HHS] W. Heinzer, C. Huneke, and J. D. Sally, *A criterion for spots*, J. Math. Kyoto Univ. **26** (1986), 667–671. MR 864469 (87k:13034)

[H] H. Hironaka, *Resolution of singularities of an algebraic variety over a field of characteristic zero. I*, Ann. of Math. (2) **79** (1964), 109–203; *II*, ibid. (2) **79** (1964), 205–326. MR 0199184 (33 #7333)

[KM] Y. Kachi and S. B. Mulay, *Local-to-global correspondence in algebraic geometry*, The Fano Conference, Univ. Torino, Turin, 2004, pp. 485–514. MR 2112588 (2006a:14018)

[K] K. Karu, *Local strong factorization of toric birational maps*, J. Algebraic Geom. **14** (2005), 165–175. MR 2092130 (2005j:14015)

[Mo] R. Morelli, *The birational geometry of toric varieties*, J. Algebraic Geom. **5** (1996), 751–782. MR 1486987 (99b:14056)

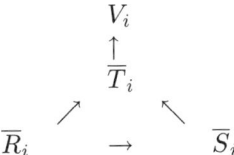

[Sa] J. Sally, *Regular overrings of regular local rings*, Trans. Amer. Math. Soc. **171** (1972), 291–300. MR 0309929 (46 #9033)

[Sh] D.L. Shannon, *Monoidal transforms*, Amer. J. Math **45** (1973), 294–320. MR0330154 (48 #8492)

[Z] O. Zariski, *Introduction to the problem of minimal models in the theory of algebraic surfaces*, Publications of the Mathematical Society of Japan, no. 4, The Mathematical Society of Japan, Tokyo, 1958. MR 0097403 (20 #3872)

Steven Dale Cutkosky, Department of Mathematics, University of Missouri, Columbia MO 65201, USA

E-mail address: cutkoskys@missouri.edu

Hema Srinivasan, Department of Mathematics, University of Missouri, Columbia MO 65201, USA

E-mail address: srinivasanh@missouri.edu,

Illinois Journal of Mathematics
Volume 51, Number 1, Spring 2007, Pages 57–65
S 0019-2082

ON EFFICIENT GENERATION OF PULL-BACK OF $T_{\mathbb{P}^n}(-1)$

S. P. DUTTA

This paper is written in honour of Phil Griffith on his retirement. The author would like to take this opportunity to express his respect for Phil's deep insight into homological methods in commutative algebra.

ABSTRACT. Let $f : X \to \mathbb{P}^n$ be a proper map such that dimension of $f(X) \geq 2$. We address the following question: Is $\dim H^o(X, f^*(T_{\mathbb{P}^n}(-1)) = n+1$? We provide an affirmative answer under standard mild restrictions on X. We also point out that this provides an affirmative answer to a similar question raised via regular alteration of a closed subvariety in a blow-up of a regular local ring at its closed point in the mixed characteristics.

Introduction

The problem that we are going to address in this note came up in the course of studying intersection multiplicity over the blow-up of a regular local ring at its closed point. Let (R, m, K) be a regular local ring of dimension n essentially of finite type over a field or over an excellent discrete valuation ring with residue field $R/m = K$. Assume K to be algebraically closed. Write $X = \operatorname{Spec} R$, $\tilde{X}$ its blow-up at the closed point $s = [m]$. Let q be a prime ideal of R. Let $\tilde{Z}$ denote the blow-up of $Z = \operatorname{Spec}(R/q)$ at its closed point and let $\tilde{Z}_s$ denote the fiber over s. Assume that $\dim Z \geq 3$. Consider the pair $(\tilde{Z}, \tilde{Z}_s)$. Let (W, W_s) be a regular alteration [J] of $(\tilde{Z}, \tilde{Z}_s)$, i.e., W is a regular scheme, $\pi : W \to \tilde{Z}$ is a dominant projective morphism, $\dim W = \dim \tilde{Z}$, and $\pi^{-1}(\tilde{Z}_s) = W_s$ is a non-reduced strict normal crossing divisor in W, i.e., reduced part of W_s is a strict normal crossing divisor. Write $\phi = \pi\big|_{(W_s)_{\text{red}}}$, $\mathcal{H} = \Omega_{\mathbb{P}}(1)$, $\mathbb{P} = \mathbb{P}_K^{n-1}$ and let $\mathcal{H}^\vee$ denote the dual of $\mathcal{H}$, i.e., $\mathcal{H}^\vee = T_{\mathbb{P}}(-1)$.

Received November 21, 2006; received in final form July 11, 2007.
2000 *Mathematics Subject Classification.* Primary 14F10, 14B05, 13H15.
Key words and phrases. sheaf cohomology, connectedness, tangent bundles, normal crossing.
Supported by NSA grant MDA H98230-04-1-0100.

Consider the exact sequence ($\mathbb{P} = \mathbb{P}^{n-1}$)

$$0 \to \mathcal{O}_{\mathbb{P}}(-1) \to \bigoplus_1^n \mathcal{O}_{\mathbb{P}} \to \mathcal{H}^\vee \to 0. \tag{1}$$

Write $\pi^*(\mathcal{O}_{\mathbb{P}}(1)) = \mathcal{O}_W(1)$, $(W_s)_{\text{red}} = \widetilde{W}$ and let

$$0 \to \mathcal{O}_{\widetilde{W}}(-1) \to \bigoplus_1^n \mathcal{O}_{\widetilde{W}} \to \mathcal{H}^\vee \otimes \mathcal{O}_{\widetilde{W}} \to 0 \tag{2}$$

denote the pull-back of (1) via ϕ. We want to address the following question.

QUESTION. *Is* $\dim H^0(\widetilde{W}, \mathcal{H}^\vee \otimes \mathcal{O}_{\widetilde{W}}) = n$*?*

An affirmative answer to the above question is the main focus of the following theorem.

THEOREM 1. *With the above set-up,*

$$H^1\left(\widetilde{W}, \mathcal{O}_{\widetilde{W}}(-1)\right) \to H^1(\widetilde{W}, \bigoplus_1^n \mathcal{O}_{\widetilde{W}})$$

is injective.

We reduce the proof of the above theorem to the following theorem. The main tool of this reduction is Zariski's Main Theorem (EGA III).

THEOREM 2. *Let $\tilde{W}$ be a reduced connected scheme of finite type over an algebraically closed field K and let $g : \tilde{W} \to \mathbb{P}^n = \mathbb{P}$ be a proper map. Let $W_1, \ldots, W_d$ be the irreducible components of $\tilde{W}$. Assume that, for $1 \le i$, $j \le d$, (i) $\dim g(W_i) \ge 2$ and (ii) if $W_i \cap W_j \ne \phi$, then $\dim g(W_i \cap W_i) \ge 1$. Consider the exact sequence*

$$o \to \mathcal{O}_{\tilde{W}}(-1) \to \bigoplus_1^{n+1} \mathcal{O}_{\tilde{W}} \to \mathcal{H}^\vee \otimes \mathcal{O}_{\tilde{W}} \to 0,$$

constructed as above. Then

$$H^1(\tilde{W}, \mathcal{O}_{\tilde{W}}(-1)) \to H^1(\tilde{W}, \bigoplus_1^{n+1} \mathcal{O}_{\tilde{W}})$$

is injective.

If $\dim g(\widetilde{W}) = 1$, the above theorem is not valid. For examples of such situations we refer the reader to [He].

I would like to thank L. Ein, R. Hartshorne, M. Hochster, and S. Katz for helpful comments. I usually don't work on this kind of question and I am rather unaware of the chronology of works in this area. I apologize in advance for my short and possibly incomplete list of references.

Section 1

We first prove Theorem 2 as stated above in the following steps.

PROPOSITION 1. *Let $\tilde{W}$ be a variety over K and let $g : \tilde{W} \to \mathbb{P}^n_K = \mathbb{P}$ be a finite map. Assume that* $\dim g(\tilde{W}) \geq 2$. *Then there exist hyperplane sections $\tilde{W}_L = g^{-1}(L) \cap \tilde{W}$ of $\tilde{W}$ such that $H^0(\tilde{W}, \mathcal{O}_{\tilde{W}_L}) = K$.*

Proof. Suppose first $\operatorname{ch} K = 0$. In this case, by Bertini's Theorem, there exist hyperplane sections L such that $\tilde{W}_L$ is a closed subvariety of $\tilde{W}$. Since g is proper, we have the required result.

Now suppose $\operatorname{ch} K = p > 0$. Let (W', h) be the normalization of $\tilde{W}$ and let g' denote the composition of $W' \xrightarrow{h} \tilde{W} \xrightarrow{g} \mathbb{P}$. Hence g' is finite and $g'^*\mathcal{O}_{\mathbb{P}}(t) = \mathcal{O}_{W'}(t)$ is very ample for $t \gg 0$. Hence W' is projective. Let $q = p^r$, $r \gg 0$. Then $\mathcal{O}_{W'}(q)$ is very ample. Let $f : W' \to W'$ denote the Frobenius map induced by $x \to x^p$ on $\mathcal{O}_{W'}$ and let f^q denote f iterated q times. By Bertini's Theorem, there exist hyperplane sections $\tilde{W}_L$ and W'_L such that they are irreducible. Consider the short exact sequences

$$0 \to \mathcal{O}_{W'}(-1) \to \mathcal{O}_{W'} \to \mathcal{O}_{W'_L} \to 0$$

and

$$0 \to \mathcal{O}_{W'}(-q) \to \mathcal{O}_{W'} \to \mathcal{O}_{W'_{Lq}} \to 0,$$

where the bottom one is obtained from the top one via f^q (the bottom sequence is also exact independently by itself). Since W' is a normal projective variety, $H^i(W', \mathcal{O}_{W'}(-q)) = 0$ for $i \leq 1$ and $q \gg 0$; thus $H^0(W', \mathcal{O}_{W'_{Lq}}) = K$. Due to normality, $\mathcal{O}_{W'_L} \to \mathcal{O}_{W'_{Lq}}$ (induced by f^q) is an injection. Hence $H^0(W', \mathcal{O}_{W'_L}) = K$. Since $h : W' \to \tilde{W}$ is finite and birational, L can be so chosen that $h_L : W'_L \to \tilde{W}_L$ is also the same and $\mathcal{O}_{\tilde{W}_L} \hookrightarrow h_*(\mathcal{O}_{W'_L})$ is an injection (Remark 3.4.11, [F-O-V]). Hence $H^0(\tilde{W}, \mathcal{O}_{\tilde{W}_L}) = K$. □

PROPOSITION 2. *Let $\tilde{W}$ be a variety over K and $g : \tilde{W} \to \mathbb{P}^n_K = \mathbb{P}$ be a finite map. Write $g^*(\mathcal{O}_{\mathbb{P}}(1)) = \mathcal{O}_{\tilde{W}}(1)$. Consider the exact sequence*

$$(*) \qquad 0 \to \mathcal{O}_{\tilde{W}}(-1) \to \bigoplus_1^{n+1} \mathcal{O}_{\tilde{W}} \to \mathcal{H}^\vee \otimes \mathcal{O}_{\tilde{W}} \to 0,$$

which is the pull-back via g of the exact sequence

$$(**) \qquad 0 \to \mathcal{O}_{\mathbb{P}}(-1) \to \bigoplus_1^{n+1} \mathcal{O}_{\mathbb{P}} \to \mathcal{H}^\vee \to 0.$$

If $\dim g(\tilde{W}) \geq 2$, *then $H^1(\tilde{W}, \mathcal{O}_{\tilde{W}}(-1)) \to H^1(\tilde{W}, \bigoplus_1^{n+1} \mathcal{O}_{\tilde{W}})$ is injective.*

Proof. By the previous Lemma, we have a hyperplane section $\tilde{W}_L$ such that $H^0(\tilde{W}, \mathcal{O}_{\tilde{W}_L}) = K$. Consider the short exact sequence

$$0 \to \mathcal{O}_{\tilde{W}}(-1) \xrightarrow{\ell} \mathcal{O}_{\tilde{W}} \to \mathcal{O}_{\tilde{W}_L} \to 0.$$

Since $H^0(\tilde{W}, \mathcal{O}_{\tilde{W}}) = K$, $H^1(\tilde{W}, \mathcal{O}_{\tilde{W}}(-1)) \xrightarrow{\tilde{\ell}} H^1(\tilde{W}, \mathcal{O}_{\tilde{W}})$ is injective.

Let L be given by $\sum_{i=0}^{n} a_i X_i = 0$ in $\mathbb{P}$, $a_o \neq 0$. Let $u : K^{n+1} \to K^{n+1}$ denote the isomorphism defined by $u(e_0) = a_0 e_0$, $u(e_i) = e_i + a_i e_0$ for $i > 0$. Let $\varphi : \mathbb{P} \to K$ denote the structure map. Then $\varphi^*(u) : \bigoplus_1^{n+1} \mathcal{O}_{\mathbb{P}} \to \bigoplus_1^{n+1} \mathcal{O}_{\mathbb{P}}$ is an isomorphism and this leads to an isomorphism $\tilde{\varphi}(u) : \bigoplus_1^{n+1} \mathcal{O}_{\tilde{W}} \to \bigoplus_1^{n+1} \mathcal{O}_{\tilde{W}}$.

We have a commutative diagram

$$\begin{array}{ccccc}
0 & \to & \mathcal{O}_{\tilde{W}}(-1) & \to & \bigoplus_1^{n+1} \mathcal{O}_{\tilde{W}} \\
 & & & & \downarrow \wr\, \tilde{\varphi}(u) \\
 & & & \searrow \ell & \bigoplus_1^{n+1} \mathcal{O}_{\tilde{W}} \\
 & & & & \downarrow \quad \text{projection onto the first component.} \\
 & & & & \mathcal{O}_{\tilde{W}}
\end{array}$$

Since $H^1(\tilde{W}, \mathcal{O}_{\tilde{W}}(-1)) \xrightarrow{\tilde{\ell}} H^1(\tilde{W}, \mathcal{O}_{\tilde{W}})$ is injective, so is $H^1(\tilde{W}, \mathcal{O}_{\tilde{W}}(-1)) \to H^1(\tilde{W}, \bigoplus_1^{n+1} \mathcal{O}_{\tilde{W}})$.

PROPOSITION 3. *Let $\tilde{W}$ be a reduced connected scheme of finite type over K and let $g : \tilde{W} \to \mathbb{P}_K^n = \mathbb{P}$ be a finite map. Let $W_1, \dots, W_d$ denote the components of $\tilde{W}$. Assume that $\dim g(W_i) \geq 2$, $1 \leq i \leq d$ and $\dim g(W_i \cap W_j) \geq 1$, whenever $W_i \cap W_j \neq \phi$.*

Consider the exact sequence (constructed as in (*))

$$0 \to \mathcal{O}_{\tilde{W}}(-1) \to \bigoplus_1^{n+1} \mathcal{O}_{\tilde{W}} \to \mathcal{H}^\vee \otimes \mathcal{O}_{\tilde{W}} \to 0.$$

Then $H^1(\tilde{W}, \mathcal{O}_{\tilde{W}}(-1)) \to H^1(\tilde{W}, \bigoplus_1^{n+1} \mathcal{O}_{\tilde{W}})$ is injective.

Proof. By the previous Lemma and Bertini's Theorem, there exist generic hyperplane sections L of $\mathbb{P}$ such that W_{iL} is irreducible and $H^0(W_{iL}, \mathcal{O}_{W_{iL}}) = K$ for $1 \leq i \leq d$, and $W_{iL} \cap W_{jL}$ is non-empty whenever $W_i \cap W_j$ is non-empty.

Consider the exact sequence

$$0 \to \mathcal{O}_{(W_i \cup W_j)} \to \mathcal{O}_{W_i} \oplus \mathcal{O}_{W_j} \to \mathcal{O}_{W_i \cap W_j} \to 0.$$

L can be so chosen that

$$0 \to \mathcal{O}_{W_{iL} \cup W_{jL}} \to \mathcal{O}_{W_{iL}} \oplus \mathcal{O}_{W_{jL}} \to \mathcal{O}_{W_{iL} \cap W_{jL}} \to 0$$

is exact. Hence $H^0(\tilde{W}, \mathcal{O}_{W_{iL} \cup W_{jL}}) = K$. Let $t \in \{1, \dots, d\}$ be such that $t \neq i, j$ and $(W_i \cup W_j) \cap W_t \neq \phi$. Consider the exact sequence

$$0 \to \mathcal{O}_{W_i \cup W_j \cup W_t} \to \mathcal{O}_{W_i \cup W_j} \bigoplus \mathcal{O}_{W_t} \to \mathcal{O}_{(W_i \cup W_j) \cap W_t} \to 0$$

L can be so chosen that

$$0 \to \mathcal{O}_{W_{iL} \cup W_{jL} \cup W_{tL}} \to \mathcal{O}_{W_{iL} \cup W_{jL}} \bigoplus \mathcal{O}_{W_{tL}} \to \mathcal{O}_{(W_{iL} \cup W_{jL}) \cap W_{tL}} \to 0$$

is also exact. Hence $H^0(\tilde{W}, \mathcal{O}_{W_{iL} \cup W_{jL} \cup W_{tL}}) = K$.

Since $\tilde{W}$ is reduced and connected, proceeding in the above manner, after a finite number steps we obtain $H^0(\tilde{W}, \mathcal{O}_{\tilde{W}_L}) = K$.

Now consider the exact sequence

$$0 \to \mathcal{O}_{\tilde{W}}(-1) \to \mathcal{O}_{\tilde{W}} \to \mathcal{O}_{\tilde{W}_L} \to 0$$

and proceed as in the proof of the previous proposition to complete the proof. □

THEOREM 2. *Let $\tilde{W}$ be a reduced connected scheme of finite type over K and let $g : \tilde{W} \to \mathbb{P}^n_K = \mathbb{P}$ be a proper map. Let $W_1, \dots, W_d$ be the irreducible components of $\tilde{W}$. Assume that, for $1 \leq i,\ j \leq d$, (i) $\dim g(W_i) \geq 2$ and (ii) if $W_i \cap W_j \neq \phi$, $\dim g(W_i \cap W_j) \geq 1$. Consider the exact sequence*

$$0 \to \mathcal{O}_{\tilde{W}}(-1) \to \bigoplus_{1}^{n+1} \mathcal{O}_{\tilde{W}} \to \mathcal{H}^{\vee} \otimes \mathcal{O}_{\tilde{W}} \to 0$$

(constructed as in (*)).

Then $H^1(\tilde{W}, \mathcal{O}_{\tilde{W}}(-1)) \to H^1(\tilde{W}, \bigoplus_{1}^{n+1} \mathcal{O}_{\tilde{W}})$ is injective.

Proof. Using Grothendieck's Theorem of Connection (Theorem 4.3.1, EGA III) we reduce the problem to the case where g is finite. The proof now follows from Proposition 2. □

Section 2

We are now ready to prove Theorem 1.

Proof of Theorem 1. Note that $\phi : \widetilde{W} \to \tilde{Z}_s$ does not necessarily satisfy the hypothesis in the statement of Theorem 2. Recall that $\pi : W \to \tilde{Z}$ is a regular alteration such that $W_s = \pi^{-1}(\tilde{Z}_s)$ is a non-reduced strict normal crossing divisor with irreducible components $W_1, \dots, W_d$. This means that for any non-empty subset $\alpha \subset \{1, \ldots, d\}$, the closed subscheme $W_\alpha = \bigcap_{i \in \alpha} W_i$ is a regular subscheme and hence a smooth ($K = \overline{K}$) subscheme of codimension $\#\alpha$ in W. Since R is regular local, $\tilde{Z}_s = \mathrm{Proj}(\mathrm{Gr}_m(R/q))$ is equidimensional. Let (W', h) be the normalization of $\tilde{Z}$ in $k(W)$. Hence we have a commutative diagram

$$\text{(3)} \qquad \begin{array}{ccc} W & \xrightarrow{f} & W' \\ & \searrow \pi \quad \swarrow h & \\ & \tilde{Z} & \end{array}$$

where h is a finite map (EGA II, 6.3.9). Since π is proper, f is also proper. $\pi_* \mathcal{O}_W = h_* f_* \mathcal{O}_W$. Since $f_* \mathcal{O}_W$ is a coherent $\mathcal{O}_{W'}$ algebra and W' is the normalization of $\tilde{Z}$ in $k(W)$, we have $\mathcal{O}_{W'} = f_* \mathcal{O}_W$. Thus $\pi_* \mathcal{O}_W = h_* \mathcal{O}_{W'}$. By Theorem (4.3.1), EGA III, for all $w' \in W'$, $\overset{-1}{f}(w')$ is connected and non-empty. Since $\dim W = \dim \tilde{Z}$, by Zariski's Main Theorem (Proposition 4.4.1, EGA III), if $V = \{w \in W \mid w \text{ is isolated in } \pi^{-1}(\pi(w))\}$, then V is non-empty open in W and $f\big|_V$ is an isomorphism of V onto an open subset of V' in W'; moreover, $f^{-1}(V') = V$.

Claim. *If $y \in W'$ is such that $\dim \mathcal{O}_{W',y} = 1$, then $y \in V'$.*

Proof. Let x be a closed point in $f^{-1}(y)$. Then, by the dimension formula, $\dim \mathcal{O}_{W,x} = \dim \mathcal{O}_{W,'y} + \mathrm{tr}_{k(W')} k(W) - \mathrm{tr}_{k(y)} k(x) = \dim \mathcal{O}_{W,'y}$ $(k(W') = k(W)) = 1$. Since f is birational and $f^{-1}(y)$ is connected, $f^{-1}(y) = \{x\}$. Thus $y \in V'$.

From (3), by taking the fiber over the closed point in Z, we obtain another commutative diagram

$$\begin{array}{ccc} W_s & \xrightarrow{f_s} & W'_s \\ & \pi_s \searrow \quad \swarrow h_s & \\ & \tilde{Z}_s & \end{array}$$

Recall that $\tilde{Z}$ is the blow-up of Z at $\{m/q\}$. We have an exact sequence

$$\mathcal{O} \to \mathcal{O}_{\tilde{Z}}(1) \to \mathcal{O}_{\tilde{Z}} \to \mathcal{O}_{\tilde{Z}_s} \to 0.$$

This leads to exact sequences

$$\mathcal{O} \to \mathcal{O}_W(1) \to \mathcal{O}_W \to \mathcal{O}_{W_s} \to 0$$

and

$$\mathcal{O} \to \mathcal{O}_{W'}(1) \to \mathcal{O}_{W'} \to \mathcal{O}_{W'_s} \to 0;$$

here $\pi^* \left(\mathcal{O}_{\tilde{Z}}(1)\right) = \mathcal{O}_W(1)$ and $h^* \left(\mathcal{O}_{\tilde{Z}}(1)\right) = \mathcal{O}_{W'}(1)$.

Applying f_* to the top sequence, we obtain an exact sequence

$$0 \to f_* \mathcal{O}_W(1) \to f_* \mathcal{O}_W \to f_* \mathcal{O}_{W_s}.$$

Since $f_*\mathcal{O}_W = \mathcal{O}_{W'}$, we obtain an injection $\mathcal{O}_{W'_s} \hookrightarrow f_*\mathcal{O}_{W_s}$ from the above sequence. Our claim shows that for every irreducible component of W'_s there exists a unique irreducible component of W_s which maps birationally onto it via the proper map f_s induced by f. Since components of W'_s are varieties over K and W_s is equidimensional, this implies that W'_s is equidimensional. Moreover, since W_s is a non-reduced strict normal crossing divisor, if W'_i and W'_j are any two components of W'_s such that $W'_i \cap W'_j \neq \phi$, then $\dim\left(W'_i \cap W'_j\right) \geq 1$. (Recall that $\dim Z \geq 3$.) To see this one may use the following observations: (a) W_s is equidimensional, $\dim(W_i \cap W_j) \geq 1$; and (b) if $\{x_1, \ldots, x_r\}$ is a finite set of points contained in an open subset U of $\operatorname{Spec} A$, then there exists an $f \in A$ s.t. $\{x_1, \ldots, x_n\} \subset \operatorname{Spec} A_f \subset U$. Since h_s is finite, $\dim h_s\left(W'_i \cap W'_j\right) \geq 1$. W_s is connected, hence so is W'_s.

Let $\widetilde{W}$, $\widetilde{W}'$ denote $W_{s\,\mathrm{red}}$ and $W'_{s\,\mathrm{red}}$, respectively, and let $\tilde{f} = f_s|_{\widetilde{W}} : \widetilde{W} \to \widetilde{W}'$. Then $\tilde{f}$ is a proper surjective map. We can factorize $\tilde{f}$ as follows (Theorem 4.3.1, EGA III):

$$
\begin{array}{ccc}
\widetilde{W} & \xrightarrow{g} & B \\
 & \tilde{f} \searrow & \downarrow h' \\
\theta \downarrow & & \widetilde{W}' \\
 & & \downarrow \theta' \\
W_s & \xrightarrow{f_s} & W'_s
\end{array}
$$

where h' is finite, $\mathcal{O}_B = g_*\mathcal{O}_{\widetilde{W}}$, and θ and θ' are the natural injections (homeomorphisms). For every $y \in \widetilde{W}'$, $\tilde{f}^{-1}(y)$ is connected. Hence, by the Theorem of Connection mentioned above, h' is a homeomorphism. By construction, there exists a dense open subset $U'(= V' \cap \widetilde{W}')$ such that if $\widetilde{U} = \tilde{f}^{-1}(U')$, then $\tilde{f}|_{\widetilde{U}} : \widetilde{U} \to U'$ is an isomorphism. By Zariski's Main Theorem (EGA III, Proposition 4.4.1), there exists a dense open subset U of B such that both $g|_{\widetilde{U}} : \widetilde{U} \to U$ and $h'|_U : U \to U'$ are isomorphisms. Moreover, by construction and our claim, each component of B has a non-empty intersection with U. Thus B is reduced, connected and equidimensional. Note that h', restricted to any component of B, is a birational homeomorphism onto a component of W'_s.

Let B_i be any irreducible component of B and let W_i and W'_i be the corresponding components of $\widetilde{W}$ and W'_s such that $g|_{W_i}$ maps W_i birationally onto B_i and B_i is birationally homeomorphic to W'_i (via $\theta' \cdot h'$). Since $\dim h_s(W'_i) \geq 2$, $\dim h_s \cdot \theta' \cdot h'(B_i) \geq 2$ and since $\dim h_s(W'_i \cap W'_j) \geq 1$ whenever $W'_i \cap W'_j \neq \phi$, $\dim h_s \cdot \theta' \cdot h'(B_i \cap B_j) \geq 1$ whenever $B_i \cap B_j \neq \phi$.

Write $\psi = h_s \cdot \theta' \cdot h'$ and $\mathcal{O}_B(1) = \psi^*(\mathcal{O}_{\mathbb{P}}(1))$. Let

$$\mathcal{O} \to \mathcal{O}_B(-1) \to \bigoplus_1^n \mathcal{O}_B \to \mathcal{H}^\vee \otimes \mathcal{O}_B \to 0$$

be the exact sequence obtained by pulling back (1) via ψ. Then, by Theorem 2,

$$H^1(B, \mathcal{O}_B(-1)) \to H^1(B, \bigoplus_1^n \mathcal{O}_B)$$

is injective. This implies that

$$H^0(B, \bigoplus_1^n \mathcal{O}_B) \xrightarrow{\sim} H^0(B, \mathcal{H}^\vee \otimes \mathcal{O}_B).$$

Since $g_*\left(\mathcal{O}_{\widetilde{W}}\right) = \mathcal{O}_B$ and $g_*\left(\mathcal{H}^\vee \otimes \mathcal{O}_{\widetilde{W}}\right) = \mathcal{H}^\vee \otimes \mathcal{O}_B$, the above isomorphism implies, via (2), that

$$H^0(\widetilde{W}, \bigoplus_1^n \mathcal{O}_{\widetilde{W}}) \xrightarrow{\sim} H^0(\widetilde{W}, \mathcal{H}^\vee \otimes \mathcal{O}_{\widetilde{W}}).$$

Hence $H^1\left(\widetilde{W}, \mathcal{O}_{\widetilde{W}}(-1)\right) \to H^1\left(\widetilde{W}, \bigoplus_1^n \mathcal{O}_{\widetilde{W}}\right)$ is injective and the proof is complete. □

References

[B] P. Berthelot, *Alteŕations de variétes algébriques [d'après A.J. de Jong]*, Séminaire Bourbaki exposé 815 (1996), Astérisque **241** (1997). MR 1472543 (98m:14021)

[Bo] N. Bourbaki, *Elements of mathematics. Commutative algebra*, Hermann, Paris, 1972. MR 0360549 (50 #12997)

[D-Gr] J. Dieudonné and A. Grothendieck, *Éléments de géométrie algébrique. I*, Inst. Hautes Études Sci. Publ. Math. **4** (1960); *II*, ibid. **8** (1961); *III*, ibid. **11** (1961), **17** (1963); *IV*, ibid. **20** (1964), **24** (1965), **28** (1966), **32** (1967).

[Du] S. P. Dutta, *Intersection multiplicity of Serre on regular schemes*, preprint.

[F-O-V] H. Flenner, L. O'Carroll, and W. Vogel, *Joins and intersections*, Springer Monographs in Mathematics, Springer-Verlag, Berlin, 1999. MR 1724388 (2001b:14010)

[Gr2] A. Grothendieck, *Cohomologie locale des faisceaux cohérents et théorèmes de Lefschetz locaux et globaux* (*SGA* 2), North-Holland Publishing Co., Amsterdam, 1968. MR 0476737 (57 #16294)

[H] R. Hartshorne, *Algebraic geometry*, 6th corrected printing, Springer-Verlag, New York, 1993, Graduate Texts in Mathematics, vol. 52. MR 0463157 (57 #3116)

[He] G. Hein, *Curves in* $\mathbf{P}^3$ *with good restriction of the tangent bundle*, Rocky Mountain J. Math. **30** (2000), 217–235. MR 1763808 (2001f:14059)

[Her] J. Herzog, *Ringe der Charakteristik p und Frobeniusfunktoren*, Math. Z. **140** (1974), 67–78. MR 0352081 (50 #4569)

[Hou] C. Houzel, *Morphisme de Frobenius et rationalite de fonctions L*, Expose XV, SGA 5, Springer Lecture Notes, vol. 589, Springer-Verlag, Berlin, 1977.

[J] A. J. de Jong, *Smoothness, semi-stability and alterations*, Inst. Hautes Études Sci. Publ. Math. (1996), 51–93. MR 1423020 (98e:14011)

[Jo] J.-P. Jouanolou, *Théorèmes de Bertini et applications*, Progress in Mathematics, vol. 42, Birkhäuser Boston Inc., Boston, MA, 1983. MR 725671 (86b:13007)

[K] E. Kunz, *Characterizations of regular local rings of characteristic p*, Amer. J. Math. **91** (1969), 772–784. MR 0252389 (40 #5609)

[M] D. Mumford, *Further pathologies in algebraic geometry*, Amer. J. Math. **84** (1962), 642–648. MR 0148670 (26 #6177)

[P-S] C. Peskine and L. Szpiro, *Dimension projective finie et cohomologie locale. Applications à la démonstration de conjectures de M. Auslander, H. Bass et A. Grothendieck*, Inst. Hautes Études Sci. Publ. Math. **42** (1973), 47–119. MR 0374130 (51 #10330)

[S] J.-P. Serre, *Algèbre locale. Multiplicités*, 3rd edition, Lecture Notes in Mathematics, vol. 11, Springer-Verlag, Berlin, 1975. MR 0201468 (34 #1352)

S. P. Dutta, Department of Mathematics, University of Illinois, 1409 West Green Street, Urbana, IL 61801, USA

E-mail address: dutta@math.uiuc.edu

Illinois Journal of Mathematics
Volume 51, Number 1, spring 2007, Pages 67–82
S 0019-2082

CYCLIC MODULES OF FINITE GORENSTEIN INJECTIVE DIMENSION AND GORENSTEIN RINGS

HANS–BJØRN FOXBY AND ANDERS J. FRANKILD

Dedicated to the achievements of Phil Griffith

ABSTRACT. The main result asserts that a local commutative Noetherian ring is Gorenstein, if it possesses a non-zero cyclic module of finite Gorenstein injective dimension. From this follows a classical result by Peskine and Szpiro stating that the ring is Gorenstein, if it admits a non-zero cyclic module of finite (classical) injective dimension. The main result applies to local homomorphisms of local rings and yields the next: if the source is a homomorphic image of a Gorenstein local ring and the target has finite Gorenstein injective dimension over the source, then the source is a Gorenstein ring. This, in turn, applies to the Frobenius endomorphism when the local ring is of prime equicharacteristic and is a homomorphic image of a Gorenstein local ring.

1. Introduction

Throughout this paper $(R, \mathfrak{m}, k)$ is a commutative Noetherian local ring.

The *Gorenstein injective dimension* in the title was introduced by Enochs, Jenda, and Xu, and it is recalled in 3.1, and for any R–module N this integer is denoted $\operatorname{Gid}_R N$. It is a *refinement* of the classical injective dimension $\operatorname{id}_R N$ in the sense that there is always the inequality $\operatorname{Gid}_R N \leq \operatorname{id}_R N$, and if $\operatorname{id}_R N$ is finite, then $\operatorname{Gid}_R N = \operatorname{id}_R N$. It turns out that the ring R is Gorenstein if and only if every R–module has finite Gorenstein injective dimension; see [6, Chapter 6].

The title refers to the next theorem, which is the main result of the paper.

THEOREM A. *If there exists a non-zero cyclic R–module N with the Gorenstein injective dimension $\operatorname{Gid}_R N$ finite, then R is Gorenstein.*

Received December 19, 2006; received in final form August 26, 2007.
1991 *Mathematics Subject Classification.* 13D05, 18E30, 18G10.
The first author was supported by FNU, the Danish Research Council.
The second author died June 10, 2007; he was the holder of an FNU Steno scolarship.

The theorem has been proved in special cases by Takahashi; see [24, (3.5)]. As an immediate consequence of our theorem it follows that R is Gorenstein if and only if R has finite Gorenstein injective dimension as an R–module; this is the local commutative version of the main result in Holm [16]. Another immediate consequence is that R is Gorenstein if and only if the residue field k has finite Gorenstein injective dimension as an R–module. Since the Gorenstein injective dimension is a refinement of the injective dimension, the above result implies the next result, which is [22, Théorème (5.5)].

THEOREM (PESKINE AND SZPIRO). *If there exists a non-zero cyclic R–module N with the injective dimension* $\mathrm{id}_R\, N$ *finite, then R is Gorenstein.*

Theorem A is restated in 4.5, and it is proved there. The proof relies on two auxiliary techniques. The first involves the Bass series $\mathrm{I}_R^M(t)$ and Poincaré series $\mathrm{P}_M^R(t)$ for an R–module M. These series are formal power series, and their coefficients are the Bass numbers and the Betti numbers, respectively. The series are presented in Section 4, where also the next key result is proved.

PROPOSITION B. *To any finitely generated R–module N of finite Gorenstein injective dimension there exists then a finitely generated R–module K of finite Gorenstein projective dimension such that there is the next equality:*

$$\mathrm{P}_N^R(t)t^{\operatorname{depth} R} = \mathrm{P}_K^R(t)\,\mathrm{I}_R^R(t)\,.$$

Moreover, if N is of finite (classical) injective dimension, then K has finite (classical) projective dimension.

The second auxiliary technique requires the ring R to admit a dualizing complex D, and it involves two categories $\mathcal{A}^{\mathrm{f}}(R)$ and $\mathcal{B}^{\mathrm{f}}(R)$ consisting of complexes R–modules; see 2.4 and 2.5. Here it is important that a finitely generated R–module is of finite Gorenstein injective dimension if and only if it belongs to $\mathcal{B}^{\mathrm{f}}(R)$; see 3.5, which is main result in [8] by Christensen, Frankild, and Holm. The next result is also used in the proof of Theorem A.

PROPOSITION C. *The two functors* $(-)^* = \mathbf{R}\mathrm{Hom}_R(-,R)$ *and* $(-)^\dagger = \mathbf{R}\mathrm{Hom}_R(-,D)$ *fit into the diagram*

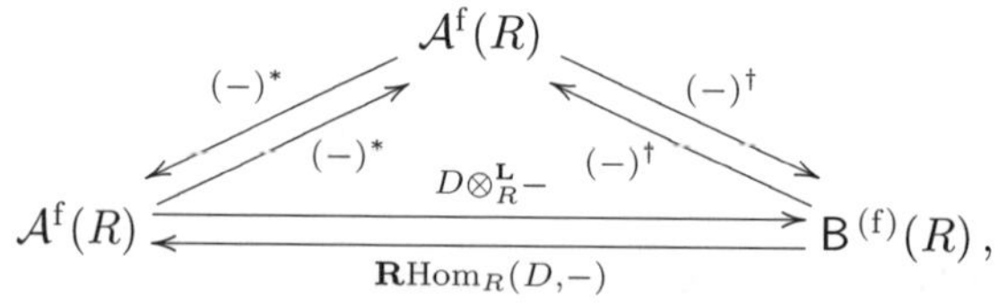

in such a way that the inner and outer triangles are both commutative, the left pair of parallel tilted arrows as well as the right pair of parallel tilted arrows provide dualities of categories, and the pair of horizontal arrows provides an equivalence of categories.

This is restated as Proposition 2.6, and it is proved there. The next result is an application of Theorem A.

THEOREM D. *Assume that $\varphi\colon R \longrightarrow S$ is local homomorphism such that the source R is a homomorphic image of a Gorenstein local ring. If the target S has finite Gorenstein injective dimension over the source R, then R is Gorenstein and S has finite Gorenstein flat dimension over R.*

Endomorphisms. Let $\varphi\colon R \longrightarrow R$ be a local endomorphism, let M be an R–module, and let n be a natural number. In this setup, ${}^{\varphi^n}\!M$ denotes M viewed as an R–module via φ^n, that is, the abelian group M equipped with the multiplication $(r, m) \longmapsto \varphi^n(r)m$.

The next result is part of Theorem 5.5.

THEOREM E. *Let $\varphi\colon (R, \mathfrak{m}) \longrightarrow (R, \mathfrak{m})$ be a local endomorphism and assume that R is a homomorphic image of a Gorenstein local ring. The following conditions are then equivalent.*

(i) *R is Gorenstein.*
(ii) $\operatorname{Gid}_R {}^{\varphi^n}\!R$ *is finite for some integer $n \geqslant 1$.*
(iii) $\operatorname{Gid}_R {}^{\varphi^n}\!R$ *is finite for all integers $n \geqslant 1$.*

If one of the above conditions is met, then $\operatorname{Gid}_R {}^{\varphi^n}\!R = \operatorname{depth} R = \dim R$.

When the local ring R is of prime characteristic p, Theorems D and E apply, in particular, to the Frobenius endomorphism $R \to R$, $r \longmapsto r^p$.

The Frobenius endomorphism is a particular instance of a local endomorphism $\varphi\colon (R, \mathfrak{m}) \longrightarrow (R, \mathfrak{m})$ such that $\varphi^i(\mathfrak{m}) \subseteq \mathfrak{m}^2$ for some integer $i \geqslant 1$; this condition is equivalent to the condition that for every element x from $\mathfrak{m}$ the sequence $(\varphi^i(x))_{i\geqslant 1}$ converges to zero in the $\mathfrak{m}$–adic topology. Such an endomorphism is called a *contraction.* Note that φ in Theorem D and E is *not* supposed to be a contraction.

Theorem E is a Gorenstein version of the next theorem, which is [4, (13.3)].

THEOREM (AVRAMOV–IYENGAR–MILLER). *Let $\varphi\colon (R, \mathfrak{m}) \longrightarrow (R, \mathfrak{m})$ be a contraction. If* $\operatorname{id}_R {}^{\varphi^n}\!R$ *is finite for some integer $n \geqslant 1$, then R is regular.*

The last result should be compared to the classical results by Kunz [20, (2.1)] and Rodicio [23] showing that the ring is regular precisely when the R–module ${}^{\varphi^n}\!R$ has finite flat dimension (or, equivalently, finite projective dimension). Moreover, the result [17, (6.5)] by Iyengar and Sather-Wagstaff implies that the ring is Gorenstein exactly when the Gorenstein flat dimension of R–module ${}^{\varphi^n}\!R$ is finite.

Organization of the paper. The main results, Theorems A, D, and E, belong to *classical homological algebra.* Their proofs, however, take—of necessity—place in the derived category $\mathcal{D}(R)$ of the category of R–modules.

For now there is no suitable description of the applications of this *hyperhomological algebra* to commutative ring theory. Thus, necessary background material is scattered throughout Sections 2–4.

2. Dualities and equivalences

2.1. Derived category. Throughout the paper, $(R, \mathfrak{m}, k)$ is a local commutative Noetherian ring. We will work within the derived category $\mathcal{D}(R)$; see, for example, Gelfand and Manin [15].

The objects in $\mathcal{D}(R)$ are complexes of R–modules. Homological notation is used, so when M is a complex, the differential has degree -1, that is, $\partial_n^M : M_n \longrightarrow M_{n-1}$. The symbol $\simeq$ denotes isomorphisms in the derived category. If n is an integer, $\Sigma^n M$ denotes the complex M shifted n degrees to the left, that is, against the direction of the differential.

The full subcategory of $\mathcal{D}(R)$ consisting of complexes with bounded homology is denoted $\mathcal{D}_\square(R)$, while $\mathcal{D}_\square^f(R)$ denotes the full subcategory of $\mathcal{D}_\square(R)$ consisting of complexes with each homology module finitely generated; the objects in $\mathcal{D}_\square^f(R)$ will be referred to as *finite complexes.* Each R–module M is viewed as a complex concentrated in degree zero. Moreover, each complex M of R–modules with homology concentrated in degree zero is isomorphic in $\mathcal{D}(R)$ to the module $\mathrm{H}_0(M)$. Thus we identify R–modules with complexes homologically concentrated in degree zero.

The homological size of a complex M is given by its *homological infimum,* its *homological supremum,* and its *amplitude:* $\inf M = \inf\{\, \ell \mid \mathrm{H}_\ell(M) \neq 0 \,\}$, $\sup M = \sup\{\, \ell \mid \mathrm{H}_\ell(M) \neq 0\}$, and $\operatorname{amp} M = \sup M - \inf M$. We set $\inf \varnothing = \infty$ and $\sup \varnothing = -\infty$, so M belongs to $\mathcal{D}_\square(R)$ if and only if $\sup M < \infty$ and $\inf M > -\infty$. Moreover, $\operatorname{amp} M = 0$ if and only if M is isomorphic (in the derived category) to $\Sigma^n K$ for some non-zero R–module K and some integer n; namely $K = \mathrm{H}_n(M)$.

If M is a homologically bounded complex, then M is said to be of *finite projective dimension, finite injective dimension,* or *finite flat dimension,* when M is isomorphic (in $\mathcal{D}(R)$) to a bounded complex consisting of, respectively, projective modules, injective modules, or flat modules, in which case we write, respectively, $\operatorname{pd}_R M < \infty$, $\operatorname{id}_R M < \infty$, or $\operatorname{fd}_R M < \infty$. For details consult [1].

The left derived tensor product functor $-\otimes_R^{\mathbf{L}} \sim$ is defined, up to isomorphism in $\mathcal{D}(R)$, by taking an appropriate projective resolution of the first argument or of the second one. The right derived homomorphism functor $\mathbf{R}\mathrm{Hom}_R(-, \sim)$ is obtained by taking an appropriate projective resolution of the first argument or by taking an appropriate injective resolution of the second one. If M and N are R–modules, then there are isomorphisms $\mathrm{H}_\ell(M \otimes_R^{\mathbf{L}} N) \cong \mathrm{Tor}_\ell^R(M, N)$ and $\mathrm{H}_\ell(\mathbf{R}\mathrm{Hom}_R(M, N)) \cong \mathrm{Ext}_R^{-\ell}(M, N)$ for all integers ℓ.

2.2. Functorial isomorphisms. The next standard isomorphisms are used throughout the paper. To facilitate the description here, also the other ring S is supposed to be commutative, and not all the boundedness conditions imposed on the complexes are strictly necessary. Let K, L, and M belong to $\mathcal{D}(R)$, let P belong to $\mathcal{D}(S)$, and let N belong to the derived category $\mathcal{D}(R,S)$ of the category of R–S–bimodules. There are then the next functorial isomorphisms in $\mathcal{D}(R,S)$.

$$\text{(Comm)}\qquad L \otimes^{\mathbf{L}}_R M \xrightarrow{\simeq} M \otimes^{\mathbf{L}}_R L\,.$$

$$\text{(Assoc)}\qquad (M \otimes^{\mathbf{L}}_R N) \otimes^{\mathbf{L}}_S P \xrightarrow{\simeq} M \otimes^{\mathbf{L}}_R (N \otimes^{\mathbf{L}}_S P)\,.$$

$$\text{(Adjoint)}\qquad \mathbf{R}\mathrm{Hom}_S(M \otimes^{\mathbf{L}}_R N, P) \xrightarrow{\simeq} \mathbf{R}\mathrm{Hom}_R(M, \mathbf{R}\mathrm{Hom}_S(N,P))\,.$$

$$\text{(Swap)}\qquad \mathbf{R}\mathrm{Hom}_R(M, \mathbf{R}\mathrm{Hom}_S(P,N)) \xrightarrow{\simeq} \mathbf{R}\mathrm{Hom}_S(P, \mathbf{R}\mathrm{Hom}_R(M,N))\,.$$

Moreover, there are the following evaluation morphisms.

(Tensor–eval)

$$\alpha_{KNP}\colon \mathbf{R}\mathrm{Hom}_R(K,N) \otimes^{\mathbf{L}}_S P \longrightarrow \mathbf{R}\mathrm{Hom}_R(K, N \otimes^{\mathbf{L}}_S P) \quad \text{and}$$

(Hom–eval)

$$\beta_{PNM}\colon P \otimes^{\mathbf{L}}_S \mathbf{R}\mathrm{Hom}_R(N,M) \longrightarrow \mathbf{R}\mathrm{Hom}_R(\mathbf{R}\mathrm{Hom}_S(P,N), M)\,.$$

- The morphism α_{KNP} is an isomorphism, if K is finite, $\mathrm{H}(N)$ is bounded, and either $\mathrm{fd}_S\, P$ or $\mathrm{pd}_R\, K$ is finite.
- The morphism β_{PNM} is an isomorphism, if P is finite, $\mathrm{H}(N)$ is bounded, and either $\mathrm{pd}_S\, P$ or $\mathrm{id}_R\, M$ is finite.

For details consult [6, A.4] and the references therein.

2.3. Dimension and depth. For a finite complex M we define its *depth* and *dimension* as follows:

$$\begin{aligned} \dim_R M &= \sup\{\, \dim_R \mathrm{H}_\ell(M) - \ell \mid \ell \in \mathbb{Z}\,\} \quad \text{and} \\ \operatorname{depth}_R M &= \inf\{\, \ell \mid \mathrm{H}_{-\ell}(\mathbf{R}\mathrm{Hom}_R(k,M)) \neq 0\,\}\,. \end{aligned}$$

Here $\dim_R \mathrm{H}_\ell(M)$ denotes the (Krull) dimension of the module $\mathrm{H}_\ell(M)$. If M is a finitely generated module, then these invariants yield the classical depth and dimension of M. It turns out that the dimension of a complex M equals the supremum of the numbers $\dim R/\mathfrak{p} - \inf M_\mathfrak{p}$ for $\mathfrak{p} \in \operatorname{Spec} R$; see [10, (16.3)].

2.4. Dagger duality. In this paragraph we assume that R admits a *normalized dualizing complex* D, that is, D is a finite R–complex, its injective dimension $\mathrm{id}_R\, D$ is finite, $\sup D = \dim R$, and the canonical morphism $\mu_D\colon R \longrightarrow \mathbf{R}\mathrm{Hom}_R(D,D)$ is an isomorphism. It follows that $k \simeq \mathbf{R}\mathrm{Hom}_R(k,D)$ and $\inf D = \operatorname{depth} R$.

When M is a finite complex, we consider the *dagger dual* $M^\dagger = \mathbf{R}\mathrm{Hom}_R(M,D)$, and the following relations hold: $\sup M^\dagger = \dim_R M$ and

$\inf M^\dagger = \operatorname{depth}_R M$; for details see [10, (16.20)]. In particular, if $\mathrm{H}(M) \neq 0$ then $\operatorname{depth}_R M \leqslant \dim_R M$, and the Cohen–Macaulay defect of M is defined to be the non-negative integer $\operatorname{cmd}_R M = \dim_R M - \operatorname{depth}_R M$.

If R is a homomorphic image of a local Gorenstein ring Q, then $\Sigma^{\dim Q}\,\mathbf{R}\mathrm{Hom}_Q(R,Q)$ is a normalized dualizing complex over R. Consequently, every complete local ring admits a dualizing complex. On the other hand, if a local ring admits a dualizing complex, then it is a homomorphic image of a Gorenstein ring by Kawasaki [19].

The contravariant functor $(-)^\dagger = \mathbf{R}\mathrm{Hom}_R(-,D)$ carries the category $\mathsf{D}^{\mathrm{f}}_\Box(R)$ into itself, and for every finite M there is the biduality isomorphism $\delta^M_D\colon M \xrightarrow{\simeq} M^{\dagger\dagger}$ from (Hom–eval) in 2.2. This induces the next duality of categories.

$$\mathsf{D}^{\mathrm{f}}_\Box(R) \underset{(-)^\dagger}{\overset{(-)^\dagger}{\rightleftarrows}} \mathsf{D}^{\mathrm{f}}_\Box(R).$$

If R is a Cohen–Macaulay ring of dimension d possessing a normalized dualizing complex D, then $\mathrm{H}_n(D) = 0$ for $n \neq d$, and $\mathrm{H}_d(D)$ is said to be the dualizing (or canonical) module for R; see [5, Chapter 3].

2.5. Dualizing equivalence. Let D be a dualizing complex for R and consider the functors $D \otimes^{\mathbf{L}}_R -$ and $\mathbf{R}\mathrm{Hom}_R(D,-)$.

The *Auslander categories* $\mathcal{A}(R)$ and $\mathcal{B}(R)$ are full subcategories of $\mathcal{D}(R)$ defined as follows.

$$\mathcal{A}(R) = \left\{ M \in \mathcal{D}_\Box(R) \;\middle|\; \begin{array}{l} \eta_M\colon M \xrightarrow{\simeq} \mathbf{R}\mathrm{Hom}_R(D, D\otimes^{\mathbf{L}}_R M) \text{ is an iso-} \\ \text{morphism in } \mathcal{D}(R), \text{ and } D\otimes^{\mathbf{L}}_R M \text{ is bounded} \end{array} \right\},$$

$$\mathcal{B}(R) = \left\{ N \in \mathcal{D}_\Box(R) \;\middle|\; \begin{array}{l} \varepsilon_N\colon D\otimes^{\mathbf{L}}_R \mathbf{R}\mathrm{Hom}_R(D,N) \xrightarrow{\simeq} N \text{ is an isomor-} \\ \text{phism in } \mathcal{D}(R), \text{ and } \mathbf{R}\mathrm{Hom}_R(D,N) \text{ is bounded} \end{array} \right\}.$$

The homomorphisms η_M and ε_N are the evaluation morphisms α_{DDM} and β_{DDN} from 2.2, respectively, composed with the isomorphism $\mathbf{R}\mathrm{Hom}_R(D,D) \simeq R$. This setup was introduced in [2], where it was also pointed out that any complex of finite flat dimension belongs to $\mathcal{A}(R)$, and that any complex of finite injective dimension belongs to $\mathcal{B}(R)$.

The Auslander categories are defined in such a way that there are the following equivalences of categories; the second equivalence then follows by restriction.

$$\mathcal{A}(R) \underset{\mathbf{R}\mathrm{Hom}_R(D,-)}{\overset{D\otimes^{\mathbf{L}}_R -}{\rightleftarrows}} \mathcal{B}(R).$$

$$\mathcal{A}^{\mathrm{f}}(R) \underset{\mathbf{R}\mathrm{Hom}_R(D,-)}{\overset{D\otimes^{\mathbf{L}}_R -}{\rightleftarrows}} \mathcal{B}^{\mathrm{f}}(R).$$

In the following result we consider the two functors $(-)^* = \mathbf{R}\mathrm{Hom}_R(-,R)$ and $(-)^\dagger = \mathbf{R}\mathrm{Hom}_R(-,D)$.

2.6. PROPOSITION. *The following statements hold for the next diagram.*

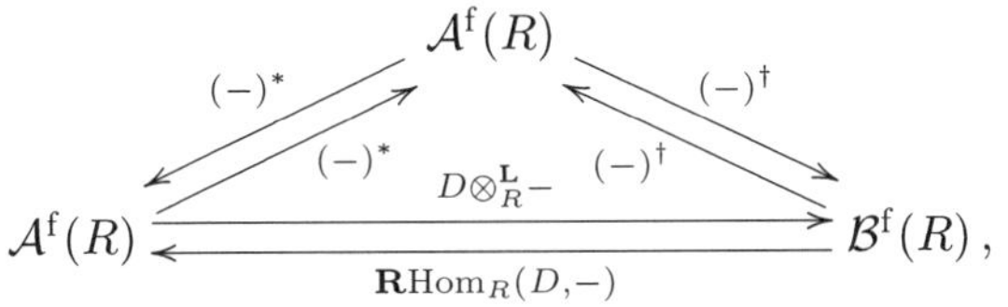

(1) *The inner triangle is commutative (up to canonical isomorphism).*
(2) *The outer triangle is commutative (up to canonical isomorphism).*
(3) *The left pair of parallel tilted arrows provides a duality of categories.*
(4) *The right pair of parallel tilted arrows provides a duality of categories.*
(5) *The pair of horizontal arrows provides an equivalence of categories.*

Proof. The assertion concerning the horizontal arrows follows by dualizing equivalence 2.5. The functor $(-)^*$ provides a duality on $\mathcal{A}^{\mathrm{f}}(R)$; see [2, (4.1.7)] or [7, (4.7)]. The corresponding assertion concerning $(-)^\dagger$ follows by similar arguments; see also the proof of [6, (3.2.9)]. The commutativity of the inner and outer triangles are up to natural isomorphisms composed by the isomorphisms in 2.2 and the isomorphism $R \simeq \mathbf{R}\mathrm{Hom}_R(D,D)$. □

3. Gorenstein homological dimensions

3.1. Gorenstein injective dimension. An R–complex I is said to be a *complete injective resolution*, if it is exact and consists of injective modules, and it is such that $\mathrm{Hom}_R(I',I)$ is exact for all injective R–modules I'. An R–module J is said to be *Gorenstein injective*, if it is a cokernel in a complete injective resolution. Thus, every injective module is Gorenstein injective.

The *Gorenstein injective dimension* $\mathrm{Gid}_R\, M$ of $M \in \mathcal{D}_\square(R)$ is defined to be the infimum of the set of integers n such that there exists a complex I consisting of Gorenstein injective modules satisfying $M \simeq I$ and $I_\ell = 0$ for $-\ell > n$. (Recall that we always use homological notation.) Thus, a complex of R–modules has finite Gorenstein injective dimension if and only if it is isomorphic in $\mathcal{D}(R)$ to a bounded complex of Gorenstein injective modules. Moreover, the Gorenstein injective dimension is a *refinement* of the (classical) injective dimension, that is, $\mathrm{Gid}_R\, M \leqslant \mathrm{id}_R\, M$ with equality if $\mathrm{id}_R\, M$ is finite; see [6, Chapter 6].

3.2. Gorenstein projective dimension. The *Gorenstein projective dimension* $\mathrm{Gpd}_R\, M$ of $M \in \mathcal{D}_\square(R)$ was introduced by Enochs and Jenda; it is defined dually to the injective one above, it is a refinement of the (classical) projective dimension, that is, $\mathrm{Gpd}_R\, M \leqslant \mathrm{pd}_R\, M$ with equality if $\mathrm{pd}_R\, M < \infty$; see [6, Chapter 4].

3.3. Gorenstein flat dimension. The definition of the *Gorenstein flat dimension* $\mathrm{Gfd}_R\, M$ of $M \in \mathcal{D}_\square(R)$ is similar to the Gorenstein injective dimension above. A complex F of R–modules is said to be a *complete flat resolution*, if it is exact and consists of flat modules, and it is such that also $I' \otimes_R F$ is exact for all injective R–modules I'. An R–module G is said to be *Gorenstein flat*, if it is a cokernel in a complete flat resolution. Thus, every flat module is Gorenstein flat.

The *Gorenstein flat dimension* $\mathrm{Gfd}_R\, M$ of $M \in \mathcal{D}_\square(R)$ is defined to be the infimum of the set of integers n such that there exists a complex F consisting of Gorenstein flat modules satisfying $M \simeq F$ and $F_\ell = 0$ for $\ell > n$. Thus, a complex of R–modules has finite Gorenstein flat dimension if and only if it is isomorphic in $\mathcal{D}(R)$ to a bounded complex of Gorenstein flat modules. Moreover, the Gorenstein flat dimension is a refinement of the (classical) flat dimension, that is, $\mathrm{Gfd}_R\, M \leqslant \mathrm{fd}_R\, M$ with equality if $\mathrm{fd}_R\, M < \infty$. For details consult [6, Chapter 5].

3.4. Auslander's G–dimension. If G is a *finitely generated* R–module, then it is Gorenstein projective if and only if it satisfies the next conditions.

- $\mathrm{Ext}^\ell_R(G,R) = 0$ and $\mathrm{Ext}^\ell_R(\mathrm{Hom}_R(G,R),R) = 0$ for $\ell > 0$, and
- the canonical map $G \longrightarrow \mathrm{Hom}_R(\mathrm{Hom}_R(G,R),R)$ is an isomorphism,

see, for example, [6, (4.4.6)]. Auslander's *Gorenstein dimension* $\mathrm{G\text{–}dim}_R\, M$ of a finite R–complex M is defined to be at most n exactly when M is isomorphic in $\mathcal{D}(R)$ to a bounded complex G such that $G_\ell = 0$ for $\ell > n$, and such that each G_ℓ is a finitely generated R–modules satisfying the above two conditions; see [6, (2.3.2)]. By [8, (3.8)] $\mathrm{G\text{–}dim}_R\, M = \mathrm{Gpd}_R\, M = \mathrm{Gfd}_R\, M$ for any finite complex M.

Moreover, [6, (3.1.11)] yields $\mathrm{G\text{–}dim}_R\, M < \infty$ if and only if $M \in \mathcal{A}^{\mathrm{f}}(R)$. The last result is extended by the next one, which is the main theorem in Christensen, Frankild and Holm [8].

3.5. Finiteness of Gorenstein dimensions. Let R be a homomorphic image of a Gorenstein ring. If M is an R–complex, then the following are equivalent.

(i) M belongs to $\mathcal{A}(R)$.
(ii) M has finite Gorenstein projective dimension, that is, $\mathrm{Gpd}_R\, M < \infty$.
(iii) M has finite Gorenstein flat dimension, that is, $\mathrm{Gfd}_R\, M < \infty$.

Furthermore, if N is an R–complex, then the following are equivalent.

(i) N belongs to $\mathcal{B}(R)$.
(ii) N has finite Gorenstein injective dimension, that is, $\mathrm{Gid}_R\, N < \infty$.

For details consult [8, (4.1) and (4.4)].

The next result on completion and Gorenstein injective dimension is due to Christensen, Frankild, and Iyengar. We thank Christensen and Iyengar for allowing us to include it here. Note that it does not require R to be a homomorphic image of a Gorenstein ring.

3.6. THEOREM. *Let R be a local ring, and let M be a finite R–complex. If M has finite Gorenstein injective dimension over R, then $M \otimes_R \widehat{R}$ has finite Gorenstein injective dimension over $\widehat{R}$.*

Proof. Let K^R be the Koszul complex on a set of generators for the maximal ideal $\mathfrak{m}$. Because the homology modules of K^R have finite length, there is a isomorphism $K^R \simeq \widehat{R} \otimes_R K^R$ in $\mathcal{D}(R)$. By flatness of the completion map $R \longrightarrow \widehat{R}$, a minimal set of generators for $\mathfrak{m}$ extends to a minimal set of generators of $\widehat{\mathfrak{m}}$; in particular, $K^{\widehat{R}} = \widehat{R} \otimes_R K^R$ is a Koszul complex on a minimal set of generators for $\widehat{\mathfrak{m}}$. Moreover, the isomorphism of R–modules $R/\mathfrak{m} \cong k \cong \widehat{R}/\widehat{\mathfrak{m}}$ together with the fact that $\mathrm{H}_\ell(K^R) \cong \mathrm{H}_\ell(K^{\widehat{R}})$ for all $\ell \in \mathbb{Z}$ shows that $K^R \simeq K^{\widehat{R}}$ (in $\mathcal{D}(R)$); we set $K = K^{\widehat{R}}$.

Under the present assumptions on M, the complex

$$N = M \otimes_R K^R \simeq M \otimes_R K \simeq (M \otimes^{\mathbf{L}}_R \widehat{R}) \otimes_{\widehat{R}} K$$

has finite Gorenstein injective dimension over R; see [9, (5.5)(c')]. Note that the homology modules of N have finite length since M is finite. Hence there is an isomorphism

$$N \xrightarrow{\simeq} \mathrm{Hom}_R(\mathrm{Hom}_R(N, \mathrm{E}_R(k)), \mathrm{E}_R(k))\,.$$

Here $\mathrm{E}_R(k)$ denotes the injective envelope of the R–module k. In particular,

$$\mathrm{Gid}_R \,\mathrm{Hom}_R(\mathrm{Hom}_R(N, \mathrm{E}_R(k)), \mathrm{E}_R(k)) < \infty$$

and, therefore, $\mathrm{Gfd}_R \,\mathrm{Hom}_R(N, \mathrm{E}_R(k))$ is finite by [6, (6.4.2)]. As the homology modules of $\mathrm{Hom}_R(N, \mathrm{E}_R(k))$ has finite length, the complex

$$(*) \qquad \mathrm{Hom}_R(N, \mathrm{E}_R(k)) \otimes_R \widehat{R} \simeq \mathrm{Hom}_R(N, \mathrm{E}_R(k))$$

has finite Gorenstein flat dimension over $\widehat{R}$. Thus, using $(*)$ and (Adjoint) from 2.2 we conclude

$$N \xrightarrow{\simeq} \mathrm{Hom}_{\widehat{R}}(\mathrm{Hom}_R(N, \mathrm{E}_R(k)), \mathrm{E}_R(k))$$

has finite Gorenstein injective dimension over $\widehat{R}$ by [8, (5.1)]; this uses the fact that the complete ring $\widehat{R}$ admits a dualizing complex D. From [8, (4.4)] it follows that the complex

$$\mathbf{R}\mathrm{Hom}_{\widehat{R}}(D, N) \simeq \mathbf{R}\mathrm{Hom}_{\widehat{R}}(D, M \otimes^{\mathbf{L}}_R \widehat{R}) \otimes^{\mathbf{L}}_{\widehat{R}} K$$

is homologically bounded. Thus, from [13, 1.3] it follows that the complex $\mathbf{R}\mathrm{Hom}_{\widehat{R}}(D, M \otimes^{\mathbf{L}}_R \widehat{R})$ is homologically bounded as well. Consider the commutative diagram

$$\begin{array}{ccc}
(M \otimes^{\mathbf{L}}_R \widehat{R}) \otimes^{\mathbf{L}}_{\widehat{R}} K & \xleftarrow[\simeq]{\varepsilon^{\widehat{R}}_{(M \otimes^{\mathbf{L}}_R \widehat{R}) \otimes^{\mathbf{L}}_{\widehat{R}} K}} & D \otimes^{\mathbf{L}}_{\widehat{R}} \mathbf{R}\mathrm{Hom}_{\widehat{R}}(D, (M \otimes^{\mathbf{L}}_R \widehat{R}) \otimes^{\mathbf{L}}_{\widehat{R}} K) \\
\Big\| & & \Big\downarrow \simeq \\
(M \otimes^{\mathbf{L}}_R \widehat{R}) \otimes^{\mathbf{L}}_{\widehat{R}} K & \xleftarrow[(\varepsilon^{\widehat{R}}_{M \otimes^{\mathbf{L}}_R \widehat{R}}) \otimes^{\mathbf{L}}_{\widehat{R}} K]{} & (D \otimes^{\mathbf{L}}_{\widehat{R}} \mathbf{R}\mathrm{Hom}_{\widehat{R}}(D, M \otimes^{\mathbf{L}}_R \widehat{R})) \otimes^{\mathbf{L}}_{\widehat{R}} K
\end{array}$$

where the rightmost vertical isomorphism is by (Tensor-eval) and (Assoc) in 2.2. Again, using [13, 1.3] and a standard mapping cone argument, it follows that $\varepsilon^{\widehat{R}}_{M \otimes^{\mathbf{L}}_R \widehat{R}}$ is a isomorphism. Whence, $M \otimes_R \widehat{R} \simeq M \otimes^{\mathbf{L}}_R \widehat{R}$ has finite Gorenstein injective dimension over $\widehat{R}$ by [8, (4.4)]. □

4. Bass and Poincaré series

4.1. Bass and Poincaré series. For a finite R–complex M and an integer ℓ, the ℓth *Bass number* and the ℓth *Betti numbers* are, respectively, the next vector space dimensions over the residue field k:

$$\mu^{\ell}_R(M) = \mathrm{rank}_k \, \mathrm{H}_{-\ell}(\mathbf{R}\mathrm{Hom}_R(k, M)) \quad \text{and} \quad \beta^R_{\ell}(M) = \mathrm{rank}_k \, \mathrm{H}_{\ell}(k \otimes^{\mathbf{L}}_R M)\,.$$

The ring of *formal Laurent series with integer coefficients* is denoted $\mathbb{Z}(|t|)$; its elements have the form $\sum_{\ell \in \mathbb{Z}} a_\ell t^\ell$ with $a_\ell \in \mathbb{Z}$ and $a_\ell = 0$ for $\ell \ll 0$. For a finite R–complex M the *Bass series* $\mathrm{I}^M_R(t)$ and the *Poincaré series* $\mathrm{P}^R_M(t)$ are elements in $\mathbb{Z}(|t|)$ defined as follows.

$$\mathrm{I}^M_R(t) = \sum_{\ell \in \mathbb{Z}} \mu^{\ell}_R(M) t^\ell \quad \text{and} \quad \mathrm{P}^R_M(t) = \sum_{\ell \in \mathbb{Z}} \beta^R_{\ell}(M) t^\ell\,.$$

The Bass series for the ring is denoted $\mathrm{I}_R(t)$, and the ring R is Gorenstein precisely when $\mathrm{I}_R(t) = t^s$ for some integer s. If D is a finite R–complex, then D is a normalized dualizing complex if and only if $\mathrm{I}^D_R(t) = 1$. The next equations for Bass and Poincaré series will be important later. The first two are proved in [2, (1.5.3)], while the last two are proved in [12, (4.3)].

4.2. Bass–Poincaré equalities. For finite complexes M and N there are the next equalities of formal Laurent series, that is, equalities in $\mathbb{Z}(|t|)$.

(PP) $$\mathrm{P}^R_{M\otimes^{\mathbf{L}}_R N}(t) = \mathrm{P}^R_M(t)\,\mathrm{P}^R_N(t)\,.$$

(PI) $$\mathrm{I}_R^{\mathbf{R}\mathrm{Hom}_R(M,N)}(t) = \mathrm{P}^R_M(t)\,\mathrm{I}^N_R(t)\,.$$

(IP) $$\mathrm{I}_R^{M\otimes^{\mathbf{L}}_R N}(t) = \mathrm{I}^M_R(t)\,\mathrm{P}^N_R(t^{-1}) \quad \text{provided} \quad \mathrm{pd}_R N < \infty\,.$$

(II) $$\mathrm{P}^R_{\mathbf{R}\mathrm{Hom}_R(M,N)}(t) = \mathrm{I}^M_R(t)\,\mathrm{I}^R_N(t^{-1}) \quad \text{provided} \quad \mathrm{id}_R N < \infty\,.$$

4.3. PROPOSITION. *Let R be a homomorphic image of a Gorenstein ring, and let N be a finite R–complex of finite Gorenstein injective dimension. There exists then a finite R–complex K of finite Gorenstein projective dimension with $\inf K = \inf N$ and $\operatorname{amp} K \leqslant \operatorname{amp} N$ such that there is the next equality of formal Laurent series.*

$$\mathrm{P}^R_N(t)t^{\operatorname{depth} R} = \mathrm{P}^R_K(t)\,\mathrm{I}_R(t)\,.$$

If N is of finite (classical) injective dimension, then K has finite (classical) projective dimension.

Proof. Throughout the proof, we let D be a normalized dualizing complex, and set $M = N^{\dagger}$, $L = \mathbf{R}\mathrm{Hom}_R(D,N)$, and $s = \operatorname{depth} R$. Note that $\operatorname{depth}_R M = \inf N$ by 2.4. As N belongs to $\mathcal{B}^{\mathrm{f}}(R)$ by 3.5, Lemma 2.6 yields that M and L belong to $\mathcal{A}^{\mathrm{f}}(R)$, and that $L^* \simeq M$ and $M^* \simeq L$. As M belongs to $\mathcal{A}^{\mathrm{f}}(R)$, the G–dimension of M is finite by 3.5. Whence, by [6, (1.2.7) and (1.4.8)] we obtain

$$\inf L = \inf M^* = -\operatorname{G-dim}_R M = \operatorname{depth}_R M - \operatorname{depth} R = \inf N - \operatorname{depth} R.$$

Next, [6, (A.4.6)] yields the inequality

$$\sup L = \sup \mathbf{R}\mathrm{Hom}_R(D,N) \leqslant -\inf D + \sup N = \sup N - \operatorname{depth} R.$$

The last equality is by 2.4. Hence, $\operatorname{amp} L \leqslant \operatorname{amp} N$. Set $K = \Sigma^s L$, which has finite Gorenstein projective dimension as it belongs to $\mathcal{A}^{\mathrm{f}}(R)$. It still remains to prove the equation for the Laurent series. Recall that $s = \operatorname{depth} R$. The computation

$$K^{*\dagger} \simeq ((\Sigma^s L)^*)^{\dagger} = (\Sigma^{-s}L^*)^{\dagger} \simeq (\Sigma^{-s}M)^{\dagger} = \Sigma^s M^{\dagger} \simeq \Sigma^s N$$

yields the next isomorphism.

(4.3.1) $$N \simeq \Sigma^{-s}K^{*\dagger}.$$

Using that $\mathrm{I}^D_R(t) = 1$ by 2.4 and that $\mathrm{P}_R(t) = 1$, the formulae (II) and (PI) in 4.2 yields the equalities

$$\mathrm{P}^R_N(t) = \mathrm{P}^R_{\Sigma^{-s}K^{*\dagger}}(t) = t^{-s}\,\mathrm{P}^R_{K^{*\dagger}}(t) = t^{-s}\,\mathrm{I}_R^{K^*}(t) = t^{-s}\,\mathrm{P}^R_K(t)\,\mathrm{I}_R(t).$$

Finally, if $\operatorname{id}_R N$ is finite, it follows that $\operatorname{pd}_R N^\dagger$ is finite as well, and this implies that $\operatorname{pd}_R N^{\dagger\,*} < \infty$, that is, $\operatorname{pd}_R K < \infty$. □

4.4. REMARK. When the finite complex N from Proposition 4.3 is a module, the finite complex K is also a module, and it has finite Gorenstein projective dimension. If the ring R is complete, then it is possible to use results in [14, Section 2] to prove that $K \cong \operatorname{Ext}^s_R(\mathrm{E}_R(k), N)$, where $s = \operatorname{depth} R$. The latter module was used by Peskine and Szpiro in the proof of their theorem mentioned in the introduction.

4.5. THEOREM. *If there exists a non-zero cyclic R–module N with the Gorenstein injective dimension $\operatorname{Gid}_R N$ finite, then R is Gorenstein.*

Proof. Note first that we may assume that R is complete and thus possesses a dualizing complex; see 3.6. As N is cyclic, we have $N \cong R/\operatorname{Ann}_R N$, and hence the constant term in $\mathrm{P}^R_N(t)$ is 1. Thus, the equlity of power series in 4.3 yields that the constant term in $\mathrm{P}^R_K(t)$ is also 1. In particular, the module K occurring in 4.3 is cyclic; whence $K \cong R/\operatorname{Ann}_R K$. The formula (4.3.1) gives that $\operatorname{Ann}_R N \supseteq \operatorname{Ann}_R K$. Applying the functor $(-)^{\dagger *}$ to (4.3.1) we obtain the equation $K \simeq \Sigma^s N^{\dagger *}$. This yields $\operatorname{Ann}_R N \subseteq \operatorname{Ann}_R K$. Thus $\operatorname{Ann}_R N = \operatorname{Ann}_R K$, and it follows that $N \cong K$, so the equation in 4.3 implies that $\mathrm{I}_R(t) = t^s$, that is, R is Gorenstein. □

4.6. Cohen–Macaulay injective dimension. Theorem 4.6 below is an immediate consequence of 4.5, and it characterizes Cohen–Macaulay rings in terms finiteness of the Cohen–Macaulay injective dimension introduced by Holm and Jørgensen [18]. Recall from [7] that a finitely generated R–module C is *semi-dualizing* if the natural homomorphism $R \longrightarrow \operatorname{Hom}_R(C,C)$ is an isomorphism and $\operatorname{Ext}^i_R(C,C) = 0$ for all $i > 0$, that is, the homothety morphism $R \longrightarrow \mathbf{R}\operatorname{Hom}_R(C,C)$ is an isomorphism in $\mathcal{D}(R)$.

The *Cohen–Macaulay injective dimension* of an R–module M is defined as

$$\operatorname{CMid}_R M = \inf\{\operatorname{Gid}_{R\ltimes C} M \mid C \text{ is a semi-dualizing module over } R\}.$$

Here $R \ltimes C$ denotes the trivial extension ring; it is the R–module $R \times C$ equipped with the multiplication $(r,c)(r',c') = (rr', rc' + r'c)$. If $(R,\mathfrak{m},k)$ is local, then so is $(R \ltimes C, \mathfrak{m} \times C, k)$. The ring homomorphism $R \ltimes C \longrightarrow R$ defined by $(r,c) \longmapsto r$ turns every R–module into an $R \ltimes C$– module; if N is cyclic over R, then it is so over $R \ltimes C$. Finally, the module C is a dualizing module precisely when the ring $R \ltimes C$ is Gorenstein; for details see Foxby [11].

4.6. THEOREM. *If there exists a non-zero cyclic R–module N with the Cohen–Macaulay injective dimension $\operatorname{CMid}_R N$ finite, then R is Cohen–-Macaulay with dualizing module.* □

5. Local homomorphisms

In this section we apply Theorem 4.5 to a local homomorphism of local rings $\varphi\colon (R,\mathfrak{m},k) \longrightarrow (S,\mathfrak{n},\ell)$.

5.1. Cohen factorizations. In this section we will use the *Cohen factorizations* of local homomorphisms introduced in Avramov, Foxby, and Herzog [3]. A Cohen factorization of φ is a commutative diagram of local homomorphisms

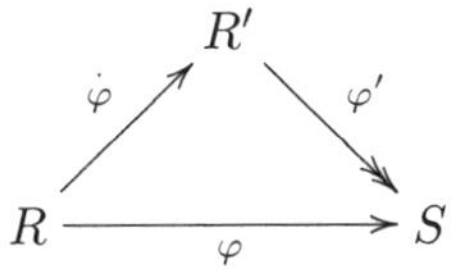

such that φ' is surjective, and $\dot{\varphi}$ is flat with the closed fiber $R'/\mathfrak{m}R'$ a regular ring and with the target R' complete.

Cohen factorizations often exist: the *semi-completion* $\grave{\varphi}\colon R \longrightarrow S \longrightarrow \widehat{S}$ always admits a Cohen factorization; see [3, (1.1)].

5.2. THEOREM. *If $\varphi\colon R \longrightarrow S$ is a local homomorphism and R is a homomorphic image of a Gorenstein ring, then* $\operatorname{Gid}_R S < \infty$ *if and only if R is Gorenstein.*

Proof. If R is Gorenstein, then $\operatorname{Gid}_R S$ is finite. Next, assume that $\operatorname{Gid}_R S$ is finite. Lemma 5.3 below implies that $\operatorname{Gid}_R \widehat{S}$ is finite as well. Choose a Cohen factorization $R \longrightarrow R' \longrightarrow \widehat{S}$ of the semi-completion $\grave{\varphi}$, and note that 5.3 below yields that also the cyclic R'–module $\widehat{S}$ has finite Gorenstein injective dimension. Thus, it follows from the main theorem 4.5 that R' is Gorenstein, and by flat descent [21, (23.4)], so is R. $\square$

5.3. LEMMA. *Assume R is a homomorphic image of a Gorenstein ring. Let $\varphi\colon R \longrightarrow S$ be a local homomorphism, let $R \longrightarrow R' \longrightarrow \widehat{S}$ be a Cohen factorization of its semi-completion, let N be a bounded complex of S–modules, and set $\widetilde{N} = N \otimes_R \widehat{S}$. The next numbers are then simultaneously finite.*

$$\operatorname{Gid}_R N, \quad \operatorname{Gid}_R \widetilde{N}, \quad \text{and} \quad \operatorname{Gid}_{R'} \widetilde{N}$$

Proof. Let D denote the normalized dualizing complex for R. According to [8, thm. 4.4] we are required to show the following two equivalences

$$N \in \mathcal{B}(R) \iff \widetilde{N} \in \mathcal{B}(R) \iff \widetilde{N} \in \mathcal{B}(R')\,.$$

The latter equivalence follows from the second part of [8, (5.3)].

To establish the former one, note that as the completion $\widehat{S}$ is flat over S there is the isomorphism $\mathbf{R}\mathrm{Hom}_R(D,N) \otimes_S^{\mathbf{L}} \widehat{S} \xrightarrow{\simeq} \mathbf{R}\mathrm{Hom}_R(D,\widetilde{N})$. Thus, the homology of $\mathbf{R}\mathrm{Hom}_R(D,\widetilde{N})$ is bounded if and only if that of $\mathbf{R}\mathrm{Hom}_R(D,N)$ is. Moreover, from the commutative diagram

$$\begin{array}{ccc}
\widetilde{N} & \xleftarrow{\ \varepsilon^R_{\widetilde{N}}\ } & D\otimes^{\mathbf{L}}_R \mathbf{R}\mathrm{Hom}_R(D,\widetilde{N}) \\
\Big\| & & \Big\uparrow \simeq \\
N\otimes^{\mathbf{L}}_S \widehat{S} & \xleftarrow[\ \varepsilon^R_N\otimes^{\mathbf{L}}_R \widehat{S}\ ]{} & D\otimes^{\mathbf{L}}_R \mathbf{R}\mathrm{Hom}_R(D,N)\otimes^{\mathbf{L}}_S \widehat{S}
\end{array}$$

it follows that $\varepsilon^R_{\widetilde{N}}$ and ε^R_N are simultaneously isomorphisms. □

The following result is [17, (8.14) and (8.15)] and it concerns contractions.

5.4. Theorem (Iyengar–Sather-Wagstaff). *Let $\varphi\colon (R,\mathfrak{m}) \longrightarrow (R,\mathfrak{m})$ be a contraction, that is, $\varphi^i(\mathfrak{m}) \subseteq \mathfrak{m}^2$ for some integer $i \geqslant 1$. The following conditions are then equivalent.*

(i) *R is Gorenstein.*
(ii) *$\mathrm{Gfd}_R\, {}^{\varphi^n}R$ is finite for all integers $n \geqslant 1$.*
(iii) *There exists a finite R–complex P with $\mathrm{H}(P) \neq 0$ and $\mathrm{pd}_R\, P$ finite such that $\mathrm{Gfd}_R\, {}^{\varphi^n}P$ is finite for some integer $n \geqslant 1$.*

If one of the above equivalent conditions is satisfied, then $\mathrm{Gfd}_R\, {}^{\varphi^n}R = 0$.

What follows may be thought of as a Gorenstein injective version of the above. However, the result is not restricted to contracting endomorphisms.

5.5. Theorem. *Let $\varphi\colon (R,\mathfrak{m}) \longrightarrow (R,\mathfrak{m})$ be a local homomorphism, and assume that R is a homomorphic image of a Gorenstein ring. The following conditions are equivalent*

(i) *R is Gorenstein.*
(ii) *$\mathrm{Gid}_R\, {}^{\varphi^n}R$ is finite for all integers $n \geqslant 1$.*
(iii) *There exists a finite R–complex P with $\mathrm{H}(P) \neq 0$ and $\mathrm{pd}_R\, P$ finite such that $\mathrm{Gid}_R\, {}^{\varphi^n}P$ is finite for some integer $n \geqslant 1$.*

Proof. The equivalence of (i) and (ii) results from Theorem 5.2. Clearly (iii) is stronger than (ii), and it is trivial that (i) implies (iii). □

If the equivalent conditions of the theorem are satisfied, then it is possible to prove that $\mathrm{Gid}_R\, {}^{\varphi^n}R = \mathrm{depth}\, R = \dim R$. However, because the only proof that the remaining author knows is quite long, it has been left out to keep this article at a reasonable length.

Acknowledgments. The authors thank Lars Winther Christensen and Sean Sather-Wagstaff for carefully reading the manuscript and for numerous thoughtful comments.

References

[1] L. L. Avramov and H.-B. Foxby, *Homological dimensions of unbounded complexes*, J. Pure Appl. Algebra **71** (1991), 129–155. MR 1117631 (93g:18017)

[2] ———, *Ring homomorphisms and finite Gorenstein dimension*, Proc. London Math. Soc. (3) **75** (1997), 241–270. MR 1455856 (98d:13014)

[3] L. L. Avramov, H.-B. Foxby, and B. Herzog, *Structure of local homomorphisms*, J. Algebra **164** (1994), 124–145. MR 1268330 (95f:13029)

[4] L. L. Avramov, S. Iyengar, and C. Miller, *Homology over local homomorphisms*, Amer. J. Math. **128** (2006), 23–90. MR 2197067 (2007e:13027)

[5] W. Bruns and J. Herzog, *Cohen-Macaulay rings*, Cambridge Studies in Advanced Mathematics, vol. 39, Cambridge University Press, Cambridge, 1993. MR 1251956 (95h:13020)

[6] L. W. Christensen, *Gorenstein dimensions*, Lecture Notes in Mathematics, vol. 1747, Springer-Verlag, Berlin, 2000. MR 1799866 (2002e:13032)

[7] ———, *Semi-dualizing complexes and their Auslander categories*, Trans. Amer. Math. Soc. **353** (2001), 1839–1883. MR 1813596 (2002a:13017)

[8] L. W. Christensen, A. Frankild, and H. Holm, *On Gorenstein projective, injective and flat dimensions—a functorial description with applications*, J. Algebra **302** (2006), 231–279. MR 2236602 (2007h:13022)

[9] L. W. Christensen and H. Holm, *Ascent properties of Auslander categories*, Canad. J. Math, to appear.

[10] H.-B. Foxby, *Hyperhomological algebra & commutative rings*, notes in preparation.

[11] ———, *Gorenstein modules and related modules*, Math. Scand. **31** (1972), 267–284 (1973). MR 0327752 (48 #6094)

[12] ———, *Isomorphisms between complexes with applications to the homological theory of modules*, Math. Scand. **40** (1977), 5–19. MR 0447269 (56 #5584)

[13] H.-B. Foxby and S. Iyengar, *Depth and amplitude for unbounded complexes*, Commutative algebra (Grenoble/Lyon, 2001), Contemp. Math., vol. 331, Amer. Math. Soc., Providence, RI, 2003, pp. 119–137. MR 2013162 (2004k:13028)

[14] A. Frankild, *Vanishing of local homology*, Math. Z. **244** (2003), 615–630. MR 1992028 (2004d:13027)

[15] S. I. Gelfand and Y. I. Manin, *Methods of homological algebra*, second ed., Springer Monographs in Mathematics, Springer-Verlag, Berlin, 2003. MR 1950475 (2003m:18001)

[16] H. Holm, *Gorenstein derived functors*, Proc. Amer. Math. Soc. **132** (2004), 1913–1923. MR 2053961 (2004m:16009)

[17] S. Iyengar and S. Sather-Wagstaff, *G-dimension over local homomorphisms. Applications to the Frobenius endomorphism*, Illinois J. Math. **48** (2004), 241–272. MR 2048224 (2005c:13016)

[18] P. Jørgensen and H. Holm, *Cohen–Macaulay homological dimensions*, Rend. Sem. Mat. Univ. Padova, to appear.

[19] T. Kawasaki, *On arithmetic Macaulayfication of Noetherian rings*, Trans. Amer. Math. Soc. **354** (2002), 123–149. MR 1859029 (2002i:13001)

[20] E. Kunz, *Characterizations of regular local rings for characteristic p*, Amer. J. Math. **91** (1969), 772–784. MR 0252389 (40 #5609)

[21] H. Matsumura, *Commutative ring theory*, second ed., Cambridge Studies in Advanced Mathematics, vol. 8, Cambridge University Press, Cambridge, 1989, Translated from the Japanese by M. Reid. MR 1011461 (90i:13001)

[22] C. Peskine and L. Szpiro, *Dimension projective finie et cohomologie locale. Applications à la démonstration de conjectures de M. Auslander, H. Bass et A. Grothendieck*, Inst. Hautes Études Sci. Publ. Math. (1973), 47–119. MR 0374130 (51 #10330)

[23] A. G. Rodicio, *On a result of Avramov*, Manuscripta Math. **62** (1988), 181–185. MR 963004 (89k:13014)

[24] R. Takahashi, *The existence of finitely generated modules of finite Gorenstein injective dimension*, Proc. Amer. Math. Soc. **134** (2006), 3115–3121. MR 2231892 (2007d:13020)

Hans–Bjørn Foxby, Afdeling for Matematiske Fag, Universitetsparken 5, DK–2100 Copenhagen Ø, Denmark

E-mail address: foxby@math.ku.dk

Illinois Journal of Mathematics
Volume 51, Number 1, Spring 2007, Pages 83–98
S 0019-2082

GENERALIZED DIVISORS AND BILIAISON

ROBIN HARTSHORNE

ABSTRACT. We extend the theory of generalized divisors so as to work on any scheme X satisfying the condition S_2 of Serre. We define a generalized notion of Gorenstein biliaison for schemes in projective space. With this we give a new proof in a stronger form of the theorem of Gaeta, that standard determinantal schemes are in the Gorenstein biliaison class of a complete intersection.

We also show, for schemes of codimension three in $\mathbb{P}^n$, that the relation of Gorenstein biliaison is equivalent to the relation of even strict Gorenstein liaison.

0. Introduction

In this paper we generalize further the theory of generalized divisors introduced in [6] by partially removing the Gorenstein hypotheses. This, we feel, puts the theory in its natural state of generality. The main difference is that instead of requiring the sheaves of ideals defining a generalized divisor to be reflexive, we require only the condition S_2 of Serre. If a scheme X satisfies G_1 and S_1, then a coherent sheaf is reflexive if and only if it satisfies S_2 [6, 1.9]. Here we show that if X satisfies S_1 only, then a coherent sheaf satisfies S_2 if and only if it is *ω-reflexive*: this means that the natural map $\mathcal{F} \to \mathcal{H}om(\mathcal{H}om(\mathcal{F}, \omega), \omega)$ is an isomorphism, where ω is the canonical sheaf. With this weaker condition we are able to establish a theory of generalized divisors on schemes X satisfying only the condition S_2.

We apply this theory to define a notion of generalized biliaison for schemes in projective space. Let D be a (generalized) divisor on an ACM scheme X in $\mathbb{P}^n_k$. If $D' \sim D + mH$, meaning D' is linearly equivalent to D plus m times the hyperplane section H of X, we say D' is obtained by an *elementary biliaison* from D. We call *biliaison* the equivalence relation generated by the elementary biliaisons.

If we do biliaisons using only complete intersection schemes X in $\mathbb{P}^n$, the resulting notion of biliaisons is equivalent to even complete intersection liaison (CI-liaison) [6, 4.4]. If we do biliaisons using an ACM scheme X satisfying

Received November 29, 2006; received in final form July 13, 2007.
2000 *Mathematics Subject Classification.* 14C20, 13C40, 14M06.

G_0, we will show (3.5) that any such biliaison (called G-biliaison) is an even Gorenstein liaison. We do not know if the converse is true. If we use arbitrary ACM schemes X, we obtain a notion of biliaison that is possibly more general than G-biliaison. Note that even this more general type of biliaison preserves the Rao modules up to shift (3.2).

As an application we give a new proof (4.1) of the theorem of KMMNP–Gaeta [11, Sec. 3] using biliaisons. I would like to thank Marta Casanellas for explaining the old proof and helping to discover the new proof. Since this paper was written in 2003, the theorem of Gaeta and the proof given here have been further generalized by E. Gorla [3], to include all determinantal schemes.

Examining the proof of (3.5) we see that the Gorenstein linkages there are all of a special kind: they use only arithmetically Gorenstein schemes of the form $M + mH$ on some ACM scheme satisfying G_0 (cf. 3.3). These we call *strict* Gorenstein linkages, and so (3.5) actually tells us that every G-biliaison is an even strict Gorenstein liaison. In Section 5 we prove a partial converse: Every even strict Gorenstein liaison of codimension 3 subschemes of $\mathbb{P}^n$ is a Gorenstein biliaison (5.1).

1. ω-reflexive modules

We will need some well-known results about the canonical module, or dualizing module as it is sometimes called, of a ring or scheme. We restrict our attention to equidimensional *embeddable* noetherian rings and schemes. For a ring A, this means that it is a quotient of a regular ring. For a scheme X, it means that it can be embedded as a closed subscheme of a regular scheme. This includes all quasiprojective schemes over a field, which will be our most common application.

An equidimensional embeddable ring or scheme always has a canonical module or sheaf unique up to isomorphism. It is finitely generated (resp. coherent). Its formation commutes with localization, and with completion of a local ring. If the ring A is a quotient of a regular ring P, and r is the difference of dimensions, then the canonical module ω of A can be obtained as $\omega = \operatorname{Ext}^r_P(A, P)$, and similarly for a closed subscheme X of a regular scheme P. If A is a Cohen–Macaulay ring, then ω is a Cohen–Macaulay module of the same dimension as A, and for any maximal Cohen–Macaulay module M, the natural map $M \to \operatorname{Hom}_A(\operatorname{Hom}_A(M, \omega), \omega)$ is an isomorphism. For references see [10] for the case of Cohen–Macaulay rings; [4, II.7] for the case of projective schemes; and see also [1].

We will expand these results somewhat by weakening their hypotheses to suit our situation. We define a module M over a ring A (as above) to be *ω-reflexive* if the natural map $M \to \operatorname{Hom}_A(\operatorname{Hom}_A(M, \omega), \omega)$ is an isomorphism. Sometimes we will denote by M^ω the module $\operatorname{Hom}_A(M, \omega)$, and call it the

ω-dual of M. We say an A-module M satisfies the condition S_r of Serre if depth $M_{\mathfrak{p}} \geq \min(r, \dim A_{\mathfrak{p}})$ for all prime ideals $\mathfrak{p} \subseteq A$.

LEMMA 1.1. *If A is a local ring of dimension 0, every finitely generated module M is ω-reflexive.*

Proof. Since A is Cohen–Macaulay, this follows from [10, 6.1]. It also follows from the local duality theorem, which says in this case that M^ω is the dual of $H^0_{\mathfrak{m}}(M) = M$, so that $M^{\omega\omega}$ is the double dual, which is isomorphic to M. Here $\mathfrak{m}$ denotes the maximal ideal of A. □

LEMMA 1.2. *For any local ring A, the canonical module ω_A satisfies the condition S_1 of Serre.*

Proof. Write A as a quotient of a regular local ring P of codimension r. Then $\omega_A = \operatorname{Ext}^r_P(A, P)$. By reason of dimension and local duality on P, the functor $\operatorname{Ext}^r_P(\cdot, P)$ is contravariant and left-exact on A modules. If $\dim A = 0$, there is nothing to prove. If $\dim A \geq 1$, let $x \in \mathfrak{m}_A$ be an element such that $\dim A/xA < \dim A$. Then from the sequence

$$A \xrightarrow{x} A \to A/xA \to 0$$

we obtain

$$0 = \operatorname{Ext}^r_P(A/xA, P) \to \omega_A \xrightarrow{x} \omega_A.$$

Thus ω_A has depth at least 1. Since formation of ω commutes with localization, we conclude that ω_A satisfies S_1. □

LEMMA 1.3. *If a local ring A satisfies S_1, then ω_A satisfies S_2.*

Proof. Write A as a quotient of a regular local ring P of codimension r, as before. Let $x \in \mathfrak{m}_A$ be a non-zero-divisor so that $B = A/xA$ has dimension one less. Then from the exact sequence

$$0 \to A \xrightarrow{x} A \to B \to 0$$

and (1.2) we obtain

$$0 \to \omega_A \xrightarrow{x} \omega_A \to \operatorname{Ext}^{r+1}_p(B, P) = \omega_B.$$

Since ω_B satisfies S_1 by (1.2), we see that if $\dim A \geq 2$, then ω_A has depth ≥ 2. Hence ω_A satisfies S_2. □

LEMMA 1.4. *Let A be a one-dimensional local Cohen–Macaulay ring. Then a finitely generated module M is ω-reflexive if and only if it has depth 1.*

Proof. Since ω has depth 1 (1.2) so does the ω-dual of any module. If M is reflexive, it is the ω-dual of M^ω and so has depth 1. The converse is [10, 6.1]. □

Proposition 1.5. *Let A be a local ring satisfying S_1. A finitely generated module M is ω-reflexive if and only if it satisfies S_2.*

Proof. First we show that the ω-dual of any module N will satisfy S_2. Write N as a cokernel of a map of free modules

$$L_1 \to L_0 \to N \to 0.$$

Taking ω-duals and the image of the second map, we obtain

$$0 \to N^\omega \to L_0^\omega \to K \to 0$$

where K is a submodule of L_1^ω. Now L_0^ω and L_1^ω are direct sums of copies of ω, so satisfy S_2 by (1.3). Hence K satisfies S_1, and then from the exact sequence it follows that N^ω satisfies S_2. In particular, any ω-reflexive module satisfies S_2.

Conversely, suppose M satisfies S_2. The map $\alpha : M \to M^{\omega\omega}$ is an isomorphism in codimension 0, by (1.1), so the kernel of α must have support of codimension ≥ 1. Since M satisfies S_1, there is no kernel. Thus we can write

$$0 \to M \xrightarrow{\alpha} M^{\omega\omega} \to R \to 0.$$

Now, since α is an isomorphism in codimension 1, by (1.4), the module R must have support of codimension ≥ 2. Since both M and $M^{\omega\omega}$ satisfy S_2, this is impossible (cf. proof of [6, 1.9]), so $R = 0$ and α is an isomorphism. □

Corollary 1.6. *Let the local ring A satisfy S_1. The ω-dual of any module is ω-reflexive.*

Proof. This follows from the first step of the proof of (1.5). □

Corollary 1.7. *If the local ring A itself satisfies S_2, then the natural map $A \to \operatorname{Hom}_A(\omega, \omega)$ is an isomorphism.*

Remark 1.8. Let X be a scheme satisfying S_1. Using the same arguments as in [6, 1.11,1.12] we see that if $\mathcal{F}$ is a coherent sheaf satisfying S_2 then $\mathcal{F}$ is *normal* in the sense of Barth [5, 1.6], namely for any open set U and any closed subset $Y \subseteq U$ of codimension ≥ 2, the restriction map $\mathcal{F}(U) \to \mathcal{F}(U - Y)$ is bijective. In fact this condition characterizes S_2, if we assume S_1.

If $\mathcal{F}$ is a coherent sheaf satisfying S_1 only, then it is easy to see that the set Y of points of X where it does not satisfy S_2 is a closed subset of codimension ≥ 2, and that the double ω-dual $\mathcal{F}^{\omega\omega}$ can be identified with $j_*(\mathcal{F}|_{X-Y})$ where $j : X - Y \to X$ is the inclusion. Thus the double ω-dual can be regarded as the S_2-ification of the sheaf.

It also follows naturally that for $Y \subseteq X$ closed of codimension ≥ 2, the category of coherent sheaves satisfying S_2 on X is equivalent by restriction to the analogous category on $X - Y$.

REMARK 1.9. To see the connection between the properties reflexive and ω-reflexive, note that the proof of [6, 1.9] shows that a reflexive module over a ring A satisfying S_2 also satisfies S_2. So we see that if A satisfies S_2, then a reflexive module is also ω-reflexive. The converse is not true without the G_1 hypothesis. For example, if X is the union of the three coordinate axes in $\mathbb{A}^3$, a scheme that satisfies G_0 but not G_1, the canonical sheaf ω is ω-reflexive by (1.4), but is easily seen not to be reflexive. On the other hand, the proof of [6, 1.9] does show that if X satisfies S_2, and $\mathcal{F}$ satisfies S_2 and is reflexive in codimension ≤ 1, then $\mathcal{F}$ is reflexive.

Note also that if A satisfies S_2, then the proofs of [6, 1.8,1.9] show that the dual of any module M satisfies S_2, and hence is ω-reflexive.

2. Generalized divisors

Let X be a noetherian, equidimensional, embeddable scheme satisfying the condition S_2 of Serre. We develop the theory of generalized divisors as in [6, §2], noting the differences in our more general setting.

Let $\mathcal{K}_X$ be the sheaf of total quotient rings on X [6, 2.1]. A *fractional ideal* is a subsheaf $\mathcal{I} \subseteq \mathcal{K}_X$ that is a coherent sheaf of $\mathcal{O}_X$-modules. It is *nondegenerate* if for each generic point $\eta \in X$, $\mathcal{I}_\eta = \mathcal{K}_{X,\eta}$.

DEFINITION. Let X be a scheme (as above) satisfying S_2. A *generalized divisor* on X is a nondegenerate fractional ideal $\mathcal{I}$ satisfying the condition S_2 as a sheaf of $\mathcal{O}_X$-modules. It is *effective* if $\mathcal{I} \subseteq \mathcal{O}_X$. We say the generalized divisor $\mathcal{I}$ is *principal* if $\mathcal{I} = (f)$ for some global section f of $\mathcal{K}_X$. We say it is *Cartier* if $\mathcal{I}$ is an invertible $\mathcal{O}_X$-module. We say it is *almost Cartier* if there exists a closed subset $Z \subseteq X$ of codimension ≥ 2 so that $\mathcal{I}|_{X-Z}$ is Cartier. We say it is *reflexive* if $\mathcal{I}$ is a reflexive $\mathcal{O}_X$-module.

Note that $\mathcal{I}$ is Cartier if and only if it is locally principal [6, 2.3]. Note that an almost Cartier divisor is reflexive (1.9) and that the sheaf $\mathcal{I}$ being reflexive implies the condition S_2.

PROPOSITION 2.1. *With X satisfying S_2, as above, the effective generalized divisors are in one-to-one correspondence with closed subschemes $Y \subseteq X$ of pure codimension one with no embedded points.*

Proof. Let Y be a closed subscheme of X, defined by a sheaf of ideals $\mathcal{I}$, so that we have an exact sequence

$$0 \to \mathcal{I} \to \mathcal{O}_X \to \mathcal{O}_Y \to 0.$$

To say that $\mathcal{I}$ is nondegenerate is equivalent to saying Y has codimension ≥ 1. Since X satisfies S_2, to say that $\mathcal{I}$ satisfies S_2 is equivalent to saying that every associated prime of Y has codimension 1 (cf. [6, 1.10]), i.e., that Y is of pure codimension 1 with no embedded points. □

DEFINITION. Let X satisfy S_2. For any coherent sheaf $\mathcal{F}$ of $\mathcal{O}_X$-modules, let us denote $\mathcal{F}^\sim$ its double ω-dual, so that $\mathcal{F}^\sim$ satisfies S_2, where ω is the canonical sheaf on X (1.8). If $\mathcal{I} \subseteq \mathcal{K}_X$ is a fractional ideal, then naturally $\mathcal{I}^\sim$ is also a fractional ideal, and will satisfy S_2. We may often denote a generalized divisor $\mathcal{I}$ by a letter D, and call $\mathcal{I}$ the ideal of D. Given two (generalized) divisors D_1 and D_2, with corresponding ideal sheaves $\mathcal{I}_1, \mathcal{I}_2$, we define the *sum* $D_1 + D_2$ by the fractional ideal $(\mathcal{I}_1 \cdot \mathcal{I}_2)^\sim$. We define the negative $-D$ by $(\mathcal{I}^{-1})^\sim$, where $\mathcal{I}^{-1}$ is the sheaf of local sections f of $\mathcal{K}_X$ for which $f \cdot \mathcal{I} \subseteq \mathcal{O}_X$ locally (cf. [6, 2.2]). We denote the divisor with ideal $\mathcal{I} = \mathcal{O}_X$ by 0.

PROPOSITION 2.2. *Let X satisfy S_2.*

(a) *Addition of divisors is associative and commutative.*
(b) *$D + 0 = D$ for all D.*
(c) *$-(-D) = D$ if and only if D is reflexive.*
(d) *$D + (-D) = 0$ if and only if D is almost Cartier.*
(e) *If D is any divisor, and E is almost Cartier, then $-(D+E) = (-D) + (-E)$.*

Proof. (a) and (b) are obvious. For (c), first note that $\mathcal{I}^{-1} \cong \mathcal{I}^\vee = \mathcal{H}om(\mathcal{I}, \mathcal{O}_X)$ by [6, 2.2]. But by (1.9), $\mathcal{I}^\vee$ is ω-reflexive, so in fact the defining sheaf of $-D$ is just $\mathcal{I}^\vee$. For (d) we follow the proof of [6, 2.5], noting that at a point of codimension 1, every ideal is ω-reflexive (1.4), so that the condition says $\mathcal{I} \cdot \mathcal{I}^{-1} = \mathcal{O}_X$, which implies $\mathcal{I}$ reflexive there [6, 2.3]. For (e) it is the same proof as [6, 2.5]. □

COROLLARY 2.3. *The set of almost Cartier divisors forms a group, containing the subgroups of Cartier divisors and of principal divisors. This group acts on the set of all divisors.*

DEFINITION. We say two divisors are *linearly equivalent* if one is obtained from the other by adding a principal divisor. We denote the equivalence classes by the group $\operatorname{Pic} X$ = Cartier divisors mod linear equivalence; the group $\operatorname{APic} X$ = almost Cartier divisors mod linear equivalence, and the set $\operatorname{GPic} X$ = generalized divisors mod linear equivalence.

PROPOSITION 2.4. *Two divisors D_1 and D_2 are linearly equivalent if and only if their ideal sheaves $\mathcal{I}_1$ and $\mathcal{I}_2$ are isomorphic as $\mathcal{O}_X$-modules. Every coherent $\mathcal{O}_X$-module that satisfies S_2 and is locally free of rank 1 at every generic point of X is isomorphic to the ideal of some divisor.*

Proof. Indeed, an isomorphism $\varphi : \mathcal{I}_1 \to \mathcal{I}_2$ of sheaves of $\mathcal{O}_X$-modules extends to $\mathcal{I}_1 \otimes \mathcal{K}_X \to \mathcal{I}_2 \otimes \mathcal{K}_X$. Each of these is isomorphic to $\mathcal{K}_X$, so the map is given by multiplication by a global section f of $\mathcal{K}_X$. If $\mathcal{F}$ is coherent satisfying S_2 and locally free of rank 1 at every generic point, then $\mathcal{F} \otimes \mathcal{K}_X \cong \mathcal{K}_X$ and the natural map $\mathcal{F} \to \mathcal{F} \otimes \mathcal{K}_X$ makes $\mathcal{F}$ into a nondegenerate fractional ideal. □

WARNING 2.5. The usual theory of the sheaf $\mathcal{L}(D) = \mathcal{I}^{-1}$ associated to a divisor D [6, 2.8] does not extend to divisors that may not be reflexive. However, we can get an analogue of [6, 2.10] using the sheaf $\mathcal{M}(D) = \mathcal{H}om(\mathcal{I}, \omega)$. Note that $\mathcal{M}(D)$ is ω-reflexive by (1.6) and therefore satisfies S_2.

PROPOSITION 2.6. *Let X be a Cohen–Macaulay scheme with canonical sheaf ω, and for any divisor D, corresponding to an ideal sheaf $\mathcal{I}$, let $\mathcal{M}(D) = \mathcal{H}om(\mathcal{I}, \omega)$. If D is an effective divisor, denoting also by D the associated closed subscheme, there are two natural exact sequences*

$$0 \to \mathcal{I} \to \mathcal{O}_X \to \mathcal{O}_D \to 0$$

and

$$0 \to \omega_X \to \mathcal{M}(D) \to \omega_D \to 0.$$

Proof. The first is the defining sequence of D. The second is obtained by applying $\mathcal{H}om(\cdot, \omega_X)$ to the first and noting (since X is Cohen–Macaulay) that $\omega_D \cong \mathcal{E}xt^1_{\mathcal{O}_X}(\mathcal{O}_D, \omega_X)$. □

DEFINITION–REMARK 2.7. Even though X has a canonical sheaf ω_X, it may not have a canonical divisor. By *canonical divisor* we mean a generalized divisor K whose ideal satisfies $\mathcal{I}_K^{-1} \cong \omega_X$. Since the ideal of any divisor is locally free at the generic points, the existence of a canonical divisor implies that X satisfies G_0. In this case we see also that ω_X must be reflexive, by [6, 1.8]. Since there are schemes satisfying G_0 and S_2 on which ω is not reflexive (1.9), we conclude that G_0 and S_2 are not sufficient conditions for the existence of a canonical divisor.

However, if X satisfies G_0 and S_2, then ω_X is locally free of rank 1 at every generic point, so is isomorphic to a fractional ideal. We choose and fix an embedding $\omega_X \subseteq \mathcal{K}_X$, and call the corresponding divisor M_X the *anticanonical* divisor. As a divisor it depends on the choice of embedding $\omega_X \subseteq \mathcal{K}_X$, but is unique up to linear equivalence. If X satisfies in addition G_1, then ω is invertible in codimension 1, so we can define a *canonical* divisor $K = -M$, which will be an almost Cartier divisor.

DEFINITION. For any two divisors D_1, D_2, we define $D_1(-D_2)$ to be the divisor whose sheaf of ideals is $\mathcal{H}om(\mathcal{I}_2, \mathcal{I}_1)^{\sim}$. In general, this operation may not be well-behaved, but we do have the following.

PROPOSITION 2.8. *The operation $D_1(-D_2)$ has the following properties.*

(a) $0(-D_2) = -D_2$ *and* $D_1(-0) = D_1$.

(b) *If E is almost Cartier,* $(D_1 + E)(-D_2) = D_1(-(D + (-E))) = D_1(-D_2) + E$.
(c) *In particular, if either D_1 or D_2 is almost Cartier, then* $D_1(-D_2) = D_1 + (-D_2)$.
(d) *If X satisfies G_0, and $D_1 \sim M + E$, where M is the anticanonical divisor and E is almost Cartier, then $D_1(-D_1(-D_2)) = D_2$ for any D_2.*

Proof. (a), (b), (c) are immediate, since an almost Cartier divisor is invertible in codimension 1, and equality of divisors can be tested in codimension 1. (d) corresponds to the fact that any divisor has an ideal sheaf satisfying S_2, and hence is ω-reflexive (1.5). □

REMARK 2.9. We take this opportunity to point out an error in [6, 2.9]. Assuming that X satisfies G_1 and S_2 as in that paper, it is true that every nondegenerate section $s \in \Gamma(X, \mathcal{L}(D))$ gives rise to an effective divisor D' in the complete linear system $|D|$, and all D' arise in this way. Two sections s_1 and s_2 give rise to the same divisor D' if and only if they differ by an isomorphism of $\mathcal{L}(D)$. If D is almost Cartier, the isomorphisms of $\mathcal{L}(D)$ are given by sections of $\Gamma(X, \mathcal{O}_X^*)$ as stated there. So in the familiar case of X integral projective, $\Gamma(X, \mathcal{O}_X^*) = k^*$ and $|D|$ is simply the projective space associated to the vector space $\Gamma(X, \mathcal{L}(D))$.

Suppose, however, that D is not almost Cartier. Then there may be more isomorphisms of $\mathcal{L}(D)$ and the statement of [6, 2.9] is not correct. For example, let $X = L_1 \cup L_2$ be the union of two lines in $\mathbb{P}^2$ meeting at a point P, and let D be the divisor P. Then one can verify that $\dim \Gamma(X, \mathcal{L}(D)) = 2$, and $\mathrm{Isom}(\mathcal{L}, \mathcal{L}) = k^* \oplus k^*$, so that the complete linear system $|D|$ consists just of the single divisor D, as we expect. (Cf. [6, 3.3] for a relevant calculation.)

How does this discussion extend to the case of the present paper, where X is only assumed to satisfy S_2? We cannot use the sheaf $\mathcal{L}(D)$. Instead, for each effective divisor $D' \sim D$, we take ω-duals of $\mathcal{I}_{D'} \subseteq \mathcal{O}_X$ to get $\omega_X \subseteq \mathcal{H}om(\mathcal{I}_{D'}, \omega_X) \cong \mathcal{M}(D)$, and this gives a section s of the sheaf $\mathcal{N}(D) = \mathcal{H}om(\omega_X, \mathcal{M}(D))$. Conversely, nondegenerate sections of $\mathcal{N}(D)$ give effective divisors $D' \sim D$ by reversing the process. The ambiguity of s is again in $\mathrm{Isom}(\mathcal{N}(D), \mathcal{N}(D)) \cong \mathrm{Isom}(\mathcal{I}_D, \mathcal{I}_D)$.

3. Biliaison

In this section we generalize the notion of biliaison introduced in [6, §4] and [12, §5.4]. Note that the word biliaison is not a synonym for even liaison. We also generalize the results of [11, §5] so as to remove the G_1 hypotheses. In fact, it was the attempt to put those results in a more natural context that led to this paper. A scheme X in $\mathbb{P}^n$ is *arithmetically Cohen–Macaulay*

(ACM) if its homogeneous coordinate ring S_X is a Cohen–Macaulay ring. It is *arithmetically Gorenstein* (AG) if S_X is a Gorenstein ring.

DEFINITION. Let V_1 and V_2 be equidimensional closed subschemes of dimension r of $\mathbb{P}^n_k$. We say that V_2 is obtained by an *elementary biliaison* of height h from V_1 if there exists an ACM scheme X in $\mathbb{P}^n$, of dimension $r+1$, containing V_1 and V_2, and so that $V_2 \sim V_1 + hH$ as generalized divisors on X, where H denotes the hyperplane class. The biliaison is *ascending* if $h \geq 0$; *descending* if $h \leq 0$. The equivalence relation generated by elementary biliaisons will be called *biliaison*.

If we restrict the schemes X in the definition all to be complete intersection schemes, we will speak of CI-*biliaison*. If we restrict the schemes X to be ACM schemes satisfying G_0, will speak of *Gorenstein biliaison* or G-*biliaison*.

REMARK 3.1. As was shown in [6, 4.4], the relation of CI-biliaison is equivalent to even CI-liaison in the usual sense.

PROPOSITION 3.2. *Suppose V_2 is obtained from V_1 by an elementary biliaison of height h on X, with* $\dim V_1 = \dim V_2 = r$.

(a) *Then reciprocally, V_1 is obtained from V_2 by an elementary biliaison of height $-h$.*

(b) *The higher Rao modules $M^i_V = H^i_*(\mathcal{I}_{V,\mathbb{P}^n})$ are related as follows:*

$$M^i_{V_2} \cong M^i_{V_1}(-h) \text{ for } 1 \leq i \leq r.$$

(c) *The Hilbert polynomials are related by*

$$\chi(\mathcal{O}_{V_2}(m)) = \chi(\mathcal{O}_X(m)) - \chi(\mathcal{O}_X(m-h)) + \chi(\mathcal{O}_{V_1}(m-h)).$$

Proof. (a) If $V_2 \sim V_1 + hH$ then $V_1 \sim V_2 - hH$.

(b) and (c) have the same proof as [6, 4.5] since only the ACM property of X was used there. □

LEMMA 3.3. *Let X be an* ACM *scheme satisfying G_0 in $\mathbb{P}^n$. Let $Y \subseteq X$ be an effective divisor, $Y \sim M + mH$, where M is the anticanonical divisor and H is the hyperplane divisor. Then Y is an arithmetically Gorenstein* (AG) *scheme in $\mathbb{P}^n$.*

Proof. Let X be of dimension $r+1$ so that Y is of dimension r. If $\dim Y \geq 1$, to show that Y is AG is equivalent to showing that Y is ACM and $\omega_Y \cong \mathcal{O}_Y(\ell)$ for some $\ell \in \mathbb{Z}$.

First to show Y is ACM, we must show $H^i_*(\mathcal{I}_{Y,\mathbb{P}^n}) = 0$ for $1 \leq i \leq r$. From the exact sequence

$$0 \to \mathcal{I}_{X,\mathbb{P}^n} \to \mathcal{I}_{Y,\mathbb{P}^n} \to \mathcal{I}_{Y,X} \to 0$$

and the fact that X is ACM, so that $H^i_*(\mathcal{I}_{X,\mathbb{P}^n}) = 0$ for $1 \leq i \leq r+1$, it is equivalent to show $H^i_*(\mathcal{I}_{Y,X}) = 0$ for $1 \leq i \leq r$. Now $Y \sim M + mH$ by hypothesis, so $\mathcal{I}_{Y,X} \cong \omega_X(-m)$. By Serre duality on X, $H^i_*(\omega_X(-m))$ is dual to $H^{r+1-i}_*(\mathcal{O}_X(m))$. These latter are 0 for $1 \leq i \leq r$ since X is ACM. Hence Y is ACM.

To study the canonical sheaf ω_Y, we use the second exact sequence of (2.6), namely

$$0 \to \omega_X \to \mathcal{M}(Y) \to \omega_Y \to 0.$$

Now since $\mathcal{I}_{Y,X} \cong \omega_X(-m)$, we have $\omega_X \cong \mathcal{I}_{Y,X}(m)$ and $\mathcal{M}(Y) = \mathcal{H}om(\mathcal{I}_{Y,X}, \omega_X) \cong \mathcal{O}_X(m)$. Thus $\omega_Y \cong \mathcal{O}_Y(m)$ and Y is arithmetically Gorenstein.

If $\dim Y = 0$, it is automatically ACM. To show it is AG, we must show that its graded canonical module Ω_Y is isomorphic to $S_Y(m)$. This follows from the graded module analogue of the argument just given using (2.6). □

REMARK 3.4. An algebraic version of this result was given in [11, 5.2], and a geometric version with the added hypothesis G_1 in [11, 5.4].

DEFINITION. Two subschemes V_1 and V_2 of $\mathbb{P}^n$, equidimensional of the same dimension and without embedded components are *linked* by a scheme Y if Y contains V_1 and V_2 and $\mathcal{I}_{V_i,Y} \cong \mathcal{H}om(\mathcal{O}_{V_j}, \mathcal{O}_Y)$ for $i, j = 1, 2, i \neq j$. If Y is a complete intersection, it is called a CI-*linkage*; if Y is arithmetically Gorenstein, it is a *Gorenstein linkage.* If Y is an arithmetically Gorenstein scheme of the form $M + mH$ on some ACM scheme X satisfying G_0 (as in (3.3) above), then we will say it is a *strict Gorenstein linkage.* (This is a slight generalization of the terminology of [7, §1], where we required that X should satisfy G_1.)

The equivalence relation generated by CI-linkages is CI-*liaison,* by Gorenstein linkages, *Gorenstein liaison,* and by strict Gorenstein linkages, *strict Gorenstein liaison.* If the liaison can be accomplished by an even number of linkages, then it is an *even* CI-liaison (resp. Gorenstein liaison, resp. strict Gorenstein liaison).

THEOREM 3.5. *Suppose that V_2 is obtained from V_1 by an elementary biliaison on an* ACM *scheme X satisfying G_0. Then V_2 can be obtained from V_1 by two strict Gorenstein linkages.*

Proof. The proof is almost the same as [6, 4.3], transposed into our context. We assume that $V_2 \sim V_1 + hH$. Thus there is a principal divisor (f) such that $V_2 = V_1 + hH + (f)$. Taking M to be the anticanonical divisor and using (2.8), we can write

$$M(-V_1) = M(-V_2) + hH + (f).$$

Now by [6, 2.11], which still holds in our case, we can find an effective Cartier divisor $E \sim mH$ such that $W = E + M(-V_1) = (M + E)(-V_1)$ is effective. Now let $Y = M + E$. Then Y is an effective divisor that is arithmetically

Gorenstein by (3.4), and I claim that V_1 and W are Gorenstein linked by Y. Indeed, the same argument as in the proof of [6, 4.1] shows that $\mathcal{I}_{W,Y} \cong \mathcal{H}om(\mathcal{O}_{V_1}, \mathcal{O}_Y)$. Since by (2.8) we also have $V_1 = (M+E)(-W)$, we obtain the reverse isomorphism $\mathcal{I}_{V_1,Y} \cong \mathcal{H}om(\mathcal{O}_W, \mathcal{O}_Y)$, so V_1 and W are linked by Y.

We also have $W = E + M(-V_2) + hH + (f)$, so if we let $Y' = E + M + hH + (f)$, then Y' will also be an effective divisor that is an arithmetically Gorenstein scheme, and as above, we see that W and V_2 are linked by Y'. Thus V_2 is obtained from V_1 by two strict Gorenstein linkages. □

COROLLARY 3.6. *Every Gorenstein biliaison is an even strict Gorenstein liaison.*

REMARK 3.7. This theorem was proved for a trivial elementary biliaison $V_2 = V_1 + hH$ (with no linear equivalence) in [11, 5.10], and with the extra hypothesis G_1 in [11, 5.14].

REMARK 3.8. In Section 5 below we will prove a converse to this theorem in codimension 3.

EXAMPLE 3.9. Let P be a point in $\mathbb{P}^3$, and let X be the union of three non-coplanar lines through P. Then X satisfies G_0 but not G_1. If H is a hyperplane section of X containing P, then $V = P + H$ is the divisor defined by the square of the ideal of P. Thus P and V are related by one G-biliaison, and hence are evenly G linked. Cf. [11, 4.1], where this was proved by a different method.

4. The theorem of Gaeta

To illustrate the theory of biliaison, we give a new proof the theorem of KMMNP–Gaeta [11, 3.6]. The statement given there is that every standard determinantal scheme is glicci. We prove a slightly stronger result.

THEOREM 4.1. *Every standard determinantal scheme in $\mathbb{P}^n$ can be obtained from a linear variety by a finite number of ascending Gorenstein biliaisons. In particular, it is glicci by* (3.3).

Proof. We follow the terminology and notation of [11, 3.6]. Let $V \subseteq \mathbb{P}^n$ be a standard determinantal scheme, i.e., a scheme of codimension $c+1$ whose ideal I_V is generated by the $t \times t$ minors of a $t \times (t+c)$ homogeneous matrix A for some $t > 0$. Let B be the matrix obtained by omitting the last column of A. Then V is contained in the determinantal scheme S defined by the $t \times t$ minors of B. By Step I of the proof of [11, 3.6], S is good determinantal. Hence it is generically a complete intersection [11, 3.2], and so satisfies G_0.

Let A' be the matrix obtained by omitting the last row of B. Then V', defined by the $(t-1) \times (t-1)$ minors of A', is also contained in S. We will show that $V \sim V' + mH$ on S for some $m > 0$, so that V is obtained by an ascending elementary Gorenstein biliaison from V'. Continuing in this manner, after a finite number of G-biliaisons, we reduce to the case $t = 1$, when V is a complete intersection. From these one can perform descending CI-biliaisons to a linear variety.

Let R be the homogeneous coordinate ring of $\mathbb{P}^n$, and let $R_S = R/I_S$ be the homogeneous coordinate ring of S. The ideal of V in S is generated by the images in R_S of the $t \times t$ minors of A that include the last column. The $t \times t$ minors that do not include the last column are just the generators of I_S. On the other hand, the ideal of V' in S is generated by the images of the $(t-1) \times (t-1)$ minors of A'. So there is a one-to-one correspondence between generators N of V in S and generators N' of V' in S, obtained by omitting the last row and column of the corresponding $t \times t$ matrix. We will show that the quotient N/N' of corresponding generators is an element of $H^0(\mathcal{K}_S(m))$, independent of the choice of N, where $\mathcal{K}_S$ is the sheaf of total quotient rings of S, and m is the difference in degrees of N and N'. This will show that $\mathcal{I}_{V,S} \cong \mathcal{I}_{V',S}(-m)$, and so we have the desired biliaison. Note that m is the degree of the element in the lower right-hand corner of the original matrix A.

To show that N/N' is independent of the choice of $N(\mod I_S)$, it will be sufficient to compare two such that differ by one column only. So let M be a $t \times t$ minor of B, let N_1 be obtained by deleting the first column of M and adding the last column of A; let N_2 be obtained by deleting the second column of M and adding the last column of A. Then N_1 and N_2 are two generators of I_V, and the corresponding generators N_1', N_2' of $I_{V'}$ are just M_{t1} and M_{t2}, where M_{ij} denotes the minor of M obtained by deleting the ith row and the jth column. We need to show that $N_1/N_1' = N_2/N_2' \mod I_S$. By making general row and column operations on A at the beginning, we may assume that all the N_i' are non-zero-divisors in R_S. So we must show that $N_1N_2' - N_2N_1' \in I_S$.

Let the last column of A be $u_1, \ldots, u_t$. We will expand N_1 and N_2 along this last column. The coefficient of u_t in $N_1N_2' - N_2N_1'$ is just $N_1'N_2' - N_2'N_1' = 0$. For $i \neq t$, the coefficient of u_i is $M_{i1}M_{t2} - M_{i2}M_{t1}$. The proof is then completed by the following identity among determinants, since $M \in I_S$. □

LEMMA 4.2. *Let M be the determinant of a $t \times t$ matrix, let M_{ij} denote the minor obtained by deleting the ith row and the jth column; let $M_{ik,jl}$ denote the minor obtained by deleting the ith and k^{th} rows and the jth and l^{th} columns. Then the determinants satisfy*

$$M_{ij} \cdot M_{kl} - M_{il} \cdot M_{kj} = \pm M_{ij,kl} \cdot M.$$

Proof. [13, p. 132ff]. □

EXAMPLE 4.3. The 4×4 minors of a general 4×6 matrix of linear forms in $\mathbb{P}^4$ define an irreducible smooth curve C of degree 20 and genus 26 which, according to the theorem, can be obtained by ascending Gorenstein biliaisons from a line. However these curves are not general in the Hilbert scheme, and it is known that a general smooth curve of degree 20 and genus 26 is ACM, but cannot be obtained by ascending biliaisons from a line. It is unknown whether it is glicci [7, 3.9].

5. Strict Gorenstein liaison

The main result of this section is a converse to (3.6) in codimension 3.

THEOREM 5.1. *For subschemes of codimension 3 in $\mathbb{P}^n$ (equidimensional and without embedded components), any even strict Gorenstein liaison is a Gorenstein biliaison.*

Proof. Suppose V and V' of codimension 3 in $\mathbb{P}^n$ are related by even strict Gorenstein liaison. Then there is a sequence

$$V = V_0, V_1, V_2, \ldots, V_{2k} = V'$$

for some k, where each V_i is related to V_{i+1} by a strict Gorenstein linkage. By composition of biliaisons, it will be sufficient to treat the case $k = 1$, i.e., when there is just one intermediary scheme Z, and V to Z is a strict Gorenstein linkage by Y of the form $M + mH$ on an ACM scheme X satisfying G_0, and Z to V' similarly is linked by a Y' of the form $M' + m'H'$ on an ACM scheme X' satisfying G_0.

Since X and X' are both ACM of codimension 2 in $\mathbb{P}^n$, they are in the same CI-biliaison class, by the classical Gaeta's theorem [12, 6.1.4]. Thus we can apply Lemma 5.2 (below) and find a chain

$$X = X_0, X_1, \ldots, X_r = X'$$

of ACM schemes satisfying G_0 and each containing Z, such that each X_i is directly CI-linked to X_{i+1}, and X_i and X_{i+1} have no common components.

Now for each $i = 1, \ldots, r$, let $D_i = X_{i-1} \cap X_i$. By Lemma 5.3 (below), D_i is an AG scheme of the form $M + mH$ on X_{i-1} and on X_i. Since the X_i all contain Z, so do the D_i. For each $i = 1, \ldots, r$, let W_i be the scheme linked to Z by D_i. We consider the chain of strict Gorenstein linkages

$$V = W_0, Z, W_1, Z, W_2, Z, \ldots, W_r, Z, V' = W_{r+1}.$$

Here, for each $i = 0, \ldots, r$, the two links W_i, Z, W_{i+1} are both strict Gorenstein links on the same ACM scheme X_i. Now, as in the proof of [6, 4.1] we see that W_i being linked to Z by $M + mH$ on X_i is equivalent to saying $Z \sim (M + mH)(-W_i)$ on X_i. Similarly, Z linked to W_{i+1} by $M + m'H$ on X_i says $W_{i+1} \sim (M + m'H)(-Z)$. Substituting the first expression for Z in the second expression for W_{i+1}, we find using (2.8) that $W_{i+1} \sim W_i + (m' - m)H$ on X_i, which is a single Gorenstein biliaison.

Thus V is joined to V' by the chain of Gorenstein biliaisons

$$V = W_0, W_1, \ldots, W_r, W_{r+1} = V'. \qquad \square$$

LEMMA 5.2. *Suppose given X, X' locally Cohen–Macaulay subschemes of codimension 2 in $\mathbb{P}^n$, both satisfying G_0, and both containing a given closed subscheme Z of codimension at least 3 in $\mathbb{P}^n$, and with X, X' in the same* CI-*liaison class in $\mathbb{P}^n$. Then there exists a chain*

$$X = X_0, X_1, \ldots, X_r = X'$$

of locally Cohen–Macaulay subschemes of $\mathbb{P}^n$, each containing Z, such that each X_{i+1} is obtained by a single geometric CI-*linkage from X_i. In particular, X_i and X_{i+1} will have no common components, so each will be generically locally complete intersection and therefore will satisfy G_0.*

Proof. Note first that the hypothesis G_0 implies that X and X' are generically locally complete intersection, since they are in codimension 2. If X to X' is an odd liaison, we can make a single general geometric liaison from X' to a new X'' also containing Z, and thus reduce to the case of an even liaison. Then, since X and X' are in the same even liaison class, by Rao's theorem, they have $\mathcal{N}$-type resolutions with stably equivalent sheaves $\mathcal{N}_i$ up to twist. By adding dissocié sheaves, we can write $\mathcal{N}$-type resolutions

$$\begin{array}{ccccccccc} 0 & \to & \mathcal{L} & \to & \mathcal{N} & \to & \mathcal{I}_X(a) & \to & 0 \\ 0 & \to & \mathcal{L}' & \to & \mathcal{N} & \to & \mathcal{I}_{X'}(a') & \to & 0 \end{array}$$

with the same locally free sheaf $\mathcal{N}$ in the middle, and $\mathcal{L}, \mathcal{L}'$ dissocié.

Now we will follow the plan of the proof of [2, 3.1] to obtain a chain

$$X = X_0, X_2, X_4, \ldots, X_{2k} = X'$$

of locally Cohen–Macaulay subschemes containing Z, such that for each i, X_{2i} and X_{2i+2} have no common components and are related by a single elementary CI-biliaison on a hypersurface S_i.

Write $\mathcal{L} = \oplus \mathcal{L}_i$ with $\mathcal{L}_i$ invertible, $i = 1, \ldots, t$. Since X is generically locally complete intersection, the rank of the map

$$\mathcal{L}(\xi) \to \mathcal{N}(\xi)$$

is $t - 1$ for each generic point ξ of X. Thus, reordering if necessary, we define $\mathcal{F}$ by

$$0 \to \bigoplus_{i \geq 2} \mathcal{L}_i \to \mathcal{N} \to \mathcal{F} \to 0$$

and $\mathcal{F}$ will be torsion-free of rank 2, and locally free at each generic point ξ of X. Now choose $b \gg 0$ so that $\mathcal{I}_Z \otimes \mathcal{N}(b)$ is generated by global sections and take $s_1 \in H^0(\mathcal{I}_Z \otimes \mathcal{N}(b))$ a sufficiently general section. Let Y_1 be defined by

$$0 \to \mathcal{O}(-b) \xrightarrow{s_1} \mathcal{F} \to \mathcal{I}_{Y_1}(a_1) \to 0.$$

Then Y_1 contains Z, and Y_1 has no component in common with X, and Y_1 is obtained from X by a single CI-biliaison [2, 3.3]. Furthermore, we can lift s_1 to $\mathcal{N}$ in such a way that $s_1(\xi_1) \neq 0$ in $\mathcal{N}(\xi_1)$ for each generic point of Y_1. In terms of $\mathcal{N}$ we now have

$$0 \to \mathcal{O}(-b) \oplus \bigoplus_{i \geq 2} \mathcal{L}_i \to \mathcal{N} \to \mathcal{I}_{Y_1}(a_1) \to 0.$$

We repeat this process with each $\mathcal{L}_i$ in turn, obtaining a sequence of biliaisons $X, Y_1, Y_2, \ldots, Y_t$, each one containing Z and having no components in common with its neighbors.

We do the same thing with X', obtaining a similar sequence $X', Y_1', \ldots, Y_t'$. Then we observe that one can take the same b in both cases, and since the sections $s_1, \ldots, s_t, s_1', \ldots, s_t'$ are all sufficiently general, we can take $s_i = s_i'$ for each i, and thus $Y_t = Y_t'$. This connects X and X' by biliaisons, all containing Z. Now just relabel Y_i and Y_i' as X_{2j} to get the sequence of biliaison above.

To conclude, let X_2 and X_4 for example be a biliaison on a hypersurface S, where X_2, X_4 both contain Z and have no common component. Then X_2 and X_4 are both generically Cartier divisors on S. When we link them both to a divisor X_3 on S, as in the proof of (3.5), we can take X_3 to have no component in common with X_2 and X_4, and by adding a complete intersection on S containing Z if necessary, we may assume X_3 contains Z. Thus the sequence of biliaisons connecting X and X' can be filled in to a sequence of geometric liaisons as required. □

LEMMA 5.3. *Let X_1, X_2 be* ACM *schemes in $\mathbb{P}^n$ that have no common component and are directly linked by an* AG *scheme S. Then $D = X_1 \cap X_2$ is arithmetically Gorenstein; moreover, it is of the form $M + \ell H$ on each of X_1, X_2, where ℓ is the integer for which $\omega_S \cong \mathcal{O}_S(\ell)$.*

Proof. (cf. [12, 4.2.1]). The fact that D is ACM follows from the exact sequence

$$0 \to \mathcal{O}_S \to \mathcal{O}_{X_1} \oplus \mathcal{O}_{X_2} \to \mathcal{O}_D \to 0.$$

Since S is AG, its dualizing sheaf ω_S is isomorphic to $\mathcal{O}_S(\ell)$ for some $\ell \in \mathbb{Z}$. Because of the linkage, $\mathcal{I}_{X_1,S} \cong \mathcal{H}om(\mathcal{O}_{X_2}, \mathcal{O}_S)$. Note that $\mathcal{I}_{X_1,S} = \mathcal{I}_{D,X_2}$ by a standard isomorphism theorem for ideals. Since $\omega_{X_2} = \mathcal{H}om(\mathcal{O}_{X_2}, \omega_S)$, we find that $\mathcal{I}_{D,X_2} \cong \omega_{X_2}(-\ell)$. This says $D \sim M + \ell H$ on X_2. The same argument shows that $D \sim M + \ell H$ on X_1 also. □

6. Conclusion

If we reflect on the outstanding problem whether every ACM subscheme of $\mathbb{P}^n$ is glicci, we can appreciate the usefulness of the extended notion of generalized divisors in this paper. It has allowed us to prove the theorem of KMMNP–Gaeta in a strengthened form, namely that any standard determinantal scheme in $\mathbb{P}^n$ can be obtained by ascending Gorenstein biliaisons from a linear space. This also makes clear the special nature of determinantal schemes, since there are known examples of other ACM schemes that cannot be obtained by ascending Gorenstein biliaisons from a linear space, even though it is still unknown whether they are glicci or not (for curves in $\mathbb{P}^4$, see [7, 3.9], and for points in $\mathbb{P}^3$ see [9, 7.2]).

We also observe that in most known proofs that some class of ACM schemes is glicci (such as the theorem of KMMNP–Gaeta discussed here) the proof could be accomplished using Gorenstein biliaisons, hence using only strict Gorenstein liaisons. Since there are AG schemes in $\mathbb{P}^n$ not of the special form $M + mH$ on some ACM scheme of one dimension higher (for curves in $\mathbb{P}^4$, see [8, 3.6, 3.11] and for points in $\mathbb{P}^3$ see [9, 3.4, 6.8]), this suggests that it would be worthwhile to investigate more deeply what kind of G-liaisons can be accomplished using AG schemes not of this special form.

References

[1] Y. Aoyama, *Some basic results on canonical modules*, J. Math. Kyoto Univ. **23** (1983), 85–94. MR 692731 (84i:13015)

[2] M. Casanellas and R. Hartshorne, *Gorenstein biliaison and ACM sheaves*, J. Algebra **278** (2004), 314–341. MR 2068080 (2005e:14074)

[3] E. Gorla, *A generalized Gaeta's theorem*, `arXiv:math.AG/0701456`.

[4] R. Hartshorne, *Algebraic geometry*, Springer-Verlag, New York, 1977, Graduate Texts in Mathematics, No. 52. MR 0463157 (57 #3116)

[5] ———, *Stable reflexive sheaves*, Math. Ann. **254** (1980), 121–176. MR 597077 (82b:14011)

[6] ———, *Generalized divisors on Gorenstein schemes*, K-Theory **8** (1994), 287–339. MR 1291023 (95k:14008)

[7] ———, *Some examples of Gorenstein liaison in codimension three*, Collect. Math. **53** (2002), 21–48. MR 1893306 (2003d:14059)

[8] ———, *Geometry of arithmetically Gorenstein curves in $\mathbb{P}^4$*, Collect. Math. **55** (2004), 97–111. MR 2028982 (2004m:14105)

[9] R. Hartshorne, T. Sabadini, and E. Schlesinger, *Codimension 3 arithmetically Gorenstein subschemes of projective space*, `arXiv:math.AG/0611478`.

[10] J. Herzog and E. Kunz, *Der kanonische Modul eines Cohen-Macaulay-Rings*, Lecture Notes in Mathematics, vol. 238, Springer-Verlag, Berlin, 1971. MR 0412177 (54 #304)

[11] J. O. Kleppe, J. C. Migliore, R. Miró-Roig, U. Nagel, and C. Peterson, *Gorenstein liaison, complete intersection liaison invariants and unobstructedness*, Mem. Amer. Math. Soc. **154** (2001). MR 1848976 (2002e:14083)

[12] J. C. Migliore, *Introduction to liaison theory and deficiency modules*, Progress in Mathematics, vol. 165, Birkhäuser Boston Inc., Boston, MA, 1998. MR 1712469 (2000g:14058)

[13] T. Muir, *A treatise on the theory of determinants*, Longmans, Green & Co., 1933.

Robin Hartshorne, Department of Mathematics, University of California, Berkeley, CA 94720-3840, USA

E-mail address: `robin@math.berkeley.edu`

Illinois Journal of Mathematics
Volume 51, Number 1, Spring 2007, Pages 99–122
S 0019-2082

BIG INDECOMPOSABLE MODULES AND DIRECT-SUM RELATIONS

WOLFGANG HASSLER, RYAN KARR, LEE KLINGLER, AND ROGER WIEGAND

Dedicated to Phillip Griffith, the master of syzygies

ABSTRACT. A commutative Noetherian local ring $(R, \mathfrak{m})$ is said to be *Dedekind-like* provided R has Krull-dimension one, R has no non-zero nilpotent elements, the integral closure $\overline{R}$ of R is generated by two elements as an R-module, and $\mathfrak{m}$ is the Jacobson radical of $\overline{R}$. A classification theorem due to Klingler and Levy implies that if M is a finitely generated indecomposable module over a Dedekind-like ring, then, for each minimal prime ideal P of R, the vector space M_P has dimension $0, 1$ or 2 over the field R_P. The main theorem in the present paper states that if R (commutative, Noetherian and local) has non-zero Krull dimension and is not a homomorphic image of a Dedekind-like ring, then there are indecomposable modules that are free of any prescribed rank at each minimal prime ideal.

1. Introduction

In a series of papers [14]–[16] Klingler and Levy proved the existence of tame-wild dichotomy for commutative Noetherian rings. They gave a complete classification of all finitely generated modules over Dedekind-like rings (cf. Definition 1.1) and showed that, over any ring that is not a homomorphic image of a Dedekind-like ring, the category of finite-length modules has wild representation type. A consequence of their classification is that if M is an indecomposable finitely generated module over a Dedekind-like ring R, then M_P is free of rank $0, 1$ or 2 at each minimal prime ideal P of R. The main theorem of the present paper complements this result of Klinger and Levy. We prove that if $(R, \mathfrak{m}, k)$ is a commutative local Noetherian ring of non-zero Krull dimension and R is not a homomorphic image of a Dedekind-like ring,

Received August 2, 2006; received in final form November 20, 2006.
2000 *Mathematics Subject Classification.* Primary 13C05, 13E05, 13D07.
The research of W. Hassler was supported by the *Fonds zur Förderung der Wissenschaftlichen Forschung*, project number P18779-N13. Wiegand's research was partially supported by NSA Grant H98230-05-1-0243.

then there are indecomposable modules that are free of any prescribed rank at each minimal prime.

This result was obtained in [9] for the case of a Cohen-Macaulay ring, using a direct but highly intricate construction. In [10] we gave a much simpler argument that handles all rings—Cohen-Macaulay or not—for which some power of the maximal ideal requires at least 3 generators. The remaining case, when $(R, \mathfrak{m}, k)$ is not Cohen-Macaulay and each power of $\mathfrak{m}$ is two-generated, was treated via an indirect argument using the bimodule structure of certain Ext modules. In this paper we apply the Ext argument, together with periodicity of resolutions over hypersurface rings, to give a unified treatment of the case when each power of $\mathfrak{m}$ is two-generated. Thus this paper does not rely on the technical construction in [9]. Our goal is to make the paper pretty much self-contained, though we do refer without proof to some of the results of [6], [10] and [14]–[16].

We actually obtain $\max\{|R/\mathfrak{m}|, \aleph_0\}$ pairwise non-isomorphic indecomposables of each rank. This refinement allows us, in dimension one, to obtain precise defining equations for the monoid of isomorphism classes of finitely generated modules that are free on the punctured spectrum. This generalizes the results of [6], which apply only to the Cohen-Macaulay case.

Our main theorem provides indecomposable modules that are free of specified rank at each prime P in a given finite set $\mathcal{P} \subseteq \operatorname{Spec}(R) - \{\mathfrak{m}\}$. In dimension greater than one we have to allow for the fact that if $M_P \cong R_P^{(n)}$ and Q is a prime ideal contained in P, then $M_Q \cong R_Q^{(n)}$. For $P_1, P_2 \in \mathcal{P}$ we write $P_1 \sim P_2$ if $P_1 \cap P_2$ contains a prime ideal of R (not necessarily in $\mathcal{P}$). (Of course "$\sim$" is not necessarily a transitive relation.)

Definition 1.1. The commutative, Noetherian local ring $(R, \mathfrak{m}, k)$ is *Dedekind-like* [14, Definition 2.5] provided R is one-dimensional and reduced, the integral closure $\overline{R}$ of R in the total quotient ring of R is generated by at most 2 elements as an R-module, and $\mathfrak{m}$ is the Jacobson radical of $\overline{R}$. We call $(R, \mathfrak{m}, k)$ an *exceptional Dedekind-like ring* provided, in addition, $\overline{R}/\mathfrak{m}$ is a purely inseparable field extension of k of degree 2.

There is a global notion of Dedekind-like, which is equivalent to Noetherian and locally Dedekind-like [16, Corollary 10.7]. In this article, "Dedekind-like" always means Dedekind-like and local, except in the last section, where we take up the question of the size of finitely generated indecomposable modules over arbitrary commutative Noetherian rings.

The classification of finitely generated modules in [15] and [16] does not apply to exceptional Dedekind-like rings. The details in the exceptional case are extremely complicated and are currently being worked out by L. Klingler, G. Piepmeyer and S. Wiegand. It appears that the indecomposable modules over an exceptional Dedekind-like ring have torsion-free rank $0, 1$ or 2, as in the non-exceptional case. Thus everything in this paper would hold without the "non-exceptional" proviso. Nonetheless, since the classification of modules in the exceptional case is still a work in progress, we have decided to restrict to non-exceptional Dedekind-like rings in the second part of our main theorem below.

THEOREM 1.2 (Main Theorem). *Let $(R, \mathfrak{m}, k)$ be a commutative Noetherian local ring.*

(i) *Suppose R is not a homomorphic image of a Dedekind-like ring. Let $\mathcal{P}$ be a finite set of non-maximal prime ideals of R, and let n_P be a non-negative integer for each $P \in \mathcal{P}$. Assume that $n_P = n_Q$ whenever $P \sim Q$. Then there exist $|k| \cdot \aleph_0$ pairwise non-isomorphic indecomposable finitely generated R-modules X such that, for each $P \in \mathcal{P}$, the localization X_P is a free R_P-module of rank n_P.*

(ii) *Conversely, assume R is not an exceptional Dedekind-like ring, but that R is a homomorphic image of some Dedekind-like ring. If X is an indecomposable finitely generated R-module and P is a non-maximal prime, then X_P either is 0 or is isomorphic to R_P or $R_P^{(2)}$.*

It is tempting to conjecture a substantial improvement of this result in higher dimensions. Let $(R, \mathfrak{m}, k)$ be a local ring of dimension at least two, and let $C_1, \ldots, C_t$ be the connected components of the punctured spectrum $\operatorname{Spec}(R) - \{\mathfrak{m}\}$. Given any sequence $n_1, \ldots, n_t$ of non-negative integers, is there necessarily an indecomposable R-module M such that $M_P \cong R_P^{(n_i)}$ for each i and each $P \in C_i$? Our methods do not seem to yield modules that are free on the entire punctured spectrum.

Part (ii) of the Main Theorem is an easy consequence of the classification theorem in [15]: Since the assertion is vacuous if $\dim(R) = 0$ and the hypotheses fail if $\dim(R) > 1$, we assume $\dim(R) = 1$. Let $R = D/J$, where D is a Dedekind-like ring. If D were an exceptional Dedekind-like ring, then, by assumption, J would have to be non-zero. But then R would be zero-dimensional, since exceptional Dedekind-like rings are domains. Therefore D is not exceptional, and we can apply the results in [15] and [16]. Write $P = Q/J$, where Q is a non-maximal, hence minimal, prime ideal of D. Viewing M as a D-module, we see, using [16, Corollary 16.4], that M_Q is either 0 or is isomorphic to D_Q or $D_Q^{(2)}$. Since the natural map $D_Q \to R_P$ is an isomorphism, the desired conclusion follows.

2. When some power of $\mathfrak{m}$ requires 3 or more generators

PROPOSITION 2.1. *Let $(R, \mathfrak{m}, k)$ be a commutative, Noetherian local ring for which some power $\mathfrak{m}^r$ of the maximal ideal requires at least three generators. Let $\mathcal{P}$ be a finite subset of $\operatorname{Spec}(R) - \{\mathfrak{m}\}$, and let n_P be a non-negative integer for each $P \in \mathcal{P}$. Assume that $n_P = n_Q$ whenever $P \sim Q$.*

Let $n_1 < \cdots < n_t$ be the distinct integers in $\{n_P \mid P \in \mathcal{P}\}$, and put $n := n_1 + \cdots + n_t$. Given any integer $q > n$, there are $|k|$ pairwise non-isomorphic indecomposable finitely generated R-modules M such that

(i) *M needs exactly $2q$ generators, and*

(ii) *$M_P \cong R_P^{(n_P)}$ for each $P \in \mathcal{P}$.*

Proof. Choose $x \in \mathfrak{m}^r - (\mathfrak{m}^{r+1} \cup (\bigcup \mathcal{P}))$, $y \in \mathfrak{m}^r - ((\mathfrak{m}^{r+1} + Rx) \cup (\bigcup \mathcal{P}))$ and $z \in \mathfrak{m}^r - ((\mathfrak{m}^{r+1} + Rx + Ry) \cup (\bigcup \mathcal{P}))$. Thus x, y and z are outside the union of the primes in $\mathcal{P}$, and their images in $\mathfrak{m}^r/\mathfrak{m}^{r+1}$ are linearly independent.

For $i = 1, \dots, t$, let $\mathcal{P}_i = \{P \in \mathcal{P} \mid n_P = n_i\}$. Put $S_i = R - \bigcup \mathcal{P}_i$, and let K_i be the kernel of the natural map $R \to S_i^{-1}R$. We claim that $0 \in S_iS_j$ if $i \neq j$. If not, there would be a prime ideal Q disjoint from the multiplicative set S_iS_j. But then Q would be contained in $P_i \cap P_j$ for some $P_i \in \mathcal{P}_i$ and $P_j \in \mathcal{P}_j$, contradicting $P_i \not\sim P_j$. It follows that $S_i^{-1}S_j^{-1}R = 0$ if $i \neq j$, that is, $K_iS_j^{-1}R = S_j^{-1}R$ if $i \neq j$. Therefore we can choose, for each $i = 1, \dots, t$, an element

$$\xi_i \in K_i\mathfrak{m}^{r+1} - \bigcup_{j \neq i} \Big(\bigcup \mathcal{P}_j\Big).$$

The image of ξ_i in $S_j^{-1}R$ is 0 if $i = j$ and a unit if $i \neq j$.

Let I_l denote the $l \times l$ identity matrix and 0_l the $l \times l$ zero matrix. Let $H = H_q$ be the $q \times q$ nilpotent Jordan block with 1's on the superdiagonal and 0's elsewhere. Given any element $u \in R$, put

$$\Delta = \Delta_{q,u} := (z + uy)I_q + yH_q.$$

Consider the following matrix:

$$A = A_{q,u} := \begin{bmatrix} \Xi & \Delta \\ 0_q & xI_q \end{bmatrix} \in \mathrm{Mat}_{2q \times 2q}(R), \tag{1}$$

where

$$\Xi := \begin{bmatrix} \xi_1 I_{n_1} & 0 & 0 & \cdots & 0 \\ 0 & \xi_2 I_{n_2} & 0 & \ddots & \vdots \\ 0 & \ddots & \ddots & \ddots & 0 \\ \vdots & \ddots & 0 & \xi_t I_{n_t} & 0 \\ 0 & \cdots & 0 & 0 & x^2 I_{q-n} \end{bmatrix} \in \mathrm{Mat}_{q \times q}(R).$$

We let A operate on $R^{(q)} \oplus R^{(q)}$ by left multiplication, and we put $M = M_{q,u} := \operatorname{coker}(A)$. Since the entries of A are in $\mathfrak{m}$, $M_{q,u}$ requires exactly $2q$ generators.

We now show that $M_{q,u}$ is indecomposable, and that $M_{q,u} \not\cong M_{q,u'}$ if $u, u' \in R$ and $u \not\equiv u' \pmod{\mathfrak{m}}$. Fix q, let $u, u' \in R$, and put $A' := A_{q,u'}$, $M' := M_{q,u'}$ and $\Delta' := \Delta_{q,u'}$. Let f be an arbitrary R-homomorphism from $M_{q,u}$ to

$M_{q,u'}$. We lift f to homomorphisms F and G making the following diagram commutative:

$$\begin{array}{ccccccc}
R^{(q)} \oplus R^{(q)} & \xrightarrow{A} & R^{(q)} \oplus R^{(q)} & \longrightarrow & M & \longrightarrow & 0 \\
{\scriptstyle G}\big\downarrow & & {\scriptstyle F}\big\downarrow & & {\scriptstyle f}\big\downarrow & & \\
R^{(q)} \oplus R^{(q)} & \xrightarrow{A'} & R^{(q)} \oplus R^{(q)} & \longrightarrow & M' & \longrightarrow & 0
\end{array}$$

When we write F and G as 2×2 block matrices, this diagram yields the equation

$$\text{(2)} \quad \begin{bmatrix} F_{11}\Xi & F_{11}\Delta + F_{12}x \\ F_{21}\Xi & F_{21}\Delta + F_{22}x \end{bmatrix} = FA = A'G$$
$$= \begin{bmatrix} \Xi G_{11} + \Delta' G_{21} & \Xi G_{12} + \Delta' G_{22} \\ G_{21}x & G_{22}x \end{bmatrix}.$$

Let stars denote the images, in $\mathfrak{m}^r/\mathfrak{m}^{r+1}$, of elements of $\mathfrak{m}^r$. Thus $\xi_i^* = 0$ for each i, $(x^2)^* = 0$, and x^*, y^* and z^* are linearly independent over k. Let bars denote reductions modulo $\mathfrak{m}$ of elements of R and of matrices over R. Comparing the 2,2 entries of the matrix equation (2), we obtain the following equation:

$$\overline{F}_{21}(\overline{u}\overline{I}_q + \overline{H})y^* + \overline{F}_{21}z^* + \overline{F}_{22}x^* = \overline{G}_{22}x^*.$$

It follows that

$$\overline{F}_{21} = 0 \text{ and } \overline{F}_{22} = \overline{G}_{22}.$$

An examination of the 1,2 entries in (2) yields the following equation:

$$\overline{F}_{11}(\overline{u}\overline{I}_q + \overline{H})y^* + \overline{F}_{11}z^* + \overline{F}_{12}x^* = G_{22}\overline{z}^* + (\overline{u'}\,\overline{I}_q + \overline{H})\overline{G}_{22}y^*$$

It follows that

$$\text{(3)} \qquad \overline{F}_{12} = 0,\ \overline{F}_{11} = \overline{G}_{22} \text{ and } \overline{F}_{11}(\overline{u}\overline{I}_q + \overline{H}) = (\overline{u'}\ \overline{I}_q + \overline{H})\overline{G}_{22}.$$

The last two equations in (3) show that

$$\text{(4)} \qquad (\overline{u} - \overline{u'})\overline{F}_{11} = \overline{H}\ \overline{F}_{11} - \overline{F}_{11}\overline{H}.$$

Suppose now that $u \not\equiv u' \pmod{\mathfrak{m}}$. Then $\overline{u} - \overline{u'} \in k^\times$, and since $\overline{H}^q = 0$ we see, by descending induction, that $\overline{H}^i\overline{F}_{11}\overline{H}^j = 0$ for $i, j = 0, \ldots, q$. Setting $i = j = 0$, we get $\overline{F}_{11} = 0$. Since $\overline{F}_{12} = 0$ too, $\overline{F}$ is not surjective, and now Nakayama's lemma implies that f is not surjective. Since f was an arbitrary element of $\operatorname{Hom}_R(M_{q,u}, M_{q,u'})$, this shows that $M_{q,u} \not\cong M_{q,u'}$.

To prove that $M = M_{q,u}$ is indecomposable, we let $u' = u$, and we assume $f \in \operatorname{End}_R(M, M)$ is idempotent but not surjective. We will show that $f = 0$. Since $\overline{H}$ is non-derogatory, (4) implies that $\overline{F}_{11} \in k[\overline{H}]$. In particular, $\overline{F}_{11}$ is upper triangular with constant diagonal. Recall that $\overline{F}_{11} = \overline{G}_{22} = \overline{F}_{22}$ and $\overline{F}_{21} = 0$, so that $\overline{F}$ is upper triangular with constant diagonal. Since $\overline{F}$ is not surjective, it must be strictly upper-triangular. Therefore $\overline{F}^q = 0$. Then

$\operatorname{im}(f) = \operatorname{im}(f^q) \subseteq \mathfrak{m}M$, whence $1 - f$ is surjective. Since f is idempotent, $f = 0$.

It remains to prove that $S_i^{-1}M \cong (S_i^{-1}R)^{(n_i)}$ for all i. Fix an index $i \leq t$, and consider the image $\tilde{A}$ in $\operatorname{Mat}_{2q\times 2q}(S_i^{-1}R)$ of the matrix A. We recall that the $\xi_j, j \neq i$, become units in $\tilde{A}$, while ξ_i maps to 0. Also, x, y and z map to units. Using these facts, one can easily do elementary row and column operations over $S_i^{-1}R$ to show that $\tilde{A}$ is equivalent to the $2q \times 2q$ matrix B with I_{2q-n_i} in the top left corner and zeros elsewhere. Thus $S_i^{-1}M \cong \operatorname{coker}(\tilde{A}) \cong \operatorname{coker}(B) \cong (S_i^{-1}R)^{(n_i)}$ as desired. □

By item (i) in the statement of the theorem, $M_{q,u} \not\cong M_{q',u'}$ if $q \neq q'$. Thus the Main Theorem is true if some power of $\mathfrak{m}$ requires at least three generators.

3. Bimodules and extensions

In this section we concoct some homological machinery to handle the more difficult case of the Main Theorem—when each power of the maximal ideal is generated by two elements.

Throughout this section let R be a commutative Noetherian ring, not necessarily local, and let A and B be module-finite R-algebras (not necessarily commutative). Let ${}_AE_B$ be an A–B-bimodule. We assume E is *R-symmetric*, that is, $re = er$ for $r \in R$ and $e \in E$. (Equivalently, E is a left $A \otimes_R B^{\mathrm{op}}$-module.) Furthermore we assume that E is finitely generated as an R-module. The Jacobson radical of a (not necessarily commutative) ring C is denoted by $\mathrm{J}(C)$, and the ring C is said to be *local* provided $C/\mathrm{J}(C)$ is a division ring; equivalently [7, Proposition 1.10], the set of non-units of C is closed under addition. The next result is [10, Theorem 3.2], and we refer the interested reader to [10] for its elementary proof.

THEOREM 3.1. *With notation above, let $\alpha : {}_AA \to {}_AE$ and $B_B \to E_B$ be module homomorphisms such that $\alpha(1_A) = \beta(1_B) \neq 0$. Assume A is local and $\ker(\beta) \subseteq \mathrm{J}(B)$. Then $C := \beta^{-1}(\alpha(A))$ is an R-subalgebra of B and is a local ring.*

Now we specialize the notation above. Still assuming that R is a commutative Noetherian ring, let M and N be finitely generated R-modules. Put $A := \operatorname{End}_R(M)$ and $B := \operatorname{End}_R(N)$. Note that each of the R-modules $\operatorname{Ext}_R^n(N, M)$ has a natural $A - B$-bimodule structure. Indeed, any $f \in B$ induces an R-module homomorphism $f^* : \operatorname{Ext}_R^n(N, M) \to \operatorname{Ext}_R^n(N, M)$. For $x \in \operatorname{Ext}_R^n(N, M)$ put $x \cdot f = f^*(x)$. The left A-module structure is defined similarly, and the fact that $\operatorname{Ext}_R^n(N, M)$ is a bimodule follows from the fact that $\mathrm{Ext}^n_R(\_,\_)$ is an additive bifunctor. Note that $\mathrm{Ext}^n_R(N,M)$ is R-symmetric, since, for $r \in R$, multiplications by r on N and on M induce the same endomorphism of $\mathrm{Ext}^n_R(N,M)$.

Put $E := \mathrm{Ext}^1_R(N,M)$, regarded as the set of equivalence classes of extensions $0 \to M \to X \to N \to 0$. Let $\alpha : {}_AA \to {}_AE$ and $\beta : B_B \to E_B$ be module homomorphisms satisfying $\alpha(1_A) = \beta(1_B) =: [\sigma]$. Then α and β are, up to signs, the connecting homomorphisms in the long exact sequences of Ext obtained by applying $\mathrm{Hom}_R(\_,M)$ and $\mathrm{Hom}_R(N,\_)$, respectively, to the short exact sequence σ. (When one computes Ext via resolutions one must adorn maps with appropriate $\pm$ signs, in order to ensure naturality of the connecting homomorphisms. In what follows, the choice of sign will not be important.)

Since it causes no extra effort, we phrase Lemma 3.2 and Theorem 3.3 in terms of a general *torsion theory* $(\mathcal{T},\mathcal{F})$ (cf., e.g., [8]). Then, in Corollary 3.4, we apply Theorem 3.3 with $\mathcal{T} = \{\text{modules of finite length}\}$ and $\mathcal{F} = \{\text{modules of positive depth}\} = \{\text{modules with zero socle}\}$.

An easy diagram chase establishes the following lemma, which is [10, Lemma 4.1]:

LEMMA 3.2. *Let R be a commutative Noetherian ring, let M and N be finitely generated R-modules, with M torsion and N torsion-free (with respect to some torsion theory). Let A, B and E be as above, and let $\alpha : A \to E$ and $\beta : B \to E$ be module homomorphisms with $\alpha(1_A) = \beta(1_B) = [\sigma] \neq 0$. Choose a short exact sequence representing σ:*

$$(\sigma) \qquad 0 \longrightarrow M \xrightarrow{i} X \xrightarrow{\pi} N \longrightarrow 0$$

Let $\rho : \mathrm{End}_R(X) \to \mathrm{End}_R(N) = B$ be the canonical homomorphism (reduction modulo torsion). Then the image of ρ is exactly the ring $C := \beta^{-1}\alpha(A) \subseteq B$.

The next result, which is [10, Theorem 4.2], follows easily from Theorem 3.1 and Lemma 3.2:

THEOREM 3.3. *Keep the notation and hypotheses of Lemma 3.2.*

(i) *Suppose C has no idempotents other than 0 and 1. If $X = U \oplus V$ (a decomposition as R-modules), then either U or V is a torsion module.*
(ii) *Suppose A is local and $\ker(\beta)$ is contained in the Jacobson radical of B. Then X is indecomposable.*

COROLLARY 3.4. *Let $(R,\mathfrak{m},k)$ be a commutative, Noetherian local ring, and let M be an indecomposable finitely generated R-module of finite length. Let N be a finitely generated R-module with $\mathrm{depth}(N) > 0$. Put $A := \mathrm{End}_R(M)$ and $B := \mathrm{End}_R(N)$. Suppose there exists a right B-module homomorphism $\beta : B_B \to E_B := \mathrm{Ext}^1_R(N,M)$ such that $\ker(\beta) \subseteq \mathrm{J}(B)$ (equivalently, assume there is an element $\xi \in E$ with $(0 :_B \xi) \subseteq \mathrm{J}(B)$). Let*

$0 \to M \to X \to N \to 0$ represent $\xi = \beta(1_B) \in E$. Then X is indecomposable.

4. Building suitable finite-length modules

To prove the Main Theorem in the remaining case, when each power of $\mathfrak{m}$ is two-generated, we need to build a sufficiently complicated indecomposable finite-length module M and then choose a suitable module N of positive depth. In this section we build the requisite finite-length modules.

The following proposition is a slightly jazzed-up version of the "Warmup" in [10]. This construction is far from new. See, for example, the papers of Higman [12], Heller and Reiner [11], and Warfield [23]. Similar constructions can be found in the classification, up to simultaneous equivalence, of pairs of matrices. (Cf. Dieudonné's discussion [3] of the work of Kronecker [17] and Weierstrass [24].)

PROPOSITION 4.1. *Let $(\Lambda, \mathfrak{m}, k)$ be a commutative Noetherian local ring with $\mathfrak{m}^2 = 0$, let q be a positive integer and let u be a unit of Λ. Let I_q denote the $q \times q$ identity matrix and H_q the $q \times q$ nilpotent Jordan block (with 1's on the superdiagonal and 0's elsewhere). Assume $\mathfrak{m}$ is minimally generated by two elements x and y, let $\Psi_{q,u} := yI_q + x(uI_q + H_q)$ and put $M_{q,u} := \operatorname{coker}(\Psi_{q,u})$.*

(i) *$M_{q,u}$ is an indecomposable Λ-module requiring exactly q generators.*
(ii) *For every non-zero element $t \in \mathfrak{m}$, $\operatorname{socle}(M_{q,u}/tM_{q,u}) \cong k^{(q)}$.*
(iii) *$M_{q,u} \cong M_{q',u'}$ if and only if $q = q'$ and $u \equiv u' \pmod{\mathfrak{m}}$.*

Proof. Clearly $M_{q,u}$ requires exactly q generators, whence $M_{q,u} \not\cong M_{q',u'}$ if $q \neq q'$. Therefore we drop the subscripts q from now on. The "if" assertion in (iii) is clear, since $\mathfrak{m}^2 = 0$. The proofs of the "only if" assertion in (iii) and of the indecomposability of the M_u are similar to (but easier than) the proofs of the analogous assertions in Proposition 2.1. Alternatively, one can note that the associated graded modules $\operatorname{gr}_{\mathfrak{m}}(M_u)$ are among the indecomposable modules in the classification of $k[X,Y]/(X^2, XY, Y^2)$-modules, found in the references above.

To prove (ii), we drop the index u and note that $M/tM = \operatorname{coker}(\Phi)$, where $\Phi = [\,\Psi \;\; tI\,]$. Suppose first that $t = by$, where b is a unit of Λ. Elementary column operations transform Φ to the matrix $[\,xH \;\; yI\,]$. Therefore $M/tM \cong k^{(q-1)} \oplus \Lambda/(y)$, and (ii) follows. The other possibility is that $t = ax + by$, where a is a unit. In this case we can do elementary column operations to replace the superdiagonal elements of Ψ by multiples of y. Further column operations transform the matrix to the form $[\,yI \;\; xI\,]$, and we have $M/tM \cong k^{(q)}$. □

If $(R, \mathfrak{m}, k)$ is Artinian and $\mathfrak{m}$ is principal, the zero-dimensional case of Cohen's Structure Theorem implies that R is a homomorphic image of a complete discrete valuation ring. Thus, if $(R, \mathfrak{m}, k)$, as in the Main Theorem, is zero-dimensional, we can apply Proposition 4.1 to the ring $R/\mathfrak{m}^2$ to get $|k|\cdot\aleph_0$ pairwise non-isomorphic indecomposable modules. Next, suppose $\dim(R) \geq 2$. Then $\mathfrak{m}$ needs three generators unless R is a two-dimensional regular local ring; and in that case $\mathfrak{m}^2$ needs three generators. By Proposition 2.1, the Main Theorem holds if $\dim(R) \geq 2$. Therefore it remains to prove the Main Theorem under the assumptions that $(R, \mathfrak{m}, k)$ is one-dimensional and each power of $\mathfrak{m}$ is at most two-generated.

DEFINITION 4.2. A commutative, Artinian local ring $(\Lambda, \mathfrak{m}, k)$ is a *Drozd ring* provided its associated graded ring is the k-algebra $\mathrm{gr}_{\mathfrak{m}}(\Lambda) \cong k[X,Y]/(X^2, XY^2, Y^3)$. (Equivalently, $\mathfrak{m}^3 = 0$, $\mathfrak{m}$ and $\mathfrak{m}^2$ each require exactly two generators, and there is an element $t \in \mathfrak{m} - \mathfrak{m}^2$ with $t^2 = 0$.)

The main result in this section is a construction, in Proposition 4.4, of suitably complex indecomposable modules over Drozd rings. The idea of the construction below originated in work of Drozd [4] and Ringel [21]. The construction was adapted by Klingler and Levy [14] to show that the category of finite-length modules over a Drozd ring has wild representation type. Drozd rings enter the picture here because of the following result, a special case of the "Ring-theoretic Dichotomy" of Klingler and Levy [16, Theorem 14.3]:

THEOREM 4.3. *Let $(\Lambda, \mathfrak{m}, k)$ be a one-dimensional local ring whose maximal ideal $\mathfrak{m}$ is generated by at most two elements. Then exactly one of the following possibilities occurs:*

(i) *Λ is a homomorphic image of a Dedekind-like ring.*
(ii) *Λ has a Drozd ring as a homomorphic image.*

Proposition 4.4 is a slight generalization of [10, Proposition 5.3]. The modification is needed to treat the case of a Cohen-Macaulay ring with multiplicity 2.

PROPOSITION 4.4. *Let $(\Lambda, \mathfrak{m}, k)$ be a Drozd ring, and let $t, y \in \Lambda$ with $(t, y) = \mathfrak{m}$ and $t^2 = 0$. There exists a family $(M_{q,\kappa})_{q\in\mathbb{N},\kappa\in k^\times}$ of pairwise non-isomorphic indecomposable Λ-modules having the following properties:*

(i) *For all $q \in \mathbb{N}$ and $\kappa \in k^\times$ we have*

$$\frac{(0 :_{M_{q,\kappa}} (t, y^2))}{tM_{q,\kappa}} \cong k^{(q)}.$$

(ii) *For every $\xi \in \mathfrak{m}$, all $\kappa \in k^\times$, and all $q \geq 1$ the k-vectorspace*

$$\frac{(0 :_{M_{q,\kappa}} \xi) + \mathfrak{m}M_{q,\kappa}}{\mathfrak{m}M_{q,\kappa}}$$

has dimension greater than or equal to q.

Proof. Given $q \in \mathbb{N}$ and $\kappa \in k^\times$, choose $u \in \Lambda^\times$ with $u + \mathfrak{m} = \kappa$. Since $\mathfrak{m}^3 = 0$, uy^2 depends only on the coset $u + \mathfrak{m}$. Therefore we can define $M_{q,\kappa}$ to be the cokernel of the $3q \times 4q$ matrix

$$\Psi_{q,\kappa} := \begin{bmatrix} yI_q & tI_q & 0 & 0 \\ 0 & -y^2 I_q & tI_q & -yI_q \\ 0 & 0 & -(uI_q + H_q)y^2 & tI_q \end{bmatrix}$$

with H_q and I_q as in Proposition 4.1. We let $\Lambda^{(3q)} \xrightarrow{\varepsilon_{q,\kappa}} M_{q,\kappa}$ denote the quotient map.

To show that $M_{q,\kappa}$ is indecomposable and that $M_{q,\kappa} \not\cong M_{q,\kappa'}$ if $\kappa \neq \kappa'$, suppose $f : M_{q,\kappa} \to M_{q,\kappa'}$ is a Λ-homomorphism. Lift κ' to $u' \in \Lambda^\times$. As in the proof of Proposition 2.1 we obtain a commutative diagram:

$$\begin{array}{ccccc} \Lambda^{(4q)} & \xrightarrow{\Psi_{q,\kappa}} & \Lambda^{(3q)} & \longrightarrow & M_{q,\kappa} \\ {\scriptstyle G}\downarrow & & {\scriptstyle F}\downarrow & & {\scriptstyle f}\downarrow \\ \Lambda^{(4q)} & \xrightarrow{\Psi_{q,\kappa'}} & \Lambda^{(3q)} & \longrightarrow & M_{q,\kappa'} \end{array}$$

In principle we could proceed as in Proposition 2.1 and derive restrictions for the entries of F from the equation $F \cdot \Psi_{q,\kappa} = \Psi_{q,\kappa'} \cdot G$; instead, we consult [14, Lemma 4.8] to shorten the argument. If we let bars denote reductions modulo $\mathfrak{m}$, this lemma implies that

$$\overline{F} = \begin{bmatrix} \overline{F}_{11} & * & * \\ 0 & \overline{F}_{11} & * \\ 0 & 0 & \overline{F}_{11} \end{bmatrix},$$

where each block is a $q \times q$ matrix and $\overline{F}_{11} \cdot (u\overline{I}_q + \overline{H}_q) = (u'\overline{I}_q + \overline{H}_q) \cdot \overline{F}_{11}$.

If $\kappa \neq \kappa'$, the argument following (4) in the proof of Proposition 2.1 shows that $M_{q,\kappa} \not\cong M_{q,\kappa'}$. Of course $M_{q,\kappa}$ requires exactly $3q$ generators, so $M_{q,\kappa} \cong M_{q',\kappa'} \implies q = q'$. Thus we assume from now on that $\kappa = \kappa'$ and omit the subscripts q and κ. The proof that M is indecomposable is essentially the same as the proof of indecomposability in Proposition 2.1.

We claim that $(0 :_M (t, y^2))$ is generated by the images, under ε, of the columns of the matrix

$$\varphi := \begin{bmatrix} tI & 0 & 0 & 0 & 0 & I \\ 0 & tI & 0 & yI & 0 & 0 \\ 0 & 0 & tI & 0 & y^2 I & -yI \end{bmatrix}$$

(where each block is $q \times q$). An easy calculation shows that both t and y^2 knock the column space of φ into the column space of Ψ, so the purported generators are, at least, in $(0 :_M (t, y^2))$.

To prove the claim, suppose $\boldsymbol{\alpha} \in \Lambda^{(3q)}$ and $t\boldsymbol{\alpha}$ and $y^2\boldsymbol{\alpha}$ are both in the image of Ψ. We will show that $\boldsymbol{\alpha} \in \operatorname{im}(\varphi)$.

We have

$$t\boldsymbol{\alpha} = \Psi \cdot \boldsymbol{\beta} \text{ and } y^2\boldsymbol{\alpha} = \Psi \cdot \boldsymbol{\gamma} \tag{5}$$

with $\boldsymbol{\beta}, \boldsymbol{\gamma} \in \Lambda^{(4q)}$. Write

$$\boldsymbol{\alpha} = \begin{bmatrix} \boldsymbol{\alpha}_1 \\ \boldsymbol{\alpha}_2 \\ \boldsymbol{\alpha}_3 \end{bmatrix}, \quad \boldsymbol{\beta} = \begin{bmatrix} \boldsymbol{\beta}_1 \\ \boldsymbol{\beta}_2 \\ \boldsymbol{\beta}_3 \\ \boldsymbol{\beta}_4 \end{bmatrix},$$

where the $\boldsymbol{\alpha}_i$ and $\boldsymbol{\beta}_j$ are in $\Lambda^{(q)}$. The first equation in (5) yields

$$\begin{bmatrix} t\boldsymbol{\alpha}_1 \\ t\boldsymbol{\alpha}_2 \\ t\boldsymbol{\alpha}_3 \end{bmatrix} = \begin{bmatrix} y\boldsymbol{\beta}_1 + t\boldsymbol{\beta}_2 \\ -y^2\boldsymbol{\beta}_2 + t\boldsymbol{\beta}_3 - y\boldsymbol{\beta}_4 \\ -y^2(uI + H) \cdot \boldsymbol{\beta}_3 + t\boldsymbol{\beta}_4 \end{bmatrix}.$$

We can write the $\boldsymbol{\alpha}_i$ and $\boldsymbol{\beta}_i$ in the form

$$\boldsymbol{\alpha}_i = \boldsymbol{u}_{i,0} + \boldsymbol{u}_{i,1}t + \boldsymbol{u}_{i,2}y + \boldsymbol{u}_{i,3}ty + \boldsymbol{u}_{i,4}y^2,$$
$$\boldsymbol{\beta}_i = \boldsymbol{v}_{i,0} + \boldsymbol{v}_{i,1}t + \boldsymbol{v}_{i,2}y + \boldsymbol{v}_{i,3}ty + \boldsymbol{v}_{i,4}y^2,$$

where the entries of $\boldsymbol{u}_{i,j}$ and $\boldsymbol{v}_{i,j}$ are either units or 0. (Cf. [14, Lemma 4.2].) Since the images of t and y in $\mathfrak{m}/\mathfrak{m}^2$ are linearly independent over k, the equation $t\boldsymbol{\alpha}_1 = y\boldsymbol{\beta}_1 + t\boldsymbol{\beta}_2$ yields $\boldsymbol{v}_{1,0} = \mathbf{0}$ and $\overline{\boldsymbol{u}}_{1,0} = \overline{\boldsymbol{v}}_{2,0}$, where bars denote reduction modulo $\mathfrak{m}$. From $t\boldsymbol{\alpha}_2 = -y^2\boldsymbol{\beta}_2 + t\boldsymbol{\beta}_3 - y\boldsymbol{\beta}_4$, it follows that $\overline{\boldsymbol{u}}_{2,0} = \overline{\boldsymbol{v}}_{3,0}$ and $\boldsymbol{v}_{4,0} = \mathbf{0}$ and, since the socle elements ty and y^2 are linearly independent over k, that $\overline{\boldsymbol{v}}_{2,0} = -\overline{\boldsymbol{v}}_{4,2}$. From $t\boldsymbol{\alpha}_3 = -y^2(uI + H) \cdot \boldsymbol{\beta}_3 + t\boldsymbol{\beta}_4$, it follows that $\overline{\boldsymbol{u}}_{3,0} = \overline{\boldsymbol{v}}_{4,0}$ and hence that $\boldsymbol{u}_{3,0} = \mathbf{0}$.

Using the equation $t\boldsymbol{\alpha}_3 = -y^2(uI + H) \cdot \boldsymbol{\beta}_3 + t\boldsymbol{\beta}_4$ again, we see that $\overline{\boldsymbol{u}}_{3,2} = \overline{\boldsymbol{v}}_{4,2}$. Further, since $uI + H$ is invertible, it follows that $\boldsymbol{v}_{3,0} = \mathbf{0}$ and hence that $\boldsymbol{u}_{2,0} = \mathbf{0}$.

To summarize, we have $\overline{\boldsymbol{u}}_{3,2} = \overline{\boldsymbol{v}}_{4,2} = -\overline{\boldsymbol{v}}_{2,0} = -\overline{\boldsymbol{u}}_{1,0}$, and $\boldsymbol{u}_{2,0} = \boldsymbol{u}_{3,0} = \mathbf{0}$. Putting $\boldsymbol{w} := \boldsymbol{u}_{1,0}$, we have $\boldsymbol{u}_{3,2} = -\boldsymbol{w} + t\boldsymbol{\mu} + y\boldsymbol{\nu}$ for suitable $\boldsymbol{\mu}, \boldsymbol{\nu} \in \Lambda^{(q)}$. Then

$$\boldsymbol{\alpha} = \begin{bmatrix} \boldsymbol{w} & + & t\boldsymbol{u}_{1,1} & + & y\boldsymbol{u}_{1,2} & + & ty\boldsymbol{u}_{1,3} & + & y^2\boldsymbol{u}_{1,4} \\ \mathbf{0} & + & t\boldsymbol{u}_{2,1} & + & y\boldsymbol{u}_{2,2} & + & ty\boldsymbol{u}_{2,3} & + & y^2\boldsymbol{u}_{2,4} \\ -y\boldsymbol{w} & + & t\boldsymbol{u}_{3,1} & + & \mathbf{0} & + & ty(\boldsymbol{u}_{3,3} + \boldsymbol{\mu}) & + & y^2(\boldsymbol{u}_{3,4} + \boldsymbol{\nu}) \end{bmatrix}. \tag{6}$$

From (6) it follows that $\boldsymbol{\alpha} \in \operatorname{im}(\boldsymbol{\varphi})$, as desired. This completes the proof of our claim.

It is easy to see, using the invertibility of $uI + H$, that the image of the left-most $3q \times 5q$ submatrix of $\boldsymbol{\varphi}$ is contained in $t\Lambda^{(3q)} + \operatorname{im}(\Psi)$. Letting $\gamma_1, \dots, \gamma_q$ be the last q columns of $\boldsymbol{\varphi}$, we see that $(0 :_M (t, y^2))/tM$ is generated by $\zeta_1 := \varepsilon(\gamma_1) + tM, \dots, \zeta_q := \varepsilon(\gamma_q) + tM$. Since $t\gamma_i,\ y\gamma_i \in t\Lambda^{(3q)} + \operatorname{im}(\Psi)$ for each i, we see that $(0 :_M (t, y^2))/tM$ is a k-vector space of dimension at most q. To complete the proof of (i), we need only show that $\zeta_1, \dots, \zeta_q$ are linearly independent. Given a relation $\sum_{i=1}^{q} \lambda_i \zeta_i = 0$, with $\lambda_i \in \Lambda$, we have

$\sum_{i=1}^{q} \lambda_i \gamma_i \in \mathrm{im}(\Psi) + t\Lambda^{(3q)} \subseteq \mathfrak{m}\Lambda^{(3q)}$. This relation obviously forces $\lambda_i \in \mathfrak{m}$ for all i, as desired.

It remains to prove assertion (ii) of the proposition. Given $\xi \in \mathfrak{m}$, write $\xi = at + by$. Suppose first that b is a unit of Λ. For each unit vector $\mathbf{e}_i \in \Lambda^{(q)}$, put

$$\boldsymbol{\sigma}_i := \begin{bmatrix} \mathbf{e}_i \\ (\frac{a^2}{b^2}t - \frac{a}{b}y)\mathbf{e}_i \\ 0 \end{bmatrix},$$

and check that

$$\xi\boldsymbol{\sigma}_i = b\begin{bmatrix} y\mathbf{e}_i \\ 0 \\ 0 \end{bmatrix} + a\begin{bmatrix} t\mathbf{e}_i \\ -y^2\mathbf{e}_i \\ 0 \end{bmatrix} \in \mathrm{im}(\Psi).$$

This shows that $\varepsilon(\boldsymbol{\sigma}_i) \in (0 :_M \xi)$ for each i, and the assertion follows easily in this case.

If b is not a unit, ξ has the form $\xi = ct + dy^2$. With $\mathbf{e}_i$ as above, put

$$\boldsymbol{\tau}_i := \begin{bmatrix} \mathbf{e}_i \\ 0 \\ -y\mathbf{e}_i \end{bmatrix}.$$

Then

$$\xi\boldsymbol{\tau}_i = dy\begin{bmatrix} y\mathbf{e}_i \\ 0 \\ 0 \end{bmatrix} + c\begin{bmatrix} t\mathbf{e}_i \\ -y^2\mathbf{e}_i \\ 0 \end{bmatrix} - cy\begin{bmatrix} 0 \\ -y\mathbf{e}_i \\ t\mathbf{e}_i \end{bmatrix} \in \mathrm{im}(\Psi).$$

As before, the assertion follows easily. □

5. When all powers of $\mathfrak{m}$ are at most 2-generated

In this section we complete the proof of the Main Theorem in the remaining case—each power of $\mathfrak{m}$ is generated by at most two elements. Recall that by Theorem 4.3 R maps onto a Drozd ring. We refer the reader to [10, Lemma 6.2] for the proof of the next result (note that $\mathrm{e}(R) = \mathrm{e}(\mathfrak{m}, R)$ denotes the multiplicity of R):

LEMMA 5.1. *Let $(R, \mathfrak{m}, k)$ be a one-dimensional local ring. Assume that $\mathfrak{m}$ and $\mathfrak{m}^2$ are two-generated and R/L is a Drozd ring for some ideal L. Write $\mathfrak{m} = Rt + Ry$, with $t^2 \in L$. Then $L = \mathfrak{m}^3$, and $\mathfrak{m}^r = y^{r-1}\mathfrak{m} = Rty^{r-1} + Ry^r$ for each $r \geq 1$. If, further, R is not Cohen-Macaulay, then the following also hold:*

(i) *$\mathfrak{m}^r = Ry^r$ for all $r \gg 1$. In particular,* $\mathrm{e}(R) = 1$.
(ii) *R has exactly one minimal prime ideal P. Moreover, R_P is a field and R/P is a discrete valuation ring.*
(iii) *P is a principal ideal, and $P \not\subseteq \mathfrak{m}^2$.*

PROPOSITION 5.2. *Let $(R, \mathfrak{m}, k)$ be a commutative local Noetherian ring, let P be a non-maximal prime ideal of R, and let n be any non-negative integer. Suppose there is an indecomposable finite-length R-module M such that $\dim_k(\operatorname{socle}_R(\operatorname{Ext}^1_R(R/P, M))) \geq n$. Then there is a short exact sequence*

$$0 \longrightarrow M \longrightarrow X \longrightarrow (R/P)^{(n)} \longrightarrow 0, \tag{7}$$

in which X is indecomposable.

Proof. Put $E_1 := \operatorname{Ext}^1_R(R/P, M)$, $N := (R/P)^{(n)}$, $A := \operatorname{End}_R(M)$, $B := \operatorname{End}_R(N) = \operatorname{Mat}_{n\times n}(R/P)$ and $E := \operatorname{Ext}^1_R(N, M) = E_1^{(n)}$. If we write elements of E as $1 \times n$ row vectors with entries in E_1, then the right B-module structure is given by matrix multiplication. Since M has finite length, A is a local ring [7, Lemmas 2.20 and 2.21].

Let $e_1, \ldots, e_n$ be linearly independent elements of $\operatorname{socle}_R(E_1)$, and put $\xi := [e_1, \ldots, e_n] \in E$. We claim that $(0 :_B \xi) \subseteq \mathrm{J}(B)$. For, suppose $\varphi := [a_{ij}] \in B$ with $\xi\varphi = 0$. Then $e_1 a_{1j} + \cdots + e_n a_{nj} = 0$ for each $j = 1, \ldots, n$. Linear independence of the e_i now implies that $a_{ij} \in \mathfrak{m}/P$ for each i, j. Then $\varphi \in \mathrm{J}(B)$, and the claim is proved.

To complete the proof, we let (7) represent the element $\xi \in E$ and apply Corollary 3.4. □

We will divide the proof of the Main Theorem into three cases.

5.1. Case 1: R is not Cohen-Macaulay. Suppose now that $(R, \mathfrak{m}, k)$ is one-dimensional and not Cohen-Macaulay, as in the Main Theorem, and assume also that each power of $\mathfrak{m}$ is generated by at most two elements. By Theorem 4.3 and Lemma 5.1, R has a unique minimal prime ideal P; moreover, P is principal, say, $P = Rt$. Given a non-negative integer n, we seek $|k| \cdot \aleph_0$ pairwise non-isomorphic indecomposable modules X such that $X_P \cong (R/P)^{(n)}$. The proof is a slight modification of the corresponding case in [10]; we give a sketch of the argument. (See [10, Proposition 6.3 and the succeeding paragraphs] for details.)

Suppose, first, that $(0 :_R t) \subseteq \mathfrak{m}^2$. Given an arbitrary integer $q \geq \max\{1, n\}$, we apply Proposition 4.1 to $R/\mathfrak{m}^2$, getting $|k| - 1$ pairwise non-isomorphic indecomposable finite-length modules M satisfying

$$\operatorname{socle}_R(M/tM) \cong k^{(q)} \text{ and } \mathfrak{m}^2 M = 0.$$

Applying $\operatorname{Hom}_R(\_, M)$ to the short exact sequence

$$0 \longrightarrow Rt \longrightarrow R \longrightarrow R/(t) \longrightarrow 0,$$

we obtain an exact sequence

$$\operatorname{Hom}_R(R, M) \longrightarrow \operatorname{Hom}_R(Rt, M) \longrightarrow E_1 \longrightarrow 0, \tag{8}$$

where $E_1 := \operatorname{Ext}^1_R(R/P, M)$. Since $Rt \cong R/(0 :_R t)$ and $(0 :_R t)M = (0)$, the map $f \mapsto f(t)$ provides an isomorphism $\operatorname{Hom}_R(Rt, M) \cong M$. Combining this isomorphism with the usual isomorphism $\mathrm{Hom}_R(R, M) \cong M$ $(g \mapsto g(1))$, we transform (8) to the exact sequence $M \xrightarrow{t} M \to E_1 \to 0$. Thus $E_1 \cong M/tM$. Now Proposition 5.2 provides, for each M, an indecomposable module X and a short exact sequence (7). Then $X_P \cong R_P^{(n)}$. Also, since $M \cong \mathrm{H}^0_{\mathfrak{m}}(X)$ (the finite-length part of X), we see that non-isomorphic M's yield non-isomorphic X's, and the proof is complete in this case.

Next, we consider the more difficult case, when $(0 :_R t) \not\subseteq \mathfrak{m}^2$. Since R maps onto a Drozd ring by Theorem 4.3, one can show easily that $t^2 \in \mathfrak{m}^3$. Also, $t \notin \mathfrak{m}^2$ by (iii) of Lemma 5.1, so we can choose y such that $\mathfrak{m} = Rt + Ry$. To summarize, we have

$$P = Rt, \ \mathfrak{m} = Rt + Ry, \text{ and } t^2 \in \mathfrak{m}^3. \tag{9}$$

We now complete the proof under the additional assumption that

$$t^2 = ty^2 = 0. \tag{10}$$

In this case, one checks easily that $(0 :_R t) = (t, y^2)$. Applying Proposition 4.4 to the Drozd ring $\Lambda := R/\mathfrak{m}^3$, we get $|k| \cdot \aleph_0$ indecomposable R-modules M such that $\mathfrak{m}^3 M = 0$ and the k-vector space $\frac{(0 :_M (t,y^2))}{tM}$ has dimension n. Again, we obtain the exact sequence (8), and since $Rt \cong R/(t, y^2)$, we see that $\mathrm{Hom}_R(Rt, M) \cong (0 :_M (t, y^2))$, and hence $E_1 \cong \frac{(0 :_M (t,y^2))}{tM}$. Thus $E_1 = \mathrm{socle}_R(E_1)$ has dimension n. As before, we can use Proposition 5.2 to produce $|k| \cdot \aleph_0$ pairwise non-isomorphic indecomposable modules X such that $X_P \cong R_P^{(n)}$.

Finally, we complete the proof when (10) is not necessarily satisfied. Since $t^2 \in \mathfrak{m}^3$ by (9), $S := R/(t^2, ty^2)$ maps onto the Drozd ring $R/\mathfrak{m}^3$. Therefore, by Theorem 4.3, S is not a homomorphic image of a Dedekind-like ring. Moreover, S is not Cohen-Macaulay, since $ty \notin Rt^2 + Rty^2$ (else $\mathfrak{m}^2$ would be principal) but $\mathfrak{m}ty \subseteq Rt^2 + Rty^2$. By the argument in the previous paragraph, we obtain $|k| \cdot \aleph_0$ pairwise non-isomorphic S-modules X such that $X_Q \cong S_Q^{(n)}$, where $Q = P/(t^2, ty^2)$. Now view these modules as R-modules and note that the natural map $R_P \to S_Q$ is an isomorphism. This completes the proof of Theorem 1.2 when R is not Cohen-Macaulay.

For the rest of Section 5, we assume that $(R, \mathfrak{m}, k)$ is a one-dimensional Noetherian local Cohen-Macaulay ring such that each power of $\mathfrak{m}$ is generated by two elements, and we assume that R is not a homomorphic image of a Dedekind-like ring, equivalently (Theorem 4.3), R has a Drozd ring as a homomorphic image. By Lemma 5.1, $\Lambda := R/\mathfrak{m}^3$ is a Drozd ring. Moreover, we have the "associativity formula" (cf. [20, Theorem 14.7] or [2, Corollary 4.6.8]):

$$2 = \mathrm{e}(R) = \sum_i \mathrm{e}(R/P_i)\ell(R_{P_i}), \tag{11}$$

where the sum ranges over all minimal prime ideals P_i of R, and $\ell(R_{P_i})$ is the length of R_{P_i} as an R_{P_i}-module. Thus, R has either one or two minimal prime ideals.

5.2. Case 2: R is Cohen-Macaulay with two minimal prime ideals. Let P_1 and P_2 denote the minimal primes of R. We are given two non-negative integers n_1 and n_2, and we want to find $|k| \cdot \aleph_0$ indecomposable modules X such that $X_{P_i} \cong R_{P_i}^{(n_i)}$ for $i = 1, 2$. By (11), each R/P_i is a discrete valuation domain and R_{P_i} is a field. Since $\mathfrak{m}$ needs two generators, it follows that each $P_i \not\subseteq \mathfrak{m}^2$, so we can choose $t_i \in P_i \not\subseteq \mathfrak{m}^2$. Then $R/(t_i)$ is one-dimensional with principal maximal ideal, i.e. a discrete valuation ring; hence $P_i = Rt_i$. Suppose r is in the kernel of the diagonal map $R \to R_{P_1} \times R_{P_2}$. Then $(0 :_R r) \not\subseteq P_1 \cup P_2$, so $(0 :_R r)$ contains a non-zerodivisor. It follows that R is reduced, with total quotient ring $R_{P_1} \times R_{P_2}$ and normalization $R/P_1 \times R/P_2$. Moreover, $(0 :_R t_1) = Rt_2$ and $(0 :_R t_2) = Rt_1$.

Given any integer $n \geq \max\{n_1, n_2\}$, let $M := M_{n,\kappa}$ be one of the indecomposable Λ-modules from Proposition 4.4. Applying $\operatorname{Hom}_R(\_, M)$ to the short exact sequence

$$0 \longrightarrow Rt_1 \longrightarrow R \longrightarrow R/(t_1) \longrightarrow 0,$$

we obtain an exact sequence

$$\operatorname{Hom}_R(R, M) \longrightarrow \operatorname{Hom}_R(Rt_1, M) \longrightarrow E_1 \longrightarrow 0,$$

where $E_1 = \operatorname{Ext}^1_R(R/P_1, M)$. Now $\operatorname{Hom}_R(Rt_1, M) \cong (0 :_M (0 :_R t_1)) = (0 :_M t_2)$. Therefore $E_1 \cong (0 :_M t_2)/t_1 M$, and, by symmetry, $E_2 := \operatorname{Ext}^1_R(R/P_2, M) \cong (0 :_M t_1)/t_2 M$. By (ii) of Proposition 4.4, E_1 and E_2 each need at least n generators.

The rest of the proof is very similar to that of Case 1. Let $N = (R/P_1)^{(n_1)} \oplus (R/P_2)^{(n_2)}$. By the annihilator relations above, $\operatorname{Hom}_R(R/P_i, R/P_j) = 0$ if $i \neq j$. Therefore $B := \operatorname{End}_R(N) = \operatorname{Mat}_{n_1 \times n_1}(R/P_1) \times \operatorname{Mat}_{n_2 \times n_2}(R/P_2)$. Put $E := \operatorname{Ext}^1_R(N, M) = E_1^{(n_1)} \times E_2^{(n_2)}$. We regard elements of E as ordered pairs (ξ_1, ξ_2), where ξ_i is a $1 \times n_i$ row vector with entries in E_i. The right action of B on E is matrix multiplication on each of the two coordinates.

Let $e_1, \dots, e_n \in E_1$ map to linearly independent elements of $E_1/\mathfrak{m}E_1$, and let $f_1, \dots, f_n \in E_2$ map to linearly independent elements of $E_2/\mathfrak{m}E_2$. Consider the elements $\xi_1 := [e_1 \dots e_{n_1}] \in E_1^{(n_1)}$ and $\xi_2 := [f_1 \dots f_{n_2}] \in E_2^{(n_2)}$, and put $\xi := (\xi_1, \xi_2) \in E$. One checks easily that $(0 :_B \xi) \in \mathfrak{m}B \subseteq \mathrm{J}(B)$ (cf., e.g., [10, Lemma 4.4]). Corollary 3.4 now provides a short exact sequence $0 \to M \to X \to N \to 0$ with X indecomposable. This completes the proof of Theorem 1.2 when R is Cohen-Macaulay and has two minimal prime ideals.

There is one remaining case, for which we will use a very different approach.

5.3. Case 3: R is Cohen-Macaulay with one minimal prime ideal P. Given a non-negative integer n, we seek $|k| \cdot \aleph_0$ indecomposable modules X with $X_P \cong R_P^{(n)}$.

Obviously no power of $\mathfrak{m}$ can be principal, so the multiplicity of R is two. Cohen's Structure Theorem implies that R is an *abstract hypersurface*, that is, the completion $\widehat{R}$ has the form $S/(f)$, where $(S, \mathfrak{n}, k)$ is a two-dimensional regular local ring and $f \in \mathfrak{n} - \{0\}$.

Again, we consider the indecomposable Λ-modules $M := M_{n,\kappa}$ provided by Proposition 4.4. This time we will take N, the torsion-free part of the desired module X, to be a suitable direct summand of the first syzygy of M.

The next three results apply more generally to any one-dimensional abstract hypersurface.

THEOREM 5.3. *Suppose $(D, \mathfrak{n}, k)$ is an abstract hypersurface of dimension 1. Let M be an indecomposable finite-length D-module whose first syzygy is isomorphic to $D^{(r)} \oplus F$, where F has no non-zero free direct summand. Let F' be an arbitrary direct summand of F. Then there is a short exact sequence*

$$0 \longrightarrow M \longrightarrow X \longrightarrow F' \longrightarrow 0, \tag{12}$$

in which X is indecomposable.

Proof. We may assume $F' \neq 0$. Put $A := \mathrm{End}_D(M)$ and $B := \mathrm{End}_D(F')$. We have a short exact sequence

$$0 \longrightarrow D^{(r)} \oplus F \longrightarrow D^{(r+s)} \longrightarrow M \longrightarrow 0, \tag{13}$$

where $s = \mathrm{rank}(F)$. Since F' is maximal Cohen-Macaulay over the Gorenstein ring D, we have $\mathrm{Ext}^i_D(F', D) = 0$ for $i > 0$ (cf. [2, Theorems 3.3.7 and 3.3.10]). Therefore, on applying the functor $\mathrm{Hom}_D(F', \_\,)$ to (13), we obtain an isomorphism

$$\mathrm{Ext}^1_D(F', M) \cong \mathrm{Ext}^2_D(F', F). \tag{14}$$

By Eisenbud's theory of matrix factorizations [5] (cf. also [26, Chapter 7]), F' has a periodic resolution with period at most 2 and with constant Betti numbers. Thus we have short exact sequences

$$0 \longrightarrow G \longrightarrow D^{(t)} \xrightarrow{\psi} F' \longrightarrow 0 \tag{15}$$

and

$$0 \longrightarrow F' \longrightarrow D^{(t)} \longrightarrow G \longrightarrow 0. \tag{16}$$

Applying $\mathrm{Hom}_D(F', \_\,)$ to (16), we get an isomorphism

$$\mathrm{Ext}^1_D(F', G) \cong \mathrm{Ext}^2_D(F', F'). \tag{17}$$

Moreover, naturality of the connecting homomorphisms in the long exact sequences of Ext implies that the isomorphisms in (14) and (17) are actually isomorphisms of right B-modules.

Next, applying $\mathrm{Hom}_D(F',\_\,)$ to (15), we get an exact sequence of right B-modules

$$\mathrm{Hom}_D(F', D^{(t)}) \xrightarrow{\psi_*} B \xrightarrow{\eta} \mathrm{Ext}^1_D(F', G) \longrightarrow 0.$$

Since F' is a direct summand of F, there is an injection of right B-modules $\mathrm{Ext}^2_D(F', F') \hookrightarrow \mathrm{Ext}^2_D(F', F)$. Composing this injection with the isomorphisms in (14) and (17), we get an injection of right B-modules $j : \mathrm{Ext}^1_D(F', G) \hookrightarrow \mathrm{Ext}^1_D(F', M)$. Putting $\beta = j\eta$, we obtain an exact sequence of right B-modules

$$\mathrm{Hom}_D(F', D^{(t)}) \xrightarrow{\psi_*} B \xrightarrow{\beta} \mathrm{Ext}^1_D(F', M)$$

We claim that $\ker(\beta)$ is contained in the Jacobson radical $\mathrm{J}(B)$ of B. To prove this, let $g \in \mathrm{Ker}(\beta) = \mathrm{im}(\psi_*)$. Then g lifts to a map $h : F' \to D^{(t)}$, with $\psi h = g$. Since F' has no non-zero free summand, $h(F') \subseteq \mathfrak{n}D^{(t)}$. This shows that $g(F') \subseteq \mathfrak{n}F'$, and the claim follows easily (cf., e.g., [10, Lemma 4.4]). The existence of the short exact sequence (12) now follows from Corollary 3.4. □

In the following, we say that a D-module M has rank s provided $M_P \cong R_P^{(s)}$ for every associated prime P of D.

PROPOSITION 5.4. *Let $(D, \mathfrak{n}, k)$ be an abstract hypersurface of dimension 1, and assume that D has a Drozd ring Λ as a homomorphic image. Let $M := M_{n,\kappa}$ be the indecomposable Λ-module built in Proposition 4.4, and let $L := \mathrm{syz}^1_D(M)$ be the first syzygy of the D-module M. Write $L = D^{(r)} \oplus F$, where F has no non-zero free direct summand. Then* $\mathrm{rank}(F) \geq \frac{n}{e-1}$, *where $e = \mathrm{e}_D(D)$ is the multiplicity of D.*

Proof. Obviously F has a rank. Put $s := \mathrm{rank}(F)$ and $m := \mu_D(F)$ (μ = minimal number of generators required). It follows, e.g., from [13, (1.6)], that $m \leq es$. (The statement of [13, (1.6)] assumes that k is infinite. This is not a problem, since none of m, e, s is changed by the flat local base change $D \to D(X) := D[X]_{\mathfrak{n}[X]}$.) Now $\mu_D(L) = r + m = 3n - s + m$, whence $\mu_D(L) - 3n \leq (e-1)s$. Therefore it will suffice to show that $\mu_D(L) \geq 4n$. Since $\mu_D(\mathfrak{n}) = 2$, the following lemma completes the proof: □

LEMMA 5.5. *Keep the notation above. There is a surjective D-homomorphism from $L = \mathrm{syz}^1_D(M)$ onto $\mathfrak{n}^{(2n)}$.*

Proof. Let χ denote the composition $D^{(3n)} \twoheadrightarrow \Lambda^{(3n)} \overset{\varepsilon}{\twoheadrightarrow} M \to 0$, so that $\ker \chi = L$, and let $\pi : D^{(3n)} \to D^{(2n)}$ be the projection onto the first two coordinates. We will show that $\pi(L) = \mathfrak{n}^{(n)} \oplus \mathfrak{n}^{(n)}$. The inclusion $\pi(L) \subseteq \mathfrak{n}^{(n)} \oplus \mathfrak{n}^{(n)}$ is obvious. For the reverse inclusion, fix $i, 1 \leq i \leq n$, and let $\mathbf{e}_i \in D^{(n)}$ be the ith unit vector. Let $\tilde{t}, \tilde{y} \in \mathfrak{n}$ lift the elements $t, y \in \Lambda$ (notation as in Proposition 4.4). It will suffice to show that the four elements

$(\tilde{t}\mathbf{e}_i, 0), (\tilde{y}\mathbf{e}_i, 0), (0, \tilde{t}\mathbf{e}_i)$ and $(0, \tilde{y}\mathbf{e}_i)$ are all in $\pi(L)$. But this follows easily from the definition of the matrix $\Psi_{n,\kappa}$. □

We now return to our special ring $(R, \mathfrak{m}, k)$ and the modules $M = M_{\kappa,n}$. As in Theorem 5.3, we write the first syzygy of M in the form $R^{(r)} \oplus F$, where F has no non-zero free summand. To complete the proof of the Main Theorem, it will suffice, by Theorem 5.3, to show that F has a direct summand F' of rank n. By Proposition 5.4 we know that F has rank at least n. By [22] F is isomorphic to a direct sum of ideals of R. (Cf. also [18, Theorem 2.1] for a more general statement and [1] for the analytically unramified case.) Each of these ideals must have rank 0 or 1. Therefore the desired module F' can be obtained from F by throwing out a few rank-one summands, if necessary.

6. The monoid of vector bundles

Let $(R, \mathfrak{m}, k)$ be a commutative Noetherian local ring. By a *vector bundle* we mean a finitely generated module M such that M_P is a free R_P-module for each prime ideal $P \neq \mathfrak{m}$. We denote by R-mod the category of finitely generated R-modules and by $\mathcal{F}(R)$ the full subcategory of vector bundles. Our goal is to obtain, in Theorem 6.3, a complete set of invariants for the monoid $\mathsf{V}(\mathcal{F}(R))$ of isomorphism classes of modules in $\mathcal{F}(R)$ when $\dim(R) = 1$, where the monoid operation is given by the direct sum. (Of course $\mathcal{F}(R) = R$-mod if each R_P is a field, e.g., if R is reduced and one-dimensional, or if R is the non-Cohen-Macaulay ring given in Lemma 5.1.) The description of these monoids was worked out in [6] for the case of a one-dimensional Cohen-Macaulay ring. We will see here that the same results hold in the one-dimensional non-Cohen-Macaulay case, thanks to our Main Theorem. We refer the reader to [6, Section 1] for the relevant terminology and basic results concerning Krull monoids, divisor homomorphisms, and the class group $\mathrm{Cl}(H)$ of a Krull monoid H.

Suppose now that $(R, \mathfrak{m}, k)$ is a one-dimensional commutative Noetherian local ring. Let $\widehat{R}$ denote the $\mathfrak{m}$-adic completion of R. The Krull-Schmidt theorem implies that $\mathsf{V}(\widehat{R}\text{-mod})$ and $\mathsf{V}(\mathcal{F}(\widehat{R}))$ are free monoids, with bases consisting of the isomorphism classes of the indecomposables. In other words, $\mathsf{V}(\mathcal{F}(\widehat{R})) \cong \mathbb{N}^{(\tau)}$, the direct sum of τ copies of the additive monoid $\mathbb{N}$ of non-negative integers, where τ is the number of isomorphism classes of indecomposable vector bundles over $\widehat{R}$. It is easy to see that if M is a finitely generated R-module, then M is a vector bundle if and only if $\widehat{R} \otimes_R M$ is a vector bundle. (This follows from the faithful flatness of $R_P \to \widehat{R}_{Q_1} \times \cdots \times \widehat{R}_{Q_t}$, where P is a minimal prime of R and the Q_j are the primes of $\widehat{R}$ lying over P.) Thus the divisor homomorphism [6, Section 1.1] $\mathsf{V}(R\text{-mod}) \to \mathsf{V}(\widehat{R}\text{-mod})$ taking $[M]$ to $[\widehat{R} \otimes_R M]$ restricts to a divisor homomorphism $\mathsf{V}(\mathcal{F}(R)) \to \mathsf{V}(\mathcal{F}(\widehat{R}))$. In particular, we can regard $\mathsf{V}(\mathcal{F}(R))$ as a submonoid of $\mathsf{V}(\mathcal{F}(\widehat{R}))$. The key is to understand exactly how $\mathsf{V}(\mathcal{F}(R))$ sits inside $\mathsf{V}(\mathcal{F}(\widehat{R}))$, that is, which modules over the $\mathfrak{m}$-adic completion $\widehat{R}$ are extended from R-modules.

PROPOSITION 6.1. *Let $(R, \mathfrak{m}, k)$ be a one-dimensional commutative Noetherian local ring with $\mathfrak{m}$-adic completion $\widehat{R}$, and let N be a vector bundle over $\widehat{R}$. Then $N \cong \widehat{R} \otimes_R M$ for some R-module M (necessarily a vector bundle) if and only if $\operatorname{rank}_{R_P}(N_P) = \operatorname{rank}_{R_Q}(N_Q)$ whenever P and Q are minimal prime ideals of $\widehat{R}$ with $P \cap R = Q \cap R$.*

Proof. The "only if" direction is clear. For the converse, let $P_1, \ldots, P_s$ be the minimal prime ideals of R, and, for each i, let n_i be the rank of N at the primes lying over P_i. Let $K = R_{P_1} \times \cdots \times R_{P_s}$, and let V be the projective K-module having rank n_i at P_i. The $K \otimes_R \widehat{R}$-module $K \otimes_R N$ is extended from the K-module V, and now [19, Theorem 3.4] implies that N is extended from an R-module. □

The next result puts an upper bound on the number of non-isomorphic vector bundles. The case of a Cohen-Macaulay ring is [6, Lemma 2.3].

PROPOSITION 6.2. *Let $(R, \mathfrak{m}, k)$ be a one-dimensional commutative Noetherian local ring. Then $|\mathsf{V}(\mathcal{F}(R))| \leq |k| \cdot \aleph_0$.*

Proof. We observe, as in the first paragraph of the proof of [6, Lemma 2.3], that each finite-length module has cardinality at most $\tau := |k| \cdot \aleph_0$, and that there are at most τ isomorphism classes of finite-length R-modules.

Let $P_1, \ldots, P_s$ be the minimal prime ideals of R. Fix a vector bundle M, and let n_i be the rank of M at P_i. Since there are only countably many sequences $(n_1, \ldots, n_s)$, it will suffice to show that there are at most τ non-isomorphic vector bundles with the same ranks as M at the minimal primes.

Let $K = R_{P_1} \times \cdots \times R_{P_s}$, the localization of R at the complement of the union of the minimal prime ideals. Given a vector bundle N with $\operatorname{rank}_{P_i}(N_i) = n_i$ for each i, one can choose a homomorphism $\varphi : M \to N$ such that $1_K \otimes_R \varphi$ is an isomorphism. Then $U := \ker(\varphi)$ and $V := \operatorname{coker}(\varphi)$ have finite length. Put $W := \operatorname{im}(\varphi) \cong \operatorname{coker}(U \hookrightarrow M)$. By the first paragraph, there are at most τ choices for U and V. Also, for each U, $\operatorname{Hom}_R(U, M)$ has finite length and therefore has cardinality at most τ. Therefore there are at most τ possibilities for W. Finally, the exact sequence $0 \to W \to N \to V \to 0$ and the fact that $\operatorname{Ext}^1_R(V, W)$ has finite length, and hence cardinality bounded by τ, show that there are at most τ possibilities for N. □

Fix a positive integer q and an infinite cardinal τ. Let B be any $q \times \tau$ integer matrix such that each element of $\mathbb{Z}^{(q)}$ occurs τ times as a column of B. We let $\mathfrak{H}(q, \tau) := \mathbb{N}^{(\tau)} \cap \ker(B : \mathbb{Z}^{(\tau)} \to \mathbb{Z}^{(q)})$, where $\mathbb{N}$ denotes the set of non-negative integers. Finally, we put $\mathfrak{H}(0, \tau) = \mathbb{N}^{(\tau)}$. These are the monoids we will obtain as $\mathsf{V}(R\text{-mod})$ for the rings that are not Dedekind-like. Not surprisingly, the isomorphism class of the monoid $H(q,\tau)$ does not depend on how the columns of B are arranged, as long as each column is repeated τ times. (Cf. [6, Lemmas 1.1 and 2.1].)

For some Dedekind-like rings, we will obtain a different monoid. Let E be the $1\times\aleph_0$ matrix $[1 \quad -1 \quad 1 \quad -1 \quad 1 \quad -1 \quad \cdots]$, and put $\mathfrak{H}_1 := \mathbb{N}^{(\aleph_0)} \cap \ker(E : \mathbb{Z}^{(\aleph_0)} \to \mathbb{Z})$.

For a one-dimensional local ring $(R, \mathfrak{m}, k)$, we define the *splitting number* $\mathrm{spl}(R)$ to be the difference $|\operatorname{Spec}(\widehat{R})| - |\operatorname{Spec}(R)|$. Thus, for example, $\mathrm{spl}(R) = 0$ means that the natural map $\operatorname{Spec}(\widehat{R}) \to \operatorname{Spec}(R)$ is bijective.

We can now state the main theorem of this section. For the proof, we refer the reader to the proof of [6, Theorem 2.2]. The only modification needed to eliminate the Cohen-Macaulay hypothesis is to replace Lemmas 2.3, 2.4 and 2.5 in [6] by, respectively, Proposition 6.2, Theorem 1.2 and Proposition 6.1 of this paper.

THEOREM 6.3. *Suppose $(R, \mathfrak{m}, k)$ is a one-dimensional commutative Noetherian local ring. Let $q := \mathrm{spl}(R)$ be the splitting number of R, and let $\tau = \tau(R) = |k| \cdot \aleph_0$.*

(1) *If R is not Dedekind-like, then $\mathsf{V}(\mathfrak{F}(R)) \cong \mathfrak{H}(q,\tau)$.*
(2) *If R is a discrete valuation ring, then $\mathsf{V}(\mathfrak{F}(R)) = \mathsf{V}(R\text{-mod}) \cong \mathbb{N}^{(\aleph_0)}$.*
(3) *If R is Dedekind-like but not a discrete valuation ring, and if $q = 0$, then $\mathsf{V}(\mathfrak{F}(R)) = \mathsf{V}(R\text{-mod}) \cong \mathbb{N}^{(\tau)}$.*
(4) *If R is Dedekind-like and $q > 0$, then $q = 1$ and $\mathsf{V}(\mathfrak{F}(R)) = \mathsf{V}(R\text{-mod}) \cong \mathbb{N}^{(\tau)} \oplus \mathfrak{H}_1$.*

In every case, $\mathrm{Cl}(\mathsf{V}(\mathfrak{F}(R))) \cong \mathbb{Z}^{(q)}$.

We remark that the yet-unpublished results on the structure of modules over exceptional Dedekind-like rings have no bearing on the validity of this theorem: If R is an exceptional Dedekind-like ring, then $\mathrm{spl}(R) = 0$, and hence all that is needed is the straightforward construction of $\tau(R)$ indecomposable modules over R, given in [6, Lemma 2.6].

7. Non-local rings

In this section only, we do not assume that Dedekind-like rings are necessarily local, calling the commutative, Noetherian ring R a *(global) Dedekind-like ring* if, for each maximal ideal $\mathfrak{m}$ of R, the localization $R_{\mathfrak{m}}$ is a (local) Dedekind-like ring [16, Corollary 10.7]. If R is a (global) Dedekind-like ring such that none of the localizations of R is exceptional, and if M is a finitely generated indecomposable R-module, then the rank of M_P is at most two for every minimal prime P of R [16, Corollary 16.9]. In this section, we prove that this result fails if at least one of the localizations of R is not a homomorphic image of a Dedekind-like ring.

THEOREM 7.1. *Let R be a connected, commutative, Noetherian ring, and suppose that R is not a homomorphic image of a (global) Dedekind-like ring. Then, for every integer $n \geq 1$, there exist infinitely many indecomposable finitely generated R-modules M such that $M_P \cong R_P^{(n)}$ for each minimal prime P of R.*

Proof. We begin by fixing a maximal ideal $\mathfrak{m}$ of R such that $R_\mathfrak{m}$ is not a homomorphic image of a (local) Dedekind-like ring. If R has dimension greater than one, then we can take $\mathfrak{m}$ to be any maximal ideal of height greater than one, since (local) Dedekind-like rings have dimension one. If R has dimension one, then the existence of such a maximal ideal $\mathfrak{m}$ follows immediately from [16, Proposition 14.1 and Corollary 13.6].

Note that a Noetherian ring A is connected if and only if, for every non-empty, proper subset $\mathcal{V}$ of the set of minimal prime ideals of A, there exist a maximal ideal $\mathfrak{m}_\mathcal{V}$ of A and minimal primes $P \in \mathcal{V}$ and $Q \notin \mathcal{V}$ such that $P + Q \subseteq \mathfrak{m}_\mathcal{V}$. Thus we can find a finite list $\mathfrak{m}_1 = \mathfrak{m}, \mathfrak{m}_2, \dots, \mathfrak{m}_t$ of maximal ideals of R such that each minimal prime of R is contained in at least one maximal ideal in the list, and such that, for every non-empty, proper subset $\mathcal{V}$ of the set of minimal prime ideals of R, there are minimal primes $P \in \mathcal{V}$ and $Q \notin \mathcal{V}$ such that $P, Q \subseteq \mathfrak{m}_i$ for some index i. Therefore, if we set $S := R - \bigcup_{i=1}^t \mathfrak{m}_i$, it follows that the localization $S^{-1}R$ is connected, with minimal primes precisely the localization of the minimal primes of R.

Suppose that we can find a finitely generated indecomposable $S^{-1}R$-module M such that $M_{S^{-1}P} \cong (S^{-1}R)^{(n)}_{S^{-1}P}$ for each minimal prime P of R. Let N be a finitely generated R-module such that $S^{-1}N \cong M$, and let $N = N_1 \oplus \cdots \oplus N_k$ be a decomposition of N into indecomposable R-modules. Since $S^{-1}N$ is indecomposable, we have $S^{-1}N_i = 0$ for all except one index i, and hence $S^{-1}N_i = M$. Then N_i is an indecomposable R-module such that $(N_i)_P \cong M_{S^{-1}P} \cong (S^{-1}R)^{(n)}_{S^{-1}P} \cong R_P^{(n)}$ for each minimal prime P of R, and the theorem is proved.

Therefore it suffices to prove the theorem under the additional hypothesis that R be semilocal. Let $\mathfrak{m}_1, \dots, \mathfrak{m}_t$ be the maximal ideals of R, where $R_{\mathfrak{m}_1}$ is not a homomorphic image of a Dedekind-like ring. Further, suppose $\mathfrak{m}_1$ has height greater than one if $\dim R > 1$. We distinguish the two cases in which R has dimension one or dimension greater than one.

Suppose first that R has dimension one. Let M_1 be an indecomposable $R_{\mathfrak{m}_1}$-module with constant rank n at the minimal primes of R contained in $\mathfrak{m}_1$ (Theorem 1.2); for $2 \leq j \leq t$, let $M_j = R_{\mathfrak{m}_j}^{(n)}$. Since R has only finitely many prime ideals, there exists, by [25, Lemma 1.11], an R-module M such that $M_{\mathfrak{m}_j} \cong M_j$ for all $j = 1, \dots, t$. If $M = U \oplus V$, then, since $M_{\mathfrak{m}_1}$ is indecomposable, we can assume that $U_{\mathfrak{m}_1} = 0$. Since $M_{\mathfrak{m}_j}$ is $R_{\mathfrak{m}_j}$-free for $2 \leq j \leq t$, $U_{\mathfrak{m}_j}$ is $R_{\mathfrak{m}_j}$-free for all $j = 1, \dots, t$, and it follows that U is R-projective.

Since R is connected and $U_{\mathfrak{m}_1} = 0$, it follows that $U = 0$. This shows that M is indecomposable. Since Theorem 1.2 produces infinitely many pairwise non-isomorphic indecomposable $R_{\mathfrak{m}_1}$-modules locally of constant rank n at the minimal primes of $R_{\mathfrak{m}_1}$, the theorem is proved in case R has dimension one.

Suppose instead that R has dimension greater than one, so that $\mathfrak{m}_1$ is a maximal ideal of height greater than one. Thus, either the maximal ideal of $R_{\mathfrak{m}_1}$ requires three or more generators, or $R_{\mathfrak{m}_1}$ is a regular local ring of dimension two, and the square of its maximal ideal requires three generators. Either way, let r be a positive integer such that $\mathfrak{m}_1^r/\mathfrak{m}_1^{r+1}$ is a vector space of dimension at least three over the residue field $R/\mathfrak{m}_1$. We adapt Proposition 2.1 to construct R-modules directly.

Let $\mathcal{P}$ be the set consisting of the minimal primes of R together with the remaining maximal ideals $\mathfrak{m}_2, \dots, \mathfrak{m}_t$, and choose x, y, and z as in the first sentence of the proof of Proposition 2.1, where $\mathfrak{m} = \mathfrak{m}_1$. As in that proof, given any integer $q > n$, set $\Delta := (z + y)I_q + yH_q$, and let

$$\Xi := \begin{bmatrix} 0_n & 0 \\ 0 & x^2 I_{q-n} \end{bmatrix} \in \mathrm{Mat}_{q\times q}(R).$$

Let A be the $2q \times 2q$ matrix over R defined by (1), and set $M := \mathrm{coker}(A)$. Since the images of x, y and z, in $\mathfrak{m}_1^r/\mathfrak{m}_1^{r+1}$, are linearly independent over $R/\mathfrak{m}_1$, while the image of x^2 in $\mathfrak{m}_1^r/\mathfrak{m}_1^{r+1}$ is 0, the proof of Proposition 2.1 shows that $M_{\mathfrak{m}_1}$ is indecomposable. Moreover, for $P \in \mathcal{P}$, localizing at P yields a matrix $\tilde{A}$ which is equivalent to $I_{2q-n} \oplus 0_n$ (because x, y, and z become units in R_P), and hence $M_P \cong R_P^{(n)}$.

To show that M is indecomposable, suppose $M = U \oplus V$. Since $M_{\mathfrak{m}_1}$ is indecomposable, we can assume that $U_{\mathfrak{m}_1} = 0$. For $2 \le j \le t$, $U_{\mathfrak{m}_j}$ is a direct summand of the free $R_{\mathfrak{m}_j}$-module $M_{\mathfrak{m}_j}$ and thus is free. Therefore U is R-projective; since $U_{\mathfrak{m}_1} = 0$ and R is connected, U must be zero. Thus M is indecomposable. As noted in the proof of Proposition 2.1, the localization $M_{\mathfrak{m}_1}$ of the R-module M just constructed requires exactly $2q$ generators as an $R_{\mathfrak{m}_1}$-module. Thus, by varying $q > n$, we get infinitely many pairwise non-isomorphic indecomposable R-modules locally of constant rank n at the minimal primes of R. □

We leave to the reader the minor adjustments required to obtain $|k| \cdot \aleph_0$ pairwise non-isomorphic indecomposable R modules of constant rank n, where k is the residue field at the maximal ideal $\mathfrak{m}_1$. One might be able to extend Theorem 7.1 to allow for *some* non-constant ranks at the minimal primes, but it is doubtful that one can obtain *arbitrary* ranks at the minimal primes. For example, if R has dimension one, two maximal ideals $\mathfrak{m}_1$ and $\mathfrak{m}_2$, and three minimal primes P_0, P_1, and P_2, such that $P_0, P_1 \subseteq \mathfrak{m}_1$ and $P_1, P_2 \subseteq \mathfrak{m}_2$, but $P_0 \not\subseteq \mathfrak{m}_2$ and $P_2 \not\subseteq \mathfrak{m}_1$, then it is not clear that there exists an indecomposable module M of rank one at P_0 and P_2 but rank zero at P_1. Moreover, in dimension greater than one, R. Wiegand's "gluing lemma" [25, Lemma 1.11] does not apply, and it is difficult to imagine how to construct a module with arbitrary localizations at finitely many maximal ideals.

References

[1] H. Bass, *On the ubiquity of Gorenstein rings*, Math. Z. **82** (1963), 8–28. MR 0153708 (27 #3669)

[2] W. Bruns and J. Herzog, *Cohen-Macaulay rings*, Cambridge Studies in Advanced Mathematics, vol. 39, Cambridge University Press, Cambridge, 1993. MR 1251956 (95h:13020)

[3] J. Dieudonné, *Sur la réduction canonique des couples de matrices*, Bull. Soc. Math. France **74** (1946), 130–146. MR 0022826 (9,264f)

[4] Yu. A. Drozd, *Representations of commutative algebras (Russian)*, Funktsional. Anal. i Priložhen. **6** (1972), 41–43; English Transl., Funct. Anal. Appl. **6** (1972), 286–288. MR 0311718 (47 #280)

[5] D. Eisenbud, *Homological algebra on a complete intersection, with an application to group representations*, Trans. Amer. Math. Soc. **260** (1980), 35–64. MR 570778 (82d:13013)

[6] A. Facchini, W. Hassler, L. Klingler and R. Wiegand, *Direct-sum decompositions over one-dimensional Cohen-Macaulay rings*, Multiplicative ideal theory in commutative algebra: a tribute to the work of Robert Gilmer (J. Brewer, S. Glaz, W. Heinzer, B. Olberding, eds.), Springer, New York, 2006, pp. 153–168. MR 2265807

[7] A. Facchini, *Module theory. Endomorphism rings and direct sum decompositions in some classes of modules.*, Progress in Mathematics, vol. 167, Birkhäuser Verlag, Basel, 1998. MR 1634015 (99h:16004)

[8] J. S. Golan, *Torsion theories*, Pitman Monographs and Surveys in Pure and Applied Mathematics, vol. 29, Longman Scientific & Technical, Harlow, 1986. MR 880019 (88c:16034)

[9] W. Hassler, R. Karr, L. Klingler, and R. Wiegand. *Indecomposable modules of large rank over Cohen-Macaulay local rings*, Trans. Amer. Math. Soc., to appear.

[10] ———, *Large indecomposable modules over local rings*, J. Algebra **303** (2006), 202–215. MR 2253659

[11] A. Heller and I. Reiner, *Indecomposable representations*, Illinois J. Math. **5** (1961), 314–323. MR 0122890 (23 #A222)

[12] D. G. Higman, *Indecomposable representations at characteristic p*, Duke Math. J. **21** (1954), 377–381. MR 0067896 (16,794c)

[13] C. Huneke and R. Wiegand, *Tensor products of modules and the rigidity of* Tor, Math. Ann. **299** (1994), 449–476. MR 1282227 (95m:13008)

[14] L. Klingler and L. S. Levy, *Representation type of commutative Noetherian rings. I. Local wildness*, Pacific J. Math. **200** (2001), 345–386. MR 1868696 (2002i:13008a)

[15] ———, *Representation type of commutative Noetherian rings. II. Local tameness*, Pacific J. Math. **200** (2001), 387–483. MR 1868697 (2002i:13008b)

[16] ———, *Representation type of commutative Noetherian rings. III. Global wildness and tameness*, Mem. Amer. Math. Soc. **176** (2005). MR 2147090 (2006g:13037)

[17] L. Kronecker. *Über die congruenten Transformationen der bilinearen Formen*, Monatsberichte Königl. Preuß. Akad. Wiss. Berlin (1874), 397–447; reprinted in: Leopold Kronecker's Werke (K. Hensel, Ed.), Vol. 1, pp. 423–483, Chelsea, New York, 1968.

[18] G. J. Leuschke and R. Wiegand, *Hypersurfaces of bounded Cohen-Macaulay type*, J. Pure Appl. Algebra **201** (2005), 204–217. MR 2158755 (2006c:13014)

[19] L. S. Levy and C. J. Odenthal, *Package deal theorems and splitting orders in dimension 1*, Trans. Amer. Math. Soc. **348** (1996), 3457–3503. MR 1351493 (96m:16006b)
[20] H. Matsumura, *Commutative ring theory*, Cambridge Studies in Advanced Mathematics, vol. 8, Cambridge University Press, Cambridge, 1986. MR 879273 (88h:13001)
[21] C. M. Ringel, *The representation type of local algebras*, Lecture Notes in Mathematics, vol. 488, Springer-Verlag, New York, 1975, pp. 282–305.
[22] D. E. Rush, *Rings with two-generated ideals*, J. Pure Appl. Algebra **73** (1991), 257–275. MR 1124788 (92j:13008)
[23] R. B. Warfield, Jr., *Decomposability of finitely presented modules*, Proc. Amer. Math. Soc. **25** (1970), 167–172. MR 0254030 (40 #7243)
[24] K. Weierstrass, *Zur Theorie der bilinearen und quadratischen Formen*, Monatsberichte Königl. Preuß. Akad. Wiss. Berlin (1868), 310–338.
[25] R. Wiegand, *Noetherian rings of bounded representation type*, Commutative algebra (Berkeley, CA, 1987), Math. Sci. Res. Inst. Publ., vol. 15, Springer, New York, 1989, pp. 497–516. MR 1015536 (90i:13010)
[26] Y. Yoshino, *Cohen-Macaulay modules over Cohen-Macaulay rings*, London Mathematical Society Lecture Note Series, vol. 146, Cambridge University Press, Cambridge, 1990. MR 1079937 (92b:13016)

Wolfgang Hassler, Institut für Mathematik und Wissenschaftliches Rechnen, Karl-Franzens-Universität Graz, Heinrichstrasse 36/IV, A-8010 Graz, Austria
E-mail address: wolfgang.hassler@uni-graz.at

Ryan Karr, Honors College, Florida Atlantic University, Jupiter, FL 33458, USA
E-mail address: rkarr@fau.edu

Lee Klingler, Department of Mathematical Sciences, Florida Atlantic University, Boca Raton, FL 33431-6498, USA
E-mail address: klingler@fau.edu

Roger Wiegand, Department of Mathematics, University of Nebraska, Lincoln, NE 68588-0323, USA
E-mail address: rwiegand@math.unl.edu

Illinois Journal of Mathematics
Volume 51, Number 1, Spring 2007, Pages 123–136
S 0019-2082

EXTENSIONS OF LOCAL DOMAINS WITH TRIVIAL GENERIC FIBER

WILLIAM HEINZER, CHRISTEL ROTTHAUS, AND SYLVIA WIEGAND

Dedicated to Phil Griffith, in honor of his contributions to commutative algebra

ABSTRACT. We consider injective local maps from a local domain R to a local domain S such that the generic fiber of the inclusion map $R \hookrightarrow S$ is trivial, that is, $P \cap R \neq (0)$ for every nonzero prime ideal P of S. We present several examples of injective local maps involving power series that have or fail to have this property. For an extension $R \hookrightarrow S$ having this property, we give some results on the dimension of S; in some cases we show $\dim S = 2$ and in some cases $\dim S = 1$.

1. Introduction and background

Our work in this paper originates with the following question raised by Melvin Hochster and Yongwei Yao.

QUESTION 1.1. Let R be a complete local domain. Can one describe or somehow classify the injective local maps of R to a complete local domain S such that $U^{-1}S$ is a field, where $U = R \setminus (0)$, i.e., such that the generic fiber of $R \hookrightarrow S$ is trivial?

By Cohen's structure theorems [4], [15, (31.6)], a complete local domain R is a finite integral extension of a complete regular local domain R_0. If R has the same characteristic as its residue field, then R_0 is a formal power series ring over a field. The generic fiber of $R \hookrightarrow S$ is trivial if and only if the generic fiber of $R_0 \hookrightarrow S$ is trivial. Thus as Hochster and Yao remark: if R is equal characteristic zero one obtains extensions as in Question 1.1 by starting with

$$R_0 := K[[x_1, \dots, x_n]] \hookrightarrow T := L[[x_1, \dots, x_n, y_1, \dots, y_m]],$$

where K is a subfield of L and the x_i, y_j are formal indeterminates. Let P be a prime ideal of T maximal with respect to being disjoint from the image

Received August 2, 2006; received in final form March 20, 2007.
2000 *Mathematics Subject Classification.* Primary 13A15, 13B02, 13F25, 13J05.
The authors are grateful for the hospitality and cooperation of Michigan State, Nebraska and Purdue, where several work sessions on this research were conducted.

of $R_0 \setminus \{0\}$. Then the composite map $R_0 \hookrightarrow T \to T/P =: S$ is an extension of this type. Of course, such prime ideals P are maximal in the generic fiber $(R_0 \setminus \{0\})^{-1}T$ of the embedding $R_0 \hookrightarrow T$.

In [11], we study the generic fiber of extensions of power series rings over the same base field. With $R = K[[x_1, \dots, x_n]]$ as above and $T = R[[y_1, \dots, y_m]]$, we show in [11, Theorem 7.2] that, if P is maximal in the generic fiber of $R \hookrightarrow T$ and $S = T/P$, then $\dim S$ is either 2 or n. This answers Question 1.1 in the case where $R = K[[x_1, \dots, x_n]]$ is a complete regular local domain with coefficient field K and S is a complete local domain that also has coefficient field K.

DEFINITION 1.2. If $R \hookrightarrow S$ is an injective map of integral domains, we say that S is a *trivial generic fiber extension*, *TGF extension*, of R if each nonzero ideal of S has a nonzero intersection with R, or equivalently, if each nonzero element of S has a nonzero multiple in R. Since ideals of S maximal with respect to not meeting the multiplicative system of nonzero elements of R are prime ideals, S is a TGF extension of R if and only if $P \cap R \neq (0)$ for each nonzero prime ideal P of S. Another condition equivalent to S is a TGF extension of R is that $U^{-1}S$ is a field, where $U = R \setminus (0)$.

Let $(R, \mathbf{m}) \hookrightarrow (S, \mathbf{n})$ be an injective local homomorphism of complete local domains, so that $\mathbf{n} \cap R = \mathbf{m}$. We say that S is a *TGF-complete extension* of R if S is a TGF extension of R.

In [12] we consider the TGF property for extensions of mixed polynomial/power series rings over the same base field and we partially characterize the prime ideal spectra of such rings. For example, we consider the nested mixed polynomial/power series rings

$$(1.1)\quad A := k[x,y] \hookrightarrow B := k[[y]]\,[x] \hookrightarrow C := k[x]\,[[y]] \hookrightarrow E := k[x, 1/x]\,[[y]],$$

$$(1.2)\quad C \hookrightarrow D_1 := k[x]\,[[y/x]] \hookrightarrow \cdots \hookrightarrow D_n := k[x]\,[[y/x^n]] \hookrightarrow \cdots \hookrightarrow E,$$

where k is a field and x and y are indeterminates over k. In Sequence (1.1) the maps are all flat. In Sequence (1.2), for n a positive integer, the map $C \hookrightarrow D_n$ is not flat, but $D_n \hookrightarrow E$ is a localization followed by an adic completion of a Noetherian ring and therefore is flat. All of the extensions in (1.1) and (1.2) except those that begin with A are TGF. The extensions that begin with A are not TGF. In dimension 3 we consider in [12] embeddings such as

$$k[x,y,z] \overset{\alpha}{\hookrightarrow} k[[z]]\,[x,y] \overset{\beta}{\hookrightarrow} k[x]\,[[z]]\,[y] \overset{\gamma}{\hookrightarrow} k[x,y]\,[[z]] \overset{\delta}{\hookrightarrow} k[x]\,[[y,z]],$$

$$k[[z]]\,[x,y] \overset{\epsilon}{\hookrightarrow} k[[y,z]]\,[x] \overset{\zeta}{\hookrightarrow} k[x]\,[[y,z]] \overset{\eta}{\hookrightarrow} k[[x,y,z]],$$

where k is a field and x, y and z are indeterminates over k. Here all of the proper inclusions fail to be TGF. Takehiko Yasuda in [18] gives additional information on the TGF property. In particular, he shows in [18, Theorem

2.7] that

$$\mathbb{C}[x,y][[z]] \hookrightarrow \mathbb{C}[x,x^{-1},y][[z]]$$

is not TGF, where $\mathbb{C}$ is the field of complex numbers.

In this article we discuss several additional topics and questions related to Question 1.1 and the TGF property. In Section 2 we record several basic facts about TGF extensions. We prove in Proposition 2.6 that if $A \hookrightarrow B$ is a TGF extension, where B is a Noetherian integral domain, then $\dim A \geq \dim B$.

We prove in Corollary 3.3 that if $(A, \mathbf{m}) \hookrightarrow (S, \mathbf{n})$ is a TGF-complete extension, where A is equicharacteristic with $\dim A = n \geq 2$ and $S/\mathbf{n}$ finite algebraic over $A/\mathbf{m}$, then either $\dim S = n$ and S is a finite integral extension of A or $\dim S = 2$. We also include in Section 3 other remarks concerning TGF-complete extensions having finite residue field extension. For each $n \geq 2$ and $R = k[[X]]$ a formal power series ring in n variables over a field k, we describe in (3.5) a TGF-complete extension $R \hookrightarrow S$, where S is a power series ring in 2 variables over k.

In Section 4 we consider a TGF-complete extension $(R, \mathbf{m}) \hookrightarrow (S, \mathbf{n})$, where $S/\mathbf{n}$ is transcendental over $R/\mathbf{m}$. We address, but do not resolve, the question of whether in this situation $\dim S \leq 1$. We prove in Theorem 4.8 that if $(A, \mathbf{m}) \hookrightarrow (B, \mathbf{n})$ is an injective local homomorphism of 2-dimensional regular local rings such that $B/\mathbf{n}$ as a field extension of $A/\mathbf{m}$ is not algebraic, then $A \hookrightarrow B$ is not TGF. We deduce that for indeterminates x, y, z, w, t over a field k, if $\varphi : R = k[[x,y]] \hookrightarrow S := k(t)[[z,w]]$ is an injective local k-algebra homomorphism, then $\varphi(R) \hookrightarrow S$ is not TGF.

There is much in the literature concerning homomorphisms of formal power series rings; see, for example, the articles of Abhyankar-Moh [2], Matsumura [13], Rotthaus [16].

2. Trivial generic fiber (TGF) extensions, general remarks

We record in Proposition 2.1 several basic facts about TGF extensions. We omit the proofs since they are straightforward.

PROPOSITION 2.1. *Let $R \hookrightarrow S$ and $S \hookrightarrow T$ be injective maps, where R, S and T are integral domains.*

1. *If $R \hookrightarrow S$ and $S \hookrightarrow T$ are TGF extensions, then so is the composite map $R \hookrightarrow T$. Equivalently, if the composite map $R \hookrightarrow T$ is not TGF, then at least one of the extensions $R \hookrightarrow S$ or $S \hookrightarrow T$ is not TGF.*
2. *If $R \hookrightarrow T$ is TGF, then $S \hookrightarrow T$ is TGF.*
3. *If the map* $\operatorname{Spec} T \to \operatorname{Spec} S$ *is surjective and $R \hookrightarrow T$ is TGF, then $R \hookrightarrow S$ is TGF.*

We consider in Proposition 2.2 the relatively easy case where the base ring has dimension one.

PROPOSITION 2.2. *Let $(R, \mathbf{m})$ be a complete one-dimensional local domain. Assume that $(S, \mathbf{n})$ is a TGF-complete extension of R. Then:*

(1) $\dim(S) = 1$ *and* $\mathbf{m}S$ *is* $\mathbf{n}$*-primary.*
(2) *If* $[S/\mathbf{n} : R/\mathbf{m}] < \infty$, *then* S *is a finite integral extension of* R.

Thus, if $R \hookrightarrow S$ *is a TGF-extension with finite residue extension and* $\dim S \geq 2$, *then* $\dim R \geq 2$.

Proof. By Krull's Principal Ideal Theorem [14, Theorem 13.5], $\mathbf{n}$ is the union of the height-one primes of S. If $\dim S > 1$, then S has infinitely many height-one primes. Each nonzero element of $\mathbf{n}$ is contained in only finitely many of these height-one primes. If $\dim S > 1$, then the intersection of the height-one primes of S is zero. Since $\dim R = 1$, every nonzero prime of S contains $\mathbf{m}$. Thus $\dim S = 1$ and $\mathbf{m}S$ is $\mathbf{n}$-primary. Moreover, if $[S/\mathbf{n} : R/\mathbf{m}] < \infty$, then S is finite over R by [14, Theorem 8.4]. □

REMARKS 2.3. (1) Notice that there exist TGF-complete extensions of R that have an arbitrarily large extension of residue field. For example, if k is a subfield of a field F and x is an indeterminate over F, then $R := k[[x]] \subseteq S := F[[x]]$ is a TGF-complete extension.

(2) Let $(R, \mathbf{m}) \hookrightarrow (T, \mathbf{q})$ be an injective local homomorphism of complete local domains. For $P \in \operatorname{Spec} T$, $S := T/P$ is a TGF-complete extension of R if and only if P is an ideal of T maximal with respect to the property that $P \cap R = (0)$.

REMARKS 2.4. Let $X = \{x_1, \ldots, x_n\}$, $Y = \{y_1, \ldots, y_m\}$ and $Z = \{z_1, \ldots, z_r\}$ be algebraically independent finite sets of indeterminates over a field k, where $n \geq 2$, $m, r \geq 1$. Set $R := k[[X]]$ and let P be a prime ideal of $k[[X, Y, Z]]$ that is maximal with respect to $P \cap R = (0)$. Then we have the inclusions

$$R := k[[X]] \overset{\sigma}{\hookrightarrow} S := k[[X, Y]]/(P \cap k[[X, Y]]) \overset{\tau}{\hookrightarrow} T := k[[X, Y, Z]]/P.$$

By Remark 2.3(2), $\tau \cdot \sigma$ is a TGF extension. By Proposition 2.1(2), $S \hookrightarrow T$ is TGF.

(1) If the map $\operatorname{Spec} T \to \operatorname{Spec} S$ is surjective, then $\sigma : R \hookrightarrow S$ is TGF by Proposition 2.1(3).
(2) If $R \hookrightarrow T$ is finite, then $R \hookrightarrow S$ is also finite, and so $\sigma : R \hookrightarrow S$ is TGF.
(3) If $R \hookrightarrow T$ is not finite, then $\dim T = 2$ by [11, Theorem 7.2].
(4) If $P \cap k[[X, Y]] = 0$, then $S = R[[Y]]$ and $R \hookrightarrow S$ is not TGF. (We show in Example 3.10 that this can occur.)

REMARKS AND QUESTION 2.5. (1) With notation as in Remarks 2.4 and with $Y = \{y\}$, a singleton set, it is always true that $\operatorname{ht}(P \cap R[[y]]) \leq n - 1$,

[11, Theorem 7.1]. Moreover, if $\operatorname{ht}(P \cap R[[y]]) = n - 1$, then $R \hookrightarrow S$ is TGF. Thus if $n = 2$ and $P \cap R[[y]] \neq 0$, then $R \hookrightarrow S$ is TGF.

(2) With notation as in (1) and $n = 3$, it can happen that $P \cap k[[X, y]] \neq (0)$ and $R \hookrightarrow R[[y]]/(P \cap R[[y]])$ is not a TGF extension. To construct an example of such a prime ideal P, we proceed as follows: Since $\dim k[[X, y]] = 4$, there exists a prime ideal Q of $k[[X, y]]$ with $\operatorname{ht} Q = 2$ and $Q \cap k[[X]] = (0)$, [11, Theorem 7.1]. Let $\mathbf{p} \subset Q$ be a prime ideal with $\operatorname{ht} \mathbf{p} = 1$. Since $\mathbf{p} \subsetneq Q$ and $Q \cap k[[X]] = (0)$, the extension $k[[X]] \hookrightarrow k[[X, y]]/\mathbf{p}$ is not a TGF extension. In particular, it is not finite. Let $P \in \operatorname{Spec} k[[X, y, Z]]$ be maximal with respect to $P \cap k[[X, y]] = \mathbf{p}$. By Corollary 3.4 below, $\dim k[[X, y, Z]]/P = 2$. Hence P is maximal in the generic fiber over $k[[X]]$.

(3) If $(R, \mathbf{m}) \hookrightarrow (S, \mathbf{n})$ is a TGF-complete extension with $S/\mathbf{n}$ finite algebraic over $R/\mathbf{m}$, can the transcendence degree of S over R be finite but nonzero?

(4) If $(R, \mathbf{m}) \hookrightarrow (S, \mathbf{n})$ is a TGF-complete extension as in (3) with R equicharacteristic and $\dim R \geq 2$, then by Corollary 3.3 below it follows that either S is a finite integral extension of R or $\dim S = 2$.

PROPOSITION 2.6. *Let $A \hookrightarrow B$ be a TGF extension, where B is a Noetherian integral domain. For each $Q \in \operatorname{Spec} B$, we have $\operatorname{ht} Q \leq \operatorname{ht}(Q \cap A)$. In particular, $\dim A \geq \dim B$.*

Proof. If $\operatorname{ht} Q = 1$, it is clear that $\operatorname{ht} Q \leq \operatorname{ht}(Q \cap A)$ since $Q \cap A \neq (0)$. Let $\operatorname{ht} Q = n \geq 2$, and assume by induction that $\operatorname{ht} Q' \leq \operatorname{ht}(Q' \cap A)$ for each $Q' \in \operatorname{Spec} B$ with $\operatorname{ht} Q' \leq n - 1$. Since B is Noetherian,

$$(0) = \bigcap \{Q' \mid Q' \subset Q \quad \text{and} \quad \operatorname{ht} Q' = n - 1\}.$$

Hence there exists $Q' \subset Q$ with $\operatorname{ht} Q' = n - 1$ and $Q' \cap A \subsetneq Q \cap A$. We have $n - 1 \leq \operatorname{ht}(Q' \cap A) < \operatorname{ht}(Q \cap A)$, so $\operatorname{ht}(Q \cap A) \geq n$. □

3. TGF-complete extensions with finite residue field extension

SETTING 3.1. Let $n \geq 2$ be an integer, let $X = \{x_1, \ldots, x_n\}$ be a set of independent variables over the field k and let $R = k[[X]]$ be the formal power series ring in n variables over the field k.

THEOREM 3.2. *Let $R = k[[X]]$ be as in Setting 3.1. Assume that $R \hookrightarrow S$ is a TGF-complete extension, where $(S, \mathbf{n})$ is a complete Noetherian local domain and $S/\mathbf{n}$ is finite algebraic over k. Then either $\dim S = n$ and S is a finite integral extension of R or $\dim S = 2$.*

Proof. It is clear that if S is a finite integral extension of R, then $\dim S = n$. Assume S is not a finite integral extension of R. Let $b_1, \ldots, b_m \in \mathbf{n}$ be such that $\mathbf{n} = (b_1, \ldots, b_m)S$, and let $Y = \{y_1, \ldots, y_m\}$ be a set of independent variables over R. Since S is complete the R-algebra homomorphism $\varphi : T :=$

$R[[Y]] \to S$ such that $\varphi(y_i) = b_i$ for each i with $1 \le i \le m$ is well defined. Let $Q = \ker \varphi$. We have

$$R \hookrightarrow T/Q \hookrightarrow S.$$

By [14, Theorem 8.4], S is a finite module over T/Q. Hence $\dim S = \dim(T/Q)$ and the map $\operatorname{Spec} S \to \operatorname{Spec} T/Q$ is surjective, so by Proposition 2.1(3), $R \hookrightarrow T/Q$ is TGF. By [11, Theorem 7.2], $\dim(T/Q) = 2$, so $\dim S = 2$. □

COROLLARY 3.3. *Let $(A, \mathbf{m})$ and $(S, \mathbf{n})$ be complete equicharacteristic local domains with $\dim A = n \ge 2$ and suppose that $A \hookrightarrow S$ is a local injective homomorphism and that the residue field $S/\mathbf{n}$ is finite algebraic over the residue field $A/\mathbf{m} := k$. If $A \hookrightarrow S$ is a TGF-complete extension, then either $\dim S = n$ and S is a finite integral extension of A or $\dim S = 2$.*

Proof. By [14, Theorem 29.4(3)], A is a finite integral extension of $R = k[[X]]$, where X is as in Setting 3.1. We have $R \hookrightarrow A \hookrightarrow S$. By Proposition 2.1(1), $R \hookrightarrow S$ is TGF. By Theorem 3.2, either $\dim S = n$ and S is a finite integral extension of A or $\dim S = 2$. □

For example, if $R = k[[x_1, \dots, x_4]]$ and $S = k[[y_1, y_2, y_3]]$, then every k-algebra embedding $R \hookrightarrow S$ fails to be TGF.

COROLLARY 3.4. *Let $R = k[[X]]$ be as in Setting 3.1. Let $Y = \{y_1, \dots, y_m\}$ be a set of m independent variables over R and let $S = R[[Y]]$. If $P \in \operatorname{Spec} R$ is such that $\dim R/P \ge 2$ and $Q \in \operatorname{Spec} S$ is maximal with respect to $Q \cap R = P$, then either*

(i) *$\dim S/Q = 2$, or*
(ii) *$R/P \hookrightarrow S/Q$ is a finite integral extension (and so $\dim R/P = \dim S/Q$).*

Proof. Let $A := R/P \hookrightarrow S/Q =: B$, and apply Corollary 3.3. □

GENERAL EXAMPLE 3.5. It is known that, for each positive integer n, the power series ring $R = k[[x_1, \dots, x_n]]$ in n variables over a field k can be embedded into a power series ring in two variables over k. The construction is based on the fact that the power series ring $k[[z]]$ in the single variable z contains an infinite set of algebraically independent elements over k. Let $\{f_i\}_{i=1}^{\infty} \subset k[[z]]$ with $f_1 \ne 0$ and $\{f_i\}_{i=2}^{\infty}$ algebraically independent over $k(f_1)$. Let $(S := k[[z, w]], \mathbf{n} := (z, w))$ be the formal power series ring in the two variables z, w. Fix a positive integer n and consider the subring $R_n := k[[f_1 w, \dots, f_n w]]$ of S with maximal ideal $\mathbf{m}_n = (f_1 w, \dots, f_n w)$. Let $x_1, \dots, x_n$ be new indeterminates over k and define a k-algebra homomorphism $\varphi : k[[x_1, \dots, x_n]] \to R_n$ by setting $\varphi(x_i) = f_i w$ for $i = 1, \dots, n$.

CLAIM 3.6. *(cf. [19, pp. 219-220]) φ is an isomorphism.*

Proof. Suppose $g = \sum_{m=0}^{\infty} g_m$, where g_m is a form of degree m in $k[x_1, \dots, x_n]$. Then

$$\varphi(g) = \sum_{m=0}^{\infty} \varphi(g_m) \quad \text{and} \quad \varphi(g_m) = g_m(f_1 w, \dots, f_n w) = w^m g_m(f_1, \dots, f_n),$$

where $g_m(f_1, \dots, f_n) \in k[[z]]$. If $\varphi(g) = 0$, then $g_m(f_1, \dots, f_n) = 0$ for each m. Thus

$$0 = g_m(f_1, \dots, f_n) = \sum_{i_1+\cdots+i_n=m} a_{i_1,\dots,i_n} f_1^{i_1} \cdots f_n^{i_n},$$

where the $a_{i_1,\dots,i_n} \in k$ and the i_j are nonnegative integers. Our hypothesis on the f_j implies that each of the $a_{i_1,\dots,i_n} = 0$, and so $g_m = 0$ for each m. □

PROPOSITION 3.7. *With notation as in Example 3.5, for each integer $n \geq 2$, the extension $(R_n, \mathbf{m}_n) \hookrightarrow (S, \mathbf{n})$ is nonfinite TGF-complete with trivial residue extension. Moreover* $\operatorname{ht}(P \cap R_n) \geq n-1$, *for each nonzero prime $P \in \operatorname{Spec} S$.*

Proof. We have $k = R_n/\mathbf{m}_n = S/\mathbf{n}$, so the residue field of S is a trivial extension of that of R_n. Since $\mathbf{m}_n S$ is not $\mathbf{n}$-primary, S is not finite over R_n. If $P \cap R_n = \mathbf{m}_n$, then $\operatorname{ht}(P \cap R_n) = n \geq n-1$. Since $\dim S = 2$, if $\mathbf{m}_n$ is not contained in P, then $\operatorname{ht} P = 1$, S/P is a one-dimensional local domain, and $\mathbf{m}_n(S/P)$ is primary for the maximal ideal $\mathbf{n}/P$ of S/P. It follows that $R_n/(P \cap R_n) \hookrightarrow S/P$ is a finite integral extension [14, Theorem 8.4]. Therefore $\dim R_n/(P \cap R_n) = 1$. Since R_n is catenary and $\dim R_n = n$, $\operatorname{ht}(P \cap R_n) = n-1$. □

COROLLARY 3.8. *Let X and $R = k[[X]]$ be as in Setting 3.1. Then there exists an infinite properly ascending chain of two-dimensional TGF-complete extensions $R =: S_0 \hookrightarrow S_1 \hookrightarrow S_2 \hookrightarrow \cdots$ such that each S_i has the same residue field as R and S_{i+1} is a nonfinite TGF-complete extension of S_i for each i.*

Proof. Example 3.5 and Proposition 3.7 imply that R can be identified with a proper subring of the power series ring in two variables so that $k[[y_1, y_2]]$ is a TGF-complete extension of R and the extension is not finite. Now Example 3.5 and Proposition 3.7 can be applied again, to $k[[y_1, y_2]]$, and so on. □

EXAMPLE 3.9. A particular case of Example 3.5.

For $R := k[[x, y]]$, the extension ring $S := k[[x, y/x]]$ has infinite transcendence over R [17]. The method used in [17] to prove that S has infinite transcendence degree over R is by constructing power series in y/x with 'special large gaps'. Since $k[[x]]$ is contained in R, it follows that S is a TGF-complete extension of R. To show this, it suffices to show $P \cap R \neq (0)$ for each $P \in \operatorname{Spec} S$ with $\operatorname{ht} P = 1$. This is clear if $x \in P$, while if $x \notin P$, then

$k[[x]] \cap P = (0)$, so $k[[x]] \hookrightarrow R/(P \cap R) \hookrightarrow S/P$ and S/P is finite over $k[[x]]$. Therefore $\dim R/(P \cap R) = 1$, so $P \cap R \neq (0)$.

Notice that the extension $k[[x,y]] \hookrightarrow k[[x,y/x]]$ is, up to isomorphism, the same as the extension $k[[x,xy]] \hookrightarrow k[[x,y]]$.

In Example 3.10 we show the situation of Remark 2.4(4) does occur.

EXAMPLE 3.10. Let k, $X = \{x_1, x_2\}$, $Y = \{y\}$, $Z = \{z\}$ and $R = k[[x_1,x_2]]$ be as in Remarks 2.4. Let $f_1, f_2 \in k[[z]]$ be algebraically independent over k. Let P denote the ideal of $k[[x_1,x_2,y,z]]$ generated by $(x_1 - f_1y, x_2 - f_2y)$, Then P is the kernel of the k-algebra homomorphism $\theta : k[[x_1,x_2,y,z]] \to k[[y,z]]$ obtained by defining $\theta(x_1) = f_1y$, $\theta(x_2) = f_2y$, $\theta(y) = y$ and $\theta(z) = z$. In the notation of Remark 2.4,

$$T = k[[x_1,x_2,y,z]]/P \cong k[[y,z]].$$

Let $\varphi := \theta|_R$ and $\tau := \theta|_{R[[y]]}$. The proof of Claim 3.6 shows that φ and τ are embeddings. Hence $P \cap R[[y]] = (0)$. By Proposition 3.7, φ and τ are TGF. We have

$$R \stackrel{\sigma}{\hookrightarrow} S = \frac{R[[y]]}{P \cap R[[y]]} = R[[y]] \stackrel{\tau}{\hookrightarrow} \frac{R[[y,z]]}{P} \cong k[[y,z]],$$

where $\sigma : R \hookrightarrow S$ is the inclusion map. Since $yS \cap R = (0)$, $\sigma : R \hookrightarrow S$ is not TGF.

QUESTIONS 3.11. (1) If $\varphi : R \hookrightarrow S$ is a TGF-complete nonfinite extension with finite residue field extension, is it always true that φ can be extended to a TGF-complete nonfinite extension $R[[y]] \hookrightarrow S$?

(2) Suppose that $R \hookrightarrow S$ is a TGF-complete extension and y is an indeterminate over S. It is natural to ask: Does $R[[y]] \hookrightarrow S[[y]]$ have the TGF property? Computing with elements, one may ask: For $s \in S \setminus R$, does $y + s$ have a multiple in $R[[y]]$? There is a $t \in S$ with $ts \in R$, but is there a $t' \in S$ with both $t't$ and $t'ts \in R$?

(3) A related question is whether the given $R \hookrightarrow S$ is extendable to an injective local homomorphism $\varphi : R[[y]] \hookrightarrow S$. For example, with k a field, $k[[x_1]][y]_{(x_1,y)} \hookrightarrow k[y][[x_1]]_{(x_1,y)}$ is TGF. Can we extend to $k[[x_1]][y][[x_2]]_{(x_1,x_2,y)} \hookrightarrow k[y][[x_1]]_{(x_1,y)}$, say by $x_2 \to \sum_{n=0}^{\infty}(yx)^n$, which is still local injective?

We show in Proposition 3.12 that the answer to Question 3.11(2) is 'no' if the answer to Question 3.11(3) is "yes", that is, the given $R \hookrightarrow S$ is extendable to an injective local homomorphism $R[[y]] \hookrightarrow S$. In Example 3.13 we present an example where this occurs.

PROPOSITION 3.12. *Let $R \hookrightarrow S$ be a TGF-complete extension and let y be an indeterminate over S. If $R \hookrightarrow S$ is extendable to an injective local homomorphism $\varphi : R[[y]] \hookrightarrow S$, then $R[[y]] \hookrightarrow S[[y]]$ is not TGF.*

Proof. Let $a := \varphi(y)$ and consider the ideal $Q = (y-a)S[[y]]$. The canonical map $S[[y]] \to S[[y]]/Q = S$ extends φ. Thus $Q \cap R[[y]] = (0)$ and $R[[y]] \hookrightarrow S[[y]]$ is not TGF. □

EXAMPLE 3.13. Let $R := R_n = k[[f_1w, \dots, f_nw]] \hookrightarrow S := k[[z,w]]$ be as in Example 3.5 with $n \geq 2$. Define the extension $\varphi : R[[y]] \hookrightarrow S$ by setting $\varphi(y) = f_{n+1}w \in S$. By Proposition 3.7, $\varphi : R[[y]] \hookrightarrow S$ is TGF-complete. Thus by Proposition 3.12, $R[[y]] \hookrightarrow S[[y]]$ is not TGF.

REMARK AND QUESTIONS 3.14. Let $(R, \mathbf{m}) \hookrightarrow (S, \mathbf{n})$ be a TGF-complete extension. Assume that $[S/\mathbf{n} : R/\mathbf{m}] < \infty$ and that S is not finite over R. By [14, Theorem 8.4], $\mathbf{m}S$ is not $\mathbf{n}$-primary. Thus $\dim S > \operatorname{ht}(\mathbf{m}S)$. Therefore $\dim S > 1$, so by Proposition 2.2, $\dim R > 1$.

(1) If $(R, \mathbf{m})$ is equicharacteristic, then by Corollary 3.3, $\dim S = 2$. Is it true in general that $\dim S = 2$?
(2) Is it possible to have $\dim S - \operatorname{ht}(\mathbf{m}S) > 1$?

EXAMPLES 3.15. (1) Let $R := k[[x, xy, z]] \hookrightarrow S := k[[x, y, z]]$. We show this is not a TGF extension. By (3.9), $\varphi : k[[x, xy]] \hookrightarrow k[[x, y]]$ is TGF-complete. By Proposition 3.12, it suffices to extend φ to an injective local homomorphism of $k[[x, xy, z]]$ to $k[[x, y]]$. Let $f \in k[[x]]$ be such that x and f are algebraically independent over k, so $(1, x, f)$ is not a solution to any nonzero homogeneous form over k. As in (3.2) and (3.5), the extension of φ obtained by mapping $z \to fy$ is an injective local homomorphism.

(2) The extension $R = k[[x, xy, xz]] \hookrightarrow S = k[[x, y, z]]$ is also not a TGF-complete extension, since $R = k[[x, xy, xz]] \hookrightarrow k[[x, xy, z]] \hookrightarrow S = k[[x, y, z]]$ is a composition of two extensions that are not TGF by part (1). Now apply Proposition 2.1.

4. The case of transcendental residue extensions

In this section we address, but do not fully resolve, the following question.

QUESTION 4.1. If $(S, \mathbf{n})$ is a TGF-complete extension of $(R, \mathbf{m})$ and if $S/\mathbf{n}$ is transcendental over $R/\mathbf{m}$ does it follow that $\dim S \leq 1$?

In Proposition 4.2 we prove every complete local domain of positive dimension has a one-dimensional TGF-complete extension.

PROPOSITION 4.2. *Let $(R, \mathbf{m})$ be a local domain of positive dimension.*

(1) *There exists a one-dimensional complete local domain $(S, \mathbf{n})$ that is a TGF extension of R.*
(2) *If R is complete, there exists a one-dimensional TGF-complete extension of R.*

Proof. It is well known that there exists a discrete rank-one valuation domain $(S, \mathbf{n})$ that dominates R (see, for example, [3]). The $\mathbf{n}$-adic completion $\widehat{S}$ of S is a one-dimensional local ring that dominates R and each minimal prime $\mathbf{p}_i$ of $\widehat{S}$ intersects S in zero, so $\widehat{S}/\mathbf{p}_i$ is a one-dimensional complete local domain that dominates R. Moreover, if $(S, \mathbf{n})$ is a one-dimensional local domain that dominates a local domain $(R, \mathbf{m})$ of positive dimension, then it is obvious that S is a TGF extension of R, so if R and S are also complete, then S is a TGF-complete extension of R. □

SETTING 4.3. Let $n \geq 2$ be an integer, let $X = \{x_1, \dots, x_n\}$ be a set of independent variables over the field k and let $R = k[[X]]$ be the formal power series ring in n variables over the field k. Let z, w, t, v be independent variables over R.

PROPOSITION 4.4. *Let notation be as in Setting 4.3.*

(1) *There exists a TGF embedding $\theta : k[[z, w]] \to k(t)[[v]]$ defined by $\theta(z) = tv$ and $\theta(w) = v$.*
(2) *Moreover, the composition $\psi = \theta \circ \varphi$ of θ with $\varphi : R \to k[[z, w]]$ given in General Example 3.5 is also TGF.*

Proof. Suppose $f \in \ker \theta$. Write $f = \sum_{n=0}^{\infty} f_n(z, w)$, where f_n is a homogeneous form of degree n with coefficients in k. We have

$$0 = \theta(f) = \sum_{n=0}^{\infty} f_n(tv, v) = \sum_{n=0}^{\infty} v^n f_n(t, 1)\,.$$

This implies $f_n(t, 1) = 0$ for each n. Since t is algebraically independent over k, we have $f_n(z, w) = 0$ for each n. Thus $f = 0$ and θ is an embedding. Since θ is a local homomorphism and $\dim k(t)[[v]] = 1$, it is clear that θ is TGF.

For the second part, we use Proposition 2.1 together with the observation in General Example 3.5 that φ is a TGF embedding. □

As a consequence of Proposition 4.4, we prove:

COROLLARY 4.5. *Let $R = k[[X]]$ be as above and let $A = k(t)[[X]]$. There exists a prime ideal $P \in \operatorname{Spec} A$ in the generic fiber over R with $\operatorname{ht} P = n - 1$. In particular, the inclusion map $R = k[[X]] \hookrightarrow A = k(t)[[X]]$ is not TGF.*

Proof. Define $\varphi : R \to k[[z, w]] := S$, by

$$\varphi(x_1) = z, \quad \varphi(x_2) = h_2(w)z, \quad \dots, \quad \varphi(x_n) = h_n(w)z,$$

where $h_2(w), \dots, h_n(w) \in k[[w]]$ are algebraically independent over k. Also define $\theta : S \to k(t)[[v]] := B$ by $\theta(z) = tv$ and $\theta(w) = v$. Consider the following diagram

$$\begin{array}{ccc} R = k[[X]] & \xrightarrow{\subset} & A = k(t)[[X]] \\ \varphi \downarrow & & \Psi \downarrow \\ S = k[[z,w]] & \xrightarrow{\theta} & B = k(t)[[v]], \end{array}$$

where $\Psi : A \to B$ is the identity map on $k(t)$ and is defined by

$$\Psi(x_1) = tv, \quad \Psi(x_2) = h_2(v)tv, \quad \dots, \quad \Psi(x_n) = h_n(v)tv.$$

Notice that $\Psi|_R = \psi = \theta \circ \varphi$. Therefore the diagram is commutative. Let $P = \ker \Psi$. Since Ψ is surjective, $\operatorname{ht} P = n-1$. Commutativity of the diagram implies that $P \cap R = (0)$. □

DISCUSSION 4.6. Let us describe generators for the prime ideal $P = \ker \Psi$ given in Corollary 4.5. Under the map Ψ, $x_1 \mapsto tv$, and so $\frac{x_1}{t} \mapsto v$. Since also $x_2 \mapsto h_2(v)tv, \dots, x_n \mapsto h_n(v)tv$, we see that

$$(x_2 - h_2(\frac{x_1}{t})x_1,\ x_3 - h_3(\frac{x_1}{t})x_1, \dots, x_n - h_n(\frac{x_1}{t})x_1)A \subseteq P$$

(that is, $\Psi(x_2 - h_2(x_1/t)x_1) = h_2(v)tv - h_2(v)tv = 0$, etc.) Since the ideal on the left-hand-side is a prime ideal of height $n-1$, the inclusion is an equality. Thus we have generators for the prime ideal $P = \ker \Psi$ resulting from the definitions of φ and θ given in the corollary.

On the other hand, in Corollary 4.5 if we change the definition of θ and we define $\theta' : k[[z,w]] \to k(t)[[v]]$ by $\theta'(z) = v$ and $\theta'(w) = tv$ (but we keep φ as above), then ψ' defined by $\psi'|_R = \theta' \cdot \varphi$ maps $x_1 \to v$, $x_2 \to h_2(tv)v$, …, $x_n \to h_n(tv)v$. In this case

$$(x_2 - h_2(tx_1)x_1,\ x_3 - h_3(tx_1)x_1, \dots, x_n - h_n(tx_1)x_1)A \subseteq \ker \Psi' = P'.$$

Again the ideal on the left-hand-side is a prime ideal of height $n-1$, so we have equality. This yields a different prime ideal P'.

In this case one can also see directly for

$$P' = (x_2 - h_2(tx_1)x_1,\ x_3 - h_3(tx_1)x_1, \dots, x_n - h_n(tx_1)x_1)A$$

that $P' \cap R = (0)$. We have $\Psi : A \to A/P' = k(t)[[v]]$. Suppose $f \in R \cap P'$. We write $f = \sum_{\ell=0}^{\infty} f_\ell(x_1, \dots, x_n)$, where $f_\ell \in k[x_1, \dots, x_n]$ is a homogeneous form of degree ℓ. We have

$$0 = \Psi'(f) = \sum_{\ell=0}^{\infty} f_\ell(v, h_2(tv)v, \dots, h_n(tv)v) = \sum_{\ell=0}^{\infty} v^\ell f_\ell(1, h_2(tv), \dots, h_n(tv)).$$

This implies $f_\ell(1, h_2(tv), \dots, h_n(tv)) = 0$ for each ℓ. Since $h_2, \dots, h_n$ are algebraically independent over k, each of the homogeneous forms $f_\ell(x_1, \dots, x_n) = 0$. Hence $f = 0$.

QUESTION 4.7. With notations as in Corollary 4.5, does every prime ideal of A maximal in the generic fiber over R have height $n-1$?

THEOREM 4.8. *Let $(A, \mathbf{m}) \hookrightarrow (B, \mathbf{n})$ be an extension of two-dimensional regular local domains. Assume that B dominates A and that $B/\mathbf{n}$ as a field extension of $A/\mathbf{m}$ is not algebraic. Then $A \hookrightarrow B$ is not TGF.*

Proof. Since $\dim A = \dim B$, the assumption that $B/\mathbf{n}$ is transcendental over $A/\mathbf{m}$ implies that B is not algebraic over A [14, Theorem 15.5]. If $\mathbf{m}B$ is $\mathbf{n}$-primary, then B is faithfully flat over A [14, Theorem 23.1], and [6, Theorem 1.12] implies that $A \hookrightarrow B$ is not TGF in this case.

If $\mathbf{m}B$ is principal, then $\mathbf{m}B = xB$ for some $x \in \mathbf{m}$ since B is local. It follows that $\mathbf{m}/x \subset B$. Localizing $A[\mathbf{m}/x]$ at the prime ideal $\mathbf{n} \cap A[\mathbf{m}/x]$ gives a local quadratic transform $(A_1, \mathbf{m}_1)$ of A. If $\dim A_1 = 1$, then $A_1 \hookrightarrow B$ is not TGF because only finitely many prime ideals of B can contract to the maximal ideal of A_1. Hence $A \hookrightarrow B$ is not TGF if $\dim A_1 = 1$. If $\dim A_1 = 2$, then $(A_1, \mathbf{m}_1)$ is a 2-dimensional regular local domain dominated by $(B, \mathbf{n})$ and the field $A_1/\mathbf{m}_1$ is finite algebraic over $A/\mathbf{m}$, and so $B/\mathbf{n}$ is transcendental over $A_1/\mathbf{m}_1$. Thus we can repeat the above analysis: If $\mathbf{m}_1 B$ is $\mathbf{n}$-primary, then as above $A \hookrightarrow B$ is not TGF. If $\mathbf{m}_1 B$ is principal, we obtain a local quadratic transform $(A_2, \mathbf{m}_2)$ of A_1. If this process does not end after finitely many steps, we have a union $V = \bigcup_{n=1}^{\infty} A_n$ of an infinite sequence A_n of quadratic transforms of a 2-dimensional regular local domains. Then V is a valuation domain of rank at most 2 contained in B, and so at most finitely many of the height-one primes of B have a nonzero intersection with V. Therefore $V \hookrightarrow B$ is not TGF and hence also $A \hookrightarrow B$ is not TGF.

Thus by possibly replacing A by an iterated local quadratic transform A_n of A, we may assume that $\mathbf{m}B$ is neither $\mathbf{n}$-primary nor principal. Let $\mathbf{m} = (x, y)A$. There exist $f, g, h \in B$ such that $x = gf, y = hf$ and g, h is a regular sequence in B. Hence $(g, h)B$ is $\mathbf{n}$-primary. Let $f = f_1^{e_1} \cdots f_r^{e_r}$, where $f_1B, \dots f_rB$ are distinct height-one prime ideals and the e_i are positive integers. Then $f_1B, \dots, f_rB$ are precisely the height-one primes of B that contain $\mathbf{m}$.

Let $t \in B$ be such that the image to t in $B/\mathbf{n}$ is transcendental over $A/\mathbf{m}$. Modifying t if necessary by an element of $\mathbf{n}$ we may assume that t is transcendental over A. We have $\mathbf{n} \cap A[t] = \mathbf{m}[t]$. Let $A(t) = A[t]_{\mathbf{m}[t]}$. Notice that $A(t)$ is a 2-dimensional regular local domain with maximal ideal $\mathbf{m}A(t)$ that is dominated by $(B, \mathbf{n})$. We have

$$A \hookrightarrow A[t] \hookrightarrow A(t) \hookrightarrow B.$$

Let $P = (xt - y)A(t)$. Then $P \cap A = (0)$. We have $PB = (gft - hf)B = f(gt - h)B$. Also $gt - h$ is a nonunit of B. Let Q be a minimal prime of $(gt-h)B$. Then $Q \notin \{f_1B, \dots, f_rB\}$. Hence $\mathbf{m}A(t) \not\subseteq Q$. Therefore $Q \cap A(t)$

has height one. Since $P \subseteq (gt - h)B \subseteq Q$, we have $Q \cap A(t) = P$. Thus $Q \cap A = (0)$. This completes the proof. □

We have the following immediate corollary to Theorem 4.8.

COROLLARY 4.9. *Let x, y, z, w, t be indeterminates over the field k and let*

$$\varphi : R = k[[x, y]] \hookrightarrow S := k(t)[[z, w]]$$

be an injective local k-algebra homomorphism. Then $\varphi(R) \hookrightarrow S$ is not TGF.

In relation to Question 4.1, Example 4.10 is a TGF extension $A \hookrightarrow B$ that is not complete for which the residue field of B is transcendental over that of A and $\dim B = 2$.

EXAMPLE 4.10. Let $A = k[x, y, z, w]_{(x,y,z,w)}$, where k is a field and $xw = yz$. Thus A is a 3-dimensional normal local domain with maximal ideal $\mathbf{m} := (x, y, z, w)A$ and residue field $A/\mathbf{m} = k$. Notice that $C := A[y/x] = k[y/x = w/z, x, z]$ is a polynomial ring in 3 variables over k. Thus $B := C_{(x,z)}$ is a 2-dimensional regular local domain with maximal ideal $\mathbf{n} = (x, z)B$. Notice that $(B, \mathbf{n})$ birationally dominates $(A, \mathbf{m})$. Hence $(A, \mathbf{m}) \hookrightarrow (B, \mathbf{n})$ is a TGF extension. Also $B = k(y/x)[x, z]_{(x,z)}$, so $k(y/x)$ is a coefficient field for B. The image t of y/x in $B/\mathbf{n}$ is transcendental over k and $B/\mathbf{n} = k(t)$. The completion of A is the normal local domain $\widehat{A} = k[[x, y, z, w]]$, where $xw = yz$. By a form of Zariski's subspace theorem [1, (10.6)], $\widehat{A}$ is dominated by $\widehat{B} = k(t)[[x, z]]$. Thus we have $\varphi : \widehat{A} \hookrightarrow \widehat{B}$, where $\varphi(x) = x, \varphi(z) = z, \varphi(y/x) = t = \varphi(w/z)$ and so also $\varphi(y) = tx, \varphi(w) = tz$, $\varphi(xw) = xtz = \varphi(yz)$.

In Example 4.10, $\widehat{A} \hookrightarrow \widehat{B}$ is not a TGF-complete extension. Equivalently, the inclusion map

$$R := k[[x, z, tx, tz]] \hookrightarrow k(t)[[x, z]] := S$$

is not a TGF-extension. We hope to expand on this in a future publication.

REFERENCES

[1] S. S. Abhyankar, *Resolution of singularities of embedded algebraic surfaces*, Pure and Applied Mathematics, Vol. 24, Academic Press, New York, 1966. MR 0217069 (36 #164)

[2] S. S. Abhyankar and T. T. Moh, *On analytic independence*, Trans. Amer. Math. Soc. **219** (1976), 77–87. MR 0414546 (54 #2647)

[3] C. Chevalley, *La notion d'anneau de décomposition*, Nagoya Math. J. **7** (1954), 21–33. MR 0067866 (16,788g)

[4] I. S. Cohen, *On the structure and ideal theory of complete local rings*, Trans. Amer. Math. Soc. **59** (1946), 54–106. MR 0016094 (7,509h)

[5] A. Grothendieck, *Élements de Géométrie Algébrique. IV*, Inst. Hautes Études Sci. Publ. Math. **24** (1965).

[6] W. Heinzer and C. Rotthaus, *Formal fibers and complete homomorphic images*, Proc. Amer. Math. Soc. **120** (1994), 359–369. MR 1189544 (94d:13020)

[7] W. Heinzer, C. Rotthaus, and S. Wiegand, *Noetherian domains inside a homomorphic image of a completion*, J. Algebra **215** (1999), 666–681. MR 1686210 (2000e:13031)

[8] ———, *Building Noetherian and non-Noetherian integral domains using power series*, Ideal theoretic methods in commutative algebra (Columbia, MO, 1999), Lecture Notes in Pure and Appl. Math., vol. 220, Dekker, New York, 2001, pp. 251–264. MR 1836605 (2002b:13031)

[9] ———, *Examples of integral domains inside power series rings*, Commutative ring theory and applications (Fez, 2001), Lecture Notes in Pure and Appl. Math., vol. 231, Dekker, New York, 2003, pp. 233–254. MR 2029829 (2004k:13037)

[10] ———, *Building Noetherian domains inside an ideal-adic completion*, Abelian groups, module theory, and topology (Padua, 1997), Lecture Notes in Pure and Appl. Math., vol. 201, Dekker, New York, 1998, pp. 279–287. MR 1651173 (99m:13035)

[11] ———, *Generic fiber rings of mixed power series/polynomial rings*, J. Algebra **298** (2006), 248–272. MR 2215127 (2006m:13013)

[12] ———, *Mixed polynomial/power series rings and relations among their spectra*, Multiplicative ideal theory in commutative algebra: a tribute to the work of Robert Gilmer, (J. Brewer, S. Glaz, W. Heinzer, B. Olberding, eds.), Springer, New York, 2006, pp. 227–242. MR 2265811
Volume in honor of Robert Gilmer, to appear.

[13] H. Matsumura, *On the dimension of formal fibres of a local ring*, Algebraic geometry and commutative algebra, Vol. I, Kinokuniya, Tokyo, 1988, pp. 261–266. MR 977763 (90a:13027)

[14] ———, *Commutative ring theory*, second ed., Cambridge Studies in Advanced Mathematics, vol. 8, Cambridge University Press, Cambridge, 1989. MR 1011461 (90i:13001)

[15] M. Nagata, *Local rings*, Interscience Tracts in Pure and Applied Mathematics, No. 13, Interscience Publishers, John Wiley & Sons, New York-London, 1962. MR 0155856 (27 #5790)

[16] C. Rotthaus, *On rings with low-dimensional formal fibres*, J. Pure Appl. Algebra **71** (1991), 287–296. MR 1117639 (92j:13006)

[17] P. B. Sheldon, *How changing* $D[[x]]$ *changes its quotient field*, Trans. Amer. Math. Soc. **159** (1971), 223–244. MR 0279092 (43 #4818)

[18] T. Yasuda, *On subschemes of formal schemes*, preprint, `arXiv:math.AG/0602542`.

[19] O. Zariski and P. Samuel, *Commutative algebra. Vol. II*, The University Series in Higher Mathematics, D. Van Nostrand Co., Inc., Princeton, N. J.-Toronto-London-New York, 1960. MR 0120249 (22 #11006)

William Heinzer, Department of Mathematics, Purdue University, West Lafayette, Indiana 47907, USA
E-mail address: `heinzer@math.purdue.edu`

Christel Rotthaus, Department of Mathematics, Michigan State University, East Lansing, MI 48824-1027, USA
E-mail address: `rotthaus@math.msu.edu`

Sylvia Wiegand, Department of Mathematics, University of Nebraska, Lincoln, NE 68588-0130, USA
E-mail address: `swiegand@math.unl.edu`

Illinois Journal of Mathematics
Volume 51, Number 1, Spring 2007, Pages 137–150
S 0019-2082

LOCAL DUALITY FOR BIGRADED MODULES

JÜRGEN HERZOG AND AHAD RAHIMI

ABSTRACT. In this paper we study local cohomology of finitely generated bigraded modules over a standard bigraded ring with respect to the irrelevant bigraded ideals and establish a duality theorem. Several applications are considered.

Introduction

Let R be a standard bigraded K-algebra with bigraded irrelevant ideals P generated by all elements of degree $(1,0)$ and Q generated by all elements of degree $(0,1)$. We want to relate the local cohomology functors $H^i_P(-)$ and $H^j_Q(-)$ via duality in the category of bigraded modules. In the ordinary local duality theorem Matlis duality establishes isomorphisms between the local cohomology modules of a module and its Ext-groups.

In our situation we have to consider Matlis duality for bigraded modules. Given a bigraded R-module M we define the bigraded Matlis-dual of M to be $M^\vee$, where the (i,j)th bigraded component of $M^\vee$ is given by $\operatorname{Hom}_K(M_{(-i,-j)}, K)$.

As the main result of our paper we have the following duality theorem:

THEOREM. *Let R be a standard bigraded K-algebra with irrelevant bigraded ideals P and Q, and let M be a finitely generated bigraded R-module. Then there exists a convergent spectral sequence*

$$E^2_{i,j} = H^{m-j}_P(H^i_{R_+}(M)^\vee) \underset{j}{\Longrightarrow} H^{i+j-m}_Q(M)^\vee$$

of bigraded R-modules, where m is the minimal number of homogeneous generators of P and R_+ is the unique graded maximal ideal of R.

Note that the above spectral sequence degenerates when M is Cohen-Macaulay and one obtains for all k the following isomorphims of bigraded

Received May 1, 2006; received in final form July 25, 2006.
2000 *Mathematics Subject Classification.* 13D45, 13C14.

R-modules

$$H^k_P(H^s_{R_+}(M)^\vee) \cong H^{s-k}_Q(M)^\vee, \tag{1}$$

where $s = \dim M$; see Corollary 2.6.

Let R_0 be the K-subalgebra of R which is generated by the elements of bidegree $(1,0)$, and let N be any bigraded R-module. Then for all j, the module $N_j = \bigoplus_i N_{(i,j)}$ is a graded R_0-module with grading $(N_j)_i = N_{(i,j)}$. Moreover, if N is finitely generated, then each N_j is a finitely generated R_0-module. In particular, if M is an s-dimensional Cohen-Macaulay module and if we set $N = H^s_{R_+}(M)^\vee$, then N is again an s-dimensional Cohen-Macaulay module and by (1) we obtain for all j the isomorphisms of graded R_0-modules

$$H^k_{P_0}(N_j) \cong (H^{s-k}_Q(M)_{-j})^\vee, \tag{2}$$

where P_0 is the graded maximal ideal of R_0. Here we used that $H^k_P(N)_j \cong H^k_{P_0}(N_j)$ for all k and j.

Brodmann and Hellus [4] raised the question whether the modules $H^k_Q(M)$ are tame if M is a finitely generated graded R-module, in other words, whether for each k there exists an integer j_0 such that either $H^k_Q(M)_j = 0$ for all $j \leq j_0$, or else $H^k_Q(M)_j \neq 0$ for all $j \leq j_0$. In various cases this problem has been answered in the affirmative; see [3], [4], [16], [10], [12] and [2] for a survey on this problem. In case M is Cohen-Macaulay the tameness problem translates, due to (2), to the following question: Given a finitely generated bigraded R-module N, does there exist an integer j_0 such that $H^k_{P_0}(N_j) = 0$ for all $j \geq j_0$, or else $H^k_{P_0}(N_j) \neq 0$ for all $j \geq j_0$? More generally, one would expect that for a finitely generated graded R_0-module W and a finitely generated bigraded R-module N there exists for all k an integer j_0 such that $\operatorname{Ext}^k_{R_0}(N_j, W) = 0$ for all $j \geq j_0$, or else $\operatorname{Ext}^k_{R_0}(N_j, W) \neq 0$ for all $j \geq j_0$. However, this is not the case as has been recently shown by Cutkosky and the first author; see [7]. Their example also provides a counterexample to the general tameness problem. To show this, Proposition 2.5 of this paper is used. On the other hand, in a recent paper [14] the second author of this paper has shown that tameness holds for all local cohomology modules of a ring with monomial relations and with respect to monomial prime ideals.

In Section 2 we use our duality to give new proofs of known cases of the tameness problem and also to add a few new cases in which tameness holds; see Corollaries 2.4, 2.8 and 2.12. The duality is also used in Corollaries 2.9 and 2.10 to prove some algebraic properties of the modules $H^k_Q(M)_j$ in case M is Cohen-Macaulay.

1. Proof of the duality theorem

Let $S = K[x_1, \ldots, x_m, y_1, \ldots, y_n]$ be the standard bigraded polynomial ring over the field K. We set $K[x] = K[x_1, \ldots, x_m]$ and $K[y] = K[y_1, \ldots, y_n]$ and consider both as standard graded polynomial rings.

If A is a standard (bi)graded K-algebra and M a (bi)graded A-module, we set $M^\vee = \operatorname{Hom}_K(M, K)$ and view $M^\vee$ as (bi)graded A-module with the (bi)grading

$$(M^\vee)_a = \operatorname{Hom}_K(M_{-a}, K)$$

for $a \in \mathbb{Z}$ (respectively $a \in \mathbb{Z}^2$ in the bigraded case).

The following simple fact is needed for the proof of the next lemma.

Lemma 1.1. *Let M be a graded $K[x]$-module and N be a graded $K[y]$-module. Then there exists a natural bigraded isomorphism of bigraded S-modules*

$$(M \otimes_K N)^\vee \cong M^\vee \otimes_K N^\vee.$$

Proof. Let $S = K[x] \otimes_K K[y] = K[x, y]$. Note that $M \otimes_K N$ is a bigraded free S-module with the natural bigrading

$$(M \otimes_K N)_{(i,j)} = M_i \otimes_K N_j.$$

Thus we see that

$$((M \otimes_K N)^\vee)_{(i,j)} = \operatorname{Hom}_K((M \otimes_K N)_{(-i,-j)}, K) = \operatorname{Hom}_K(M_{-i} \otimes_K N_{-j}, K).$$

By using the universal property of the tensor product one has the following natural isomorphism of K-vector spaces

$$\operatorname{Hom}_K(M_{-i} \otimes_K N_{-j}, K) \cong \operatorname{Hom}_K(M_{-i}, K) \otimes_K \operatorname{Hom}_K(N_{-j}, K).$$

Thus we have

$$\begin{aligned}((M \otimes_K N)^\vee)_{(i,j)} &\cong \operatorname{Hom}_K(M_{-i}, K) \otimes_K \operatorname{Hom}_K(N_{-j}, K)\\ &= (M^\vee)_i \otimes_K (N^\vee)_j\\ &= (M^\vee \otimes_K N^\vee)_{(i,j)}.\end{aligned}$$

So the desired isomorphism follows. □

Lemma 1.2. *Let $S = K[x_1, \ldots, x_m, y_1, \ldots, y_n]$ be the standard bigraded polynomial ring over the field K with the irrelevant bigraded ideals $P = (x_1, \ldots, x_m)$ and $Q = (y_1, \ldots, y_n)$. Then we have the following isomorphism of bigraded S-modules*

$$H^m_P(\omega_S) \cong H^n_Q(S)^\vee,$$

where ω_S is the bigraded canonical module of S.

Proof. We denote by P_0 the graded maximal ideal of $K[x]$ and by Q_0 the graded maximal ideal of $K[y]$. First we notice that there is a natural isomorphism of bigraded S-modules

$$H_P^m(S) \cong H_{P_0}^m(K[x]) \otimes_K K[y].$$

By the graded version of the local duality theorem (see [5, Example 13.4.6]) we have

$$H_{P_0}^m(K[x])^\vee \cong K[x](-m).$$

Thus we see that

$$\begin{aligned} H_P^m(\omega_S) = H_P^m(S(-m,-n)) &= H_P^m(S)(-m,-n) \\ &\cong (K[x](-m)^\vee \otimes_K K[y])(-m,-n) \\ &= K[x]^\vee \otimes_K K[y](-n). \end{aligned}$$

On the other hand, using again the local duality theorem, Lemma 1.1 yields

$$\begin{aligned} H_Q^n(S)^\vee \cong (K[x] \otimes_K H_{Q_0}^n(K[y]))^\vee &\cong (K[x] \otimes_K K[y](-n)^\vee)^\vee \\ &\cong K[x]^\vee \otimes_K K[y](-n)^{\vee\vee} \\ &\cong K[x]^\vee \otimes_K K[y](-n), \end{aligned}$$

as desired. □

COROLLARY 1.3. *Let F be a finitely generated bigraded free S-module, and set $F^* = \operatorname{Hom}_S(F, \omega_S)$. Then there exists a natural isomorphism of bigraded S-modules*

$$H_P^m(F^*) \cong H_Q^n(F)^\vee.$$

Proof. Let $F = \bigoplus_{k=1}^t S(-a_k, -b_k)$. Thus $F^* = \bigoplus_{k=1}^t (\omega_S)(a_k, b_k)$ and hence by Lemma 1.2 we have

$$\begin{aligned} H_P^m(F^*) \cong \bigoplus_{k=1}^t H_P^m(\omega_S)(a_k,b_k) &\cong \bigoplus_{k=1}^t H_Q^n(S)^\vee(a_k,b_k) \\ &\cong H_Q^n(\bigoplus_{k=1}^t S(-a_k,-b_k))^\vee \\ &\cong H_Q^n(F)^\vee. \end{aligned}$$ □

The previous result can easily be extended as follows.

LEMMA 1.4. *Let $\mathbb{F}$ be a bounded complex of bigraded free S-modules. We set $\mathbb{F}^* = \operatorname{Hom}_S(\mathbb{F}, \omega_S)$. Then we have a functorial isomorphism*

$$H_P^m(\mathbb{F}^*) \cong H_Q^n(\mathbb{F})^\vee$$

of complexes of bigraded modules.

Proof. In order to prove that the complexes of $H_P^m(\mathbb{F}^*)$ and $H_Q^n(\mathbb{F})^\vee$ are isomorphic, we observe that for any bihomogeneous linear map $\varphi : G \to F$ between finitely generated free bigraded S-modules we obtain the following commutative diagram

$$\begin{array}{ccc} H_Q^n(F)^\vee & \xrightarrow{\psi_1^\vee} & H_Q^n(G)^\vee \\ \downarrow & & \downarrow \\ H_P^m(F^*) & \xrightarrow{\psi_2} & H_P^m(G^*), \end{array}$$

where $\psi_1 = H_P^n(\varphi)$ and $\psi_2 = H_Q^m(\varphi^*)$ and where the vertical maps are the isomorphisms given in Corollary 1.3. The commutativity of the diagram results from the fact that all maps in the diagram are functorial. $\square$

PROPOSITION 1.5. *Let M be a finitely generated bigraded S-module, P and Q be the irrelevant bigraded ideals of S. Then we have the following convergent spectral sequence*

$$E_{i,j}^2 = H_P^{m-j}(\mathrm{Ext}_S^{n+m-i}(M, \omega_S)) \underset{j}{\Longrightarrow} H_Q^{i+j-m}(M)^\vee.$$

Proof. Let $(\mathbb{F}, d)$ be a bigraded free resolution of M of length $n+m$, and let $\mathbb{G}$ be the complex of bigraded S-modules with $G_i = \mathrm{Hom}_S(F_{m+n-i}, \omega_S)$ and differential $\partial_i = \mathrm{Hom}_S(d_{m+n-i}, \omega_S)$. Next we choose a bigraded free resolution $\mathbb{C}$ of the complex $\mathbb{G}$. In other words, $\mathbb{C}$ is a double complex C_{ij} of finitely generated bigraded free S-modules with $i, j \geq 0$ such that:

(i) The ith column of $\mathbb{C}$ is a free resolution of G_i for all i, i.e.,

$$H_j(C_{i\bullet}) = \begin{cases} G_i & \text{for } j = 0, \\ 0 & \text{for } j > 0. \end{cases}$$

(ii) For each row the image of $C_{i-1,j} \longleftarrow C_{i,j}$ is a bigraded free direct summand of the kernel of $C_{i-2,j} \longleftarrow C_{i-1,j}$. In particular, the homology of

$$C_{i-2,j} \longleftarrow C_{i-1,j} \longleftarrow C_{i,j}$$

is a bigraded free S-module for all i and j.

(iii) For each i the complex

$$0 \longleftarrow H_i(C_{\bullet,0}) \longleftarrow H_i(C_{\bullet,1}) \longleftarrow H_i(C_{\bullet,2}) \longleftarrow \cdots$$

is a bigraded free resolution of $H_i(\mathbb{G})$.

Now we compute the total homology of the double complex $H_P^m(\mathbb{C})$: Since all G_i are free S-modules, it follows that the complexes

$$0 \longleftarrow G_i \longleftarrow C_{i,0} \longleftarrow C_{i,1} \longleftarrow \cdots$$

are all split exact. Hence the complexes

$$0 \longleftarrow H_P^m(G_i) \longleftarrow H_P^m(C_{i,0}) \longleftarrow H_P^m(C_{i,1}) \longleftarrow \cdots$$

are again exact.

This implies that the E^1-terms of the double complex $H_P^m(\mathbb{C})$ with respect to the column filtration are

$$E^1_{i,j} = \begin{cases} H_P^m(G_i) & \text{for } j = 0, \\ 0 & \text{for } j > 0. \end{cases}$$

As a consequence, for the E^2-terms of $H_P^m(\mathbb{C})$ we have that $E^2_{i,j} = 0$ for $j > 0$, and that $E^2_{i,0}$ is the ith homology of the complex $H_P^m(\mathbb{G})$. Now we use Lemma 1.4 as well as [13, Theorem 1.1] and obtain

$$E^2_{i,j} = \begin{cases} H_Q^{i-m}(M)^\vee & \text{for } j = 0, \\ 0 & \text{for } j > 0, \end{cases}$$

since $H_i(H_P^m(\mathbb{G})) = (H_{n+m-i}(H_Q^n(\mathbb{F})))^\vee$. From this it follows that the $(i+j)$th total homology of $H_P^m(\mathbb{C})$ is equal to $H_Q^{i+j-m}(M)^\vee$.

Now we compute the homology of $H_P^m(\mathbb{C})$ using the row filtration. Each row $H_P^m(C_{\bullet j})$ of $H_P^m(\mathbb{C})$ is split exact with homology $H_i(H_P^m(C_{\bullet j})) = H_P^m(H_i(C_{\bullet j}))$. In other words, $E^1_{i,j} = H_P^m(H_i(C_{\bullet j}))$. Hence by property (iii) of the complex $\mathbb{C}$ and by [13, Theorem 1.1] it follows that $E^2_{i,j} = H_P^{m-j}(\operatorname{Ext}_S^{m+n-i}(M, \omega_S))$. This yields the desired conclusion. □

Now our main theorem is an easy consequence of Proposition 1.5:

Proof. As R is a standard bigraded K-algebra, it is the homomorphic image of a standard bigraded polynomial ring $S = K[x_1, \ldots, x_m, y_1, \ldots, y_n]$. We may consider R and S as well as standard graded K-algebras with the unique graded maximal ideal R_+ (resp. S_+), and M as a graded R-module (resp. S-module). Then by the graded local duality theorem we have

$$\operatorname{Ext}_S^{m+n-i}(M, \omega_S) \cong H^i_{S_+}(M)^\vee.$$

Since $H^i_{S_+}(M) \cong H^i_{R_+}(M)$, it follows that

$$H_P^{m-j}(H^i_{R_+}(M)^\vee) = H_P^{m-j}(\operatorname{Ext}_S^{m+n-i}(M, \omega_S)).$$

Let $(x) = (x_1, \ldots, x_m)$ and $(y) = (y_1, \ldots, y_n)$ be the irrelevant ideals of S. We note that $H_P^{m-j}(\operatorname{Ext}_S^{m+n-i}(M, \omega_S)) = H_{(x)}^{m-j}(\operatorname{Ext}_S^{m+n-i}(M, \omega_S))$ and that $H_Q^{i+j-m}(M)^\vee = H_{(y)}^{i+j-m}(M)^\vee$. Therefore, Proposition 1.5 yields the desired convergent spectral sequence. □

Corollary 1.6. *Let R be a standard bigraded d-dimensional Cohen-Macaulay K-algebra with irrelevant bigraded ideals P and Q, and let M be a finitely generated bigraded R-module. Then there exists a convergent spectral sequence*

$$E^2_{i,j} = H_P^{m-j}(\operatorname{Ext}_R^{d-i}(M, \omega_R)) \underset{j}{\Longrightarrow} H_Q^{i+j-m}(M)^\vee$$

of bigraded R-modules, where m is the minimal number of homogeneous generators of P.

Proof. The assertion follows from our main theorem by using the fact that $H^i_{R_+}(M)^\vee = \mathrm{Ext}^{d-i}_R(M, \omega_R)$. □

2. Some applications

In this section, unless otherwise stated, R denotes a standard bigraded K-algebra of dimension d, and M a finitely generated bigraded R-module.

We note that for the E^2-terms in the spectral sequence of our main theorem we have $E^2_{i,j} = H^{m-j}_P(H^i_{R_+}(M)^\vee) = 0$ if $i < \operatorname{depth} M$ or $i > \dim M$ or $j < 0$ or $j > m$. Thus the possible non-zero E^2-terms are in the shadowed region of the following picture.

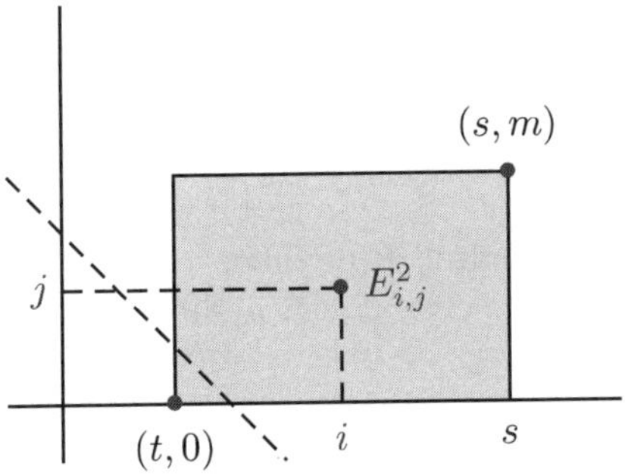

FIGURE 1

Here $t = \operatorname{depth} M$, $s = \dim M$ and $E^2_{i,j} = H^{m-j}_P(H^i_{R_+}(M)^\vee)$.

We first observe that the graded local duality theorem is a special case of our main theorem. In fact, if we assume that $P = (0)$, then $m = 0$, and $\mathfrak{m} = Q$ is the unique graded maximal ideal of R. Moreover, $E^2_{i,j} = E^\infty_{i,j} = 0$ for $j \neq 0$ and all i, since $H^k_{(0)}(-) = 0$ if $k \neq 0$. Therefore we have

$$\mathrm{Ext}^{d-i}_R(M, \omega_R) = H^0_{(0)}(\mathrm{Ext}^{d-i}_R(M, \omega_R)) \cong H^i_{\mathfrak{m}}(M)^\vee.$$

Considering Figure 1 we immediately obtain the following corner isomorphisms.

PROPOSITION 2.1. *Let* $\dim M = s$ *and* $\operatorname{depth} M = t$. *Then there are natural isomorphisms*

$$H^m_P(H^t_{R_+}(M)^\vee) \cong H^{t-m}_Q(M)^\vee \quad \text{and} \quad H^0_P(H^s_{R_+}(M)^\vee) \cong H^s_Q(M)^\vee.$$

Moreover, for $i < t - m$ *we have* $H^i_Q(M) = 0$.

Definition 2.2. Let R_0 be a commutative Noetherian ring, R a graded R_0-algebra and N a graded R-module. The R-module N is called *tame*, if there exists an integer j_0 such that

$$N_j = 0 \text{ for all } j \leq j_0, \quad \text{or } N_j \neq 0 \text{ for all } j \leq j_0.$$

In case of a standard bigraded K-algebra R we let R_0 be the K-subalgebra of R generated by all elements of degree $(1,0)$. Then R is a graded R_0-algebra with components $R_j = R_{(*,j)} = \bigoplus_i R_{(i,j)}$. Let N be a bigraded R-module. We may view N as a graded R-module with graded components $N_j = N_{(*,j)} = \bigoplus_i N_{(i,j)}$. Each of the modules N_j is a graded R_0-module, and if N is a finitely generated R-module then each N_j is a finitely generated R_0-module.

Now let M be a finitely generated bigraded R-module. Then $H^i_P(M)_j = H^i_{P_0}(M_j)$, where P_0 is the graded maximal ideal of R_0. Since M_j is a finitely generated R_0-module it follows that $H^i_{P_0}(M_j)$ is a graded Artinian R_0-module. Hence we see that $(H^i_P(M)^\vee)_j = (H^i_P(M)_{-j})^\vee = H^i_{P_0}(M_{-j})^\vee$ is a finitely generated graded R_0-module for all j. Of course this does not imply that $H^i_P(M)^\vee$ is a finitely generated R-module.

We denote by $\operatorname{cd}(M)$ the cohomological dimension of M with respect to Q, i.e., the number

$$\operatorname{cd}(M) = \sup\{i \in \mathbb{N}_0 : H^i_Q(M) \neq 0\}.$$

Corollary 2.3. *Let* $N = H^s_{R_+}(M)^\vee$. *Then the following statements hold:*

(a) $\operatorname{cd}(M) < \dim(M)$ *if and only if* $\operatorname{depth}_{R_0} N_j > 0$ *for all* j.
(b) *If* $\operatorname{cd}(M) < \dim(M) - 1$, *then* $\operatorname{depth}_{R_0} N_j > 1$ *for all* j.

Proof. We note that $\operatorname{cd}(M) < \dim(M)$, if and only if $H^s_Q(M) = 0$. Hence Proposition 2.1 yields part (a) of the corollary.

For the proof of (b) we notice that $H^i_P(N) = E^\infty_{s,m-i}$ for $i = 0, 1$, and that $E^\infty_{s,m-i}$ is a submodule of $H^{s-i}_Q(M)^\vee$ for all i. Thus our assumption implies that $H^i_{P_0}(N_j) = H^i_P(N)_j = 0$ for $i = 0, 1$ and all j. This yields the desired conclusion. □

The second statement of the next corollary is well-known (see [2, Theorem 4.8 (e)]).

Corollary 2.4. *Let M be a finitely generated bigraded R-module of dimension s and depth t. Then* $H^{t-m}_Q(M)$ *and* $H^s_Q(M)$ *are tame.*

Proof. We first prove $H^{t-m}_Q(M)$ is tame. By [1, Proposition 2.5] the dimension of N_j as an R_0-module is constant for $j \gg 0$. We set $N = H^t_{R_+}(M)^\vee$ and $s_0 = \dim_{R_0} N_j$ for $j \gg 0$. Note that $s_0 \leq \dim R_0 \leq m$. Thus we have

$H_P^m(N)_j = H_{P_0}^m(N_j) = 0$ for $j \gg 0$ if $s_0 < m$ and $H_P^m(N)_j = H_{P_0}^m(N_j) \neq 0$ for $j \gg 0$ if $s_0 = m$. Therefore by Proposition 2.1 there exists an integer j_0 such that

$$H_Q^{t-m}(M)_j = 0 \text{ for all } j \leq j_0, \quad \text{or } H_Q^{t-m}(M)_j \neq 0 \text{ for all } j \leq j_0,$$

as desired. In order to prove that $H_Q^s(M)$ is tame, we set $N = H_{R_+}^s(M)^\vee$. Since $H_{R_+}^s(M)$ is a graded Artinian R-module, N is a finitely generated graded R-module. Thus N_j is a finitely generated R_0-module. By [1, Proposition 2.5] the set of associated prime ideals of $\operatorname{Ass}_{R_0}(N_j)$ is constant for large j. If $P_0 \in \operatorname{Ass}_{R_0}(N_j)$, it follows that $H_P^0(N)_j = H_{P_0}^0(N_j) \neq 0$ for large j, and if $P_0 \notin \operatorname{Ass}_{R_0}(N_j)$, then $H_P^0(N)_j = H_{P_0}^0(N_j) = 0$ for large j. Thus in view of Proposition 2.1, $H_Q^s(M)$ is also tame. □

We say that M is a generalized Cohen-Macaulay R-module if $H_{R_+}^i(M)$ has finite length for all $i \neq \dim M$.

PROPOSITION 2.5. *Let M be a generalized Cohen-Macaulay R-module of dimension s. Then we have the following long exact sequence of bigraded R-modules*

$$\begin{aligned} 0 \to H_P^1(H_{R_+}^s(M)^\vee) &\to H_Q^{s-1}(M)^\vee \to H_{R_+}^{s-1}(M)^\vee \to \\ H_P^2(H_{R_+}^s(M)^\vee) &\to H_Q^{s-2}(M)^\vee \to H_{R_+}^{s-2}(M)^\vee \to \\ \cdots &\to H_Q^{s-m}(M)^\vee \to H_{R_+}^{s-m}(M)^\vee \to 0. \end{aligned}$$

Moreover, we have the following isomorphisms

$$H_{R_+}^i(M) \cong H_Q^i(M) \quad \textit{for all } i < s-m.$$

Proof. Since M is a generalized Cohen-Macaulay module, we have that $H_{R_+}^i(M)^\vee$ is of finite length for $i \neq s$. Thus by Grothendieck's vanishing theorem [5, Theorem 6.1.2] we see that $E_{i,j}^2 = E_{i,j}^\infty = 0$ for $j = 0, \ldots, m-1$ and $i \neq s$. The following picture will make this clear.
Therefore for all k with $s \leq k < s+m$ we get the following exact sequences

$$0 \to E_{s,r}^\infty \to H_Q^l(M)^\vee \to E_{l,m}^\infty \to 0,$$

$$0 \to E_{l,m}^\infty \to E_{l,m}^2 \to E_{s,r-1}^2 \to E_{s,r-1}^\infty \to 0,$$

where l and r are defined by the equations $s + r = l + m = k$.

Composing these two exact sequences we get the long exact sequence

$$\begin{aligned} \cdots \to E_{s,r}^2 \to H_Q^l(M)^\vee &\to E_{l,m}^2 \to E_{s,r-1}^2 \to \\ H_Q^{l-1}(M)^\vee &\to E_{l-1,m}^2 \to E_{s,r-2}^2 \to \cdots, \end{aligned}$$

which yields the desired exact sequence, observing that

$$H_P^0(H_{R_+}^i(M)^\vee) = H_{R_+}^i(M)^\vee$$

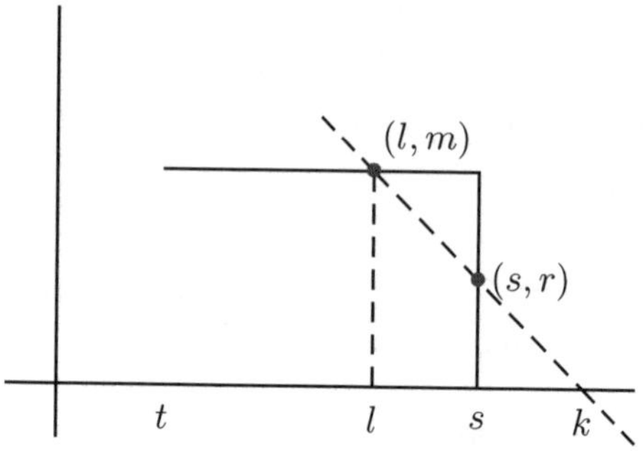

FIGURE 2

for $i \neq s$, since for such i the modules $H^i_{R_+}(M)^\vee$ have finite length. The last statement of the proposition follows similarly. □

COROLLARY 2.6. *Suppose M is a generalized Cohen-Macaulay module of dimension s. Then the following conditions are equivalent:*

(a) *M is Cohen-Macaulay.*

(b) *$H^k_P(H^s_{R_+}(M)^\vee) \cong H^{s-k}_Q(M)^\vee$ for all k.*

Proof. (a) $\Longrightarrow$ (b): Since M is Cohen-Macaulay we have $H^i_{R_+}(M)^\vee = 0$ for all $i \neq s$. Therefore it follows from the long exact sequence in Proposition 2.5 that $H^k_P(H^s_{R_+}(M)^\vee) \cong H^{s-k}_Q(M)^\vee$ for $k = 1, \dots, m$. The assertion for $k = 0$ follows from Proposition 2.1. The assertion is also clear when $k < 0$. Now assume that $k > m$. Then $s-k < s-m$, and hence by Proposition 2.5 it follows that $H^{s-k}_Q(M)^\vee = H^{s-k}_{R_+}(M)^\vee = 0$. On the other hand, we also have $H^k_P(H^s_{R_+}(M)^\vee) = 0$ because $k > m$.

(b) $\Longrightarrow$ (a): This is proved is the same way. □

As a generalization of Lemma 1.2 we obtain as an immediate consequence of Corollary 2.6 the following

REMARK 2.7. Let R be a bigraded Cohen-Macaulay K-algebra of dimension d. Then

$$H^k_P(\omega_R) \cong H^{d-k}_Q(R)^\vee \quad \text{for all } k.$$

Recall that for a finitely generated graded R-module N one has that $\dim_{R_0} N_j$ as well as $\operatorname{depth}_{R_0} N_j$ is constant for large j; see [1, Proposition 2.5]. In fact, if N is Cohen-Macaulay, then $\lim_{j\to\infty} \operatorname{depth}_{R_0} N_j = \dim N - \dim N/P_0N$ as shown in [9]. We call these constants the limit depth and limit dimension, respectively. Using this fact we have:

COROLLARY 2.8. *Let M be a bigraded Cohen-Macaulay R-module of dimension s. We set $N = H^s_{R_+}(M)^\vee$, and put $t_0 = \lim_{j\to\infty} \operatorname{depth}_{R_0} N_j$ and*

$s_0 = \lim_{j\to\infty} \dim_{R_0} N_j$. *Then the* R*-modules* $H_Q^j(M)$ *are tame for all* $j \leq s - s_0$ *and* $j \geq s - t_0$.

Proof. We see that $H_P^{s-i}(N)_j = H_{P_0}^{s-i}(N_j) \neq 0$ for $j \gg 0$ if $i = s - s_0$ and $i = s - t_0$, and also $H_P^{s-i}(N)_j = H_{P_0}^{s-i}(N_j) = 0$ for $j \gg 0$ if $i < s - s_0$ and $i > s - t_0$. Therefore by Corollary 2.6 we have the desired conclusion. □

Corollary 2.9. *Assume* R_0 *is Cohen-Macaulay and* M *is a bigraded Cohen-Macaulay* R*-module of dimension* s. *We set* $N = H_{R_+}^s(M)^\vee$. *Then:*

(a) *For all* k *and* j *we have the following isomorphism of graded* R_0*-modules*
$$\operatorname{Ext}_{R_0}^{d-k}(N_j, \omega_{R_0}) \cong H_Q^{s-k}(M)_{-j},$$
where $d = \dim R_0$.

(b) $\dim H_Q^{s-k}(M)_{-j} \leq k$ *for all* k *and* j.

Proof. Corollary 2.6 implies that
$$(H_Q^{s-k}(M)_{-j})^\vee \cong (H_Q^{s-k}(M)^\vee)_j = H_{P_0}^k(N_j).$$
Thus the local duality theorem yields
$$H_Q^{s-k}(M)_{-j} \cong H_{P_0}^k(N_j)^\vee \cong \operatorname{Ext}_{R_0}^{d-k}(N_j, \omega_{R_0}),$$
as desired.

Finally by [6, Corollary 3.5.11(c)] one has $\dim_{R_0} \operatorname{Ext}_{R_0}^{d-k}(N_j, \omega_{R_0}) \leq k$. This proves statement (b). □

Let $N \neq 0$ be a graded R_0-module. We set $a(N) = \inf\{i\colon N_i \neq 0\}$ and $b(N) = \sup\{i\colon N_i \neq 0\}$. If $N = 0$, we set $a(N) = \infty$ and $b(N) = -\infty$.

Recall that the *regularity* of N is defined to be
$$\operatorname{reg} N = \max\{b(H_{P_0}^k(N)) + k\colon k = 0, 1, \ldots\}.$$

With the assumptions and notation introduced in Corollary 2.6 we therefore have
$$\operatorname{reg}(N_j) = -\min\{a(H_Q^{s-k}(M)_{-j}) - k\colon k = 0, 1, \ldots\}.$$
In [8] and [11] it is shown that $\operatorname{reg}(N_j)$ is bounded above by a linear function of j. Thus in view of the preceding formula we get:

Corollary 2.10. *Let* M *be a Cohen-Macaulay* R*-module. Then there exist integers* c *and* d *such that* $a(H_Q^k(M)_j) \geq cj + d$ *for all* k *and all* j.

If the dimension and the depth of M differ at most by 1 or $\dim R_0 \leq 1$ one obtains:

Proposition 2.11. *The following statements hold:*

(a) *If* $\dim M = s$ *and* $\operatorname{depth} M = s-1$, *then we obtain the long exact sequence*

$$\cdots \to H_P^{m-j-2}(H_{R_+}^{s-1}(M)^\vee) \to H_P^{m-j}(H_{R_+}^{s}(M)^\vee) \to H_Q^{s-m+j}(M)^\vee \to$$
$$H_P^{m-j-1}(H_{R_+}^{s-1}(M)^\vee) \to H_P^{m-j+1}(H_{R_+}^{s}(M)^\vee) \to \cdots.$$

(b) *If* $\dim R_0 = 0$, *then* $H_{R_+}^i(M) \cong H_Q^i(M)$ *for all* i.

(c) *If* $\dim R_0 = 1$, *then for all* i *we have the short exact sequence*

$$0 \to H_P^1(H_{R_+}^{i+1}(M)^\vee) \to H_Q^i(M)^\vee \to H_P^0(H_{R_+}^i(M)^\vee) \to 0.$$

Proof. We first prove (a). Our hypotheses imply the following exact sequences

$$0 \to E_{s,j}^\infty \to H_Q^{s-m+j}(M)^\vee \to E_{s-1,j+1}^\infty \to 0,$$
$$0 \to E_{s-1,j+1}^\infty \to E_{s-1,j+1}^2 \to E_{s,j-1}^2 \to E_{s,j-1}^\infty \to 0.$$

Putting these two exact sequences together we get the long exact sequence

$$\cdots \to E_{s-1,j+2}^2 \to E_{s,j}^2 \to H_Q^{s-m+j}(M)^\vee \to E_{s-1,j+1}^2 \to$$
$$E_{s,j-1}^2 \to H_Q^{s-m-1+j}(M)^\vee \to E_{s-1,j}^2 \to E_{s,j-2}^2 \to \cdots,$$

which yields the desired exact sequence.

For the proof of (b) we set $N = H_{R_+}^i(M)^\vee$. Since $\dim R_0 = 0$, it follows that N_k is a finitely generated R_0-module of finite length. Thus $H_P^{m-j}(N)_k = H_{P_0}^{m-j}(N_k) = 0$ for all k and $j < m$ and hence $E_{i,j}^2 = 0$ for all i and $j \neq m$. Therefore we have $N = H_P^0(N) \cong H_Q^i(M)^\vee$ for all i, and so $H_{R_+}^i(M) \cong H_Q^i(M)$ for all i. In order to prove (c) we again set $N = H_{R_+}^i(M)^\vee$. Since $\dim R_0 = 1$, it follows that $H_P^{m-j}(N)_k = H_{P_0}^{m-j}(N_k) = 0$ for all k and $j < m-1$ and hence $E_{i,j}^2 = 0$ for all i and $j \neq m, m-1$. Thus for all i we get the exact sequence

$$0 \to E_{i+1,m-1}^\infty \to H_Q^i(M)^\vee \to E_{i,m}^\infty \to 0.$$

Since $E_{i+1,m-1}^\infty = E_{i+1,m-1}^2$ and $E_{i,m}^\infty = E_{i,m}^2$ for all i, the result follows. □

As a simple consequence of Proposition 2.11 (b),(c) we obtain the following tameness result due to [2, Theorem 4.5].

COROLLARY 2.12. *Let* $\dim R_0 \leq 1$. *Then* $H_Q^i(M)$ *is tame for all* i.

Proof. First we assume that $\dim R_0 = 0$. Since any Artinian graded R-module is tame, the result follows from Proposition 2.11 (b).

Now we assume that $\dim R_0 = 1$. Let N be a finitely generated bigraded R-module. By Proposition 2.11 (c) it is enough to prove that there exists an integer j_0 such that for $i = 0, 1$ one has:

$$H_{P_0}^i(N_j) = 0 \text{ for all } j \geq j_0, \quad \text{or } H_{P_0}^i(N_j) \neq 0 \text{ for all } j \geq j_0.$$

We set $t_0 = \lim_{j\to\infty} \operatorname{depth}_{R_0} N_j$ and $s_0 = \lim_{j\to\infty} \dim_{R_0} N_j$. Then $H^0_{P_0}(N_j) \neq 0$ for $j \gg 0$ if $t_0 = 0$, and $H^0_{P_0}(N_j) = 0$ for $j \gg 0$ if $t_0 \neq 0$. Similarly, $H^1_{P_0}(N_j) \neq 0$ for $j \gg 0$ if $s_0 = 1$, and $H^1_{P_o}(N_j)$ for $j \gg 0$ if $s_0 = 0$. □

Finally we want to mention two standard 5-term exact sequences arising from our spectral sequence.

PROPOSITION 2.13. *There is a 5-term exact sequence for the corner* $(t, 0)$

$$H_Q^{t+2-m}(M)^\vee \to H_P^{m-2}(H^t_{R_+}(M)^\vee) \to H_P^m(H^{d+1}_{R_+}(M)^\vee) \to H_Q^{t+1-m}(M)^\vee \to H_P^{m-1}(H^t_{R_+}(M)^\vee) \to 0,$$

and a 5-term exact sequence for the corner (s, m)

$$H^s_Q(M)^\vee \to H^0_P(H^{s-1}_{R_+}(M)^\vee) \to H^2_P(H^s_{R_+}(M)^\vee) \to H_Q^{s-1}(M)^\vee \to H^0_P(H^{s-1}_{R_+}(M)^\vee) \to 0.$$

REFERENCES

[1] M. Brodmann, *Asymptotic depth and connectedness in projective schemes*, Proc. Amer. Math. Soc. **108** (1990), 573–581. MR 1031674 (90m:13010)

[2] ———, *Asymptotic behaviour of cohomology: tameness, supports and associated primes*, Commutative algebra and algebraic geometry, Contemp. Math., vol. 390, Amer. Math. Soc., Providence, RI, 2005, pp. 31–61. MR 2187323 (2007b:13025)

[3] M. Brodmann, S. Fumasoli, and C. S. Lim, *Low-codimensional associated primes of graded components of local cohomology modules*, J. Algebra **275** (2004), 867–882. MR 2052643 (2005d:13028)

[4] M. Brodmann and M. Hellus, *Cohomological patterns of coherent sheaves over projective schemes*, J. Pure Appl. Algebra **172** (2002), 165–182. MR 1906872 (2003f:13017)

[5] M. P. Brodmann and R. Y. Sharp, *Local cohomology: an algebraic introduction with geometric applications*, Cambridge Studies in Advanced Mathematics, vol. 60, Cambridge University Press, Cambridge, 1998. MR 1613627 (99h:13020)

[6] W. Bruns and J. Herzog, *Cohen-Macaulay rings*, Cambridge Studies in Advanced Mathematics, vol. 39, Cambridge University Press, Cambridge, 1993. MR 1251956 (95h:13020)

[7] S. D. Cutkosky and J. Herzog, *Failure of tameness for local cohomology*, J. Pure Appl. Algebra **211** (2007), 428–432.

[8] S. D. Cutkosky, J. Herzog, and N. V. Trung, *Asymptotic behaviour of the Castelnuovo-Mumford regularity*, Compositio Math. **118** (1999), 243–261. MR 1711319 (2000f:13037)

[9] J. Herzog and T. Hibi, *The depth of powers of an ideal*, J. Algebra **291** (2005), 534–550. MR 2163482 (2006h:13023)

[10] M. Katzman and R. Y. Sharp, *Some properties of top graded local cohomology modules*, J. Algebra **259** (2003), 599–612. MR 1955534 (2004a:13011)

[11] V. Kodiyalam, *Asymptotic behaviour of Castelnuovo-Mumford regularity*, Proc. Amer. Math. Soc. **128** (2000), 407–411. MR 1621961 (2000c:13027)

[12] C. S. Lim, *Graded local cohomology modules and their associated primes: the Cohen-Macaulay case*, J. Pure Appl. Algebra **185** (2003), 225–238. MR 2006428 (2004k:13032)

[13] A. Rahimi, *On the regularity of local cohomology of bigraded algebras*, J. Algebra **302** (2006), 313–339. MR 2236605 (2007h:13024)

[14] ———, *Tameness of local cohomology of monomial ideals with respect to monomial prime ideals*, J. Pure Appl. Algebra **211** (2007), 83–93.

[15] J. Rotman, *An introduction to homological algebra*, Pure and Applied Mathematics, vol. 85, Acadamic Press, New York, 1979. MR 0538169 (80k:18001)

[16] C. Rotthaus and L. M. Şega, *Some properties of graded local cohomology modules*, J. Algebra **283** (2005), 232–247. MR 2102081 (2005h:13029)

Jürgen Herzog, Fachbereich Mathematik und Informatik, Universität Duisburg-Essen, Campus Essen, 45117 Essen, Germany
E-mail address: juergen.herzog@uni-essen.de

Ahad Rahimi, Fachbereich Mathematik und Informatik, Universität Duisburg-Essen, Campus Essen, 45117 Essen, Germany
E-mail address: ahad.rahimi@uni-essen.de

Illinois Journal of Mathematics
Volume 51, Number 1, Spring 2007, Pages 151–169
S 0019-2082

HOMOLOGICAL CONJECTURES, OLD AND NEW

MELVIN HOCHSTER

This paper is written to honor the many contributions of Phil Griffith to commutative algebra.

ABSTRACT. We discuss a number of the local homological conjectures, many of which are now theorems in equal characteristic and conjectures in mixed characteristic. One focus is the syzygy theorem of Evans and Griffith, and its connection with the direct summand conjecture, the existence of big Cohen-Macaulay modules and algebras, and tight closure theory.

1. Introduction

All given rings in this paper are commutative, associative with identity, and Noetherian. Our objective is to discuss some conjectures and theorems related to the local homological conjectures—for example, almost all of the results have some connection with the direct summand conjecture. Some of the conjectures have been around for decades. Others are quite recent, and some have grown out of our understanding of the related subjects of tight closure theory and existence of big Cohen-Macaulay algebras.

In Section 2 we discuss the syzygy theorem of Evans and Griffith, and several related intersection theorems. We are naturally led to consider the direct summand conjecture: other results, which either are equivalent to or imply the direct summand conjecture are considered in Section 3. Some of these other results involve the behavior of the local cohomology of the *absolute integral closure*, defined in Section 3. See [23], [26], and [21].

In Section 4 we give a characteristic p proof of the key lemma in the proof of the syzygy theorem that uses a little known variant of tight closure theory, following [30, Section 10]. One of our reasons for giving the argument is that this particular variant of tight closure is almost entirely unexplored, despite the fact that it has important applications.

Received August 22, 2006; received in final form June 7, 2007.
2000 *Mathematics Subject Classification.* Primary 13D02, 13D22, 13D45, 13A35, 13C14.
The author was supported by NSF grant DMS-0400633.

In Section 5 we describe recent progress in dimension 3: there was a spectacular breakthrough by Heitmann in [20], and the result of that paper was used in [25] to establish the existence of big Cohen-Macaulay algebras in dimension 3 in mixed characteristic.

Section 6 treats the strong direct summand conjecture, which was shown in [41] to be equivalent to a conjecture on vanishing of maps of Tor that can be proved in equal characteristic either using tight closure theory [33, Theorem 4.1] or the existence of big Cohen-Macaulay algebras in a weakly functorial sense as described in [34, Section 4].

In Section 7 we describe some recent results of G. Dietz [11] on when algebras over a local ring of characteristic $p > 0$ can be mapped to a big Cohen-Macaulay algebra. Finally, in Section 8 we treat quite briefly some other homological questions about local rings that remain open.

2. The syzygy theorem and some intersection theorems

A remarkable theorem of Evans and Griffith [15] and [16, Corollary 3.16] asserts the following:

THEOREM 2.1 (Syzygy Theorem). *Let R be a Cohen-Macaulay local ring* that contains a field *and let M be a finitely generated k th module of syzygies that has finite projective dimension over R. If M is not free, then M has rank at least k.*

This is not the most general result known: the condition that the ring be Cohen-Macaulay may be relaxed. However, the theorem is of great interest, and very hard to prove, even when the ring is regular! The original proof of Evans and Griffith used the existence of big Cohen-Macaulay modules [22], and while it is now known that the result can be deduced from *a priori* weaker statements, such as the direct summand conjecture [21], [23], it remains an open question in mixed characteristic, even when the ring is regular.

In fact, the theorem has the following amazing consequence [16, Theorem 4.4]:

COROLLARY 2.2. *Let R be a regular local ring of dimension at least three* such that R contains a field. *Let I be an ideal of R that is generated by three elements and is unmixed of height 2. Then R/I is Cohen-Macaulay or, equivalently, the projective dimension of I over R is 1.*

This result was so unexpected that I did not believe it. I tried repeatedly to give counterexamples, and on one occasion called Phil at an inconvenient time claiming that I had one. This turned out to be wrong, of course.

The method of proof of Evans and Griffith in essence showed that the theorem follows from an intersection theorem of the type introduced by Peskine and Szpiro [39], [40] and also studied by Roberts [42], [43], [44], [45], [46], Dutta [12], [13] and in my papers [22], [23]. Here is the result:

THEOREM 2.3 (Evans-Griffith improved new intersection theorem). *Let* (R, m) *be a local ring of Krull dimension* d that contains a field *and let*

$$0 \longrightarrow G_n \longrightarrow \cdots \longrightarrow G_1 \longrightarrow G_0 \longrightarrow 0$$

be a finite complex of finitely generated free modules such that

(1) $H_i(G_\bullet)$ *has finite length for* $i \geq 1$, *and*

(2) $H_0(G_\bullet)$ *has a minimal generator that is killed by a power of* m.

Then $d \leq n$.

The *new intersection theorem* of Peskine and Szpiro is the case where $H_0(G_\bullet)$ is nonzero of finite length. To see that this really is an intersection theorem, we deduce the Krull height theorem from it: consider a minimal prime P of an ideal generated by d elements, $x_1, \ldots, x_d$. We may pass to R_P, and we want to show that $\dim(R_P) \leq d$. This follows by applying the new intersection theorem to the Koszul complex $\mathcal{K}_\bullet(x_1, \ldots, x_d; R)$, which has finite length homology with nonzero $H_0\big(\mathcal{K}_\bullet(x_1, \ldots, x_d; R)\big) \cong R/(x_1, \ldots, x_d)R$.

By the main results of [23], the direct summand conjecture, which asserts that regular local rings are direct summands of their module-finite extensions, implies the improved new intersection theorem. In fact, this work of the author and the results of S. P. Dutta [13] combine to show that the direct summand conjecture, the improved new intersection theorem, and the canonical element conjecture,[1] are equivalent. What we want to get across is that the direct summand conjecture appears to be central, and I conjecture that whatever method leads to its solution will also yield the existence of big Cohen-Macaulay algebras.

With the exception of the new intersection conjecture, these conjectures all remain open in mixed characteristic in dimension four or more. Progress in dimension 3 is discussed in Section 5. The new intersection conjecture was proved in complete generality by P. Roberts [45], [46], [47]. This work made use of an idea of S. P. Dutta [12] as well as techniques from intersection theory developed in [5]: an exposition of the latter material is given in [17]. For further background see also [50], and my review of [47] in [25]. Note that

[1]We shall not discuss this conjecture [23] in detail here, but here is one statement: a free resolution of the residue class field K of the local ring (R, m, K) may be truncated so as to end at a dth module of syzygies S of K, where $d = \dim(R)$: call the complex one obtains $G_\bullet$. If $x_1, \ldots, x_d$ is any system of parameters, the obvious surjection $R/(x_1, \ldots, x_d)R \twoheadrightarrow K$ lifts to a map of the Koszul complex $\mathcal{K}_\bullet(x_1, \ldots, x_d; R) \longrightarrow G_\bullet$ and so there is a map at the dth spots, $R \longrightarrow S$. The conjecture asserts that no matter what choices are made, the image of $1 \in R$ in S (and even in $S/(x_1, \ldots, x_d)S$) is not 0. In fact, it turns out to be equivalent to assert that even after one takes a direct limit as the system of parameters varies, one has that the image of 1 in $H^d_m(S)$ (this image is *the canonical element*) is not 0.

the results of [39] showed that an affirmative answer to Bass's question [2] (if a local ring has a finitely generated nonzero module of finite injective dimension, must it be Cohen-Macaulay) and M. Auslander's zerodivisor conjecture [1] (which asserts that a zerodivisor on a nonzero finitely generated module M of finite projective dimension over a local ring R must be a zerodivisor in the ring) both follow from the version of the intersection theorem stated in [39], and this follows easily from the new intersection theorem. Thus, Bass's question is affirmatively answered, and the zerodivisor conjecture is true, even in mixed characteristic.

3. The direct summand conjecture and local cohomology of R^+

We mentioned in Section 1 that the direct summand conjecture is equivalent to a host of other conjectures. In fact, even to give all the known forms of the conjecture would require a lengthy manuscript. In this section we want to discuss primarily one form, which is connected with the behavior of the local cohomology of rings of the form R^+, where this notation is defined in the next paragraph.

By the *absolute integral closure* of a domain R, denoted R^+, we mean the integral closure of R in an algebraic closure of its fraction field. The domain R^+ is unique up to non-unique isomorphism, since this is true of the algebraic closure of the fraction field. It is a maximal integral extension of R that is still a domain. Note that every monic polynomial over R (or R^+) factors into monic linear factors over R^+, and this characterizes R^+. Evidently, if $R \subseteq S$ are domains, we may identify R^+ with a subring of S^+, so that there is a commutative diagram

$$\begin{array}{ccc} R^+ & \longrightarrow & S^+ \\ \uparrow & & \uparrow \\ R & \longrightarrow & S \end{array}$$

where the horizontal arrows are inclusions. Likewise, given a surjection $S \longrightarrow T$ of domains, so that $T \cong S/Q$, there is a prime ideal Q' of S^+ lying over Q. Since S^+/Q' is an integral extension of T in which every monic polynomial factors into linear factors, we have that S^+/Q' may be identified with T^+, and so there is a commutative diagram

$$\begin{array}{ccc} S^+ & \longrightarrow & T^+ \\ \uparrow & & \uparrow \\ S & \longrightarrow & T \end{array}$$

where the horizontal arrows are now surjections. Since every homomorphism of domains $R \longrightarrow T$ is the composition of an injection and a surjection, it follows that there is a commutative diagram

$$\begin{array}{ccc} R^+ & \longrightarrow & T^+ \\ \uparrow & & \uparrow \\ R & \longrightarrow & T \end{array}$$

for *every* homomorphism $R \longrightarrow T$ of domains.

The rings R^+ have attracted great interest recently for several reasons. In [32] it was shown that when R is an excellent local domain of characteristic $p > 0$, R^+ is a big Cohen-Macaulay algebra over R, that is, every system of parameters of R is a regular sequence on R^+, and the maximal ideal of R expands to a proper ideal. This is false, in general, in equal characteristic 0. However, recent work of Heitmann [20] has used the properties of the local cohomology of these rings to prove the direct summand conjecture in dimension 3 in mixed characteristic: see Section 5.

The equivalence given in the following result on the direct summand conjecture in mixed characteristic is, I feel, somewhat surprising. It was first observed in [23, Theorem (6.1)], but does not seem to be well known except to experts on the problem.

THEOREM 3.1. *Let V be a complete discrete valuation ring of mixed characteristic p, and let $A = V[[x_2, \ldots, x_d]]$. Let m be the maximal ideal of A. Then the direct summand conjecture holds for regular rings of dimension d in mixed characteristic if and only if for every such A, $H^d_m(A^+) \neq 0$.*

Before giving the proof, we remark that standard and relatively straightforward manipulations reduce the problem of proving the direct summand conjecture to the case where the regular ring A is a complete regular local ring. A subtler argument given in [23] makes the reduction to the case where A has the form above.

Proof. The fact that the direct summand conjecture implies that $H^d_m(A^+) \neq 0$ is easy: it suffices to show that the map $H^d_m(A) \longrightarrow H^d_m(A^+)$ is injective, and since A^+ is the direct limit of module-finite extension domains B of A, it suffices to see that for each such B the map $H^d_M(A) \longrightarrow H^d_m(B)$ is injective. But this is immediate if $B = A \oplus_A W$ for some A-module W.

To prove the other direction we want to show that every ring A as above is a direct summand of every module-finite extension ring B, and it suffices to consider domains, for we may first kill a minimal prime P of B disjoint from $A - \{0\}$. (A splitting of $A \hookrightarrow B/P$ composed with $B \twoheadrightarrow B/P$ gives a splitting of $A \hookrightarrow B$.) In fact, we shall show that under the condition $H^d_m(A^+) \neq 0$ considered in Theorem 3.1, A is a direct summand of A^+.

We may identify $H^d_m(A^+) \cong H^d_m(A) \otimes_A A^+$, and $H^d_m(A) = E$ is the injective hull of $K = A/m$ over A. If $H^d_m(A^+) \neq 0$, we can conclude that

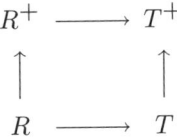

$\mathrm{Hom}_A(H^d_m(A^+), E) \neq 0$, and this may be written

$$\mathrm{Hom}_A(A^+ \otimes_A E, E) \cong \mathrm{Hom}_A\big(A^+, \mathrm{Hom}_A(E, E)\big)$$

by the adjointness of tensor and Hom. We have that $A \cong \mathrm{Hom}_A(E, E)$ by Matlis duality, since A is complete. Thus, our hypothesis implies that there is a nonzero A-linear map $f : A^+ \longrightarrow A$. Let π be the generator of the maximal ideal of V. The image of this map is a nonzero ideal of A, and we can write the image in the form $\pi^t J$, where t is a nonnegative integer chosen as large as possible, so that $J \not\subseteq \pi A$. We may compose f with $\pi^t J \cong J \subseteq A$ to obtain a map $f_1 : A^+ \longrightarrow A$ whose image is J. Hence, for some $a \in A^+$ we know that $f_1(a) \notin \pi A$. Define $g : A^+ \longrightarrow A$ via $g(r) = f_1(ar)$ for all $r \in A^+$. Then g is an A-linear map $A^+ \longrightarrow A$ such that $g(1) \notin \pi A$. Since $\bigcap_{N=1}^\infty (\pi, x_2^N, \dots, x_d^N)A = \pi A$, we can fix $N > 0$ such that $g(1) \notin (\pi, x_2^N, \dots, x_d^N)A$.

Let $A_0 = V[[x_2^N, \dots, x_d^N]] \subseteq A$, which has maximal ideal

$$m_0 = (\pi, x_2^N, \dots, x_d^N)A_0.$$

Then A is A_0-free over A_0, and $g(1) \in A - m_0 A$ is part of a free basis. Thus, there is an A_0-linear map $h : A \longrightarrow A_0$ that sends $g(1)$ to 1. Then $h \circ g : A^+ \longrightarrow A_0$ is an A_0-linear map sending 1 to 1, and it follows that A_0 is a direct summand of A^+ as an A_0-module. Since A is module-finite over A_0, we have that A^+ is also A_0^+, and so $A_0 \hookrightarrow A_0^+$ splits over A_0. Since $A_0 \cong A$, it follows that $A \hookrightarrow A^+$ splits over A. $\square$

We next want to describe a conjecture, which we refer to as the *Galois conjecture*, that implies the direct summand conjecture in all characteristics. This observation was first made in [26].

The Galois conjecture, made explicit below, asserts that a certain module is "small" in a sense that we shall make precise. This is true both in equal characteristic $p > 0$ and in equal characteristic 0: in fact, in both equicharacteristic cases, the module is not just small—it is 0. It is striking that the reasons why it is zero in those two cases appear to be completely different.

Let V be a complete discrete valuation ring, which may be either equal characteristic or mixed characteristic. In the mixed characteristic case assume that the residual characteristic p is the generator of the maximal ideal. In either case, denote the generator of the maximal ideal by $x = x_1$. Let $A = V[[x_2, \dots, x_d]]$ be a formal power series ring over V. Let $\mathcal{F}$ denote the fraction field of A, and then the fraction field of A^+ is an algebraic closure $\overline{\mathcal{F}}$ of $\mathcal{F}$. Let G be the Galois group of $\mathcal{F}$-automorphisms of $\overline{\mathcal{F}}$, which also acts on A^+. Note that ${A^+}^G = A$ when $\mathcal{F}$ has characteristic zero, which includes the case where A has equal characteristic zero and the case where A has mixed characteristic.

We let $E = H^d_m(A)$, the highest (in fact, the only) nonzero local cohomology module of A with support in $m = m_A$, since it is also an injective hull $E_A(K)$ for the residue field $K = A/m$ of A over A. We write $M^\vee$ for $\mathrm{Hom}_A(M, E)$.

If (B, n, L) is any complete local ring, we shall call a B-module Q *small* if $E_B(L)$, the injective hull of $L = B/n$ over B, cannot be injected into Q. Note that if $E_B(L)$ is a submodule of Q, then it is actually a direct summand of Q, since $E_B(L)$ is an injective B-module. Thus, the condition that a module be small does not seem unduly restrictive.

Theorem 3.1 above implies that in order to prove the direct summand conjecture, it suffices to show that the modules $H^d_m(A^+)$ are not zero.

Now $x = x_1$ is a regular parameter in A, and we have a short exact sequence

$$0 \longrightarrow A^+ \xrightarrow{\ x\ } A^+ \longrightarrow A^+/xA^+ \longrightarrow 0.$$

If we contradict the direct summand conjecture and assume that $H^d_m(A^+) = 0$, part of the corresponding long exact sequence for local cohomology gives:

$$H^{d-1}_m(A^+) \xrightarrow{\ x\ } H^{d-1}_m(A^+) \longrightarrow H^{d-1}_m(A^+/xA^+) \longrightarrow 0.$$

This implies an isomorphism

$$H^{d-1}_m(A^+/xA^+) \cong H^{d-1}_m(A^+)/xH^{d-1}_m(A^+).$$

The regular ring A/xA injects into A^+/xA^+ (because A is normal, the principal ideal xA is contracted from A^+). Suppose that A provides a counterexample to the direct summand conjecture of smallest dimension or that A has mixed characteristic, provides a counterexample, and $x = p$. Under either hypothesis, A provides a counterexample, but the direct summand conjecture holds for the regular ring A/xA. Then A/xA is a direct summand of A^+/xA^+ as an (A/xA)-module, and it follows that $H^{d-1}_m(A/xA)$ injects into $H^{d-1}_m(A^+/xA^+)$. Evidently, since G acts on A^+, m is contained in the ring of invariants of this action, and the element x is an invariant, we have that G acts on $H^{d-1}_m(A^+)/xH^{d-1}_m(A^+)$, and it is clear that $H^{d-1}_m(A/xA)$ injects into

$$\left(H^{d-1}_m(A^+)/xH^{d-1}_m(A^+)\right)^G \subseteq H^{d-1}_m(A^+)/xH^{d-1}_m(A^+).$$

We therefore will have a contradiction that establishes the direct summand conjecture if we can prove the following:

CONJECTURE 3.2 (Galois Conjecture). *Let (A, m, K) be a complete regular local ring of dimension d with fraction field $\mathcal{F}$, let G be the automorphism group of the algebraic closure $\overline{\mathcal{F}}$ over $\mathcal{F}$, and let x be a regular parameter in A. Then $\left(H^{d-1}_m(A^+)/xH^{d-1}_m(A^+)\right)^G$ is a small (A/xA)-module.*

THEOREM 3.3. *The Galois Conjecture holds if* $\dim A \leq 2$ *or if A contains a field. In fact, in all of these cases* $\left(H^{d-1}_m(A^+)/xH^{d-1}_m(A^+)\right)^G = 0$.

Proof. The explanation when A contains a field is quite different depending on whether the field has characteristic 0 or positive characteristic. In the first case, it turns out that ^{G} is an exact functor here, so that what we have is

$$(H_m^{d-1}(A^{+^G})/xH_m^{d-1}(A^{+^G})),$$

and since $A^{+^G} = A$, this is $H_m^{d-1}(A)/xH_m^{d-1}(A)$, and $H_m^{d-1}(A) = 0$. In the positive characteristic case we know from the main result of [32] that A^+ is a big Cohen-Macaulay algebra, so that $H_m^{d-1}(A^+) = 0$, and the result follows again. The same argument shows that the conjecture is true when A has dimension at most two. □

From the discussion above, we have the following:

THEOREM 3.4. *If the Galois Conjecture is true whenever A is a formal power series ring $V[[x_2, \dots, x_d]]$ over a complete discrete valuation domain (V, pV, K) of mixed characteristic and residual characteristic $p > 0$, and $x = p$, then the direct summand conjecture is true.* □

4. A "tight closure" proof of the syzygy theorem in characteristic p

In this section, we give a "tight closure" proof of a key lemma in the proof of the syzygy theorem. The usual notion of tight closure in a Noetherian ring R of characteristic $p > 0$ uses a fixed multiplier c that is required not to be in any minimal prime of R. The reader is referred to [30], [29], [31], [35], [37], [28], and [8] for background. The argument below, based on a treatment in [30, Section 10], utilizes a variant notion of tight closure in which the only restriction placed on c is that it be nonzero. This has certain disadvantages. The operation one gets, when iterated, typically produces a larger closure. However, the usefulness of this variant notion in the argument below argues that it deserves further study. One of our main motivations in presenting the argument here is to encourage investigation of this variant notion.

We first present a family of variant notions of tight closure:

Consider a non-empty family of nonzero ideals $\mathcal{C}$ in a Noetherian ring R of characteristic $p > 0$ with the property:

(∗) if $C, C' \in \mathcal{C}$ then there exists $C'' \in \mathcal{C}$ such that $C'' \subseteq C \cap C'$.

Then we can define the tight closure with respect to $\mathcal{C}$: an element $u \in N \subseteq M$ is in the *tight closure with respect to* $\mathcal{C}$ of N in M if there exists an ideal $C \in \mathcal{C}$ such that $Cu^q \in N^{[q]}$ for all $q = p^e \gg 0$.[2] We can also define the *small tight*

[2]Let R be Noetherian of prime characteristic $p > 0$. We abbreviate $q = p^e$, and let $\mathbf{F}^e$ denote the eth iteration of the Frobenius functor: this is the base change functor $S_e \otimes_R \_$, where S_e is the R-algebra whose structural homomorphism is $F^e : R \longrightarrow R$, with $F^e(r) = r^q$ for all $r \in R$. If $N \subseteq M$, let $N^{[q]}$ be the image of $\mathbf{F}^e(N) \longrightarrow \mathbf{F}^e(M)$. If $u \in M$, let u^q denote $1 \otimes u$ in $\mathbf{F}^e(M)$. Then $N^{[q]}$ is the R-span within M of the elements $\{u^q : u \in N\}$. In particular, for ideals $I \subseteq R$, $I^{[q]} = (i^q : i \in I)R$.

closure of N in M with respect to $\mathcal{C}$: for this we require that for some $C \in \mathcal{C}$, $Cu^q \in N^{[q]}$ for all q (which includes $q = 1$). The property $(*)$ is needed so that the tight closure of N will be closed under addition.

If we take the family $\mathcal{C}$ to consist of all principal ideals generated by an element of R°, we obtain the usual notion of tight closure.

If the family consists of only the unit ideal R, tight closure with respect to this family is Frobenius closure, while the small tight closure of N is the submodule N itself.

If R has a test element, tight closure with respect to the family consisting of the single ideal it generates gives ordinary tight closure, as does small tight closure with respect to the family consisting of the single ideal it generates.

We note that iterating one of these variant tight closure operations may give a larger result than performing it once. One can show that iterating the operation gives the same result if the family of ideals has the property that for all C, $C' \in \mathcal{C}$, there exists $C'' \in \mathcal{C}$ such that $C'' \subseteq CC'$. To see this, suppose that the tight closure Q of N with respect to $\mathcal{C}$ has generators u_i. For every i, we can choose $C_i \in \mathcal{C}$ such that $C_i u^q \in N^{[q]}$ for all $q \gg 0$. By property $(*)$ we can choose $C \subseteq \bigcap_i C_i$ with $C \in \mathcal{C}$. It follows that $CQ^{[q]} \subseteq N^{[q]}$ for all $q \gg 0$. If v is in the tight closure of Q with respect to $\mathcal{C}$, then we can choose C' such that $C'v^q \subseteq Q^{[q]}$ for all $q \gg 0$. Then $C'Cv^q \subseteq N^{[q]}$ for all $q \gg 0$, and so if $C'' \in \mathcal{C}$ and $C'' \subseteq C'C$, then $C''v^q \subseteq N^{[q]}$ for all $q \gg 0$. An entirely similar argument establishes the corresponding fact for small tight closure.

We now want to show how one of these variant notions of tight closure can be used to prove the Evans-Griffith syzygy theorem. We want to make two remarks. First, it is immediate from the definition that $u \in M$ is in the tight closure (respectively, small tight closure) with respect to $\mathcal{C}$ of N in M if and only if the image of u in M/N is in the tight closure (respectively, small tight closure) of 0 in M/N with respect to $\mathcal{C}$. The second remark we state as:

LEMMA 4.1. *If (R, m, K) is local, $\mathcal{C}$ is a non-empty family of nonzero ideals of R, and x is a minimal generator of a finitely generated module M, then x is not in the tight closure (nor in the small tight closure) of 0 in M with respect to $\mathcal{C}$.*

Proof. If u is in the tight closure of 0 in M we have that $Cx^q = 0$ in $F^e(M)$ for all $q \gg 0$. We can map $M \twoheadrightarrow K$ so that $x \mapsto 1$. We get an induced surjection $F^e(M) \longrightarrow R/m^{[q]}$. It follows that $C \subseteq m^{[q]}$ for all $q \gg 0$, which implies that $C = (0)$, a contradiction. □

We shall need to make use of the notion of order ideal.

DEFINITION 4.2. Let x be an element of M, a finitely generated module over a Noetherian ring R. We define the *order ideal* $\mathfrak{O}_M(x) = \mathfrak{O}(x)$ to be $\{f(x) : f \in \mathrm{Hom}_R(M, R)\}$.

For finitely generated modules over a Noetherian ring R, the formation of the order ideal commutes with localization. The map $R \longrightarrow M$ sending $1 \mapsto x$ evidently splits if and only if $\mathfrak{O}_M(x) = R$.

Also note that for any finitely generated free R-module G, any R-linear map $M \longrightarrow G$ takes x into $\mathfrak{O}_M(x)G$.

The Evans-Griffith syzygy theorem asserts that a k*th* module of syzygies over a regular local ring, if not free, has rank at least k. They prove more general statements, in which the conditions on the ring are weakened but the module is assumed to have finite projective dimension. However, the key point in their proof is the following:

THEOREM 4.3 (Evans-Griffith). *Let R be a local ring* such that R contains a field, *let M be a kth module of syzygies of a finitely generated module of finite projective dimension, and suppose that M_P is R_P-free for every prime P of R except the maximal ideal, i.e., M is locally free on the punctured spectrum of R. Let $x \in M$ be a minimal generator. Then $\mathfrak{O}(x)$ is either the unit ideal or else has height at least k.*

In fact, they show that this is true by using the fact that the improved new intersection theorem is true when R contains a field, which they deduce from the existence of big Cohen-Macaulay modules in the equal characteristic case. We shall not duplicate their argument here. But we shall prove a better result in characteristic p, with depth replacing height and without the assumption that M is locally free on the punctured spectrum. This is where we use a variant notion of tight closure. The following is [30, Theorem 10.8, p. 103].

THEOREM 4.4. *Let (R, m, K) be a local ring of prime characteristic $p > 0$ and let N be a finitely generated module of finite projective dimension over R. Let M be a finitely generated k th module of syzygies of N, and let $x \in M$ be a minimal generator of M. Let $I = \mathfrak{O}_M(x)$. Then either $I = R$ or else* $\operatorname{depth}_I R \geq k$.

Proof. If not, let $y_1, \dots, y_d$ be a maximal regular sequence in the proper ideal I, where $d < k$, and let $J = (y_1, \dots, y_d)R$. Then we can choose $c \in R - J$ such that $cI \subseteq J$. Let c' denote the image of c in $R' = R/J$. Let $G_\bullet$ be a resolution of N by finitely generated free modules over R such that $G_k \longrightarrow G_{k-1}$ factors $G_k \twoheadrightarrow M \hookrightarrow G_{k-1}$, which we know exists because M is a kth module of syzygies of N over R. Let B denote the image of G_{k+1} in G_k. Let G'_j denote $R' \otimes_R G_j$, while M' denotes $R' \otimes_R M$ and B' denotes the image of $R' \otimes_R B$ in G'_k. Choose an element z of G_k that maps onto $x \in M$. We shall obtain a contradiction by showing that the image z' of z in G'_k is in the tight closure of B' in G'_k with respect to the family $\{c'R'\}$. This implies that the image x' of x in M' is in the tight closure of 0 in M' with respect to the family $\{c'R'\}$, a contradiction using the Lemma 4.1 above, because x' is a minimal generator of M'.

To see this, note that $F_R^e(G_\bullet)$ remains acyclic for all e: the determinantal ranks of the maps and the depths of the ideals of minors do not change. Thus, this is a free resolution of $F_R^e(N)$, and it follows that $R' \otimes_R F_R^e(G_\bullet)$ has homology $\mathrm{Tor}_\bullet^R(R', F_R^e(N))$. Since $\mathrm{pd}_R R' = d < k$, we have that $\mathrm{Tor}_k^R(R', F_R^e(N)) = 0$. But the complex $R' \otimes_R F_R^e(G_\bullet)$ may be identified with $F_{R'}^e(G'_\bullet)$. Let δ denote the map $G'_k \longrightarrow G'_{k-1}$. Now consider the value of the R-linear map $F_{R'}^e(\delta)$ evaluated on $c'(z')^q$. This is evidently $c'F_{R'}^e(\delta)\big((z')^q)\big)$. Since the map $G_k \longrightarrow G_{k-1}$ factors through M, the image of z, which maps to $x \in M$, is in IG_{k-1}. It follows that the image of z' under δ in IG'_{k-1}, and, hence, that the image of $(z')^q$ under $F^e(\delta)$ is in

$$I^{[q]}F^e(G'_{k-1}) \subseteq IF_{R'}^e(G'_{k-1}).$$

Since $cI \subseteq J$ and J becomes 0 in R', we have that $F_{R'}^e(\delta)\big(c'(z')^q\big) = 0$. Since $c'(z')^q$ is a cycle and the homology at this spot is 0, it follows that $c'(z')^q$ is a boundary, which means that it is in the image $(B')^{[q]}$ of $F_{R'}^e(B')$. Thus, z' is in the tight closure with respect to the family $\{c'R'\}$ of B' in G'_k, and this means that x' is in the tight closure with respect to $\{cR'\}$ of 0 in M'. Since x is a minimal generator of M and $J \subseteq m$, it follows that x' is a minimal generator of M', and we have obtained the contradiction of Lemma 4.1 mentioned earlier. □

It is quite easy to deduce the syzygy theorem from Theorem 4.3: for details see, for example, [30, Cor. 10.10, p. 105].

Finally, note that the equal characteristic 0 cases of both Theorem 4.3 and of the syzygy theorem can be deduced from the equal characteristic $p > 0$ case by standard methods of reduction to characteristic p.

5. Recent progress in dimension 3

Ray Heitmann [20] recently proved the direct summand conjecture in dimension 3. The argument is very difficult. The main result of Heitmann's paper (although stated differently from the version in [20]) is:

THEOREM 5.1 (Heitmann). *Let (R, m, K) be a complete local domain of mixed characteristic p and dimension 3. Then every element of $H^m(R^+)$ is killed by arbitrarily small powers of p, i.e., by $p^{1/N}$ for arbitrarily large values of the positive integer N.*

This result can be expressed more concretely as follows: let x, y, z be a system of parameters for a three dimensional complete local domain R of mixed characteristic p, and suppose that $zu \in (x, y)R$. Then for every positive integer N there exists a module-finite extension domain S of R such that $p^{1/N} \in S$ and $p^{1/N}u \in (x, y)S$.

We make the curious observation that, in general, the direct summand conjecture seems to follow if we can show that, for all complete local domains R of dimension d, $H^d_m(R^+)$ is "big" (in fact, simply nonzero) or if we can show that for all such R of dimension d, $H^{d-1}_m(R^+)$ is "small" (in one of several senses that are rather technical).

In [27] Heitmann's result is used to prove that every complete local domain of dimension at most 3 has a big Cohen-Macaulay algebra. In characteristic p, one has a "weakly functorial" version of the existence of big Cohen-Macaulay algebras: if $R \longrightarrow S$ is a local homomorphism of complete local domains, one has a commutative diagram:

$$\begin{array}{ccc} R^+ & \longrightarrow & S^+ \\ \uparrow & & \uparrow \\ R & \longrightarrow & S \end{array}$$

and, hence, a commutative diagram:

$$\begin{array}{ccc} B & \longrightarrow & C \\ \uparrow & & \uparrow \\ R & \longrightarrow & S \end{array}$$

where B and C are big Cohen-Macaulay algebras. A corresponding result for the equal characteristic 0 case is proved in [34] by reduction to characteristic p. It is shown in [27] that one has:

THEOREM 5.2 (weakly functorial big Cohen-Macaulay algebras in dimension at most 3). *Let $R \longrightarrow S$ be a local homomorphism of complete local domains, both of mixed characteristic, and both of dimension at most 3. Then there exists a commutative diagram*

$$\begin{array}{ccc} B & \longrightarrow & C \\ \uparrow & & \uparrow \\ R & \longrightarrow & S \end{array}$$

where B is a big Cohen-Macaulay algebra over R and C is a big Cohen-Macaulay algebra over S.

I conjecture that a sufficiently good result on the weakly functorial existence of big Cohen-Macaulay algebras in mixed characteristic is equivalent to the existence of tight closure theory in mixed characteristic. This is an admittedly vague statement. A much more precise version of this statement is made in [11]. Some of the results of [11] are discussed in Section 7.

We note that the existence of big Cohen-Macaulay algebras in a weakly functorial sense implies the vanishing conjecture for maps of Tor discussed in Section 6 just below.

6. The strong direct summand conjecture and vanishing of maps of Tor

In this section we discuss one of the new homological conjectures, the vanishing conjecture for maps of Tor. It is a theorem in equal characteristic but remains open in mixed characteristic. In a reasonably non-technical version, the conjecture may be phrased as follows:

CONJECTURE 6.1 (Vanishing conjecture for maps of Tor). *Suppose that* $A \longrightarrow R \longrightarrow T$ *are homomorphisms of Noetherian rings such that:*

(1) *A is regular.*
(2) *R is a module-finite and torsion-free extension of A.*
(3) *T is regular.*

Then for every A module M and $i \geq 1$, the map

$$\mathrm{Tor}_i^A(M, R) \longrightarrow \mathrm{Tor}_i^A(M, T)$$

is 0.

It is reasonably straightforward to reduce to the case where M is finitely generated, $i = 1$, T and A are complete regular local rings, and R is a complete local domain. As mentioned earlier, proofs for the equicharacteristic case are given in [33, Section 4] and [32, Theorem (4.1)].

This conjecture, when it is true, is a very powerful tool. For example, it implies the direct summand conjecture [33, Section 4] and the statement that a ring R that is a direct summand of a regular ring T is Cohen-Macaulay. The latter statement was originally proved for certain rings of invariants [36], and stronger statements are true in equal characteristic 0 [7].

The proof of the statement that direct summands of regular rings are Cohen-Macaulay may be sketched as follows. One can come down to the case where R is complete local and module-finite over a regular local ring (A, m, K). But then the maps

$$\mathrm{Tor}_i^A(K, R) \longrightarrow \mathrm{Tor}_i^A(K, T),$$

which are injective because R is a direct summand of T, are 0 for $i \geq 1$ by the Vanishing Conjecture (6.1), and so the modules $\mathrm{Tor}_i(K, R) = 0$ for $i \geq 1$, which implies that R is Cohen-Macaulay.

Although Conjecture 6.1 does not appear to be purely a result about "splitting," N. Ranganathan proved that it is equivalent to the following conjecture in [41]:

Conjecture 6.2 (Strong direct summand conjecture). *Let $A \longrightarrow R$ be a local homomorphism where A is a regular local ring and R is a domain module-finite over A. Let Q be a height one prime ideal of R lying over xA, where x is a regular parameter for A (i.e., A/xA is again a regular local ring). Then the inclusion map $xA \longrightarrow Q$ splits as a map of A-modules.*

Note that this result implies the direct summand conjecture, for $xA \subseteq xR \subseteq Q$, and so xA is a direct summand of xR, which means that A is a direct summand of R. The direct summand conjecture has an easy proof in equal characteristic 0, using a trace argument, but Conjecture 6.2 appears subtle and difficult in all characteristics.

7. Algebras that map to big Cohen-Macaulay algebras

A central and challenging open question is this: given an algebra S over a complete local domain R, when can S be mapped to a big Cohen-Macaulay algebra over R? Following [11], we call such an R-algebra S a *seed* over R. Some remarkable results about seeds in characteristic $p > 0$ are obtained in [11]. Whether there are corresponding results in equal characteristic 0 is an open question: in these instances, standard methods of reduction to characteristic p do not succeed.

Theorem 7.1 (G. Dietz). *Let (R, m) be a complete local domain of characteristic $p > 0$. Then every seed S over R maps to a seed T with all of the following properties:*

(1) *T is a domain.*
(2) *T is absolutely integrally closed, i.e., $T = T^+$.*
(3) *T is m-adically complete and separated.*

Theorem 7.2 (G. Dietz). *If $R \longrightarrow R'$ is a local homomorphism of complete local domains and S is a seed over R then $R' \otimes_R S$ is a seed over R'. That is, for every big Cohen-Macaulay algebra B over R there is a big Cohen-Macaulay C over R' and a commutative diagram*

$$\begin{array}{ccc} B & \longrightarrow & C \\ \uparrow & & \uparrow \\ R & \longrightarrow & R' \end{array}$$

Remark 7.3. Note that before the work of [11], it was known that there exist big Cohen-Macaulay algebras B and C such that the diagram commutes: one can take, for example, $B = R^+$ and $C = S^+$. But it was not known that one can construct such a diagram for every given B.

Theorem 7.4 (G. Dietz). *Let R be a complete local domain and let S and S' be seeds over R. Then $S \otimes_R S'$ is a seed over R. Equivalently, given big*

Cohen-Macaulay algebras B and B' over R there exists a big Cohen-Macaulay algebra C and a commutative diagram

$$\begin{array}{ccc} B & \longrightarrow & C \\ \uparrow & & \uparrow \\ R & \longrightarrow & B' \end{array}$$

Of course, in mixed characteristic in dimension 4 or more, we do not even know whether the complete local domain R is itself a seed over R: this question is the existence of big Cohen-Macaulay algebras.

Even in characteristic p, we do not fully understand how to characterize seeds. First, we recall:

DEFINITION 7.5. If R is a domain, an R algebra S is called *solid* if there exists a nonzero R-linear module homomorphism $f : S \longrightarrow R$.

If (R, m) is a complete local domain of dimension d, it turns out that R is solid if and only if $H^d_m(R) \neq 0$ [24, Corollary (2.4)]. We refer the reader to [24] for a detailed treatment. It is an open question in characteristic p whether an R-algebra S is solid if and only if it is a seed. However, this is known to be false in equal characteristic 0.[3]

8. Other questions

In this final section we want to make some brief comments on other homological questions. The conjecture of Buchsbaum and Eisenbud [3], reported in [4] as a question by Horrocks, that the ith Betti number of a module of finite length over a local ring is at least $\binom{n}{i}$ remains open in dimension 5 or more.

Serre's conjecture [49] that if (R, m) is regular local of Krull dimension d and M, N are finitely generated nonzero modules such that $M \otimes_R N$ has finite length then

$$\chi(M, N) = \sum_{i=0}^{d} \ell(\mathrm{Tor}_i^R(M, N))$$

[3]For example, let $X = (x_{ij})$ be a 2×3 matrix of indeterminates over a field K of characteristic 0, and let $\Delta_1, \Delta_2, \Delta_3$ be the 2×2 minors. Then $R = K[\Delta_1, \Delta_2, \Delta_3] \subseteq K[x_{ij}] = S$ splits, since $R = S^G$, where $G = \mathrm{SL}(2, K)$ acting so that $\gamma \in G$ maps the entries of X to the corresponding entries of γX. The splitting is a consequence of the fact that G is reductive: these ideas originate in [51], but see also [36, p. 119]. We still have the splitting after applying $\widehat{R} \otimes_R \_$. Thus, $T = \widehat{R} \otimes_R S$ is solid over $\widehat{R}$, but T does not map to a big-Cohen Macaulay algebra over $\widehat{R}$. In any such algebra the relations $\sum_{j=1}^{3}(-1)^{j-1} x_{ij}\Delta_j = 0$ will force each x_{ij} into the ideal J generated by the Δ_j, which, since each Δ_j is a quadratic form in the x_{ij}, implies that $\Delta_j \in J^2$, so that $J = J^2$. This is not possible when J is generated by a proper regular sequence.

is nonnegative and vanishes if and only if $\dim(M)+\dim(N) < \dim(R)$ (Serre showed that one must have $\dim(M)+\dim(N) \leq \dim(R)$) remains open. The conjecture was already proved by Serre if $\widehat{R}$ is formal power series over a field or a discrete valuation ring. That $\chi(M, N) = 0$ if $\dim(M) + \dim(N) < \dim(R)$ was proved by P. Roberts [44] and Gillet-Soulé [18] independently. See also [47, Corollary 13.1.1]. Non-negativity in the remaining cases was proved by O. Gabber (there is an exposition in [6]) using results of De Jong [10] on alterations. Strict positivity in the mixed characteristic case when $\dim(M) + \dim(N) = \dim(R)$ remains an open question.

This intersection multiplicity is defined more generally when the local ring is not necessarily regular, but M, N are finitely generated modules such that $M \otimes_R N$ has finite length and one of them, say M, has finite projective dimension. But in this generality $\chi(M, N)$ can be negative [14], although [40] has an affirmative result for the graded equicharacteristic case.

With the same hypotheses as in the preceding paragraph, one may ask whether it must be true that $\dim(M) + \dim(N) \leq \dim(R)$. See [39], [40]. This is very much an open question. In fact, Peskine and Szpiro [39] even raised the following question: under the same hypotheses, with $I = \mathrm{Ann}_R M$, must it be true that $\dim(N) \leq \mathrm{depth}_I R$ (which is always $\leq \dim(R) - \dim(M)$). This remains open.

A conjecture of M. Auslander on the rigidity of Tor for modules of finite projective dimension was disproved by Ray Heitmann [19] using "generic" modules of projective dimension two introduced in [22].

Finally, we mention that [9] has results over local complete intersections connecting a version of rigidity and the dimension inequality

$$\dim(M) + \dim(N) \leq \dim(R).$$

References

[1] M. Auslander, *Modules oever unramified regular local rings*, Proc. Internat. Congr. Mathematicians (Stockholm, 1962), Inst. Mittag-Leffler, Djursholm, 1963, pp. 230–233. MR 0175930 (31 #206)

[2] H. Bass, *On the ubiquity of Gorenstein rings*, Math. Z. **82** (1963), 8–28. MR 0153708 (27 #3669)

[3] D. A. Buchsbaum and D. Eisenbud, *Algebra structures for finite free resolutions, and some structure theorems for ideals of codimension* 3, Amer. J. Math. **99** (1977), 447–485. MR 0453723 (56 #11983)

[4] R. Hartshorne, *Algebraic vector bundles on projective spaces: a problem list*, Topology **18** (1979), 117–128. MR 544153 (81m:14014)

[5] P. Baum, W. Fulton, and R. MacPherson, *Riemann-Roch for singular varieties*, Inst. Hautes Études Sci. Publ. Math. (1975), 101–145. MR 0412190 (54 #317)

[6] P. Berthelot, *Alteŕations de variétes algébriques [d'après A.J. de Jong]*, Séminaire Bourbaki exposé 815 (1996), Astérisque **241** (1997). MR 1472543 (98m:14021)

[7] J.-F. Boutot, *Singularités rationnelles et quotients par les groupes réductifs*, Invent. Math. **88** (1987), 65–68. MR 877006 (88a:14005)

[8] W. Bruns, *Tight closure*, Bull. Amer. Math. Soc. (N.S.) **33** (1996), 447–457. MR 1397097 (97d:13003)

[9] H. Dao, *Homological properties of modules over complete intersections*, Thesis, University of Michigan, 2006.

[10] A. J. de Jong, *Smoothness, semi-stability and alterations*, Inst. Hautes Études Sci. Publ. Math. (1996), 51–93. MR 1423020 (98e:14011)

[11] G. Dietz, *Closure operations in positive characteristic and big Cohen-Macaulay algebras,* Thesis, University of Michigan, 1995.

[12] S. P. Dutta, *Frobenius and multiplicities*, J. Algebra **85** (1983), 424–448. MR 725094 (85f:13022)

[13] ———, *On the canonical element conjecture*, Trans. Amer. Math. Soc. **299** (1987), 803–811. MR 869233 (87m:13028)

[14] S. P. Dutta, M. Hochster, and J. E. McLaughlin, *Modules of finite projective dimension with negative intersection multiplicities*, Invent. Math. **79** (1985), 253–291. MR 778127 (86h:13023)

[15] E. G. Evans and P. Griffith, *The syzygy problem*, Ann. of Math. (2) **114** (1981), 323–333. MR 632842 (83i:13006)

[16] ———, *Syzygies*, London Mathematical Society Lecture Note Series, vol. 106, Cambridge Univ. Press, Cambridge, 1985. MR 811636 (87b:13001)

[17] W. Fulton, *Intersection theory*, Ergebnisse der Mathematik und ihrer Grenzgebiete (3), vol. 2, Springer-Verlag, Berlin, 1984. MR 732620 (85k:14004)

[18] H. Gillet and C. Soulé, *K-théorie et nullité des multiplicités d'intersection*, C. R. Acad. Sci. Paris Sér. I Math. **300** (1985), 71–74. MR 777736 (86k:13027)

[19] R. C. Heitmann, *A counterexample to the rigidity conjecture for rings*, Bull. Amer. Math. Soc. (N.S.) **29** (1993), 94–97. MR 1197425 (93m:13007)

[20] ———, *The direct summand conjecture in dimension three*, Ann. of Math. (2) **156** (2002), 695–712. MR 1933722 (2003m:13008)

[21] M. Hochster, *Contracted ideals from integral extensions of regular rings*, Nagoya Math. J. **51** (1973), 25–43. MR 0349656 (50 #2149)

[22] ———, *Topics in the homological theory of modules over commutative rings*, CBMS Regional Conference, Lincoln, Neb., 1974, Conference Board of the Mathematical Sciences Regional Conference Series in Mathematics, vol. 24, American Mathematical Society, Providence, R.I., 1975. MR 0371879 (51 #8096)

[23] ———, *Canonical elements in local cohomology modules and the direct summand conjecture*, J. Algebra **84** (1983), 503–553. MR 723406 (85j:13021)

[24] ———, *Solid closure*, Commutative algebra: syzygies, multiplicities, and birational algebra (South Hadley, MA, 1992), Contemp. Math., vol. 159, Amer. Math. Soc., Providence, RI, 1994, pp. 103–172. MR 1266182 (95a:13011)

[25] ———, *Review of the book "Multiplicities and Chern classes in local algebra" by P. Roberts*, Bull. Amer. Math. Soc. **38** (2001), 83–92.

[26] ———, *Parameter-like sequences and extensions of tight closure*, Commutative ring theory and applications (Fez, 2001), Lecture Notes in Pure and Appl. Math., vol. 231, Dekker, New York, 2003, pp. 267–287. MR 2029831 (2005e:13004)

[27] ———, *Big Cohen-Macaulay algebras in dimension three via Heitmann's theorem*, J. Algebra **254** (2002), 395–408. MR 1933876 (2004c:13011)

[28] ———, *Tight closure theory and characteristic p methods. With an appendix by Graham J. Leuschke*, Trends in commutative algebra, Math. Sci. Res. Inst. Publ., vol. 51, Cambridge Univ. Press, Cambridge, 2004, pp. 181–210. MR 2132652 (2006e:13004)

[29] M. Hochster and C. Huneke, *Tight closure and strong F-regularity*, Mém. Soc. Math. France (N.S.) **38** (1989), 119–133. MR 1044348 (91i:13025)

[30] ———, *Tight closure, invariant theory, and the Briançon-Skoda theorem*, J. Amer. Math. Soc. **3** (1990), 31–116. MR 1017784 (91g:13010)

[31] ———, *F-regularity, test elements, and smooth base change*, Trans. Amer. Math. Soc. **346** (1994), 1–62. MR 1273534 (95d:13007)

[32] ———, *Infinite integral extensions and big Cohen-Macaulay algebras*, Ann. of Math. (2) **135** (1992), 53–89. MR 1147957 (92m:13023)

[33] ———, *Phantom homology*, Mem. Amer. Math. Soc. **103** (1993). MR 1144758 (93j:13020)

[34] ———, *Applications of the existence of big Cohen-Macaulay algebras*, Adv. Math. **113** (1995), 45–117. MR 1332808 (96d:13014)

[35] ———, *Tight closure in equal characteristic zero*, preprint, available at `http://www.math.lsa.umich.edu/~hochster/msr.html`.

[36] M. Hochster and J. L. Roberts, *Rings of invariants of reductive groups acting on regular rings are Cohen-Macaulay*, Advances in Math. **13** (1974), 115–175. MR 0347810 (50 #311)

[37] C. Huneke, *Tight closure and its applications*, CBMS Regional Conference Series in Mathematics, vol. 88, American Mathematical Society, Providence, R.I., 1996. MR 1377268 (96m:13001)

[38] E. Kunz, *Characterizations of regular local rings for characteristic p*, Amer. J. Math. **91** (1969), 772–784. MR 0252389 (40 #5609)

[39] C. Peskine and L. Szpiro, *Dimension projective finie et cohomologie locale. Applications à la démonstration de conjectures de M. Auslander, H. Bass et A. Grothendieck*, Inst. Hautes Études Sci. Publ. Math. **42** (1973), 47–119. MR 0374130 (51 #10330)

[40] ———, *Syzygies et multiplicités*, C. R. Acad. Sci. Paris Sér. A **278** (1974), 1421–1424. MR 0349659 (50 #2152)

[41] N. Ranganathan, *Splitting in module-finite extension rings and the vanishing conjecture for maps of Tor*, Thesis, University of Michigan, 2000.

[42] P. Roberts, *Two applications of dualizing complexes over local rings*, Ann. Sci. École Norm. Sup. (4) **9** (1976), 103–106. MR 0399075 (53 #2926)

[43] ———, *Cohen-Macaulay complexes and an analytic proof of the new intersection conjecture*, J. Algebra **66** (1980), 220–225. MR 591254 (82e:14009)

[44] ———, *The vanishing of intersection multiplicities of perfect complexes*, Bull. Amer. Math. Soc. (N.S.) **13** (1985), 127–130. MR 799793 (87c:13030)

[45] ———, *Le théorème d'intersection*, C. R. Acad. Sci. Paris Sér. I Math. **304** (1987), 177–180. MR 880574 (89b:14008)

[46] ———, *Intersection theorems*, Commutative algebra (Berkeley, CA, 1987), Math. Sci. Res. Inst. Publ., vol. 15, Springer, New York, 1989, pp. 417–436. MR 1015532 (90j:13024)

[47] ———, *Multiplicities and Chern classes in local algebra*, Cambridge Tracts in Mathematics, vol. 133, Cambridge University Press, Cambridge, 1998. MR 1686450 (2001a:13029)

[48] ———, *Heitmann's proof of the direct summand conjecture in dimension 3*, expository e-print, available at `http://www.math.utah.edu/~roberts/eprints.html`.

[49] J.-P. Serre, *Algèbre locale. Multiplicités*, Lecture Notes in Mathematics, vol. 11, Springer-Verlag, Berlin, 1965. MR 0201468 (34 #1352)

[50] L. Szpiro, *Sur la théorie des complexes parfaits*, Commutative algebra: Durham 1981 (Durham, 1981), London Math. Soc. Lecture Note Ser., vol. 72, Cambridge Univ. Press, Cambridge, 1982, pp. 83–90. MR 693628 (84m:13015)

[51] H. Weyl, *The Classical Groups*, Princeton University Press, Princeton, 1946.

Melvin Hochster, Department of Mathematics, University of Michigan, Ann Arbor, MI 48109, USA
E-mail address: hochster@umich.edu
URL: http://www.math.lsa.umich.edu/~hochster

Illinois Journal of Mathematics
Volume 51, Number 1, Spring 2007, Pages 171–183
S 0019-2082

FINE BEHAVIOR OF SYMBOLIC POWERS OF IDEALS

MELVIN HOCHSTER AND CRAIG HUNEKE

This paper is written in appreciation of the many contributions of Phil Griffith to commutative algebra.

ABSTRACT. A fundamental property connecting the symbolic powers and the usual powers of ideals in regular rings was discovered by Ein, Lazarsfeld, and Smith in 2001, and later extended by Hochster and Huneke in 2002. In this paper we give further generalizations which give better results in case the quotient of the regular ring by the ideal is F-pure or F-pure type. Our methods also give insight into a conjecture of Eisenbud and Mazur concerning the existence of evolutions. The methods used come from tight closure and reduction to positive characteristic.

1. Introduction

All given rings in this paper are commutative, associative with identity, and Noetherian. In [3], L. Ein, R. Lazarsfeld, and K. Smith discovered the following remarkable fact about the behavior of symbolic powers of ideals in affine regular rings of equal characteristic 0: if c is the largest height[1] of an associated prime of I, then $I^{(cn)} \subseteq I^n$ for all $n \geq 0$. Here, if W is the complement of the union of the associated primes of I, $I^{(t)}$ denotes the contraction of $I^t R_W$ to R, where R_W is the localization of R at the multiplicative system W. Their proof depended on the theory of multiplier ideals (see [1], [16], [17], [23], and [24] for background), including an asymptotic version, and, in particular, needed resolution of singularities as well as vanishing theorems.

Stronger results were obtained in [11, Theorem 1.1] with proofs by methods that were, in some ways, more elementary. The results of [11] are valid in both equal characteristic 0 and in positive prime characteristic p, but depend on reduction to characteristic p. The paper [11] used tight closure methods

Received July 1, 2006; received in final form May 8, 2007.
2000 *Mathematics Subject Classification.* Primary 13A35, 13B40, 13H05.
Both authors were supported in part by grants from the National Science Foundation. The first author was supported by DMS-0400633 and the second author was supported by grant DMS-0244405.

[1]See Discussion 1.1 and Theorem 2.1.

and, in consequence, needed neither resolution of singularities nor vanishing theorems that may fail in positive characteristic.

In this paper we use ideas related to those of [11] to prove further results in this direction that are more subtle.

E.g., in the local case the conclusion may be that a certain symbolic power of I is contained in the product of the maximal ideal with a certain other symbolic power of I. It appears to be very difficult to establish these theorems without tight closure theory or some correspondingly intricate method: we do not know whether the techniques of [3] can be used to obtain the main theorems here even in equal characteristic 0.[2]

The papers [2], [12], and [13] study the existence of evolutions, which is equivalent to the question of whether, for a prime P in a regular local ring $(R, \mathfrak{m})$, $P^{(2)} \subseteq \mathfrak{m}P$. Eisenbud and Mazur [2] ask whether this is always true in a regular local ring of equal characteristic 0. Kunz gave a counterexample in characteristic 2 which is extended to all positive characteristics in [2]: these counterexamples are codimension three primes in regular local rings of dimension four.[3] Our Theorems 3.5 and 4.2 prove that in a regular local ring containing a field, if P is a prime of codimension three, then $P^{(4)} \subseteq \mathfrak{m}P$. In dimension 3 regular local rings, codimension 2 primes are in the linkage class of a complete intersection (this is true in regular local rings of any dimension whenever the quotient by the codimension two ideal is Cohen-Macaulay), and the fact that $P^{(2)} \subseteq \mathfrak{m}P$ is established for primes in the linkage class of a complete intersection in [2]. Our Theorems 3.5 and 4.2 prove that in a regular local ring of arbitrary dimension containing a field, if P is a prime of codimension two, then $P^{(3)} \subseteq \mathfrak{m}P$. In general we are able to prove that if P is a prime of codimension c in a regular local ring containing a field, then $P^{(c+1)} \subseteq \mathfrak{m}P$ (see Theorem 3.5).

We note that to prove the most basic form of our results, all that we need to know about tight closure is the definition and the fact that, in a regular ring, every ideal is tightly closed. The results of Section 4 require more of the theory, and we refer the reader to [7]–[10], [14], and [22] for further background on tight closure theory.

Although all of our proofs initially take place in positive characteristic, relatively standard methods show that whenever, roughly speaking, the statements "make sense," corresponding results hold for rings containing a field of characteristic 0. We need a definition before stating our results.

---

[2]Since we finished this research, Shunsuke Takagi [26] has announced that he can obtain similar results, and in some cases stronger results, using an interpretation of multiplier ideals via tight closure.

[3]Specifically, map $K[[x_1, x_2, x_3, x_4]] \twoheadrightarrow K[[t]]$ by sending x_1, x_2, x_3, x_4 to t^a, t^b, t^c, t^d resp. Let P be the kernel. If K has characteristic $p > 0$ and the values of a, b, c, d are p^2, $p(p+1)$, $p(p+1)+1$, and $(p+1)^2$, resp., then $f = x_1^{p+1}x_2 - x_2^{p+1} - x_1x_3^p + x_4^p \in P^{(2)} \setminus \mathfrak{m}P$.

DISCUSSION 1.1. Our results are typically expressed in terms of a number associated with I which we call the *key number* of I. We shall give here several possible definitions of this term: the reader may choose any one of them. Our reason for proceeding in this way is that the results are sharpest for a rather technical version of "key number" based on analytic spread, but are correct and a lot less technical if one simply uses instead the largest number of generators after localizing at an associated prime of the ideal (or the largest height of an associated prime of the ideal). We note that all of the definitions that we give agree for radical ideals in regular rings.

Thus, in the sequel, the reader may use any of the following as the definition of the key number of I: the number obtained will, in general, depend on which definition is used, but the theorems will be correct for any of these numbers (or any larger number).

(1) The *key number* of the ideal I is the largest number of generators of IR_P for any associated prime P of I.
(2) The *key number* of the ideal I is the largest height (or codimension) of any associated prime P of I, where the height of P is the Krull dimension of R_P.
(3) The *key number* of the ideal I is the largest analytic spread[4] of IR_P for any associated prime P of I.

Because the analytic spread of I in a local ring R is bounded both by the Krull dimension of R and by the least number of generators of I, the notion of key number given in (3) is never larger than either of the other two. As already mentioned, the three notions coincide in the case of a radical ideal of a regular ring.

We note that by the main results of [11], if c is the key number for I, then $I^{(cn)} \subseteq I^n$ for all nonnegative integers n: see Theorem 2.1 of Section 2. On first perusal of this paper the reader may well want to focus on the case where I is radical or even prime.

It is a natural question to ask whether the result that $I^{(cn)} \subseteq I^n$ for all $n \geq 1$ can be improved, perhaps by assuming more about the singularities of R/I. The following example, which we learned from L. Ein, shows that one will not be able to improve this result too much.

EXAMPLE 1.2. In $K[x_1, \ldots, x_n]$, consider the primes (x_i, x_j) for $j \neq i$. Let I denote their intersection, which is generated by all monomials consisting of

---

[4]For a discussion of *analytic spread* and related ideas we refer the reader to [25], and [19]. A summary of what is needed here is given in [11], §2.3. We note that when $I \subseteq (R, \mathfrak{m})$ with $(R, \mathfrak{m})$ local, the analytic spread of I is the same as the Krull dimension of the ring $(R/\mathfrak{m}) \otimes_R \operatorname{gr}_I R$, where $\operatorname{gr}_I R = R/I \oplus I/I^2 \oplus I^2/I^3 \oplus \cdots$ is the associated graded ring of R with respect to the I-adic filtration. When $R/\mathfrak{m}$ is infinite, this is the same as thc smallest number of generators of an ideal $J \subseteq I$ such that I is contained in the integral closure of the ideal J.

products of $n-1$ of the variables. This is a radical ideal of pure height 2 with key number $c = 2$.

For every integer k, $x_1^k \cdots x_n^k \in I^{(2k)}$, but if $k < n-1$, it is not in I^{k+1}: since each generator omits one variable, for a product of $k+1 < n$ generators there must be some variable that occurs in all $k+1$ factors, and that variable will have an exponent of at least $k+1$ in the product. Note that the main result of [11] implies that $I^{(2k)} \subseteq I^k$.

In giving proofs in Section 3 we first give the argument in the case where the key number c is defined as in (1). We then later explain the modifications in the arguments needed for case (3). The point in case (3) is that by the introduction of an additional fixed multiplier, we can, in essence, replace I, in certain localizations, by an ideal with c generators on which it is integrally dependent. We do not need to give an argument for the case where the key number is defined as in (2), since the number defined in (3) is never larger.

The result that follows, Theorem 1.3, is a composite of Theorems 3.5, 3.6, and 4.2 containing our main results.

THEOREM 1.3. *Let $(R, \mathfrak{m})$ be a regular local ring containing a field and let I be a proper ideal of R with key number c.*

(1) *For every $s \geq 1$, $I^{(cs+1)} \subseteq \mathfrak{m}I^{(s)}$.*

(2) *If R is characteristic p and R/I is F-pure, then $I^{(rc-1)} \subseteq II^{(r-1)}$.*

We shall also discuss a version of part (2) for local rings of finitely generated algebras over a field of characteristic 0 in Section 3: one needs to replace the notion of "F-pure" by a suitable characteristic 0 notion ("F-pure type").

We do not know how to prove (1) or (2) by elementary means, even when $(R, \mathfrak{m})$ is local and has dimension 3, $s = 1$, and $I = P$ is a prime of codimension 2: in that case (1) gives that $P^{(3)} \subseteq \mathfrak{m}P$.

Our results take a simpler form when $s = 1$ (in (1)) or $r = 2$ (in (2)) because $I^{(1)} = I$, and we make this explicit:

THEOREM 1.4. *Let $(R, \mathfrak{m})$ be a regular local ring containing a field and let I be a proper ideal of R with key number c.*

(1) *$I^{(c+1)} \subseteq \mathfrak{m}I$.*

(2) *If R is characteristic p and R/I is F-pure, then $I^{(2c-1)} \subseteq I^2$.*

In the next section we recall the main results of [11]. Section 3 contains the proofs of our results for regular rings in positive prime characteristic, and Section 4 contains the equal characteristic 0 versions of our results.

2. Prior comparison results

The main results of [11] in all characteristics are summarized in the following theorem. Note that I^* denotes the tight closure of the ideal I. The characteristic zero notion of tight closure used in this paper is the *equational tight closure* of [10] (see, in particular, Definition (3.4.3) and the remarks in (3.4.4) of [10]). This is the smallest of the characteristic zero notions of tight closure, and therefore gives the strongest result.

THEOREM 2.1. *Let R be a Noetherian ring containing a field. Let I be any ideal of R, and let c be the key number*[5] *of I.*

(1) *If R is regular, $I^{(cn+kn)} \subseteq (I^{(k+1)})^n$ for all positive n and nonnegative k. In particular, $I^{(cn)} \subseteq I^n$ for all positive integers n.*
(2) *If I has finite projective dimension, then $I^{(cn)} \subseteq (I^n)^*$ for all positive integers n.*
(3) *If R is finitely generated, geometrically reduced (in characteristic 0, this simply means that R is reduced) and equidimensional over a field K, and locally I is either 0 or contains a nonzerodivisor (this is automatic if R is a domain), then, with J equal to the Jacobian ideal of R over K, for every nonnegative integer k and positive integer n, we have that $J^n I^{(cn+kn)} \subseteq ((I^{(k+1)})^n)^*$ and $J^{n+1}I^{(cn+kn)} \subseteq (I^{(k+1)})^n$. In particular, we have that $J^n I^{(cn)} \subseteq (I^n)^*$ and $J^{n+1}I^{(cn)} \subseteq I^n$ for all positive integers n.*

The following result is implicit but not explicit in [11].

THEOREM 2.2. *Let R be a regular ring of positive prime characteristic p. Let I be an ideal of R.*

(1) *Let c be the largest number of generators of I after localizing at an associated prime of I. Then $I^{(cnq)} \subseteq (I^{(n)})^{[q]}$ for every $n \in \mathbb{N}$ and $q = p^e$.*
(2) *Let c be the key number of I in any of the senses of Discussion 1.1. Then there is a positive integer s such that $I^s I^{(cnq)} \subseteq (I^{(n)})^{[q]}$ for every $n \in \mathbb{N}$ and $q = p^e$.*

Proof. Let W be the multiplicative system that is the complement of the union of the associated primes of I. The elements of W are not zerodivisors on any symbolic power of I, by construction of the symbolic powers, nor on any bracket power of a symbolic power of I, since the Frobenius endomorphism is flat: see, for example, Lemma 2.2(d) of [11] (and [5], [15], [20] for several related results).

(1) Thus, it will suffice to show that $I^{(cnq)}R_W \subseteq (I^{(n)})^{[q]}R_W$, i.e., that $I^{cnq}R_W \subseteq (I^n)^{[q]}R_W$. Since R_W is semilocal, it suffices to show this after localizing at a maximal ideal, and this will be a (maximal) associated prime P of I. Thus, we need only show that $(I_P)^{cnq} \subseteq \left((I_P)^n\right)^{[q]}$. Since I_P has at most c generators, say $f_1, \dots, f_c$ (we may take some of these to be 0), by

[5]See Discussion 1.1.

definition of c, any monomial of degree cnq in these must be such that at least one of the f_j has exponent $\geq nq$, which means that it is in $\big((I_P)^n\big)^{[q]}$, and the result follows.

(2) We replace R by $R[t]$, where t is an indeterminate, and I by its expansion to this ring. The new associated primes of I are the expansions of the original associated primes. The crucial effect of this trick is that now the localization of the ring at any associated prime of the ideal has infinite residue field. We go back to our original notation and call the ring R. Then for each associated prime P of I we may choose an ideal J of R_P with c generators such that IR_P is integral over J. Then there is an integer s_P such that $I^{s_P}I^NR_P \subseteq J^N$ for every $N \in \mathbb{N}$. Choose s to be at least the maximum of these finitely many s_P. We shall show that $I^sI^{(cnq)} \subseteq (I^{(n)})^{[q]}$. Since no element of W is a zerodivisor on $(I^{(n)})^{[q]}$, it suffices to show this after localizing at W and hence after localizing at each of the associated primes P of I. We replace R by R_P and I by IR_P. But then all we need to show is that $I^sI^{cnq} \subseteq (I^n)^{[q]}$, and we have that $I^sI^{cnq} \subseteq J^{cnq} \subseteq (J^n)^{[q]}$ (exactly as in part (1), since J has c generators), and this is contained in $(I^n)^{[q]}$. □

3. The main results for characteristic p regular rings

Let R be a regular domain of positive characteristic $p > 0$, let I be an ideal of R, let c be the key number for I, and let r be an integer with $r \geq 2$. We define a sequence of ideals as follows:

Definition 3.1. Fix an integer k, $0 \leq k \leq r(c-1)$. Set $I_{0,k} = I$, and inductively set $I_{n,k} = \big(I_{n-1}I^{(r-1)} : I^{(rc-k)}\big)$. Set $J_k = \bigcup_j I_{j,k}$. We will usually simplify notation when k is fixed and write $I_n = I_{n,k}$.

Observe that I_j form an increasing sequence of ideals (depending on k). Since $k \leq r(c-1)$ it follows that $rc - k \geq rc - r(c-1) = r$ and thus $I_{n-1}I^{(rc-k)} \subseteq I_{n-1}I^{(r-1)}$, and $I_{n-1} \subseteq I_{n-1}I^{(rc-k)} : I^{(r-1)} = I_n$. Hence $J_k = I_N$ for all large N.

Theorem 3.2. *Let R be a regular domain of positive characteristic $p > 0$, let I be an ideal of R, let c be the key number for I, and let r be an integer with $r \geq 2$. For all $q = p^e$, and any integer k, $0 \leq k \leq r(c-1)$,*

$$I^{(q((n+1)k-nc))} \subseteq I_{n+1}^{[q]}.$$

Proof. We first assume that key number is defined as in Discussion 1.1(1) and prove the case $n = 0$. We need to prove that

$$I^{(kq)} \subseteq \big(II^{(r-1)} : I^{(rc-k)}\big)^{[q]}.$$

Since we are in a regular ring,

$$\big(II^{(r-1)} : I^{(rc-k)}\big)^{[q]} = (II^{(r-1)})^{[q]} : (I^{(rc-k)})^{[q]}.$$

Suppose that $u \in I^{(qk)}$ and $v \in I^{(rc-k)}$. It will suffice to show that $uv^q \in (II^{(r-1)})^{[q]}$. Evidently, we may assume that $v \neq 0$.

Say $b \geq rc - k$. Fix any power Q of p. We shall show that $v^b(uv^q)^Q \in \bigl((II^{(r-1)})^{[q]}\bigr)^{[Q]}$. Since b is independent of Q, this shows that $uv^q \in \bigl((II^{(r-1)})^{[q]}\bigr)^*$ (the tight closure), and since R is regular, this will establish that $uv^q \in (II^{(r-1)})^{[q]}$.

By the division algorithm, $qQ = (a-1)(rc-k) + \rho$, where $a - 1 \in \mathbb{N}$ and $0 \leq \rho < rc - k$. Then $\rho + b \geq rc - k$. Clearly

$$(*) \qquad a(rc-k) \geq qQ.$$

Then

$$v^b(uv^q)^Q = u^Q v^{b+qQ} \in \bigl(u^Q v^{ac} v^{a\left((r-1)c-k\right)}\bigr) = (v^{ac})(u^Q v^{a((r-1)c-k)}).$$

Now, since $v \in I^{(rc-k)}$, we have that

$$v^{ac} \in I^{((rc-k)ac)} \subseteq I^{(qQc)} \subseteq I^{[qQ]}$$

by Theorem 2.2, and since $u \in I^{(qk)}$, we find that

$$u^Q v^{a((r-1)c-k)} \subseteq I^{(qQk+(rc-k)a((r-1)c-k))},$$

and, using $(*)$, the exponent is $\geq qQk + qQ\bigl((r-1)c-k\bigr) = qQ(r-1)c$. Thus,

$$u^Q v^{a((r-1)c-k)} \subseteq I^{(qQ(r-1)c)} \subseteq (I^{(r-1)})^{[qQ]}$$

by Theorem 2.2. Multiplying, we have that

$$v^b(uv^q)^Q \in I^{[qQ]}(I^{(r-1)})^{[qQ]} = (II^{(r-1)})^{[qQ]} = \bigl((II^{(r-1)})^{[q]}\bigr)^{[Q]},$$

as required.

We next describe the changes needed in the argument in the case where the key number is defined as in Discussion 1.1(3). If $I = (0)$, there is nothing to prove. If not, choose s as in Theorem 2.2(b) and let y be any nonzero element of I. The argument is almost the same, but we show instead that $y^{2s}v^b(uv^q)^Q \in \bigl((II^{(r-1)})^{[q]}\bigr)^{[Q]}$ for all Q, which again allows us to conclude that $uv^q \in (II^{(r-1)})^{[q]})^*$. In the final paragraph of the argument we get that $y^s v^{ac} \subseteq y^s I^{(qQc)} \subseteq I^{[qQ]}$ by Theorem 2.2(b), and, similarly, that $y^s u^Q v^{a((r-1)c-k)} \subseteq I^{(qQ(r-1)c)} \subseteq (I^{(r-1)})^{[qQ]}$ by Theorem 2.2(b). We can multiply: the argument is otherwise unchanged. This completes the case $n = 0$.

We now assume the result for $n - 1$ and prove it for n. This induction works for every choice of key number as defined in Discussion 1.1.

We need to prove that

$$I^{(q((n+1)k-nc))} \subseteq I_{n+1}^{[q]} = (I_n I^{(r-1)} : I^{(rc-k)})^{[q]}.$$

Suppose that $u \in I^{(q((n+1)k-nc)}$ and $v \in I^{(rc-k)}$. It will suffice to show that $uv^q \in (I_n I^{(r-1)})^{[q]}$. Evidently, we may assume that $v \neq 0$.

First suppose that $c \leq k$. Then $(n+1)k - nc = k + n(k-c) \geq k$ and the result follows from the case $n = 0$, since by that case

$$I^{(q((n+1)k-nc))} \subseteq I^{(qk)} \subseteq (II^{(r-1)} : I^{(rc-k)})^{[q]} \subseteq (I_n I^{(r-1)} : I^{(rc-k)})^{[q]}.$$

We can therefore assume that $c > k$. Fix any power Q of p. By the division algorithm,

$$(c-k)qQ = (a-1)(rc-k) + \rho,$$

where $a - 1 \in \mathbb{N}$ and $0 \leq \rho < rc - k$. We shall show that $v^\rho (uv^q)^Q \in \big((I_n I^{(r-1)})^{[q]}\big)^{[Q]}$ for all Q. Since ρ is independent of Q, this shows that $uv^q \in \big((I_n I^{(r-1)})^{[q]}\big)^*$ (the tight closure), and since R is regular, this will establish that $uv^q \in (I_n I^{(r-1)})^{[q]}$. Note that $a(rc-k) \geq (c-k)qQ$. We break the product into two terms:

$$v^\rho (uv^q)^Q = (u^Q v^a) \cdot v^{\rho + qQ - a}.$$

Now, since $v \in I^{(rc-k)}$, we have that

$$v^a \in I^{((rc-k)a)} \subseteq I^{((c-k)qQ)}$$

and hence

$$u^Q v^a \in I^{((c-k)qQ)} I^{(qQ((n+1)k-nc))} \subseteq I^{(qQ(nk-(n-1)c)} \subseteq I_n^{[qQ]}.$$

The last step follows from our induction.

Observe that

$$\begin{aligned}(\rho + qQ - a)(rc-k) &\geq qQ(rc-k) - ((a-1)(rc-k) + \rho)) \\ &= qQ(rc-k) - (c-k)qQ = qQ(r-1)c,\end{aligned}$$

so that

$$v^{\rho+qQ-a} \in I^{((r-1)cqQ)} \subseteq (I^{(r-1)})^{[qQ]}$$

by Theorem 2.2.

Hence

$$v^\rho (uv^q)^Q = (u^Q v^a) \cdot v^{\rho+qQ-a} \in (I_n I^{(r-1)})^{[qQ]},$$

which proves that $uv^q \in (I_n I^{(r-1)})^*$ and finishes the proof of the theorem. □

COROLLARY 3.3. *Let $(R, \mathfrak{m})$ be a regular local ring of positive characteristic $p > 0$, let I be an ideal of R, and let c be the key number.*[6] *Then either $I^{(rc-1)} \subseteq II^{(r-1)}$ for every $r \geq 2$ or $I^{(q)} \subseteq \mathfrak{m}^{[q]}$ for every $q = p^e$.*

Proof. Take $k = 1$ and apply Theorem 3.2 to the ideal I_1. Either $I_1 = R$, in which case $I^{(rc-1)} \subseteq II^{(r-1)}$, or else it is a proper ideal, in which case $I^{(q)} \subseteq I_1^{[q]} \subseteq \mathfrak{m}^{[q]}$. □

---

[6] See Discussion 1.1.

REMARK 3.4. The second of the alternative conclusions is a very strong form of the Zariski-Nagata theorem[7], which asserts that $P^{(N)} \subseteq \mathfrak{m}^N$ for every prime of the regular local ring $(R, \mathfrak{m})$. For $q = p^e$, $\mathfrak{m}^{[q]}$ tends to be quite a bit smaller than $\mathfrak{m}^q$. Note that the second of the alternate conclusions does imply that the Zariski-Nagata theorem holds even when N is not a power of p. For suppose $u \in P^{(N)}$. For any $q \geq N$ we have $q = aN + b$, where $a = \lfloor \frac{q}{N} \rfloor$ and $0 \leq b < N$. Then $u^b u^a \in P^{(q)} \subseteq \mathfrak{m}^q$, which shows that $(b+a) \operatorname{ord}_{\mathfrak{m}} u \geq Na+b$ for all $q \geq N$, and so $\operatorname{ord}_{\mathfrak{m}} u \geq (Na+b)/(b+a) = (N + \frac{b}{a})/(1+\frac{b}{a})$. Since $b < N$ while $a \longrightarrow \infty$ as $q \longrightarrow \infty$, this is $> N-1$ for $q \gg 0$. Thus, $\operatorname{ord}_{\mathfrak{m}} u \geq N$, i.e., $u \in \mathfrak{m}^N$.

THEOREM 3.5. *Let $(R, \mathfrak{m})$ be a regular local ring of positive characteristic p, let I be an ideal of R, and let c be the key number for I. For all $s \geq 1$,*

$$I^{(cs+1)} \subseteq \mathfrak{m} I^{(s)}.$$

Proof. We claim that $J_{c-1} = R$. By Theorem 3.2,

$$I^{(q((n+1)(c-1)-nc))} \subseteq I_{n+1}^{[q]}$$

for all n. But for all large n, $(n+1)(c-1) - nc = c - n + 1 < 0$, and hence $I_{n+1} = R$. Since $I_{n+1} \subseteq J_{c-1}$ we obtain $J_{c-1} = R$.

Choose n least with the property that $I_n = R$ (for $k = c-1$). Note that $I_0 = I \neq R$, so that $I_{n-1} \neq R$. But then $R = I_n = I_{n-1} I^{(r-1)} : I^{(rc-c+1)}$ implies that

$$I^{(rc-c+1)} \subseteq I_{n-1} I^{(r-1)} \subseteq \mathfrak{m} I^{(r-1)}.$$

Setting $s = r-1$ gives the conclusion of the theorem. □

THEOREM 3.6. *Let R be a regular ring of positive characteristic $p > 0$, let I be a proper ideal of R, let $r \geq 2$ be an integer, and let c be the key number for I. If R/I is F-pure, then $I^{(rc-1)} \subseteq I I^{(r-1)}$.*

Proof. The proof reduces at once to the local case, so we may assume that $(R, \mathfrak{m})$ is a regular local ring. If $I^{(c-1)} \subseteq I I^{(r-1)}$ we are done, since $rc - 1 \geq c - 1$. Thus, we may assume that $I^{(c-1)} \not\subseteq I I^{(r-1)}$.

By Fedder's criterion for F-purity (cf. [4]), $I^{[q]} : I \not\subseteq \mathfrak{m}^{[q]}$. If $I^{(rc-1)} \not\subseteq I I^{(r-1)}$, then we may set $k = 1$ in Theorem 3.2. But then $I^{(kq)} = I^{(q)} \subseteq (I I^{(r-1)} : I^{(rc-1)})^{[q]} \subseteq \mathfrak{m}^{[q]}$, a contradiction, for then $I^{(q)} \subseteq \mathfrak{m}^{[q]}$, while $I^{[q]} : I \not\subseteq \mathfrak{m}^{[q]}$. □

---

[7]The theorem is equivalent to Theorem 1 of [6], where Zariski's proof is given; for Nagata's proof see [18, p. 143].

4. The main results in characteristic 0

In this section we give the extensions of the various positive characteristic results to the equal characteristic case. As mentioned in Section 2, the notion of tight closure that we use here is that of equational tight closure from [10, Sections 3.4.3–4]. The main results of this section are contained in Theorem 4.2 below. We need to do some groundwork before we can prove that theorem, however. The proof of the main results depends on three steps: one is to localize and complete, the second is to descend from the complete case to the affine case, and the third is to use reduction to positive characteristic in the affine case. We follow the same path as was done in detail in [11]. To state the main theorem we need the definition of F-pure type.

DEFINITION 4.1. Let R be a ring which is finitely generated over a field k of characteristic 0. The ring R is said to be of *F-pure type* if there exists a finitely generated $\mathbb{Z}$-algebra $A \subseteq k$ and a finitely generated A-algebra R_A which is free over A such that $R \cong R_A \otimes_A K$ and such that for all maximal ideals μ in a Zariski dense subset of $\mathrm{Spec}(A)$, with $\kappa = A/\mu$, the fiber rings $R_A \otimes_A \kappa$ are F-pure.

THEOREM 4.2. *Let R be a regular ring containing a field of characteristic* 0*, and let I be a proper ideal of R with key number c.*

(1) *If R is local with maximal ideal $\mathfrak{m}$, then for all $s \geq 1$,*

$$I^{(cs+1)} \subseteq \mathfrak{m}I^{(s)}.$$

(2) *If R is of finite type over a field of equal characteristic* 0 *and R/I is of F-pure type, then $I^{(rc-1)} \subseteq II^{(r-1)}$.*

Proof. We prove a stronger statement to prove (1), which does not require the ring to be local. Recall the definition of I_j, which makes sense in all characteristics: Fix an integer k, $0 \leq k \leq r(c-1)$. Set $I_0 = I$, and inductively set $I_n = \left(I_{n-1}I^{(r-1)} : I^{(rc-k)}\right)$. Set $J_k = \bigcup_j I_j$.

Recall that I_j form an increasing sequence of ideals (depending on k), since $I_{n-1}I^{(rc-k)} \subseteq I_{n-1}I^{(r-1)}$ as $k \leq (r-1)c$, so that $rc - k \geq rc - r(c-1) = r$. Hence $J_k = I_N$ for all large N.

We replace (1) by

$$(1') \qquad J_{c-1} = R.$$

We first prove that $(1')$ implies (1). Assume $(1')$ and assume that R is local with maximal ideal $\mathfrak{m}$. The proof is entirely similar to the proof of Theorem 3.5 in positive characteristic. We let $k = c - 1$, and consider the ascending chain of ideal $\{I_j\}$. By $(1')$, for some large n, $I_n = R$. Choose n least with this property. Note that $I_0 = I \neq R$, so that $I_{n-1} \neq R$ and is a proper ideal. But then $R = I_n = I_{n-1}I^{(r-1)} : I^{(rc-c+1)}$ implies that

$$I^{(rc-c+1)} \subseteq I_{n-1}I^{(r-1)} \subseteq \mathfrak{m}I^{(r-1)}.$$

Setting $s = r - 1$ gives (1).

We prove (1′) for finitely generated algebras over a field K of characteristic 0, and at the same time prove (2). Assume there is a counterexample in either of these cases. We use the standard descent theory of Chapter 2 of [10] to replace the field K by a finitely generated Z-subalgebra A, so that we have a counterexample in an affine algebra R_A over A with $R_A \subseteq R$ and $R \cong K \otimes_A R_A$. In particular, R_A will be reduced. In doing so we descend I to an ideal I_A of R_A as well as the ideals and their prime radicals in its primary decomposition. We define the ideals I_{jA} and J_{kA} in a similar manner. In case (1′), we have the stable ideal J_{kA}, which is the union of the ideals I_{jA}, and we are assuming that $1 \notin J_{(c-1)A}$. In case (3) we have an element u_A that fails to satisfy the containment we are trying to prove. Since R is regular, we can localize at a nonzero element of A to make R_A smooth over A. In either case, we can localize at a nonzero element of A to make A smooth over $\mathbb{Z}$.

For variable maximal ideals of A, we denote the residue field by κ. Notice that as we pass to fibers $\kappa \longrightarrow R_\kappa = R_A \otimes_A \kappa$ we may assume that each minimal prime P_A of I_A becomes a radical ideal whose minimal primes in R_κ are all of the same height as the original. Thus, in the fiber, the primary decomposition of I_κ may have more components, but each of these will be obtained from the image of one of the original components by localization. The biggest analytic spread after localizing at an associated prime will not change. The ring R_κ will be regular, and in part (2) we know that for a dense set of maximal ideals of A, the fiber R_κ/I_κ is F-pure. Both (2) and (1′) now follow since both results hold in characteristic p for a dense set of closed fibers.

We now consider the general case for (1). The problem reduces to the local case, and then it suffices to prove the required containment after completion. Although $I\widehat{R}$ may have more associated primes, Discussion 1.1 shows that the biggest analytic spread as one localizes at these cannot increase.

Since R is regular, note that for all integers j, $\widehat{I^{(j)}} = I^{(j)}\widehat{R}$, so that $(\widehat{I^{(j)}})^n = (I^{(j)})^n\widehat{R}$. (The associated primes of $\widehat{R}/I^{(j)}\widehat{R}$ are among those associated to $\widehat{R}/P\widehat{R}$ for some associated prime P of I, by Proposition 15 in IV B.4 of [21], since any associated prime of $I^{(j)}$ must be an associated prime of I, and by another application of Proposition 15 in IV B.4 of [21] these in turn are associated primes of $I\widehat{R}$.) Thus,

$$I^{(cs+1)} \subseteq \widehat{I}^{(cs+1)} \subseteq \widehat{\mathfrak{m}}\widehat{I}^{(s)} \cap R = \mathfrak{m}I^{(s)}\widehat{R} \cap R = \mathfrak{m}I^{(s)},$$

as required, by the faithful flatness of $\widehat{R}$ over R.

We may now use Theorem 4.3 of [11] to descend to a suitable affine algebra over a coefficient field for the complete local ring, and the results follow from what we have already proved in the affine case. This reduction is entirely similar to the reduction done in [11], and we refer the reader to that paper for full details. □

REMARK 4.3. We still do not know the best possible conclusion, even in codimension two. For example, in $K[x, y, z]$, for homogeneous primes $P_1, \ldots, P_s$ of height two, we do not know whether or not $(P_1 \cap \cdots \cap P_s)^{(3)} \subseteq (P_1 \cap \cdots \cap P_s)^2$. All evidence suggests this is always true, but we have not been able to extend our methods to decide this question.

It is also possible that for regular local rings $(R, \mathfrak{m})$ in all characteristics $I^{(cs)} \subseteq \mathfrak{m}I$, where c is the key number for I. If so, this would prove the Eisenbud-Mazur conjecture (even in positive characteristic) for prime ideals of codimension two. It would also show that the counterexamples in positive characteristic are sharp.

REFERENCES

[1] L. Ein and R. Lazarsfeld, *A geometric effective Nullstellensatz*, Invent. Math. **137** (1999), 427–448. MR 1705839 (2000j:14028)

[2] D. Eisenbud and B. Mazur, *Evolutions, symbolic squares, and Fitting ideals*, J. Reine Angew. Math. **488** (1997), 189–201. MR 1465370 (98h:13035)

[3] L. Ein, R. Lazarsfeld, and K. E. Smith, *Uniform bounds and symbolic powers on smooth varieties*, Invent. Math. **144** (2001), 241–252. MR 1826369 (2002b:13001)

[4] R. Fedder, *F-purity and rational singularity*, Trans. Amer. Math. Soc. **278** (1983), 461–480. MR 701505 (84h:13031)

[5] J. Herzog, *Ringe der Charakteristik p und Frobeniusfunktoren*, Math. Z. **140** (1974), 67–78. MR 0352081 (50 #4569)

[6] H. Hironaka, *Resolution of singularities of an algebraic variety over a field of characteristic zero. I*, Ann. of Math. (2) **79** (1964), 109–203; *II*, ibid. **79** (1964), 205–326. MR 0199184 (33 #7333)

[7] M. Hochster and C. Huneke, *Tight closure and strong F-regularity*, Mém. Soc. Math. France (N.S.) **38** (1989), 119–133. MR 1044348 (91i:13025)

[8] ———, *Tight closure, invariant theory, and the Briançon-Skoda theorem*, J. Amer. Math. Soc. **3** (1990), 31–116. MR 1017784 (91g:13010)

[9] ———, *F-regularity, test elements, and smooth base change*, Trans. Amer. Math. Soc. **346** (1994), 1–62. MR 1273534 (95d:13007)

[10] ———, *Tight closure*, Commutative algebra (Berkeley, CA, 1987), Math. Sci. Res. Inst. Publ., vol. 15, Springer, New York, 1989, pp. 305–324. MR 1015524 (91f:13022)

[11] ———, *Comparison of symbolic and ordinary powers of ideals*, Invent. Math. **147** (2002), 349–369. MR 1881923 (2002m:13002)

[12] R. Hübl, *Evolutions and valuations associated to an ideal*, J. Reine Angew. Math. **517** (1999), 81–101. MR 1728546 (2000j:13047)

[13] R. Hübl and C. Huneke, *Fiber cones and the integral closure of ideals*, Collect. Math. **52** (2001), 85–100. MR 1833088 (2002h:13007)

[14] C. Huneke, *Tight closure and its applications*, CBMS Regional Conference Series in Mathematics, vol. 88, American Mathematical Society, Providence, R.I., 1996. MR 1377268 (96m:13001)

[15] E. Kunz, *Characterizations of regular local rings for characteristic p*, Amer. J. Math. **91** (1969), 772–784. MR 0252389 (40 #5609)

[16] R. Lazarsfeld, *Positivity in Algebraic Geometry. I*, Ergebnisse der Mathematik und ihrer Grenzgebiete, vol. 48, Springer-Verlag, Berlin, 2004. MR 2095471 (2005k:14001a)

[17] J. Lipman, *Adjoints of ideals in regular local rings*, Math. Res. Lett. **1** (1994), 739–755. MR 1306018 (95k:13028)

[18] M. Nagata, *Local rings*, Interscience Tracts in Pure and Applied Mathematics, No. 13, John Wiley & Sons, New York-London, 1962. MR 0155856 (27 #5790)

[19] D. G. Northcott and D. Rees, *Reductions of ideals in local rings*, Proc. Cambridge Philos. Soc. **50** (1954), 145–158. MR 0059889 (15,596a)

[20] C. Peskine and L. Szpiro, *Dimension projective finie et cohomologie locale. Applications à la démonstration de conjectures de M. Auslander, H. Bass et A. Grothendieck*, Inst. Hautes Études Sci. Publ. Math. (1973), 47–119. MR 0374130 (51 #10330)

[21] J.-P. Serre, *Algèbre locale. Multiplicités*, Cours au Collège de France, 1957–1958, rédigé par Pierre Gabriel. Seconde édition, 1965. Lecture Notes in Mathematics, vol. 11, Springer-Verlag, Berlin, 1965. MR 0201468 (34 #1352)

[22] K. E. Smith, *Tight closure of parameter ideals*, Invent. Math. **115** (1994), 41–60. MR 1248078 (94k:13006)

[23] ———, *The multiplier ideal is a universal test ideal*, Comm. Algebra **28** (2000), 5915–5929. MR 1808611 (2002d:13008)

[24] I. Swanson, *Linear equivalence of ideal topologies*, Math. Z. **234** (2000), 755–775. MR 1778408 (2001f:13037)

[25] I. Swanson and C. Huneke, *Integral closure of ideals, rings, and modules*, London Mathematical Society Lecture Note Series, vol. 336, Cambridge University Press, Cambridge, 2006. MR 2266432

[26] S. Takagi, *Formulas for multiplier ideals on singular varieties*, Amer. J. Math. **128** (2006), 1345–1362. MR 2275023

Melvin Hochster, Department of Mathematics, University of Michigan, Ann Arbor, MI 48109, USA

E-mail address: hochster@umich.edu

URL: http://www.math.lsa.umich.edu/~hochster

Craig Huneke, Department of Mathematics, University of Kansas, Lawrence, KS 66045, USA

E-mail address: huneke@math.ku.edu

URL: http://www.math.ku.edu/~huneke

Illinois Journal of Mathematics
Volume 51, Number 1, Spring 2007, Pages 185–208
S 0019-2082

SOCLE DEGREES OF FROBENIUS POWERS

ANDREW R. KUSTIN AND ADELA N. VRACIU

In honor of Phillip Griffith, on the occasion of his retirement

ABSTRACT. Let k be a field of positive characteristic p, R be a Gorenstein graded k-algebra, and $S = R/J$ be an artinian quotient of R by a homogeneous ideal. We ask how the socle degrees of S are related to the socle degrees of $F_R^e(S) = R/J^{[q]}$. If S has finite projective dimension as an R-module, then the socles of S and $F_R^e(S)$ have the same dimension and the socle degrees are related by the formula $D_i = qd_i - (q-1)a(R)$, where $d_1 \le \cdots \le d_\ell$ and $D_1 \le \cdots \le D_\ell$ are the socle degrees of S and $F_R^e(S)$, respectively, and $a(R)$ is the a-invariant of the graded ring R, as introduced by Goto and Watanabe. We prove the converse when R is a complete intersection.

Let $(R, \mathfrak{m})$ be a Noetherian graded algebra over a field of positive characteristic p, with irrelevant ideal $\mathfrak{m}$. We usually let $R = P/C$ with P a polynomial ring, and C a homogeneous ideal. Let J be an $\mathfrak{m}$-primary homogeneous ideal in R. Recall that if $q = p^e$, then the eth Frobenius power of J is the ideal $J^{[q]}$ generated by all i^q with $i \in J$. The basic question is:

QUESTION. *How do the degrees of the minimal generators of $(J^{[q]} : \mathfrak{m})/J^{[q]}$ vary with q?*

The largest of the degrees of a generator of the socle $(J : \mathfrak{m})/J$ will be called the *top socle degree* of R/J. The question of finding a linear bound for the top socle degree of $R/J^{[q]}$ has been considered by Brenner in [3] from a different point of view; his main motivation there is finding inclusion-exclusion criteria for tight closure.

The answer to the Question is well-known (although not explicitly stated in the existing literature) in the case when J has finite projective dimension; see Observation 2.1. We prove that the converse holds when $R = P/C$ is a complete intersection.

Received June 13, 2006; received in final form January 3, 2007.
2000 *Mathematics Subject Classification.* 13A35.
The second author was supported in part by a Research And Productive Scholarship Award from the University of South Carolina.

THEOREM A. *Let k be a field of positive characteristic p, $q = p^e$ for some positive integer e, P be a positively graded polynomial ring over k, and $R = P/C$ be a complete intersection ring with C generated by a homogeneous regular sequence. Let $\mathfrak{m}$ be the maximal homogeneous ideal of R, J be a homogeneous $\mathfrak{m}$-primary ideal in R, and I be a lifting of J to P. Let ℓ be the dimension of the socle $(J : \mathfrak{m})/J$ of R/J and $d_1, \dots, d_\ell$ be the degrees of the generators of the socle. Then the following statements are equivalent:*

(a) $\operatorname{pd}_R R/J < \infty$.

(b) *The socle $(J^{[q]} : \mathfrak{m})/J^{[q]}$ of $R/J^{[q]}$ has dimension ℓ and the degrees of the generators are $qd_i - (q-1)a(R)$, for $1 \le i \le \ell$, where $a(R)$ is the a-invariant of R.*

(c) $(C + I)^{[q]} : (C^{[q]} : C) = C + I^{[q]}$.

(d) $I^{[q]} \cap C = (I \cap C)^{[q]} + CI^{[q]}$.

Of course, the following general question remains wide open and very compelling:

QUESTION. *How do the socle degrees of Frobenius powers $J^{[q]}$ encode homological information about the ideals $J^{[q]}$?*

The proof of Theorem A appears in Section 2.

1. Preliminary notions

In this paper, ring means commutative noetherian ring with one. Let k be a field of positive characteristic p. We say that the ring R is a *graded k-algebra* if

(1.1) R is non-negatively graded, $R_0 = k$, and R is finitely generated as a ring over k.

Every ring that we study in this paper is a graded k-algebra. In particular, "Let P be a polynomial ring" means $P = k[x_1, \dots, x_n]$, for some n, and each variable has positive degree. Every calculation in this paper is homogeneous: all elements and ideals that we consider are homogeneous, all ring or module homomorphisms that we consider are homogeneous of degree zero. If r is a homogeneous element of the ring R, then $|r|$ is the degree of r. The graded k-algebra R has a unique homogeneous maximal ideal

$$\mathfrak{m} = \mathfrak{m}_R = R_+ = \bigoplus_{i>0} R_i;$$

furthermore, R has a unique graded canonical module K_R, which is equal to the graded dual of the graded local cohomology module $\mathrm{H}^d_{\mathfrak{m}}(R)$, where d is the Krull dimension of R; that is,

$$K_R = \operatorname{Hom}_R(\mathrm{H}^d_{\mathfrak{m}}(R), E_R),$$

for $E_R = \mathrm{Hom}_k(R, R/\mathfrak{m})$ the injective envelope of $R/\mathfrak{m}$ as a graded R-module. (See, for example, [7, Def. 2.1.2].) The a-invariant of R is defined to be

$$a(R) = -\min\{m \mid (K_R)_m \neq 0\} = \max\{m \mid (\mathrm{H}^d_{\mathfrak{m}}(R))_m \neq 0\}.$$

The definition of the a-invariant is rigged so that if R is a Gorenstein graded k-algebra, then $K_R = R(a(R))$. When the ring R is Cohen-Macaulay, there are many ways to compute $a(R)$. The main tool for these calculations, Proposition 1.2 below, may be found as Proposition 2.2.9 in [7] or Proposition 3.6.12 in [4].

PROPOSITION 1.2. *If $R \to S$ is a graded surjection of graded k-algebras, and R is Cohen-Macaulay, then*

$$K_S = \mathrm{Ext}^c_R(S, K_R),$$

where $c = \dim R - \dim S$. In particular, if $S = R/C$ and the ideal C is generated by the homogeneous regular sequence $f_1, \ldots, f_c$, then

$$K_{R/C} = (K_R/CK_R)\left(\sum_{i=1}^{c} |f_i|\right).$$

COROLLARY 1.3.

(a) *If P is the polynomial ring $k[x_1, \ldots, x_n]$, then $a(P) = -\sum_{i=1}^n |x_i|$.*

(b) *If R is the complete intersection ring P/C, where P is the polynomial ring $k[x_1, \ldots, x_n]$ and C is the ideal in P generated by the homogeneous regular sequence $f_1, \ldots, f_c$, then $a(R) = \sum_{i=1}^c |f_i| - \sum_{i=1}^n |x_i|$.*

(c) *If $R \to S$ is a surjection of graded Cohen-Macaulay k-algebras, and S has finite projective dimension as an R-module, then $a(S) = a(R) + N$, where N is the largest back twist in the minimal homogeneous resolution of S by free R-modules. In other words, if*

$$0 \to \bigoplus_i R(-b_{c,i}) \to \cdots \to \bigoplus_i R(-b_{1,i}) \to R \to S \to 0$$

is the minimal homogeneous resolution of S by free R-modules, then $N = \max_i\{b_{c,i}\}$.

DEFINITION. If S is an artinian graded k-algebra, then the socle of S,

$$\mathrm{soc}\, S = 0 : \mathfrak{m}_S = \{s \in S \mid s\mathfrak{m}_S = 0\},$$

is a finite dimensional graded k-vector space: $\mathrm{soc}\, S = \bigoplus_{i=1}^{\ell} k(-d_i)$. We refer to the numbers $d_1 \leq d_2 \leq \cdots \leq d_\ell$ as the *socle degrees of S*.

OBSERVATION 1.4. *Let R be an artinian Gorenstein graded k-algebra with socle degree δ, and let J be a homogeneous ideal of R. If the socle degrees of R/J are $\{d_i\}$, then the minimal generators of* $\mathrm{ann}\, J$ *have degrees $\{\delta - d_i\}$.*

Proof. Choose minimal generators $g_1, \dots, g_s$ of $\operatorname{ann} J$. Gorenstein duality (see Lemma 1.11) implies that

$$\operatorname{ann}(g_1, \dots, \hat{g}_i, \dots, g_s) \not\subseteq \operatorname{ann}(g_i);$$

and thus, for each i, we can choose an element $u_i \in \operatorname{ann}(g_1, \dots, \hat{g}_i, \dots, g_s)$, which represents a generator for the socle of $R/\operatorname{ann}(g_i)$. The ideals J and $\operatorname{ann}(g_1, \dots, g_s)$ are equal and the socle of $R/\operatorname{ann}(g_1, \dots, g_s)$ is minimally generated by $u_1, \dots, u_s$. On the other hand, $u_i g_i$ generates the socle of R, so the degree of u_i is equal to $\delta - |g_i|$. □

PROPOSITION 1.5. *If S is an artinian graded k-algebra and $d_1 \le \cdots \le d_\ell$ are the socle degrees of S, then the minimal generators of the canonical module K_S have degrees $-d_\ell \le \cdots \le -d_1$.*

Proof. Let $P = k[x_1, \dots, x_n]$ be a polynomial ring which maps onto S. One may compute the degrees of the generators of K_S as well the socle degrees of S in terms of the back twists in the minimal homogeneous resolution of S as a P-module:

$$0 \to \bigoplus_i P(-b_{n,i}) \to \cdots \to \bigoplus_i P(-b_{1,i}) \to P \to S \to 0.$$

The canonical module K_S is equal to $\operatorname{Ext}_P^n(S, K_P)$, where $K_P = P(a(P))$ and $a(P) = -\sum_{i=1}^n |x_i|$. It follows that the minimal homogeneous resolution of K_S is

$$0 \to P(a(P)) \to \cdots \to \bigoplus_i P(a(P) + b_{n,i}) \to K_S \to 0; \tag{1.6}$$

therefore, the minimal generators of K_S (over either S or P) have degrees $\{-a(P) - b_{n,i}\}$. On the other hand, one may compute $\operatorname{Tor}_n^P(S, P/\mathfrak{m}_P)$ in each coordinate (see, for example, [9, Lemma 1.3]) in order to conclude that

$$\bigoplus_i k(-b_{n,i}) = \operatorname{Tor}_n^P(S, k) = \operatorname{soc} S(a(P)).$$

Thus, the socle degrees of S are equal to $\{a(P) + b_{n,i}\}$. □

COROLLARY 1.7. *Let $R \to S$ be a surjection of graded k-algebras with S artinian, and R Gorenstein. If $\operatorname{pd}_R S$ finite, then the socle degrees of S are $\{b_i + a(R)\}$, where the back twists in the minimal homogeneous resolution of S by free R-modules are $\{b_i\}$.*

Proof. We know, from Proposition 1.2, that $K_S = \operatorname{Ext}_R^{\dim R}(S, K_R)$, with $K_R = R(a(R))$; therefore,

$$0 \to R(a(R)) \to \cdots \to \bigoplus_i R(a(R) + b_i) \to K_S \to 0$$

is a minimal resolution of K_S and the minimal generators of K_S as an R-module, or as an S-module, have degrees $\{-a(R) - b_i\}$. Apply Proposition 1.5. $\square$

Let R be a graded k-algebra. We write ${}^eR$ to represent the ring R endowed with an R-module structure given by the eth iteration of the Frobenius endomorphism $\phi_R : R \to R$. (If r is a scalar in R and s is a ring element in ${}^eR$, then $r \cdot s$ is equal to $r^q s \in {}^eR$, for $q = p^e$.) The Frobenius functor $F_R^e(\_\_) = \_\_ \otimes_R {}^eR$ is base change along the homomorphism ϕ_R^e.

Notation 1.8. Let R be a graded k-algebra. We use the notation $(\ )^{[q]}$ in three ways.

(a) If $\boldsymbol{g}$ is a matrix with entries in R, then $\boldsymbol{g}^{[q]}$ is the matrix in which each entry of $\boldsymbol{g}$ is raised to the power q. In particular, if z is an element of the free module $\bigoplus_{i=1}^m R(-b_i)$, then z is an $m \times 1$ matrix and $z^{[q]}$ is the matrix in which each entry of z is raised to the power q.

(b) If $\mathbb{G}_1$ is the free module $\bigoplus_{i=1}^m R(-b_i)$, then $\mathbb{G}_1^{[q]}$ is the free module $\bigoplus_{i=1}^m R(-qb_i)$.

(c) If J is the R-ideal $(a_1, \dots, a_m)$, then $J^{[q]}$ is the R-ideal $(a_1^q, \dots, a_m^q)$.

In particular, if $\mathbb{N}$ is the homogeneous complex of graded free R-modules

$$\cdots \to \mathbb{N}_3 \xrightarrow{\boldsymbol{n}_3} \mathbb{N}_2 \xrightarrow{\boldsymbol{n}_2} \mathbb{N}_1 \xrightarrow{\boldsymbol{n}_1} \cdots,$$

then $\mathbb{N}^{[q]}$ is a very clean way to write the homogeneous complex

$$F_R^e(\mathbb{N}): \quad \cdots \to \mathbb{N}_3^{[q]} \xrightarrow{\boldsymbol{n}_3^{[q]}} \mathbb{N}_2^{[q]} \xrightarrow{\boldsymbol{n}_2^{[q]}} \mathbb{N}_1^{[q]} \xrightarrow{\boldsymbol{n}_1^{[q]}} \cdots. \tag{1.9}$$

Furthermore, the Frobenius functor is always right exact; so, $F_R^e(R/J) = R/J^{[q]}$.

We conclude this section by gathering a few properties of Gorenstein ideals, Gorenstein duality, and linkage. All of these tricks appear elsewhere in the literature, usually in more generality. We are likely to use them at any moment, without any further ado. Theorem 1.10 is due to Bass [2].

Theorem 1.10. *Let R be a local artinian ring. The following statements are equivalent:*

(1) *R is a Gorenstein ring.*

(2) *The socle of R is principal.*

(3) *The ideal (0) is irreducible (in the sense that (0) is not equal to the intersection of two non-zero ideals).*

(4) *The ring R is self-injective.*

When the conditions of Theorem 1.10 are in effect, then the functor $\operatorname{Hom}_R(\ \ , R) = (\ \ )^*$ is exact and if M is a finitely generated R-module, then the modules M and M^* have the same length.

LEMMA 1.11. *Let M be an ideal in the artinian local Gorenstein ring $(R, \mathfrak{m})$.*

(1) *The ideals M and $\operatorname{ann}(\operatorname{ann}(M))$ are equal.*
(2) *If N is an ideal of R with $M \subseteq N$, then $\operatorname{ann}(M)/\operatorname{ann}(N) \cong \operatorname{Hom}(N/M, R)$.*
(3) *If R/M is a Gorenstein ring, then there exists an element y of R with $\operatorname{ann}(y) = M$ and $\operatorname{ann}(M) = (y)$.*
(4) *If y is a non-zero element of R, then $R/\operatorname{ann}(y)$ is a Gorenstein ring.*

Proof. The ideal M is contained in $\operatorname{ann}(\operatorname{ann}(M))$ and the two ideals have the same length because $M^* = R/\operatorname{ann}(M)$. Assertion (1) follows. Apply $\operatorname{Hom}(\_, R)$ to the short exact sequence

$$0 \to N/M \to R/M \to R/N \to 0$$

to obtain (2). If $\operatorname{ann}(M) = (y_1, \dots, y_s)$, then (1) shows that

$$M = \operatorname{ann}(\operatorname{ann}(M)) = \operatorname{ann}(y_1) \cap \dots \cap \operatorname{ann}(y_s).$$

Under the hypothesis of (3), the ideal M is irreducible and $M = \operatorname{ann}(y_i)$, for some i. Apply (1) again to complete the proof of (3). For (4), it is not difficult to see that the socle of $R/\operatorname{ann}(y)$ is a principal ideal. □

If the conditions of (1) in Proposition 1.12 hold, then the ideal C is called a *Gorenstein ideal.* Notice that when we use this term, we automatically assume the ideal C to have finite projective dimension.

PROPOSITION 1.12. *Let R be a graded k-algebra. Assume that R is a Gorenstein ring. Let C be a homogeneous ideal of R of grade c and finite projective dimension, and $(\mathbb{G}, \boldsymbol{g}_\bullet)$ be the minimal homogeneous resolution of R/C by free R-modules.*

(1) *The following statements are equivalent:*
  (a) *The ring R/C is Gorenstein.*
  (b) *The ring R/C is Cohen-Macaulay and $\operatorname{Ext}_R^c(R/C, R)$ is a cyclic R/C-module.*
  (c) *The ring R/C is Cohen-Macaulay and $\operatorname{Ext}_R^c(R/C, R)$ is isomorphic to a shift of R/C.*
  (d) *The complex $\operatorname{Hom}_R(\mathbb{G}, R)$ is isomorphic to a shift of $\mathbb{G}$.*
(2) *If C is a Gorenstein ideal, then the entries of any matrix representation for $\boldsymbol{g}_c$ form a minimal generating set for C.*
(3) *If the field k has characteristic p and the ideal C is a Gorenstein ideal, then $C^{[p]}$ is a Gorenstein ideal.*

Proof. The equivalence of (a), (b), and (c) is Section 5 of [2]. The ring R/C is Cohen-Macaulay if and only if R/C is a perfect R-module of projective dimension equal to c; see, for example, Section 16.C of [5]. It is now obvious that (c) and (d) are equivalent. Assertion (2) follows from (d). Assertion (3) follows from Theorem 1.7 of [10], which guarantees that $F_R(\mathbb{G})$ is the minimal homogeneous resolution of $R/C^{[p]}$. □

Peskine and Szpiro popularized the concept of linkage by complete intersection ideals. Only slight modifications need be made to [11, Proposition 2.6] in order prove assertion (1) in the following result about linkage by Gorenstein ideals; see, for example, Section 1 of [8].

PROPOSITION 1.13. *Let R be a Gorenstein graded k-algebra, and let $L \subseteq M$ be homogeneous Gorenstein ideals (in the sense of Proposition 1.12) of R of grade c.*

(1) *Let $\mathbb{F}_\bullet$ and $\mathbb{G}_\bullet$ be the minimal homogeneous resolutions of R/L and R/M, respectively, and let $\alpha_\bullet : \mathbb{F}_\bullet \to \mathbb{G}_\bullet$ be a homogeneous map of resolutions which extends the natural map $R/L \to R/M$. The map $\alpha_c : \mathbb{F}_c \to \mathbb{G}_c$ is multiplication by a homogeneous element y of R. Then*
$$L : M = (L, y) \quad \text{and} \quad L : y = M.$$
(2) *If M is generated by the homogeneous regular sequence $f_1, \dots, f_c$ and L is generated by $f_1^{r_1}, \dots, f_c^{r_c}$, then the conclusion of (1) holds for the product $y = f_1^{r_1-1} \cdots f_c^{r_c-1}$.*

Proof. One can prove (2) directly, or deduce it from (1). □

If L, M, and N are ideals in a ring R then a quick calculation yields the remarkably useful formula

$$L : MN = (L : M) : N. \tag{1.14}$$

The proof of the next result may be read from the proof of Proposition 2 in [13].

PROPOSITION 1.15. *Let L and M be homogeneous ideals of a ring R of positive prime characteristic p. If L and $L : M$ have finite projective dimension, then $L^{[p]} : M^{[p]} = (L : M)^{[p]}$.* □

2. The plan of attack and a few examples

We first establish "(a) implies (b)" from Theorem A.

OBSERVATION 2.1. *Let k be a field of positive characteristic p, $R \to S$ be a surjection of graded k-algebras in the sense of (1.1), with R Gorenstein and S artinian. If S has finite projective dimension as an R-module, then the socles of S and $F_R^e(S)$ have the same dimension; furthermore, if the socle degrees of S and $F_R^e(S)$ are given by*
$$d_1 \le d_2 \le \cdots \le d_\ell \quad \text{and} \quad D_1 \le D_2 \le \cdots \le D_\ell,$$

respectively, then

$$D_i = qd_i - (q-1)a(R), \tag{2.2}$$

for all i.

Proof. Consider the minimal homogeneous resolution $\mathbb{F}$ of S by free R-modules. We know from [10] that $F_R^e(\mathbb{F}) = \mathbb{F}^{[q]}$ is the minimal homogeneous resolution of $F_R^e(S)$. If the back twists of $\mathbb{F}$ are $\{b_i \mid 1 \le i \le L\}$, then the back twists of $\mathbb{F}^{[q]}$ are $\{qb_i\}$. Use Corollary 1.7 to see that $L = \ell$, $d_i = b_i + a(R)$, and $D_i = qb_i + a(R)$, for all i. □

REMARK. The hypothesis that R is Gorenstein is necessary in Observation 2.1. Let $R = S$ be a non-Gorenstein artinian graded k-algebra whose socle lives in at least two distinct degrees. The ring S has finite projective dimension as an R-module, and $d_i = D_i$, for all i, since $F_R^e(R) = R$. The a-invariant of R is equal to the top socle degree of R; so, $a(R) = d_\ell$ and (2.2) holds for $d_i = d_\ell$; but does not hold for $d_i < d_\ell$.

We prove the converse of Observation 2.1 under the assumption that R is a complete intersection. Our main result is the following statement.

THEOREM 2.3. *Let k be a field of positive characteristic p, $R \to S$ be a surjection of graded k-algebras in the sense of (1.1), with R a complete intersection and S artinian. Let e be a positive integer, $q = p^e$, and $d_1 \le \cdots \le d_\ell$ be the socle degrees of S. If the socle of $F_R^e(S)$ has the same dimension as the socle of S, and the socle degrees of $F_R^e(S)$ are given by $D_1 \le D_2 \le \cdots \le D_\ell$ as in (2.2), then $\operatorname{Tor}_1^R(S, {}^eR) = 0$.*

STANDING NOTATION 2.4. We express $R = P/C$, where P is the polynomial ring $P = k[x_1, \dots, x_n]$, each variable has positive degree, and C is a homogeneous Gorenstein ideal in P of grade c. Let I be a homogeneous $\mathfrak{m}_P$-primary ideal in P, $S = P/(I+C)$, $T = P/I$, and let Z be the $(c-1)$-syzygy of the P-module $K_T(-a(P))$.

Proof of Theorem 2.3. Adopt the notation of 2.4. In Corollary 3.2, we convert numerical information about the socle degrees of S and $F_R^e(S)$ into numerical information about $\operatorname{Tor}_c^P(K_T, R)$ and $\operatorname{Tor}_c^P(K_{F_P^e(T)}, R)$. In Proposition 4.1, the numerical information about Tor_c's is converted into the statement

$$\operatorname{Tor}_1^R(Z \otimes_P R, {}^eR) = 0.$$

This homological statement is expressed as a statement about ideals:

$$(C^{[q]} + I^{[q]}) : (C^{[q]} : C) = C + I^{[q]}$$

in Proposition 5.1. In Proposition 6.1 we deduce

$$I^{[q]} \cap C = (I \cap C)^{[q]} + CI^{[q]}.$$

This result is equivalent to $\operatorname{Tor}_1^R(S, {}^eR) = 0$, as is recorded in Proposition 3.5. ☐

We would like to prove that the conclusion of Theorem 2.3 continues to hold after one replaces the hypothesis that R is a complete intersection with the weaker hypothesis that R is Gorenstein. Three of our five steps (3.2, 5.1, and 3.5) work when R is Gorenstein. The arguments that we use in the other two steps (4.1 and 6.1) require that R be a complete intersection, although in Proposition 7.4 we prove the ideal theoretic version of (4.1) under the hypothesis that R is Gorenstein and F-pure. At any rate, if R is a complete intersection and the conclusion of Theorem 2.3 holds, then the Theorem of Avramov and Miller [1] (see also [6]) guarantees that S has finite projective dimension as an R-module. We are very curious to know if some form of the Avramov-Miller result,

$$\operatorname{Tor}_1^R(M, {}^eR) = 0 \implies \operatorname{pd}_R M < \infty,$$

for finitely generated R-modules M, can be proven when R is Gorenstein, but not necessarily a complete intersection.

Proof of Theorem A. Adopt the notation of 2.4.

(a)$\Longrightarrow$(b): This is Observation 2.1.

(b)$\Longrightarrow$(c): Assume (b). In Corollary 3.2 and Observation 3.3, we show that if the generator degrees of $\operatorname{Tor}_1^P(Z, R)$ are $\{\gamma_i\}$, then the generator degrees of $\operatorname{Tor}_1^P(F_P^e(Z), R)$ are $\{q\gamma_i\}$. Proposition 4.1 shows that $\operatorname{Tor}_1^R(Z \otimes_P R, {}^eR) = 0$. Proposition 5.1 yields (c).

(c)$\Longrightarrow$(d): This is Proposition 6.1.

(d)$\Longrightarrow$(a): If (d) holds, then Proposition 3.5 shows that $\operatorname{Tor}_1^R(T \otimes_P R, {}^eR) = 0$. The Theorem of Avramov and Miller [1] guarantees that $T \otimes_P R$ has finite projective dimension as an R-module. ☐

If the hypothesis of Theorem 2.3 is weakened to say only that the socles of S and $F_R^e(S)$ have the same dimension (with no mention about how the socle degrees are related), then the conclusion of Theorem 2.3 fails to hold; see Example 2.9. It is curious, however, that if S is Gorenstein, and the defining ideal of S is contained in a proper ideal of R of finite projective dimension (for example, a parameter ideal of R), then one need only verify that the socles of $F_R^e(S)$ and S both have dimension one in order to conclude that S has finite projective dimension over R.

THEOREM 2.5. *Let $(R, \mathfrak{m})$ be a local k-algebra, with R a complete intersection, and let $J \subset R$ be an $\mathfrak{m}$-primary ideal. Assume that $J \subseteq \mathfrak{b}$ for some proper ideal $\mathfrak{b}$ with $\operatorname{pd}_R \mathfrak{b} < \infty$. If R/J and $R/J^{[q]}$ both are Gorenstein, then $\operatorname{pd}_R J < \infty$.*

Proof. Choose $\mathfrak{a} \subseteq J$ a parameter ideal. Use Lemma 1.11 and (1.14) to write $J = \mathfrak{a} : f$, and write $\mathfrak{b} = J : K = \mathfrak{a} : fK$. Since $\mathfrak{b}$ has finite projective dimension, we know, from Proposition 1.15, that

$$\mathfrak{b}^{[q]} = \mathfrak{a}^{[q]} : f^q K^{[q]}. \tag{2.6}$$

On the other hand, we have

$$\mathfrak{b}^{[q]} \subseteq J^{[q]} : K^{[q]}. \tag{2.7}$$

Since $J^{[q]} \subseteq \mathfrak{a}^{[q]} : f^q$ are both irreducible, we can write $\mathfrak{a}^{[q]} : f^q = J^{[q]} : b$ for some $b \in R$. Plugging this into (2.6) yields

$$\mathfrak{b}^{[q]} = J^{[q]} : bK^{[q]}.$$

Comparing this with (2.7) we get $J^{[q]} : bK^{[q]} = J^{[q]} : K^{[q]}$ (equation (2.7) gives one inclusion; the other inclusion is always true), which is equivalent to $K^{[q]} \subseteq J^{[q]} + bK^{[q]}$. If b is not a unit, this means that $K^{[q]} \subseteq J^{[q]}$, and so $\mathfrak{b}^{[q]} = J^{[q]} : K^{[q]} = R$, which is a contradiction.

Thus, b must be a unit, so

$$J^{[q]} = \mathfrak{a}^{[q]} : f^q. \tag{2.8}$$

Consider the short exact sequence

$$0 \to R/J \to R/\mathfrak{a} \to R/(\mathfrak{a}, f) \to 0.$$

Equation (2.8) implies that its tensorization with ${}^eR$ is exact, and thus

$$\operatorname{Tor}_1^R(R/(\mathfrak{a}, f), {}^eR) = 0,$$

and $(\mathfrak{a}, f)$ has finite projective dimension by the Avramov-Miller result. Consequently, J has finite projective dimension as well. □

EXAMPLE 2.9. Note that the conclusion of Theorem 2.5 no longer holds without the assumption that J is contained in a proper ideal of finite projective dimension. Consider, for instance, $R = k[x, y, z]/(x^3 + y^3 + z^3)$, $J = (x, y, z^2)$, where k is a field of characteristic $p \equiv 2 \pmod 3$. Clearly J is irreducible, $J^{[p]} = (x^p, y^p)$ is also irreducible, but J does not have finite projective dimension.

3. Convert socle degrees into $\operatorname{Tor}_1^R(Z \otimes_P R, {}^eR)$

Adopt the notation of 2.4. The first three results in this section convert hypothesis (2.2) about the socle degrees of S and $F_R^e(S)$ into a statement about the generator degrees of $\operatorname{Tor}_1^P(Z, R)$ and $\operatorname{Tor}_1^P(F_P^e(Z), R)$. Observation 3.4 shows that $\operatorname{Tor}_1^R(Z \otimes_P R, {}^eR)$ is a quotient of $\operatorname{Tor}_1^P(F_P^e(Z), R)$ by a submodule which is built from the generators of $\operatorname{Tor}_1^P(Z, R)$. The notation that is introduced in the proof of Observation 3.4 will be used again in the proofs of Propositions 4.1 and 5.1.

All of the calculations in Section 3 which we have described so far pertain to the proof of (b)$\Longrightarrow$(c) in Theorem A. It is curious, however, that Observation 3.4 also may be applied (see Proposition 3.5) to give an ideal theoretic interpretation of $\mathrm{Tor}_1^R(T \otimes_P R, {}^eR)$, which is our contribution to the proof of (d)$\Longrightarrow$(a) in Theorem A.

LEMMA 3.1. *Adopt the notation of 2.4. If the socle degrees of S are*

$$\{d_i \mid 1 \le i \le \ell\},$$

then the minimal generators of $\mathrm{Tor}_c^P(K_T(-a(P)), R)$ *have degrees*

$$\{a(R) - d_i \mid 1 \le i \le \ell\}.$$

Proof. Let $\mathbb{G}$ be the minimal homogeneous resolution of R by free P-modules. Corollary 1.3 (c) tells us that $\mathbb{G}_c = P(a(P) - a(R))$. It follows that

$$\begin{aligned}\mathrm{Tor}_c^P(K_T(-a(P)), R) &= \mathrm{H}_c(K_T(-a(P)) \otimes_P \mathbb{G})\\ &= \{\alpha \in K_T(-a(R)) \mid C\alpha = 0\}\\ &= \mathrm{Hom}_P(R, K_T(-a(R)).\end{aligned}$$

On the other hand, we have a surjection $T \to S$; so Proposition 1.2 guarantees

$$K_S = \mathrm{Hom}_T(S, K_T) = \mathrm{Hom}_P(R, K_T).$$

Thus,

$$K_S(-a(R)) = \mathrm{Hom}_P(R, K_T(-a(R))) = \mathrm{Tor}_c^P(K_T(-a(P)), R).$$

Apply Proposition 1.5. □

Lemma 3.1 also applies when the ideal I is replaced by the ideal $I^{[q]}$; consequently, if the socle degrees of $F_R^e(S)$ are $\{D_i \mid 1 \le i \le L\}$, then the minimal generators of $\mathrm{Tor}_c^P(K_{F_P^e(T)}(-a(P)), R)$ have degrees $\{a(R) - D_i \mid 1 \le i \le L\}$. We have established the following conversion of the original hypothesis about socle degrees into a statement about generator degrees of Tor_c.

COROLLARY 3.2. *Retain the notation of 2.4. Assume that the socles of S and $F_R^e(S)$ have the same dimension. Let*

$$d_1 \le \cdots \le d_\ell \quad \textit{and} \quad D_1 \le \cdots \le D_\ell$$

be the socle degrees of S and $F_R^e(S)$, respectively, and

$$\gamma_1 \le \cdots \le \gamma_\ell \quad \textit{and} \quad \Gamma_1 \le \cdots \le \Gamma_\ell,$$

be the minimal generator degrees of

$$\mathrm{Tor}_c^P(K_T(-a(P)), R) \quad \textit{and} \quad \mathrm{Tor}_c^P(K_{F_P^e(T)}(-a(P)), R),$$

respectively. Then

$$D_i = qd_i - (q-1)a(R) \text{ for all } i \iff \Gamma_i = q\gamma_i \text{ for all } i. \qquad \square$$

We interpret the homological objects of Corollary 3.2 as Tor_1 of the appropriate modules. This process involves index shifting and keeping careful track of the twists.

OBSERVATION 3.3. *In the notation of 2.4:*

(a) $\operatorname{Tor}_c^P(K_T(-a(P)), R) = \operatorname{Tor}_1^P(Z, R)$, *and*
(b) $\operatorname{Tor}_c^P(K_{F_P^e(T)}(-a(P)), R) = \operatorname{Tor}_1^P(F_P^e(Z), R)$.

Proof. We prove (b). Let $\mathbb{F}$ be the minimal homogeneous resolution of $K_T(-a(P))$ by free P-modules. The functor $F_P^e(\_\_)$ is exact; so, $\mathbb{F}^{[q]}$, which is equal to $F_P^e(\mathbb{F})$ (see (1.9)), resolves $F_P^e(K_T(-a(P)))$. On the other hand, it is not hard to see that $\mathbb{F}^{[q]}$ resolves some twist of $K_{F_P^e(T)}$. Indeed, if $(\ )^* = \operatorname{Hom}_P(\_\_, P)$, then it is clear that $(F_P^e(\mathbb{F}^*))^*$ is equal to a shift of $\mathbb{F}^{[q]}$; furthermore, it is also clear that $(F_P^e(\mathbb{F}^*))^*$ resolves some shift of $K_{F_P^e(T)}$; see, for example, Proposition 1.12. There are many ways to keep track of the shifts. One pain free approach is to apply the technique of (1.6) to K_T and to $K_{F_P^e(T)}$ in order to nail down the fact that $\mathbb{F}^{[q]}$ is the minimal homogeneous resolution of $K_{F_P^e(T)}(-a(P))$; hence,

$$F_P^e(K_T(-a(P))) = K_{F_P^e(T)}(-a(P)),$$
$$\operatorname{Tor}_i^P(K_{F_P^e(T)}(-a(P)), R) = \mathrm{H}_i(\mathbb{F}^{[q]} \otimes_P R),$$

for all i. The beginning of the minimal homogeneous resolution of Z is

$$\cdots \to \mathbb{F}_{c+1} \to \mathbb{F}_c \to \mathbb{F}_{c-1} \to Z \to 0.$$

The functor $F_P^e(\ )$ is exact; so,

$$\operatorname{Tor}_1^P(F_P^e(Z), R) = \mathrm{H}_c(\mathbb{F}^{[q]} \otimes_P R) = \operatorname{Tor}_c^P(K_{F_P^e(T)}(-a(P)), R). \qquad \square$$

OBSERVATION 3.4. *Let $P \to R$ be a surjection of graded k-algebras, with P a polynomial ring, and let M be a finitely generated graded P-module. Then there is an exact sequence of graded R-modules:*

$$F_R^e\left(\operatorname{Tor}_1^P(M, R)\right) \to \operatorname{Tor}_1^P(F_P^e(M), R) \to \operatorname{Tor}_1^R(M \otimes_P R, {}^eR) \to 0.$$

Proof. Let $(\mathbb{N}, \boldsymbol{n})$ be the minimal homogeneous resolution of M by free P-modules. The functor $F_P^e(\_\_)$ is exact; so, $\mathbb{N}^{[q]}$ is the minimal homogeneous resolution of $F_P^e(M)$ by free P-modules, and $\operatorname{Tor}_1^P(F_P^e(M), R)$ is equal to

$$\mathrm{H}_1(F_P^e(\mathbb{N}) \otimes_P R) = \mathrm{H}_1(F_R^e(\mathbb{N} \otimes_P R)).$$

The functors $F_P^e(\_\_)\otimes_P R$ and $F_R^e(\_\_\otimes_P R)$ are equal because the homomorphisms

$$\begin{array}{ccc} P & \xrightarrow{\phi_P^e} & P \\ {\scriptstyle \text{quot. map}}\downarrow & & \downarrow{\scriptstyle \text{quot. map}} \\ R & \xrightarrow{\phi_R^e} & R \end{array}$$

commute. Let $^-$ denote the functor $\_\_\otimes_P R$. Select elements $z_1,\dots,z_\ell$ of $\mathbb{N}_1$ so that $\overline{z}_1,\dots,\overline{z}_\ell$ are cycles in $\mathbb{N}\otimes_P R$ and the homology classes $[\overline{z}_1],\dots,[\overline{z}_\ell]$ form a minimal generating set for $\mathrm{H}_1(\mathbb{N}\otimes_P R)=\mathrm{Tor}_1^P(M,R)$. It is clear that $z_i^{[q]}$ is an element of $F_P^e(\mathbb{N}_1)$ with $\overline{z_i^{[q]}}=\overline{z}_i^{[q]}$ a cycle in $F_P^e(\mathbb{N})\otimes_P R=F_R^e(\mathbb{N}\otimes_P R)$, for each i.

The technique of killing cycles (see, for example, Section 2 of [12]) tells us that

$$\mathbb{M}:\quad \overline{\mathbb{N}}_2\oplus\bigoplus_{i=1}^{\ell}R(-|z_i|)\xrightarrow{[\overline{\boldsymbol{n}}_2\,\overline{z}_1\,\dots\,\overline{z}_\ell]}\overline{\mathbb{N}}_1\xrightarrow{\overline{\boldsymbol{n}}_1}\overline{\mathbb{N}}_0\to\overline{M}\to 0$$

is the beginning of a homogeneous resolution of $\overline{M}$ by free R-modules. It follows that

$$\mathrm{Tor}_1^R(\overline{M},{}^eR)=\mathrm{H}_1(\mathbb{M}\otimes_R{}^eR)=\frac{\mathrm{H}_1(F_R^e(\overline{\mathbb{N}}))}{([\overline{z}_1^{[q]}],\dots,[\overline{z}_\ell^{[q]}])}=\frac{\mathrm{Tor}_1^P(F_P^e(M),R)}{([\overline{z}_1^{[q]}],\dots,[\overline{z}_\ell^{[q]}])}.\quad\square$$

The following result is an application of the technique of Observation 3.4.

Proposition 3.5. *Let $R=P/C$ and $T=P/I$, where I and C are homogeneous ideals in the polynomial ring P. Then*

$$\mathrm{Tor}_1^R(T\otimes_P R,{}^eR)=\frac{I^{[q]}\cap C}{(I\cap C)^{[q]}+I^{[q]}C}.$$

Proof. Apply Observation 3.4 to see that

$$\mathrm{Tor}_1^R(T\otimes_P R,{}^eR)=\frac{\mathrm{H}_1(\mathbb{N}^{[q]}\otimes_P R)}{([\overline{z}_1^{[q]}],\dots,[\overline{z}_\ell^{[q]}])},$$

where $(\mathbb{N},\boldsymbol{n})$ is the minimal homogeneous resolution of T by free P-modules, $^-$ is the functor $\_\_\otimes_P R$, and $z_1,\dots,z_\ell$ are homogeneous elements of $\mathbb{N}_1$ with $[\overline{z}_1],\dots,[\overline{z}_\ell]$ a minimal generating set for

$$\mathrm{H}_1(\mathbb{N}\otimes_P R)=\mathrm{Tor}_1^P(P/I,P/C)=(I\cap C)/IC.$$

Observe that $I\cap C=(\boldsymbol{n}_1(z_1),\dots,\boldsymbol{n}_1(z_\ell))+IC$. Observe also, that

$$\mathrm{H}_1(\mathbb{N}^{[q]}\otimes_P R)=\mathrm{Tor}_1^P\left(P/I^{[q]},P/C\right)=(I^{[q]}\cap C)/I^{[q]}C.$$

The isomorphism $\mathrm{H}_1(\mathbb{N}^{[q]}\otimes_P R)\to(I^{[q]}\cap C)/I^{[q]}C$ carries $[\overline{z}_i^{[q]}]$ to the class of $(\boldsymbol{n}_1(z_i))^q$. $\square$

4. Degree considerations concerning Tor_1

Proposition 4.1, followed by [1], is a general statement that says that if the degrees of the minimal generators of

$$\mathrm{Tor}_1^P(M,R) \quad \text{and} \quad \mathrm{Tor}_1^P(F_P^e(M),R)$$

are related in the appropriate manner, then $M \otimes_P R$ has finite projective dimension as an R-module. When the notation of 2.4 and the hypothesis of Theorem 2.3 are in effect, then Corollary 3.2 and Observation 3.3 show that Proposition 4.1 may be applied with $M = Z$.

PROPOSITION 4.1. *Let $P \to R$ be a surjection of graded k-algebras, with P a polynomial ring and R a complete intersection, and let M be a finitely generated graded P-module. Suppose that the minimal generators of $\mathrm{Tor}_1^P(M,R)$ have degrees $\{\gamma_i \mid 1 \le i \le \ell\}$. If the minimal generators of $\mathrm{Tor}_1^P(F_P^e(M),R)$ have degrees $\{q\gamma_i \mid 1 \le i \le \ell\}$, then $\mathrm{Tor}_1^R(M \otimes_P R, {}^eR) = 0$.*

Proof. Inflation of the base field $k \to K$ gives rise to faithfully flat extensions $P \to P \otimes_k K$ and $R \to R \otimes_k K$. Consequently, we may assume that k is a perfect field. Let C be the ideal in P with $R = P/C$, and let $f_1, \dots, f_c$ be a homogeneous regular sequence in P that generates C. We retain the notation from the proof of Observation 3.4. So,

$$\mathbb{N}: \quad \mathbb{N}_2 \xrightarrow{n_2} \mathbb{N}_1 \xrightarrow{n_1} \mathbb{N}_0 \to M \to 0$$

is the beginning of the minimal homogeneous resolution of M by free P-modules, $\overline{}$ is the functor $\_\_ \otimes_P R$, and $z_1, \dots, z_\ell$ are homogeneous elements of $\mathbb{N}_1$ with $[\overline{z}_1], \dots, [\overline{z}_\ell]$ a minimal generating set for $\mathrm{H}_1(\overline{\mathbb{N}}) = \mathrm{Tor}_1^P(M,R)$. Let Y be the subset $\{[\overline{z}_1^{[q]}], \dots, [\overline{z}_\ell^{[q]}]\}$ of $\mathrm{H}_1(\overline{\mathbb{N}}^{[q]}) = \mathrm{Tor}_1^P(F_P^e(M),R)$. We will prove that

$$Y \text{ generates } \mathrm{H}_1(\overline{\mathbb{N}}^{[q]}). \tag{4.2}$$

As soon as (4.2) is established, then the proof is complete by Observation 3.4.

Fix an integer δ. Define W'_δ and W''_δ to be the P-submodules of $\mathrm{H}_1(\overline{\mathbb{N}}^{[q]})$ which are generated by

$$\sum_{\delta < i} \left[\mathrm{H}_1(\overline{\mathbb{N}}^{[q]})\right]_i \quad \text{and} \quad \sum_{i<\delta} \left[\mathrm{H}_1(\overline{\mathbb{N}}^{[q]})\right]_i,$$

respectively. Let V_δ be the vector space $V_\delta = \mathrm{H}_1(\overline{\mathbb{N}}^{[q]})/(W'_\delta + W''_\delta)$. Let X be a homogeneous minimal generating set for the P-module $\mathrm{H}_1(\overline{\mathbb{N}}^{[q]})$. Let X_δ be the subset of X which consists of those generators which have degree equal to δ. Notice that the set X_δ is a basis for the vector space V_δ over k. Let

$$Y_\delta = \{[\overline{z}_i^{[q]}] \in Y \mid |z_i^{[q]}| = \delta\}.$$

Our hypothesis guarantees that the dimension of the vector space V_δ is exactly equal to the number of elements in Y_δ. We prove that

(4.3) the elements of Y_δ are linearly independent in V_δ.

Once (4.3) is established, then we will know that Y_δ is a basis for V_δ and that the elements of Y_δ are part of a minimal generating set of $\mathrm{H}_1(\overline{\mathbb{N}}^{[q]})$. We repeat this calculation for all δ to see that Y is part of a minimal generating set for $\mathrm{H}_1(\overline{\mathbb{N}}^{[q]})$. The hypothesis tells us that the elements of Y have the correct degrees to be a complete minimal generating set for $\mathrm{H}_1(\overline{\mathbb{N}}^{[q]})$. We conclude that Y is a complete minimal generating set for $\mathrm{H}_1(\overline{\mathbb{N}}^{[q]})$. At that point the proof will be complete by (4.2).

We establish (4.3) by induction on δ. When δ is small, then Y_δ is the empty set, V_δ is zero, and everything is fine. Now we work on the inductive step. There is nothing to do unless δ is a multiple of q. Relabel the z_i, if necessary, and select the integer λ so that $z_1, \dots, z_\lambda$ have degree less than δ/q, and $z_{\lambda+1}, \dots, z_\ell$ have degree at least δ/q. Consider a non-trivial k-linear combination of the elements of Y_δ. The field k is closed under the taking of qth roots; so this linear combination is equal to $[\overline{z}^{[q]}]$, where z is a non-trivial k-linear combination of those z_i that have degree equal to δ/q. So

$$(4.4) \qquad [\overline{z}] \text{ is not zero in } \frac{\mathrm{H}_1(\overline{\mathbb{N}})}{P([\overline{z}_1], \dots, [\overline{z}_\lambda])}.$$

We assume that

$$(4.5) \qquad [\overline{z}^{[q]}] = 0 \text{ in } \frac{\mathrm{H}_1(\overline{\mathbb{N}}^{[q]})}{W''_\delta},$$

and we will show that this assumption leads to a contradiction. Keep in mind that the induction hypothesis guarantees that

$$(4.6) \qquad W''_\delta = P([\overline{z}_1^{[q]}], \dots, [\overline{z}_\lambda^{[q]}]).$$

We know that $\overline{z}$ and $\overline{z}_1, \dots, \overline{z}_\lambda$ all are cycles in $\overline{\mathbb{N}}$; so

$$(4.7) \qquad \boldsymbol{n}_1 z \in C\mathbb{N}_0 \quad \text{and} \quad \boldsymbol{n}_1 z_i \in C\mathbb{N}_0 \text{ for } 1 \le i \le \lambda.$$

We introduce a notational convenience. Let

$$\mathbb{G}_2 = \mathbb{N}_2 \oplus \bigoplus_{i=1}^{\lambda} P(-|z_i|)$$

and let $\boldsymbol{g}_2 : \mathbb{G}_2 \to \mathbb{N}_1$ be the map

$$\boldsymbol{g}_2 = \begin{bmatrix} \boldsymbol{n}_2 & z_1 & \dots & z_\lambda \end{bmatrix}.$$

Of course, the map

$$\boldsymbol{g}_2^{[q]} : \mathbb{G}_2^{[q]} \to \mathbb{N}_1^{[q]}$$

also now has meaning. Assumption (4.5), together with the induction hypothesis (4.6), tells us that

$$z^{[q]} \in \operatorname{im} \boldsymbol{g}_2^{[q]} + C\mathbb{N}_1^{[q]},$$

which is the base case for the following induction. We will prove that if $1 \le t \le c(q-1)$, then

$$z^{[q]} \in \operatorname{im} \boldsymbol{g}_2^{[q]} + C^{[q]}\mathbb{N}_1^{[q]} + C^t\mathbb{N}_1^{[q]} \implies z^{[q]} \in \operatorname{im} \boldsymbol{g}_2^{[q]} + C^{[q]}\mathbb{N}_1^{[q]} + C^{t+1}\mathbb{N}_1^{[q]}. \tag{4.8}$$

As soon as (4.8) is established, then

$$z^{[q]} \in \operatorname{im} \boldsymbol{g}_2^{[q]} + C^{[q]}\mathbb{N}_1^{[q]}$$

because $C^{c(q-1)+1} \subseteq C^{[q]}$. Then the proof will be complete because we may apply Observation 4.10 to

$$z^{[q]} \in \operatorname{im} \begin{bmatrix} \boldsymbol{g}_2 & f_1 I & \cdots & f_c I \end{bmatrix}^{[q]}$$

to conclude that

$$z \in \operatorname{im} \begin{bmatrix} \boldsymbol{g}_2 & f_1 I & \cdots & f_c I \end{bmatrix},$$

and this violates (4.4).

We prove (4.8). For each c-tuple $\alpha = (\alpha_1, \dots, \alpha_c)$ of non-negative integers, with $\alpha_i < q$ for all i, and $\sum \alpha_i = t$, there exists $y_\alpha \in \mathbb{N}_1^{[q]}$ such that

$$z^{[q]} - \sum_\alpha f_1^{\alpha_1} \cdots f_c^{\alpha_c} y_\alpha \in C^{[q]}\mathbb{N}_1^{[q]} + \operatorname{im} \boldsymbol{g}_2^{[q]}. \tag{4.9}$$

Each element y_α is homogeneous of degree $\delta - t$. Fix a c-tuple α. Apply $\boldsymbol{n}_1^{[q]}$ to (4.9) and use (4.7) to see that

$$f_1^{\alpha_1} \cdots f_c^{\alpha_c} \boldsymbol{n}_1^{[q]} y_\alpha \in (f_1^{\alpha_1+1}, \dots, f_c^{\alpha_c+1})\mathbb{N}_0^{[q]}.$$

It follows, from the fact that $f_1, \dots, f_c$ is a regular sequence, that

$$\boldsymbol{n}_1^{[q]} y_\alpha \in C\mathbb{N}_0^{[q]}.$$

So, $\overline{y}_\alpha$ is a one-cycle, of degree less than δ, in $\mathbb{N}^{[q]}$. The induction hypothesis (4.6) tells us that

$$y_\alpha \in \operatorname{im} \boldsymbol{g}_2^{[q]} + C\mathbb{N}_1^{[q]},$$

and (4.8) is established. □

We close this section with a quick application of the flatness of the Frobenius functor for regular rings. The result is well known and is a critical step in the proof of the previous result.

Observation 4.10. *Let P be a polynomial ring, $\boldsymbol{g} : \mathbb{G}_2 \to \mathbb{G}_1$ be a homomorphism of graded P-modules, and z be an element of $\mathbb{G}_1$. If $z^{[q]}$ is in the image of $\boldsymbol{g}^{[q]}$, then z is in the image of $\boldsymbol{g}$.*

Proof. Let $\widetilde{\mathbb{G}}_2$ be the graded free module $\mathbb{G}_2 \oplus P(-|z|)$ and $\widetilde{\boldsymbol{g}}$ be the map of graded free modules

$$\widetilde{\boldsymbol{g}} = \begin{bmatrix} \boldsymbol{g} & z \end{bmatrix} : \widetilde{\mathbb{G}}_2 \to \mathbb{G}_1.$$

The hypothesis ensures the existence of $h \in \mathbb{G}_2^{[q]}$ with $\begin{bmatrix} -h & 1 \end{bmatrix}^{\mathrm{T}}$ in the kernel of $\widetilde{\boldsymbol{g}}^{[q]}$, where $(\ )^{\mathrm{T}}$ means "transpose". The Frobenius functor $\_ \otimes_P {}^eP$ is flat; so, there exist $\begin{bmatrix} t_i & b_i \end{bmatrix}^{\mathrm{T}}$ in $\ker \widetilde{\boldsymbol{g}}$ and $a_i \in P$ such that $\sum_i a_i \begin{bmatrix} t_i^{[q]} & b_i^q \end{bmatrix}^{\mathrm{T}} = \begin{bmatrix} -h & 1 \end{bmatrix}^{\mathrm{T}}$. Degree considerations tell us that b_i is a unit, for some i. For this i, $\boldsymbol{g}(t_i) + b_i z = 0$ and $\boldsymbol{g}(-t_i/b_i) = z$. □

5. We interpret Tor_1 of the syzygy in terms of ideals

Recall, from the beginning of Section 4, that if the notation of 2.4 and the hypotheses of Theorem 2.3 are in effect, then $\operatorname{Tor}_1^R(Z \otimes_P R, {}^eR) = 0$. In this section, we interpret this Tor-module in terms of ideals. Our interpretation holds even when the hypotheses of Theorem 2.3 are not in effect.

PROPOSITION 5.1. *Adopt the notation of 2.4. If $A \subseteq I$ is any homogeneous $\mathfrak{m}_P$-primary Gorenstein ideal and N is equal to $a(P/A^{[q]}) - a(R)$, then*

$$\operatorname{Tor}_1^R(Z \otimes_P R, {}^eR) = \operatorname{Hom}_P\left(\frac{(C^{[q]} + I^{[q]}) : (C^{[q]} : C)}{C + I^{[q]}}, \frac{P}{A^{[q]}}(N)\right); \tag{5.2}$$

furthermore,

$$\operatorname{Tor}_1^R(Z \otimes_P R, {}^eR) = 0 \iff \frac{(C^{[q]} + I^{[q]}) : (C^{[q]} : C)}{C + I^{[q]}} = 0.$$

Proof. In our argument, we assume that $A \subseteq I$ are perfect ideals of the same grade without assuming that these ideals are $\mathfrak{m}_P$-primary, and we prove that

$$\operatorname{Tor}_1^R(Z \otimes_P R, {}^eR) = \frac{A^{[q]} : (C + I^{[q]})}{(A : (C + I))^{[q]}(C^{[q]} : C) + A^{[q]}}(N). \tag{5.3}$$

Once (5.3) is established, then, when A is an $\mathfrak{m}_P$-primary ideal, Gorenstein duality (see, for example, Lemma 1.11 and (1.14)) may be employed to show that the right side of (5.3) and the right side of (5.2) are both equal to

$$\frac{A^{[q]} : (C + I^{[q]})}{A^{[q]} : ((C + I)^{[q]} : (C^{[q]} : C))}(N).$$

Let $(\mathbb{G}, \boldsymbol{g}_\bullet)$ be the minimal homogeneous resolution of R by free P-modules and $\boldsymbol{y}_\bullet : \mathbb{G}^{[q]} \to \mathbb{G}$ be a map of complexes which lifts the natural quotient map $P/C^{[q]} \to R$. The ideal C is Gorenstein of grade c; so $\mathbb{G}_c = P(a(P) - a(R))$. The map $\boldsymbol{y}_c : \mathbb{G}_c^{[q]} \to \mathbb{G}_c$ is multiplication by y, for some element y in P of degree $(q-1)(a(R)-a(P))$. We know, from linkage theory (see Proposition 1.13), that

$$C^{[q]}:C=(y,C^{[q]}). \tag{5.4}$$

Use the surjections $P/A \to T$ and $P/A^{[q]} \to F_P^e(T)$ to calculate

$$K_T=\frac{A:I}{A}(a(P/A)) \quad \text{and} \quad K_{F_P^e(T)}=\frac{A^{[q]}:I^{[q]}}{A^{[q]}}(a(P/A^{[q]})).$$

It follows that $\mathrm{H}_c(K_T(-a(P))\otimes_P \mathbb{G})$ is equal to

$$\left\{\alpha\in\frac{A:I}{A}(a(P/A)-a(R))\,\middle|\,\alpha C=0\right\}=\frac{A:(I+C)}{A}\,(a(P/A)-a(R))$$

and

$$\mathrm{H}_c(K_{F_P^e(T)}(-a(P))\otimes_P\mathbb{G})=\frac{A^{[q]}:(I^{[q]}+C)}{A^{[q]}}(N).$$

Let $(\mathbb{F},\boldsymbol{f})$ be the minimal homogeneous resolution of $K_T(-a(P))$ by free P-modules. We saw in Observation 3.4 that

$$\mathrm{Tor}_1^R(Z\otimes_P R,{}^eR)=\frac{\mathrm{H}_c(\mathbb{F}^{[q]}\otimes_P R)}{([\overline{z}_1^{[q]}],\dots,[\overline{z}_\ell^{[q]}])},$$

where $z_1,\dots,z_\ell$ are elements in $\mathbb{F}_c$ with $[\overline{z}_1],\dots,[\overline{z}_\ell]$ a minimal generating set for $\mathrm{H}_c(\mathbb{F}\otimes_P R)$. We use the isomorphisms

$$\begin{aligned}\mathrm{H}_c(\mathbb{F}\otimes_P R)&=\mathrm{H}_c(\mathrm{Tot}(\mathbb{F}\otimes_P\mathbb{G}))=\mathrm{H}_c(K_T(-a(P))\otimes_P\mathbb{G})\\&=\frac{A:(C+I)}{A}(a(P/A)-a(R))\end{aligned}\tag{5.5}$$

and

$$\begin{aligned}\mathrm{H}_c(\mathbb{F}^{[q]}\otimes_P R)=\mathrm{H}_c(\mathrm{Tot}(\mathbb{F}^{[q]}\otimes_P\mathbb{G}))\quad&=\mathrm{H}_c(K_{F_P(T)}(-a(P))\otimes_P\mathbb{G})\\=\frac{A^{[q]}:(C+I^{[q]})}{A^{[q]}}(N)&\end{aligned}\tag{5.6}$$

to re-express $\mathrm{Tor}_1^R(Z\otimes_P R,{}^eR)$ as a subquotient of $P(N)$. First, we give an explicit description of the isomorphism (5.5). Let w_c be an element of $\mathbb{F}_c$ with $\boldsymbol{f}_c(w_c)\in C\mathbb{F}_{c-1}$. Each row and column of the double complex $\mathbb{F}\otimes\mathbb{G}$ is acyclic; therefore, for each i, with $0\le i\le c$, there exists $w_i\in\mathbb{F}_i\otimes\mathbb{G}_{c-i}$ such that

$$(\boldsymbol{f}_i\otimes 1)(w_i)=(1\otimes\boldsymbol{g}_{c-i+1})(w_{i-1}). \tag{5.7}$$

In particular, w_0 is an element of $\mathbb{F}_0\otimes\mathbb{G}_c$. The isomorphism (5.5) sends the homology class $[\overline{w}_c]$ to the image of w_0 in

$$\frac{A:(C+I)}{A}(a(P/A)-a(R)).$$

In a similar manner, if W_c is an element of $\mathbb{F}_c^{[q]}$ with $\boldsymbol{f}_c^{[q]}(W_c) \in C\mathbb{F}_{c-1}^{[q]}$, then the isomorphism (5.6) sends the homology class $[\overline{W}_c]$ in $\mathrm{H}_c(\mathbb{F}^{[q]} \otimes_P R)$ to the image of W_0 in

$$\frac{A^{[q]} : (C + I^{[q]})}{A^{[q]}}(N),$$

where $W_i \in \mathbb{F}_i^{[q]} \otimes \mathbb{G}_{c-i}$ and

$$(\boldsymbol{f}_i^{[q]} \otimes 1)(W_i) = (1 \otimes \boldsymbol{g}_{c-i+1})(W_{i-1}). \tag{5.8}$$

We finish the argument by showing that the submodule $([\overline{z}_1^{[q]}], \dots, [\overline{z}_\ell^{[q]}])$ of $\mathrm{H}_c(\mathbb{F}^{[q]} \otimes_P R)$ is mapped onto the submodule

$$\frac{(A : (C + I))^{[q]}y + A^{[q]}}{A^{[q]}}(N) \quad \text{of} \quad \frac{A^{[q]} : (C + I^{[q]})}{A^{[q]}}(N)$$

under the isomorphism (5.6). Let $\overline{w}_c$ be a cycle in $\mathbb{F} \otimes R$, for some element w_c of $\mathbb{F}_c$. We are given the family $\{w_i\}$ with $w_i \in \mathbb{F}_i \otimes \mathbb{G}_{c-i}$ such that (5.7) holds. If $w_i = \sum_j u_{i,j} \otimes v_{c-i,j}$ with $u_{i,j} \in \mathbb{F}_i$ and $v_{c-i,j} \in \mathbb{G}_{c-i}$, then let

$$W_i = \sum_j u_{i,j}^{[q]} \otimes \boldsymbol{y}_{c-i}(v_{c-i,j}^{[q]}) \in \mathbb{F}_i^{[q]} \otimes \mathbb{G}_{c-i}.$$

A short calculation shows that (5.8) holds for $\{W_i\}$ and we conclude that if a in

$$\frac{A : (C + I)}{A}(a(P/A) - a(R))$$

is the image of the homology class $[\overline{w}_c]$ under the isomorphism (5.5), then ya^q in

$$\frac{A^{[q]} : (C + I^{[q]})}{A^{[q]}}(N)$$

is the image of the homology class $[\overline{w}_c^{[q]}]$ under the isomorphism (5.6). □

6. The key calculation

Retain the notation of 2.4. In Section 4, we proved that if the socle hypothesis (2.2) is in effect, then $\mathrm{Tor}_1^R(Z \otimes_P R, {}^eR) = 0$. Our goal in Theorem 2.3 is to prove that $\mathrm{Tor}_1^R(T \otimes_P R, {}^eR) = 0$. Homological arguments in Sections 3 and 5 connect these Tor-modules to quotients of ideals. In the present section we show how information about the Tor-module of Z gives information about the Tor-module of T, when R is a complete intersection.

PROPOSITION 6.1. *Let P be a regular ring of positive characteristic p, and let C and I be ideals in P. Assume that C is generated by the regular sequence $f_1, \dots, f_c$ and that*

$$(C + I)^{[q]} : y = C + I^{[q]}, \tag{6.2}$$

where $y = (f_1 \cdots f_c)^{q-1}$. *Then*

$$I^{[q]} \cap C = (I \cap C)^{[q]} + CI^{[q]}.$$

Proof. Recall, from Proposition 1.13, that $C^{[q]} : C = (y, C^{[q]})$. Take $\xi \in I^{[q]} \cap C$. We prove that if $1 \le t \le c(q-1)$, then

$$\xi \in C^t + C^{[q]} + CI^{[q]} \implies \xi \in C^{t+1} + C^{[q]} + CI^{[q]}. \tag{6.3}$$

Of course, we know that the hypothesis of (6.3) holds for $t = 1$. Once we have established (6.3), then, since $C^{c(q-1)+1} \subseteq C^{[q]}$, we know that

$$\xi \in I^{[q]} \cap C^{[q]} + CI^{[q]} = (I \cap C)^{[q]} + CI^{[q]},$$

because the Frobenius functor on P is flat.

Now we prove (6.3). Write ξ as an element of $C^{[q]} + CI^{[q]}$ plus

$$\sum_{\alpha} b_\alpha f_1^{\alpha_1} \cdots f_c^{\alpha_c},$$

where $\alpha = (\alpha_1, \dots, \alpha_c)$ varies over all c-tuples of non-negative integers with $\alpha_i < q$ for all i and $\sum_{i=1}^{c} \alpha_i = t$. Fix an index α. Observe that

$$f_1^{q-1-\alpha_1} \cdots f_c^{q-1-\alpha_c} \xi$$

is equal to $b_\alpha y$ plus an element of $C^{[q]} + I^{[q]}$. Hypothesis (6.2) tells us that b_α is in $C + I^{[q]}$; (6.3) is established, and the proof is complete. □

7. The Gorenstein F-pure case

The question of whether the conclusion of Theorem 2.3 holds when R is Gorenstein is still open. In this section, we include partial results in this direction. Recall that the ring R of positive prime characteristic p is F-pure if whenever J is an ideal of R and x is an element of R with $x \notin J$, then $x^q \notin J^{[q]}$ for all $q = p^e$.

First note that the top socle degree (tsd) of a Frobenius power is always at least equal to the "expected" top socle degree, in the sense of Observation 2.1:

PROPOSITION 7.1. *Let k be a field of positive characteristic p, $R \to S$ be a surjection of graded k-algebras with S artinian. Assume that either R is a complete intersection or R is Gorenstein and F-pure. If d is the top socle degree of S, then the top socle degree of $F_R^e(S)$ is at least $qd - (q-1)a(R)$.*

Proof. Write $S = R/J$, with $J \subset R$ an $\mathfrak{m}$-primary ideal, where $\mathfrak{m}$ is the unique homogeneous maximal ideal of R.

We first assume that R is Gorenstein and F-pure. Let $\mathfrak{a}$ be an $\mathfrak{m}$-primary ideal of R, generated by a regular sequence, with $\mathfrak{a} \subset J$. Let $g_1, \dots, g_s$, with $|g_1| \le \cdots \le |g_s|$, be elements in R which represent a minimal generating set for $(\mathfrak{a} : J)/\mathfrak{a}$. The hypothesis that R is F-pure ensures that

$g_i^q \notin (g_1^q, \dots, \widehat{g_i^q}, \dots, g_s^q, \mathfrak{a}^{[q]})$ for all i; therefore, $g_1^q, \dots, g_s^q$ represents a minimal generating set for $(g_1^q, \dots, g_s^q, \mathfrak{a}^{[q]})/\mathfrak{a}^{[q]}$. It follows that the minimum generator degree (min gen degree) of $(g_1^q, \dots, g_s^q, \mathfrak{a}^{[q]})/\mathfrak{a}^{[q]}$ is $q|g_1|$. Apply Observation 1.4 to the ideal $J/\mathfrak{A}$ of the ring $R/\mathfrak{a}$ to see that

$$\begin{aligned} \operatorname{tsd} R/J &= \operatorname{tsd}\left(\frac{R/\mathfrak{a}}{J/\mathfrak{a}}\right) \\ &= \text{socle degree}\, R/\mathfrak{a} - \text{min gen degree}\,(\operatorname{ann}(J/\mathfrak{a})) \\ &= \text{socle degree}\, R/\mathfrak{a} - \text{min gen degree}\,((\mathfrak{a} : J)/\mathfrak{a}) \\ &= \text{socle degree}\, R/\mathfrak{a} - |g_1|. \end{aligned}$$

Duality (see Lemma 1.11) gives

$$\mathfrak{a}^{[q]} : (\mathfrak{a}^{[q]} : (g_1^q, \dots, g_s^q)) = (g_1^q, \dots, g_s^q, \mathfrak{a}^{[q]})$$

in R; so the annihilator of the ideal

$$\frac{\mathfrak{a}^{[q]} : (g_1^q, \dots, g_s^q)}{\mathfrak{a}^{[q]}} \tag{7.2}$$

in the ring $R/\mathfrak{a}^{[q]}$ is $(g_1^q, \dots, g_s^q, \mathfrak{a}^{[q]})/\mathfrak{a}^{[q]}$. Apply Observation 1.4 to the ideal (7.2) to see that

$$\begin{aligned} &\operatorname{tsd} \frac{R}{\mathfrak{a}^{[q]} : (g_1^q, \dots, g_s^q)} = \operatorname{tsd} \frac{R/\mathfrak{a}^{[q]}}{(\mathfrak{a}^{[q]} : (g_1^q, \dots, g_s^q))/\mathfrak{a}^{[q]}} \\ &= \text{socle degree}\, \frac{R}{\mathfrak{a}^{[q]}} - \text{min gen degree}\left(\operatorname{ann} \frac{\mathfrak{a}^{[q]} : (g_1^q, \dots, g_s^q)}{\mathfrak{a}^{[q]}}\right) \\ &= \text{socle degree}\, \frac{R}{\mathfrak{a}^{[q]}} - \text{min gen degree}\left(\frac{(g_1^q, \dots, g_s^q, \mathfrak{a}^{[q]})}{\mathfrak{a}^{[q]}}\right) \\ &= \text{socle degree}\, \frac{R}{\mathfrak{a}^{[q]}} - q|g_1|. \end{aligned}$$

The R-module $R/\mathfrak{a}$ has finite projective dimension; so Observation 2.1 yields

$$\text{socle degree}\, \frac{R}{\mathfrak{a}^{[q]}} = q\, \text{socle degree}\, \frac{R}{\mathfrak{a}} - (q-1)a(R).$$

Duality gives $J = \mathfrak{a} : (g_1, \dots, g_s)$. It follows that $J^{[q]} \subseteq \mathfrak{a}^{[q]} : (g_1^q, \dots, g_s^q)$; therefore,

$$\begin{aligned}\operatorname{tsd}\frac{R}{J^{[q]}} &\geq \operatorname{tsd}\frac{R}{\mathfrak{a}^{[q]}:(g_1^q,\dots,g_s^q)} = \text{socle degree}\,\frac{R}{\mathfrak{a}^{[q]}} - q|g_1| \\ &= \text{socle degree}\,\frac{R}{\mathfrak{a}^{[q]}} - q\left(\text{socle degree}\,\frac{R}{\mathfrak{a}} - \operatorname{tsd}\frac{R}{J}\right) \\ &= q\operatorname{tsd}\frac{R}{J} + \left(\text{socle degree}\,\frac{R}{\mathfrak{a}^{[q]}} - q\,\text{socle degree}\,\frac{R}{\mathfrak{a}}\right) \\ &= q\operatorname{tsd}\frac{R}{J} - (q-1)a(R).\end{aligned}$$

The proof is complete if R is Gorenstein and F-pure. Throughout the rest of the argument, R is a complete intersection. We begin by reducing to the case where J is an irreducible ideal. Assume, for the time being, that the result has been established for irreducible ideals. Let $J = J_1 \cap \dots \cap J_n$, with each J_i irreducible. Recall that $\operatorname{tsd} R/J$ is the largest integer d with $R_d \not\subseteq J$. It follows that the $\operatorname{tsd} R/J$ is equal to the maximum of the set $\{\operatorname{tsd} R/J_k\}$. Fix a subscript k with $\operatorname{tsd} R/J = \operatorname{tsd} R/J_k$. We know that $J^{[q]} \subseteq J_k^{[q]}$; therefore,

$$q\operatorname{tsd}\frac{R}{J} - (q-1)a(R) = q\operatorname{tsd}\frac{R}{J_k} - (q-1)a(R) \leq \operatorname{tsd}\frac{R}{J_k^{[q]}} \leq \operatorname{tsd}\frac{R}{J^{[q]}}.$$

Henceforth, the ideal J is irreducible. Write $R = P/C$, where P is a polynomial ring and the ideal C is generated by the homogeneous regular sequence $f_1, \dots, f_c$. Let I be the pre-image of J in P. In particular, $C \subseteq I$. The rings $R/J = P/I$ and $P/I^{[q]}$ are Gorenstein, so Observation 1.4 gives

$$\text{socle degree}\,\frac{P}{I^{[q]}} - M = \operatorname{tsd}\frac{P}{I^{[q]}+C},$$

where M is the least degree among homogeneous non-zero elements of $(I^{[q]} : C)/I^{[q]}$. The P-module P/I has finite projective dimension; so Observation 2.1 yields

$$q\,\text{socle degree}(P/I) - (q-1)a(P) = \text{socle degree}(P/I^{[q]}).$$

Recall the formula $a(P/C) = a(P) + \sum_{i=1}^{c}|f_i|$. The inequality

$$q\,\text{socle degree}(P/I) - (q-1)a(P/C) \leq \operatorname{tsd}\left(\frac{P}{I^{[q]}+C}\right)$$

is equivalent to the inequality

$$M \leq (q-1)(|f_1| + \dots + |f_c|). \tag{7.3}$$

We establish (7.3). There exists an integer t, with $0 \leq t \leq c(q-1)$, such that $C^t \not\subseteq I^{[q]}$; but $C^{t+1} \subseteq I^{[q]}$. Thus, some element $f_1^{t_1}\cdots f_c^{t_c}$ of C^t, with $\sum t_i = t$ and $0 \leq t_i \leq q-1$ for all i, represents a non-zero element of $(I^{[q]} : C)/I^{[q]}$; therefore,

$$M \leq |f_1^{t_1}\cdots f_c^{t_c}| \leq (q-1)(|f_1| + \dots + |f_c|).$$ □

The next result shows that we can get most of the way through the proof of Theorem 2.3 under the assumption that R is Gorenstein and F-pure. The conclusion of Proposition 7.4 is exactly the same as the conclusion that is obtained when Proposition 5.1 is used in the proof of Theorem 2.3. Only the last step (the analogue of Proposition 6.1) is still missing.

PROPOSITION 7.4. *Let k be a field of positive characteristic p, $R \to S$ be a surjection of graded k-algebras with R Gorenstein and S artinian. Assume, in addition, that R is F-pure. Assume that $d_1 \leq \cdots \leq d_\ell$ are the socle degrees of S and the socle of $F_R^e(S)$ has the same dimension as the socle of S, with degrees of the generators given by $D_1 \leq D_2 \leq \cdots \leq D_\ell$, with*

$$D_i = qd_i - (q-1)a(R),$$

for all i. Let $R = P/C$, and $S = R/IR$, with P a polynomial ring, $I \subset P$. Then we have

$$(C^{[q]} + I^{[q]}) : (C^{[q]} : C) = C + I^{[q]}.$$

Proof. Let $A = C + (x_1, \ldots, x_d)$, where the images of $x_1, \ldots, x_d$ are a system of parameters in R. Let $K = A : (I + C)$, so that we also have $I + C = A : K$. We have

$$(I^{[q]} + C^{[q]}) : (C^{[q]} : C) = (A^{[q]} : K^{[q]}) : (C^{[q]} : C) = (A^{[q]} : (C^{[q]} : C)) : K^{[q]}.$$

We claim that $A^{[q]} : (C^{[q]} : C) = A^{[q]} + C$. We see this by looking at the comparison map of resolutions induced by the projection $P/A^{[q]} \to P/(A^{[q]} + C)$. If $\mathbb{F}$ is the resolution of P/C, and $\mathbb{K}$ is the Koszul complex on $x_1, \ldots, x_d$, then the resolution of $P/A^{[q]}$ is given by $\mathbb{F}^{[q]} \otimes \mathbb{K}^{[q]}$, the resolution of $P/(A^{[q]} + C)$ is given by $\mathbb{F} \otimes \mathbb{K}^{[q]}$, and the comparison map between them is given by the comparison map $\mathbb{F}^{[q]} \to \mathbb{F}$, tensored with $\mathbb{K}^{[q]}$. Thus, the last map is multiplication by an element of P which represents the generator of $(C^{[q]} : C)/C^{[q]}$.

It follows that $(I^{[q]} + C^{[q]}) : (C^{[q]} : C) = (A^{[q]} + C) : K^{[q]}$. It is clear that

$$(A^{[q]} + C) : K^{[q]} \supseteq I^{[q]} + C.$$

We next show that the rings defined by these ideals have the same socle degrees. Let δ and Δ be the socle degrees of the Gorenstein rings P/A and $P/(A^{[q]} + C)$, respectively. The P/C-module P/A has finite projective dimension; so

$$\Delta = q\delta - (q-1)a(P/C).$$

Let $g_1, \ldots, g_s$ be elements of K which represent a minimal generating set for K/A. Observation 1.4 gives that the socle degrees of $P/(I + C)$ are $\{\delta - |g_i|\}$. So, our hypothesis tells us that the socle degrees of $P/(I^{[q]} + C)$ are

$$\{q(\delta - |g_i|) - (q-1)a(P/C)\} = \{\Delta - q|g_i|\}.$$

It is clear that $g_1^q, \dots, g_s^q$ represents a generating set for $(K^{[q]}+C)/(A^{[q]}+C)$. Apply the hypothesis that P/C is F-pure to each of the ideals $(g_1^q, \dots, \widehat{g_i^q}, \dots, g_s^q, A^{[q]})$ of P/C to conclude that $(K^{[q]}+C)/(A^{[q]}+C)$ is minimally generated by $g_1^q, \dots, g_s^q$. Observation 1.4 yields that the socle degrees of

$$\frac{P}{(A^{[q]}+C):K^{[q]}}$$

are exactly the same as the socle degrees of $P/(I^{[q]}+C)$; therefore, Lemma 1.1 of [9] shows that

$$I^{[q]}+C = (A^{[q]}+C):K^{[q]} = (I^{[q]}+C^{[q]}):(C^{[q]}:C). \qquad \square$$

References

[1] L. L. Avramov and C. Miller, *Frobenius powers of complete intersections*, Math. Res. Lett. **8** (2001), 225–232. MR 1825272 (2002b:13022)

[2] H. Bass, *On the ubiquity of Gorenstein rings*, Math. Z. **82** (1963), 8–28. MR 0153708 (27 #3669)

[3] H. Brenner, *A linear bound for Frobenius powers and an inclusion bound for tight closure*, Michigan Math. J. **53** (2005), 585–596. MR 2207210 (2006k:13010)

[4] W. Bruns and J. Herzog, *Cohen-Macaulay rings*, Cambridge Studies in Advanced Mathematics, vol. 39, Cambridge University Press, Cambridge, 1993. MR 1251956 (95h:13020)

[5] W. Bruns and U. Vetter, *Determinantal rings*, Lecture Notes in Mathematics, vol. 1327, Springer-Verlag, Berlin, 1988. MR 953963 (89i:13001)

[6] S. P. Dutta, *On modules of finite projective dimension over complete intersections*, Proc. Amer. Math. Soc. **131** (2003), 113–116 (electronic). MR 1929030 (2003j:13016)

[7] S. Goto and K. Watanabe, *On graded rings. I*, J. Math. Soc. Japan **30** (1978), 179–213. MR 494707 (81m:13021)

[8] A. R. Kustin and M. Miller, *Deformation and linkage of Gorenstein algebras*, Trans. Amer. Math. Soc. **284** (1984), 501–534. MR 743730 (85k:13015)

[9] A. R. Kustin and B. Ulrich, *If the socle fits*, J. Algebra **147** (1992), 63–80. MR 1154674 (93e:13017)

[10] C. Peskine and L. Szpiro, *Dimension projective finie et cohomologie locale. Applications à la démonstration de conjectures de M. Auslander, H. Bass et A. Grothendieck*, Inst. Hautes Études Sci. Publ. Math. **42** (1973), 47–119. MR 0374130 (51 #10330)

[11] ———, *Liaison des variétés algébriques. I*, Invent. Math. **26** (1974), 271–302. MR 0364271 (51 #526)

[12] J. Tate, *Homology of Noetherian rings and local rings*, Illinois J. Math. **1** (1957), 14–27. MR 0086072 (19,119b)

[13] A. Vraciu, *Tight closure and linkage classes in Gorenstein rings*, Math. Z. **244** (2003), 873–885. MR 2000463 (2004k:13008)

Andrew R. Kustin, Mathematics Department, University of South Carolina, Columbia, SC 29208, USA

E-mail address: kustin@math.sc.edu

Adela N. Vraciu, Mathematics Department, University of South Carolina, Columbia, SC 29208, USA

E-mail address: vraciu@math.sc.edu

Illinois Journal of Mathematics
Volume 51, Number 1, Spring 2007, Pages 209–236
S 0019-2082

QUASI-PERFECT SCHEME-MAPS AND BOUNDEDNESS OF THE TWISTED INVERSE IMAGE FUNCTOR

JOSEPH LIPMAN AND AMNON NEEMAN

To Phillip Griffith, on his 65th birthday

ABSTRACT. For a map $f\colon X \to Y$ of quasi-compact quasi-separated schemes, we discuss *quasi-perfection*, i.e., the right adjoint $f^\times$ of $\mathbf{R}f_*$ respects small direct sums. This is equivalent to the existence of a functorial isomorphism $f^\times\mathcal{O}_Y \otimes^{\mathbf{L}} \mathbf{L}f^*(-) \xrightarrow{\sim} f^\times(-)$; to *quasi-properness* (preservation by $\mathbf{R}f_*$ of pseudo-coherence, or just *properness* in the noetherian case) plus boundedness of $\mathbf{L}f^*$ (finite tor-dimensionality), or of the functor $f^\times$; and to some other conditions. We use a globalization, previously known only for divisorial schemes, of the local definition of pseudo-coherence of complexes, as well as a refinement of the known fact that the derived category of complexes with quasi-coherent homology is generated by a single perfect complex.

1. Introduction

This paper, inspired by [V, p. 396, Lemma 1 and Corollary 2], deals with matters raised there, but not yet fully treated in the literature.

Throughout, *scheme* will mean *quasi-compact quasi-separated scheme* (see [GD, §6.1, p. 290ff]), though weaker assumptions would sometimes suffice. Unless otherwise indicated, a *map* $f\colon X \to Y$ will be a *scheme-morphism,* necessarily quasi-compact and quasi-separated.

For a scheme X, $\mathbf{D}(X)$ is the (unbounded) derived category of the category of (sheaves of) $\mathcal{O}_X$-modules, and $\mathbf{D}_{\mathsf{qc}}(X)$ is the full subcategory whose objects are the $\mathcal{O}_X$-complexes whose homology sheaves are all quasi-coherent. For any map $f\colon X \to Y$, the derived functor $\mathbf{R}f_*\colon \mathbf{D}(X) \to \mathbf{D}(Y)$ takes $\mathbf{D}_{\mathsf{qc}}(X)$ to $\mathbf{D}_{\mathsf{qc}}(Y)$ [Lp, Prop. (3.9.2)]. Grothendieck Duality theory asserts, to begin, that *the restriction* $\mathbf{R}f_*\colon \mathbf{D}_{\mathsf{qc}}(X) \to \mathbf{D}_{\mathsf{qc}}(Y)$ *has a right adjoint* $f^\times$, the "twisted inverse image functor" in our title.[1]

Received November 24, 2006; received in final form March 16, 2007.
2000 *Mathematics Subject Classification.* Primary 14A15.
The first author is partially supported by National Security Agency award H98230-06-1-0010. The second author is partially supported by the Australian Research Council.

A proof for maps of *separated* schemes, suggested by Deligne's appendix to [H], is described in [Lp, §4.1]. This proof depends ultimately on the Special Adjoint Functor Theorem, applied to categories of sheaves. A more direct approach, via Brown Representability—which applies immediately to derived categories—is given in [N1]. Originally this too required separability, but now that assumption can be dropped because of [BB, p. 9, Thm. 3.3.1], which gives that $\mathbf{D}_{\mathsf{qc}}(X)$ is compactly generated, and because $\mathbf{R}f_*$ commutes with $\mathbf{D}_{\mathsf{qc}}$-coproducts (= direct sums) [Lp, (3.9.3.3)].[2]

The functor $f^\times$ emerging from these proofs commutes with translation (=suspension) of complexes, and is *bounded-below* (*way-out right* in the sense of [H, p. 68]), i.e., there exists an integer m such that for every $F \in \mathbf{D}_{\mathsf{qc}}(Y)$ with $H^iF = 0$ for all i less than some integer $n(F)$, it holds that $H^if^\times F = 0$ for all $i < n(F) - m$ (see [Lp, (4.1.8) and the remarks preceding it]).

"Bounded-below" has a similar meaning for any functor between derived categories. *Bounded-above* is defined in an analogous way, with $>$ (resp. $+$) in place of $<$ (resp. $-$). A functor is *bounded* if it is bounded both above and below. Boundedness enables a potent form of induction in derived categories, expressed by the "way-out Lemmas" [H, p. 68, Prop. 7.1 and p. 73, Prop. 7.3].

For example, the left adjoint $\mathbf{L}f^*$ of $\mathbf{R}f_*$ is always bounded-above; and $\mathbf{L}f^*$ is bounded iff f has *finite tor-dimension* (a.k.a. *finite flat dimension*), that is, there is an integer $d \geq 0$ such that for each $x \in X$ there exists an exact sequence of $\mathcal{O}_{Y,f(x)}$-modules

$$0 \to P_d \to P_{d-1} \to \cdots \to P_1 \to P_0 \to \mathcal{O}_{X,x} \to 0$$

with P_i flat over $\mathcal{O}_{Y,f(x)}$ $(0 \leq i \leq d)$.

We will be concerned with the relation between boundedness of the right adjoint $f^\times$ and the left adjoint $\mathbf{L}f^*$, especially in the context of *quasi-perfection,* a property of maps to be discussed at length now and in §2.

DEFINITION 1.1. We say a map $f\colon X \to Y$ is *quasi-perfect* if $f^\times$ respects direct sums in $\mathbf{D}_{\mathsf{qc}}$, i.e., for any small $\mathbf{D}_{\mathsf{qc}}(Y)$-family (E_α) the natural map is an isomorphism

$$\bigoplus_\alpha f^\times E_\alpha \xrightarrow{\sim} f^\times(\bigoplus_\alpha E_\alpha).$$

As will be explained below, quasi-perfection is also characterized by the existence of a canonical isomorphism

$$f^\times\mathcal{O}_Y \otimes^{\mathbf{L}} \mathbf{L}f^*F \xrightarrow{\sim} f^\times F \qquad (F \in \mathbf{D}_{\mathsf{qc}}(Y)).$$

---

[1]Warning: for nonproper maps of noetherian schemes, the usual twisted inverse image $f^!$ differs from $f^\times$, and is not covered by this paper. For that, see, e.g., [Lp, §4.9].

[2]Subsequently, a slightly simpler proof was given in [BB, p. 14, 3.3.4]. (In that proof one needs to replace "flabby" by "quasi-flabby," see [Kf, §2].)

More characterizations are given in §2—for instance, via compatibility of $f^\times$ with tor-independent base change (Theorem 2.7). That section also brings in the related condition on maps of being *perfect*, i.e., pseudo-coherent and of finite tor-dimension. (Pseudo-coherence will be reviewed in §2. It holds, for instance, for all finite-type maps of noetherian schemes; and then descent to the noetherian case yields that every flat, finitely presentable map is pseudo-coherent.) For example, for a proper map f of noetherian schemes, *f is quasi-perfect $\Leftrightarrow$ f is perfect $\Leftrightarrow$ $f^\times$ is bounded.*

It is stated in [V, p. 396, Lemma 1] that any proper map f of finite-dimensional noetherian schemes is quasi-perfect. In general, however, this fails even for closed immersions. But $f^\times$ *does* respect direct sums when the summands E_α are uniformly homologically bounded below, i.e., there exists an integer n such that for all α, $H^iE_\alpha = 0$ whenever $i < n$ [Lp, (4.7.6)(b)]. Consequently, *if the functor $f^\times$ is bounded, then f is quasi-perfect.*

Our main results say more. But first, call a map $f\colon X \to Y$ *quasi-proper* if $\mathbf{R}f_*$ takes pseudo-coherent $\mathcal{O}_X$-complexes to pseudo-coherent $\mathcal{O}_Y$-complexes. (Again, pseudo-coherence is explained in §2. In particular, if X is noetherian then $E \in \mathbf{D}(X)$ is pseudo-coherent iff the homology sheaves $H^n(E)$ are all coherent, and vanish for $n \gg 0$.) Kiehl showed that *every proper pseudo-coherent map is quasi-proper.* Consequently, any flat, finitely presentable, proper map, being pseudocoherent, is quasi-proper (and perfect and quasi-perfect as well). Moreover, *when Y is noetherian, every finite-type separated quasi-proper $f\colon X \to Y$ is proper.*

Here are the main results.

THEOREM 1.2. *For a map $f\colon X \to Y$, the following are equivalent:*
(i) *f is quasi-perfect (resp. perfect).*
(ii) *f is quasi-proper (resp. pseudo-coherent) and has finite tor-dimension.*
(iii) *f is quasi-proper (resp. pseudo-coherent) and $f^\times$ is bounded.*

Hence, by Kiehl's theorem, *every proper perfect map is quasi-perfect.*

The implication (i) $\Rightarrow$ (iii) is worked out in §4. The proofs in §4 are based on Theorems 4.1 and 4.2, which are of independent interest.

Theorem 4.1 states that for a scheme X, any pseudo-coherent $\mathcal{O}_X$-complex can be "arbitrarily-well approximated," *globally,* by a perfect complex. (*Local* approximability is essentially the definition of pseudo-coherence. The global result was previously known only for divisorial schemes.)

This leads to quasi-proper maps being characterized as those f such that $\mathbf{R}f_*$ takes perfect complexes to pseudo-coherent ones. Since by Prop. 2.1, quasi-perfect maps are those f such that $\mathbf{R}f_*$ takes perfect complexes to perfect ones, it follows at once that quasi-perfect maps are quasi-proper.

Theorem 4.2 refines a theorem of Bondal and van den Bergh [BB, p. 9, Thm. 3.1.1] which states that the triangulated category $\mathbf{D_{qc}}(X)$ is generated by a single perfect complex. With this in hand, one can prove Corollary 4.3.1, which says that for any quasi-perfect *or* perfect f as above, $f^\times$ is bounded.

The implication (iii) $\Rightarrow$ (ii) results from Theorem 3.1, which says, for any $f\colon X \to Y$ as above, *if $f^\times$ is bounded then f has finite tor-dimension.*

Finally, the implication (ii) $\Rightarrow$ (i) holds by definition for the resp. case, and is proved for the other case in §2, Example 2.2(a).

Let us call a map $f\colon X \to Y$ *locally embeddable* if every $y \in Y$ has an open neighborhood V over which the induced map $f^{-1}V \to V$ factors as $f^{-1}V \xrightarrow{i} Z \xrightarrow{p} V$ where i is a closed immersion and p is smooth. (For instance, any quasi-projective f satisfies this condition.) Proposition 2.5 asserts that *any quasi-proper locally embeddable map is pseudo-coherent.* A similar proof shows that any quasi-perfect locally embeddable map is perfect. By 1.2, then, *a locally embeddable map is quasi-perfect iff it is quasi-proper and perfect.*

The equivalence of (i) and (ii) in Theorem 1.2 generalizes [V, p. 396, Cor. 2], in view of the following characterization (mentioned above) of quasi-perfection.

For a map $f\colon X \to Y$, and for any $E \in \mathbf{D}_{\mathsf{qc}}(X)$, $F \in \mathbf{D}_{\mathsf{qc}}(Y)$, with $\underline{\otimes} := \otimes^{\mathbf{L}}$, the derived tensor product, the "projection map"

$$\pi\colon (\mathbf{R}f_*E) \underline{\otimes} F \to \mathbf{R}f_*(E \underline{\otimes} \mathbf{L}f^*F),$$

defined to be adjoint to the natural composite map

$$\mathbf{L}f^*\big((\mathbf{R}f_*E) \underline{\otimes} F\big) \xrightarrow{\;\sim\;} (\mathbf{L}f^*\mathbf{R}f_*E) \underline{\otimes} \mathbf{L}f^*F \to E \underline{\otimes} \mathbf{L}f^*F,$$

is an *isomorphism.* (This is well-known under more restrictive hypotheses; for a proof in the stated generality, see [Lp, Prop. (3.9.4)].) There results a natural map

$$\chi_F\colon f^\times\mathcal{O}_Y \underline{\otimes} \mathbf{L}f^*F \to f^\times F \qquad \big(F \in \mathbf{D}_{\mathsf{qc}}(Y)\big), \tag{1.3}$$

adjoint to the natural composite map

$$\mathbf{R}f_*(f^\times\mathcal{O}_Y \underline{\otimes} \mathbf{L}f^*F) \underset{\pi^{-1}}{\xrightarrow{\;\sim\;}} \mathbf{R}f_*f^\times\mathcal{O}_Y \underline{\otimes} F \to \mathcal{O}_Y \underline{\otimes} F = F.$$

It is clear (since $\underline{\otimes}$ and $\mathbf{L}f^*$ both respect direct sums, see e.g., [Lp, 3.8.2]) that *if χ_F is an isomorphism for all $F \in \mathbf{D}_{\mathsf{qc}}(Y)$ then f is quasi-perfect;* and Proposition 2.1 gives the converse.

2. quasi-perfect maps

For surveying quasi-perfection in more detail, starting with Proposition 2.1, we need some preliminaries.

First, a brief review of the notion of pseudo-coherence of complexes. (Details can be found in the primary source [I, Exposé III], or, perhaps more accessibly, in [TT, pp. 283*ff*, §2]; a summary appears in [Lp, §4.3].) The idea is built up from that of *strictly perfect* $\mathcal{O}_X$-complex, i.e., bounded complex of finite-rank free $\mathcal{O}_X$-modules.

For $n \in \mathbb{Z}$, a map $\xi\colon P \to E$ in $\mathbf{K}(X)$, the homotopy category of $\mathcal{O}_X$-complexes, (resp. in $\mathbf{D}(X)$), is said to be an *n-quasi-isomorphism* (resp. *n-isomorphism*) if the following two equivalent conditions hold:

(1) The homology map $H^j(\xi)\colon H^j(P) \to H^j(E)$ is bijective for all $j > n$ and surjective for $j = n$.

(2) For any $\mathbf{K}(X)$- (resp. $\mathbf{D}(X)$-)triangle

$$P \xrightarrow{\ \xi\ } E \longrightarrow Q \longrightarrow P[1],$$

it holds that $H^j(Q) = 0$ for all $j \geq n$.

Then E is said to be *n-pseudo-coherent* if X has an open covering (U_α) such that for each α there exists a strictly perfect $\mathcal{O}_{U_\alpha}$-complex P_α and an n-quasi-isomorphism (or equivalently, an n-isomorphism) $P_\alpha \to E|_{U_\alpha}$, see [I, p. 98, Définition 2.3]; and E is *pseudo-coherent* if E is n-pseudo-coherent for every n. If $\mathcal{O}_X$ is coherent, this means simply that F has coherent homology sheaves, vanishing in all sufficiently large degrees [*ibid.*, p. 116, top]. When X is noetherian and finite-dimensional, it means that F is globally $\mathbf{D}$-isomorphic to a bounded-above complex of coherent $\mathcal{O}_X$-modules [*ibid.*, p. 168, Cor. 2.2.2.1].

A complex $E \in \mathbf{D}(X)$ (X a scheme) is said to be *perfect* if it is locally $\mathbf{D}$-isomorphic to a strictly perfect $\mathcal{O}_X$-complex. More precisely, E is said to have *perfect amplitude in* $[a, b]$ ($a \leq b \in \mathbb{Z}$) if locally on X, E is $\mathbf{D}$-isomorphic to a bounded complex of finite-rank free $\mathcal{O}_X$-modules vanishing in all degrees $< a$ or $> b$. Thus E is perfect iff it has perfect amplitude in some interval $[a, b]$.

By [I, p. 134, 5.8], E has perfect amplitude in $[a, b]$ iff E is $(a-1)$-pseudo-coherent and has tor-amplitude in $[a, b]$ (i.e., is globally $\mathbf{D}$-isomorphic to a flat complex vanishing in all degrees $< a$ and $> b$). So *E is perfect iff it is pseudo-coherent and has finite tor-dimension* (the latter meaning that it is $\mathbf{D}$-isomorphic to a bounded flat complex).

A map $f\colon X \to Y$ is *pseudo-coherent* if every $x \in X$ has an open neighborhood U such that the restriction $f|_U$ factors as $U \xrightarrow{i} Z \xrightarrow{p} Y$, where i is a closed immersion such that $i_*\mathcal{O}_U$ is pseudo-coherent on Z, and p is smooth [I, p. 228, Défn. 1.2]. Pseudo-coherent maps are finitely presentable. Compositions of pseudo-coherent maps are pseudo-coherent [I, p. 236, Cor. 1.14].

A map is *perfect* if it is pseudo-coherent and has finite tor-dimension [I, p. 250, Défn. 4.1]. Any smooth map is perfect, any regular immersion (= closed immersion corresponding to a quasi-coherent ideal generated locally by a regular sequence) is perfect, and compositions of perfect maps are perfect [I, p. 253, Cor. 4.5.1(a)].

For noetherian Y, any finite-type $f\colon X \to Y$ is pseudo-coherent. Pseudo-coherence (resp. perfection) of maps survives tor-independent base change [I, p. 233, Cor. 1.10; p. 257, Cor. 4.7.2]. Hence, by descent to the noetherian case [EGA, IV, (11.2.7)], *every flat finitely-presentable map is perfect.*

A map $f\colon X \to Y$ is *quasi-proper* if $\mathbf{R}f_*$ takes pseudo-coherent $\mathcal{O}_X$-complexes to pseudo-coherent $\mathcal{O}_Y$-complexes.

Kiehl's Finiteness Theorem [Kl, p. 315, Thm. 2.9′] (first proved by Illusie for projective maps [I, p. 236, Thm. 2.2]) generalizes preservation of coherence by higher direct images under proper maps of noetherian schemes. It states that *every proper pseudo-coherent map is quasi-proper.*

This theorem (or its special case [I, p. 240, Cor. 2.5]), plus [Lp, Ex. (4.3.9)]) implies that *if Y is noetherian then a finite-type separated $f\colon X \to Y$ is quasi-proper iff it is proper.*

For details in the proof of the following Proposition, and for some subsequent considerations, recall that an object C in a triangulated category $\mathcal{T}$ is *compact* if for every small $\mathcal{T}$-family (E_α) the natural map is an isomorphism

$$\bigoplus_\alpha \operatorname{Hom}(C, E_\alpha) \xrightarrow{\sim} \operatorname{Hom}(C, \bigoplus_\alpha E_\alpha).$$

For any scheme X, the compact objects of $\mathbf{D}_{\mathsf{qc}}(X)$ are just the perfect complexes, of which one is a generator [BB, p. 9, Thm. 3.1.1].

PROPOSITION 2.1. *For a map $f\colon X \to Y$, the following are equivalent:*
(i) *f is quasi-perfect (Definition 1.1).*
(ii) *The functor $\mathbf{R}f_*$ takes perfect complexes to perfect complexes.*
(ii)′ *If S is a perfect generator of $\mathbf{D}_{\mathsf{qc}}(X)$ then $\mathbf{R}f_*S$ is perfect.*
(iii) *The twisted inverse image functor $f^\times$ has a right adjoint.*
(iv) *For all $F \in \mathbf{D}_{\mathsf{qc}}(Y)$, the map in (1.3) is an* isomorphism

$$\chi_F\colon f^\times\mathcal{O}_Y \underset{=}{\otimes} \mathbf{L}f^*F \xrightarrow{\sim} f^\times F.$$

Proof. (i) ⇔ (ii) ⇔ (ii)′: [N1, p. 224, Thm. 5.1].
(i) ⇒ (iii): [N1, p. 223, Thm. 4.1].
(iii) ⇒ (i): simple.
(i) ⇒ (iv) ⇒ (i): See [N1, p. 226, Thm. 5.4].
To be precise, the results in [N1] are proved for *separated* schemes; but with the remark preceding Prop. 2.1, one readily verifies that the proofs survive without any separability requirement. □

EXAMPLES 2.2. (a) *Any quasi-proper map f of finite tor-dimension—in particular, by Kiehl's theorem, any proper perfect map—is quasi-perfect.* Indeed, $\mathbf{R}f_*$ preserves pseudo-coherence, and by [I, p. 250, 3.7.2] (a consequence of the projection isomorphism mentioned near the end of the above Introduction), $\mathbf{R}f_*$ preserves finite tor-dimensionality of complexes; so Prop. 2.1(ii) holds.

(b) Let $f\colon X \to Y$ be a map with X *divisorial*—i.e., X has an ample family $(\mathcal{L}_i)_{i\in I}$ of invertible $\mathcal{O}_X$-modules [I, p. 171, Défn. 2.2.5]. Then [N1, p. 212, Example 1.11 and p. 224, Theorem 5.1] show that *f is quasi-perfect ⇔ for each $i \in I$, the $\mathcal{O}_Y$-complex $\mathbf{R}f_*(\mathcal{L}_i^{\otimes -n_i})$ is perfect for all $n_i \gg 0$.*

(c) Let f be quasi-projective and let $\mathcal{L}$ be an f-ample invertible sheaf. Then *f is quasi-perfect $\Leftrightarrow$ the $\mathcal{O}_Y$-complex $\mathbf{R}f_*(\mathcal{L}^{\otimes -n})$ is perfect for all $n \gg 0$.*

Indeed, condition (ii) in Prop. 2.1, together with the compatibility of $\mathbf{R}f_*$ and open base change, implies that quasi-perfection is a property of f which can be checked locally on Y, and the same holds for perfection of $\mathbf{R}f_*(\mathcal{L}^{\otimes -n})$; so we may assume Y affine, and apply (b).

(d) For a *finite* map $f\colon X \to Y$ the following are equivalent:

(i) f is quasi-perfect.

(ii) f is perfect.

(iii) The complex $f_*\mathcal{O}_X \cong \mathbf{R}f_*\mathcal{O}_X$ is perfect.

This follows quickly from (a) and from Proposition 2.1(ii).

A *tor-independent square* is a fiber square of maps

$$\begin{array}{ccc} X' & \xrightarrow{\ v\ } & X \\ {\scriptstyle g}\big\downarrow & & \big\downarrow{\scriptstyle f} \\ Y' & \xrightarrow[\ u\ ]{} & Y \end{array} \tag{2.3}$$

(that is, the natural map is an isomorphism $X' \xrightarrow{\sim} X \times_Y Y'$) such that for all $x \in X$, $y' \in Y'$ and $y \in Y$ with $f(x) = u(y') = y$, and all $i > 0$, $\operatorname{Tor}_i^{\mathcal{O}_{Y,y}}(\mathcal{O}_{X,x}, \mathcal{O}_{Y',y'}) = 0$.

The following stability properties will be useful.

PROPOSITION 2.4. *For any tor-independent square* (2.3),

(i) *If the functor $f^\times$ is bounded then so is $g^\times$.*

(ii) *If f is quasi-perfect then so is g.*

(iii) *If f is quasi-proper then so is g.*

Proof. (i) and (ii) are proved in [Lp, (4.7.3.1)]; and (iii) is treated in Prop. 4.4 below (a slight change in whose proof gives another proof of (ii)). $\square$

Since perfection (resp. pseudo-coherence) is a local property of complexes, and $\mathbf{R}f_*$ is compatible with open base change on Y, we deduce:

COROLLARY 2.4.1. *Let $f\colon X \to Y$ be a map, and let $(Y_i)_{i\in I}$ be an open cover of Y. Then f is quasi-perfect (resp. quasi-proper) $\Leftrightarrow$ for all i, the same is true of the induced map $f^{-1}Y_i \to Y_i$.*

PROPOSITION 2.5. *Let $f\colon X \to Y$ be a* locally embeddable *map, i.e., every $y \in Y$ has an open neighborhood V over which the induced map $f^{-1}V \to V$ factors as $f^{-1}V \xrightarrow{i} Z \xrightarrow{p} V$ where i is a closed immersion and p is smooth. (For instance, any quasi-projective f satisfies this condition* [EGA, II, (5.3.3)].)

(i) *If f is quasi-proper then f is pseudo-coherent.*

(ii) *If f is quasi-perfect then f is perfect.*

Proof. By Corollary 2.4.1, quasi-properness (resp. quasi-perfection) of f is a property local over Y; and since they are compatible with tor-independent base change, the same is true of pseudo-coherence and perfection. So we may as well assume that $X = f^{-1}V$. Then it suffices to show that the complex $i_*\mathcal{O}_X$ is pseudo-coherent when f is quasi-proper, (resp., by [I, p. 252, Prop. 4.4], that $i_*\mathcal{O}_X$ is perfect when f is quasi-perfect).

But i factors as $X \xrightarrow{\gamma} X \times_Y Z \xrightarrow{g} Z$ with γ the graph of i and g the projection. The map γ is a local complete intersection [EGA, IV, (17.12.3)], so the complex $\gamma_*\mathcal{O}_X$ is perfect. Also, g arises from f by flat base change, so by Proposition 2.4, g is quasi-proper (resp. quasi-perfect). Hence $i_*\mathcal{O}_X = \mathbf{R}i_*\mathcal{O}_X = \mathbf{R}g_*\gamma_*\mathcal{O}_X$ is indeed pseudo-coherent (resp. perfect). □

(2.6). For any tor-independent square (2.3), the map

$$\theta(E)\colon \mathbf{L}u^*\mathbf{R}f_*E \to \mathbf{R}g_*\mathbf{L}v^*E \qquad \big(E \in \mathbf{D}_{\mathsf{qc}}(X)\big) \tag{2.6.1}$$

adjoint to the natural composition

$$\mathbf{R}f_*E \to \mathbf{R}f_*\mathbf{R}v_*\mathbf{L}v^*E \cong \mathbf{R}u_*\mathbf{R}g_*\mathbf{L}v^*E$$

(equivalently, to $\mathbf{L}g^*\mathbf{L}u^*\mathbf{R}f_*E \cong \mathbf{L}v^*\mathbf{L}f^*\mathbf{R}f_*E \to \mathbf{L}v^*E$) is an isomorphism, so that one has a *base-change map*

$$\beta(F)\colon \mathbf{L}v^*f^\times F \to g^\times\mathbf{L}u^*F \qquad \big(F \in \mathbf{D}_{\mathsf{qc}}(Y)\big) \tag{2.6.2}$$

adjoint to the natural composition

$$\mathbf{R}g_*\mathbf{L}v^*f^\times F \underset{\theta^{-1}}{\xrightarrow{\ \sim\ }} \mathbf{L}u^*\mathbf{R}f_*f^\times F \to \mathbf{L}u^*F.$$

The fundamental *independent base-change* theorem states that:

Let there be given a tor-independent square (2.3) *and an* $F \in \mathbf{D}_{\mathsf{qc}}(Y)$. *If* f *is quasi-proper,* u *has finite tor-dimension, and* $H^nF = 0$ *for all* $n \ll 0$, *then* $\beta(F)$ *is an isomorphism.*

This theorem is well-known, at least under more restrictive hypotheses. For a treatment in full generality, see [Lp, §§4.4–4.6].

One consequence, in view of Proposition 2.4(i), is:

COROLLARY 2.6.3. *Let* $f\colon X \to Y$ *be a quasi-proper map and let* $(Y_i)_{i\in I}$ *be an open cover of* Y. *Then* $f^\times$ *is bounded* $\Leftrightarrow$ *for all* i, *the same is true of the induced map* $f^{-1}Y_i \to Y_i$.

For quasi-perfect f, a stronger base-change theorem holds—which, together with boundedness of $f^\times$ (Corollary 4.3.1), characterizes quasi-perfection:

THEOREM 2.7 ([Lp, Thm. 4.7.4]). *Let*

$$\begin{array}{ccc} X' & \xrightarrow{\;v\;} & X \\ {\scriptstyle g}\big\downarrow & & \big\downarrow{\scriptstyle f} \\ Y' & \xrightarrow[\;u\;]{} & Y \end{array}$$

be a tor-independent square, with f quasi-perfect. Then for all $F \in \mathbf{D}_{\mathsf{qc}}(Y)$ the base-change map of (2.6.2) *is an* isomorphism

$$\beta(F)\colon v^*f^\times F \xrightarrow{\;\sim\;} g^\times u^*F.$$

The same holds, with no assumption on f, whenever u is finite and perfect.

Conversely, the following conditions on a map $f\colon X \to Y$ are equivalent; and if Y is separated and $f^\times$ bounded above, they imply that f is quasi-perfect:

(i) *For any flat affine universally bicontinuous map $u\colon Y' \to Y$,*[3] *the base-change map associated to the (tor-independent) square*

$$\begin{array}{ccc} Y' \times_Y X = X' & \xrightarrow{\;v\;} & X \\ {\scriptstyle g}\big\downarrow & & \big\downarrow{\scriptstyle f} \\ Y' & \xrightarrow[\;u\;]{} & Y \end{array}$$

is an isomorphism

$$\beta(\mathcal{O}_Y)\colon v^*f^\times \mathcal{O}_Y \xrightarrow{\;\sim\;} g^\times u^*\mathcal{O}_Y.$$

(ii) *The map in* (1.3) *is an isomorphism*

$$\chi_F\colon f^\times\mathcal{O}_Y \underset{=}{\otimes} \mathbf{L}f^*F \xrightarrow{\;\sim\;} f^\times F$$

whenever F is a flat quasi-coherent $\mathcal{O}_Y$-module.

Keeping in mind Corollary 4.3.1 below (f quasi-perfect $\Rightarrow$ $f^\times$ bounded), we can deduce:

COROLLARY 2.7.1. *When Y is separated, a map $f\colon X \to Y$ is quasi-perfect iff $f^\times$ is bounded and the following two conditions hold:*

(i) *If $u\colon Y' \to Y$ is an open immersion, and if $v\colon Y' \times_Y X \to X$ and $g\colon Y' \times_Y X \to Y$ are the projection maps, then the base-change map is an isomorphism*

$$\beta(\mathcal{O}_Y)\colon v^*f^\times \mathcal{O}_Y \xrightarrow{\;\sim\;} g^\times u^*\mathcal{O}_Y.$$

Equivalently (see [Lp, §4.6, subsection V]), *for all $E \in \mathbf{D}_{\mathsf{qc}}(X)$ the natural composite map*

$$\mathbf{R}f_*\mathbf{R}\mathcal{H}om^\bullet_X(E, f^\times\mathcal{O}_Y) \to \mathbf{R}\mathcal{H}om^\bullet_Y(\mathbf{R}f_*E, \mathbf{R}f_*f^\times\mathcal{O}_Y) \to \mathbf{R}\mathcal{H}om^\bullet_Y(\mathbf{R}f_*E, \mathcal{O}_Y)$$

is an isomorphism.

---

[3] *Universally bicontinuous* means that for any $Y'' \to Y$ the resulting projection map $Y' \times_Y Y'' \to Y''$ is a homeomorphism onto its image [GD, p. 249, Défn. (3.8.1)].

(ii) *If* (F_α) *is a filtered direct system of flat quasi-coherent* $\mathcal{O}_Y$*-modules, then for all* $n \in \mathbb{Z}$ *the natural map is an isomorphism*

$$\varinjlim_\alpha H^n(f^\times F_\alpha) \xrightarrow{\sim} H^n(f^\times \varinjlim_\alpha F_\alpha).$$

Remarks. 1. Conditions (i) and (ii) in Theorem 2.7 are connected via the flat, affine, and universally bicontinuous natural map $\mathrm{Spec}(S_{\le 1}(F)) \to Y$, where $S_{\le 1}(F)$ is the $\mathcal{O}_Y$-algebra $\mathcal{O}_Y \oplus F$ with $F^2 = 0$.

2. The idea behind the proof of Corollary 2.7.1 is to use Lazard's theorem that over a commutative ring A any flat module is a $\varinjlim$ of finite-rank free A-modules [GD, p. 163, (6.6.24)], to show that (i) and (ii) imply condition (ii) in Theorem 2.7.

3. Boundedness of $f^\times$ implies finite tor-dimension

THEOREM 3.1. *Let* $f\colon X \to Y$ *be a map. If* $f^\times$ *is bounded then* f *has finite tor-dimension.*

The proof uses the following two Lemmas.

An $\mathcal{O}_X$-complex E is *a_locally projective* ($a \in \mathbb{Z}$) if there is a $b \ge a$ and an affine open covering $\big(U_i := \mathrm{Spec}(A_i)\big)_{i\in I}$ of X such that for each $i \in I$, the restriction $E|_{U_i}$ is $\mathbf{D}$-isomorphic to a quasi-coherent direct summand of a complex F of free $\mathcal{O}_{U_i}$-modules, with F vanishing in all degrees outside $[a, b]$.

Every complex with perfect amplitude in $[a, b]$ (§2) is a_locally projective.

LEMMA 3.2. *For any scheme* X, *there is an integer* $s > 0$ *such that for all* $a \in \mathbb{Z}$ *and* a*_locally projective* $E \in \mathbf{D}(X)$, *if* $G \in \mathbf{D}_{\mathsf{qc}}(X)$ *and* $H^jG = 0$ *for all* $j > a - s$ *then* $\mathrm{Hom}_{\mathbf{D}(X)}(E, G) = 0$.

LEMMA 3.3. *Let* $f\colon X \to Y$ *be a perfect map, of tor-dim* $d < \infty$. *Then there exists an integer* $t > 0$ *such that for any* a*_locally projective* $E \in \mathbf{D}_{\mathsf{qc}}(X)$, $\mathbf{R}f_*E \in \mathbf{D}_{\mathsf{qc}}(Y)$ *is* $(a - d - t)$*_locally projective.*

These Lemmas are proved below.

Proof of Theorem 3.1.

Part (i) of Proposition 2.4 gives an immediate reduction to the case where Y is affine, say $Y = \mathrm{Spec}(A)$. We need to show in this case that for any open immersion $\iota\colon U \hookrightarrow X$ with U affine the $\mathcal{O}_Y$-module $f_*\iota_*\mathcal{O}_U$ has finite tor-dimension.

Since U is affine, there are natural isomorphisms

$$f_*\iota_*\mathcal{O}_U = (f\iota)_*\mathcal{O}_U \xrightarrow{\sim} \mathbf{R}(f\iota)_*\mathcal{O}_U \xrightarrow{\sim} \mathbf{R}f_*\mathbf{R}\iota_*\mathcal{O}_U.$$

So for any $G \in \mathbf{D_{qc}}(Y)$ there are natural isomorphisms

$$\mathrm{Hom}_{\mathbf{D}(Y)}(f_*\iota_*\mathcal{O}_U, G) \cong \mathrm{Hom}_{\mathbf{D}(Y)}(\mathbf{R}f_*\mathbf{R}\iota_*\mathcal{O}_U, G) \cong \mathrm{Hom}_{\mathbf{D}(X)}(\mathbf{R}\iota_*\mathcal{O}_U, f^\times G).$$

Lemma 3.3 provides an integer t such that if U is *any* quasi-compact open subscheme of X, with inclusion $\iota\colon U \subset X$, then $\mathbf{R}\iota_*\mathcal{O}_U$ is $(-t)$-locally projective. By Lemma 3.2 and the boundedness of $f^\times$, it follows, for U affine, G a quasi-coherent $\mathcal{O}_Y$-module, and some $j \gg 0$ not depending on G, that

$$\begin{aligned}\mathrm{Ext}^j(f_*\iota_*\mathcal{O}_U, G) &= \mathrm{Hom}_{\mathbf{D}(Y)}(f_*\iota_*\mathcal{O}_U, G[j])\\ &\cong \mathrm{Hom}_{\mathbf{D}(X)}(\mathbf{R}\iota_*\mathcal{O}_U, f^\times G[j]) = 0.\end{aligned}$$

The natural equivalences $\mathbf{D}(A) \xrightarrow{\approx} \mathbf{D}(Y_{\mathsf{qc}}) \xrightarrow{\approx} \mathbf{D_{qc}}(Y)$ (where Y_{qc} is the category of quasi-coherent $\mathcal{O}_Y$-modules—see [BN, p. 30, Cor. 5.5]) show then that $f_*\iota_*\mathcal{O}_U$ has a resolution by the sheafification of a bounded projective A-complex, and thus has finite tor-dimension, as desired. □

Proof of Lemma 3.2.

Let us call an open $U \subset X$ *E-good* if U is affine, say $U = \mathrm{Spec}(A)$, and if there is a $b \ge a$ such that the restriction $E|_U$ is $\mathbf{D}$-isomorphic to the sheafification of a projective A-complex E vanishing in all degrees outside $[a, b]$.

Clearly, every quasi-compact open subset of X is a finite union of E-good open subsets. Hence, as in the proof of [BB, p. 13, Prop. 3.3.1], it will suffice to show that Lemma 3.2 holds for X if X itself is E-good, or if $X = X_1 \cup X_2$ with X_1 and X_2 quasi-compact open subsets such that Lemma 3.2 holds for X_1, X_2 and $X_1 \cap X_2$ (which is also quasi-compact, since X is quasi-separated).

Suppose first that $X = \mathrm{Spec}(A)$ is E-good. Let E be as in the definition of E-good, and let $G \in \mathbf{D_{qc}}(X)$ be such that $H^jG = 0$ for all $j > a - 1$. The natural equivalence of categories $\mathbf{D}(X_{\mathsf{qc}}) \xrightarrow{\approx} \mathbf{D_{qc}}(X)$ (where X_{qc} is the category of quasi-coherent $\mathcal{O}_X$-modules) allows us to assume G quasi-coherent, so that G is the sheafification of an A-complex G; and further, after applying the well-known truncation functor (see e.g., [Lp, §1.10]) we can assume that G vanishes in all degrees $> a - 1$.

The dual versions of [Lp, (2.3.4) and (2.3.8)(v)], and the equivalences $\mathbf{D}(A) \xrightarrow{\approx} \mathbf{D}(X_{\mathsf{qc}})$, $\mathbf{D}(X_{\mathsf{qc}}) \xrightarrow{\approx} \mathbf{D_{qc}}(X)$, yield natural isomorphisms, with $\mathbf{K}(A)$ the homotopy category of A-complexes:

$$\mathrm{Hom}_{\mathbf{K}(A)}(\mathrm{E}, \mathrm{G}) \cong \mathrm{Hom}_{\mathbf{D}(A)}(\mathrm{E}, \mathrm{G}) \cong \mathrm{Hom}_{\mathbf{D}(X_{\mathsf{qc}})}(E, G) \cong \mathrm{Hom}_{\mathbf{D}(X)}(E, G).$$

So since E vanishes in all degrees $< a$ and G vanishes in all degrees $> a - 1$, therefore $\mathrm{Hom}_{\mathbf{D}(X)}(E, G) = 0$, proving Lemma 3.2 in this case.

Suppose next that $X = X_1 \cup X_2$ as above. Let $s > 0$ be such that Lemma 3.2 holds with this s for all three of X_1, X_2, and $X_1 \cap X_2$. Let $G \in \mathbf{D_{qc}}(X)$ satisfy $H^jG = 0$ for all $j > a - (s + 1)$. Let $i\colon X_1 \hookrightarrow X$, $j\colon X_2 \hookrightarrow X$, and $k\colon X_1 \cap X_2 \hookrightarrow X$ be the inclusion maps. One gets the natural triangle

$$G \longrightarrow \mathbf{R}i_*i^*G \oplus \mathbf{R}j_*j^*G \longrightarrow \mathbf{R}k_*k^*G \longrightarrow G[1],$$

by applying the usual exact sequence, holding for any flasque $\mathcal{O}_X$-module F,

$$0 \to F \to i_*i^*F \oplus j_*j^*F \to k_*k^*F \to 0$$

to an injective q-injective resolution[4] of E^+. There results an exact sequence [H, p, 21, 1.1(b)], with $\mathrm{Hom} := \mathrm{Hom}_{\mathbf{D}(X)}$,

$$\mathrm{Hom}(E, \mathbf{R}k_*k^*G[-1]) \to \mathrm{Hom}(E, G) \to \mathrm{Hom}(E, \mathbf{R}i_*i^*G) \oplus \mathrm{Hom}(E, \mathbf{R}j_*j^*G).$$

Adjointness of $\mathbf{R}k_*$ and $\mathbf{L}k^* = k^*$ gives that

$$\mathrm{Hom}_{\mathbf{D}(X)}(E, \mathbf{R}k_*k^*G[-1]) \cong \mathrm{Hom}_{\mathbf{D}(X_1 \cap X_2)}(k^*E, k^*G[-1]);$$

and Lemma 3.2 makes these groups vanish. Similarly, $\mathrm{Hom}(E, \mathbf{R}j_*j^*G) = 0$ and $\mathrm{Hom}(E, \mathbf{R}i_*i^*G) = 0$. Hence $\mathrm{Hom}(E, G) = 0$. □

Proof of Lemma 3.3.

The question is local on Y, so we may assume Y affine, say $Y = \mathrm{Spec}(B)$.

Arguing as in the preceding proof, suppose first that X is E-good. We begin with the case $E = \mathcal{O}_X$. Then for some $t > 0$, f factors as

$$X \xrightarrow{\iota} Y_t := \mathrm{Spec}(B[T_1, T_2, \dots, T_t]) \xrightarrow{\pi} \mathrm{Spec}(B),$$

where $T_1, \dots, T_t$ are independent indeterminates, ι is a closed immersion, and π is the natural map. By [I, p. 252, Prop. 4.4(ii) and p. 174, Prop. 2.2.9(b)], the sheaf $\iota_*\mathcal{O}_X$ is $\mathbf{D}(Y_n)$-isomorphic to a bounded quasi-coherent complex G of direct summands of finite-rank free $\mathcal{O}_{Y_n}$-modules, vanishing in all degrees $< -d-t$. Hence $\mathbf{R}f_*\mathcal{O}_X \cong \pi_*\iota_*\mathcal{O}_X \cong \pi_*G$ is $(-d-t)$_locally projective.

Since $\mathbf{R}f_*$ commutes with direct sums in $\mathbf{D}_{\mathsf{qc}}$ (because $\mathbf{R}f_*$ has a right adjoint, or more directly, by [Lp, 3.9.3.3]), it follows that for any free $\mathcal{O}_X$-module E, $\mathbf{R}f_*E$ is $(-d-t)$_locally projective. Finally, to show that for any a_locally projective E, $\mathbf{R}f_*E$ is $(a-d-t)$_locally projective, one reduces easily to where E is a bounded free complex, and then argues by induction on the number of degrees in which E is nonvanishing, using the following observation:

(∗): *In a* $\mathbf{D}(X)$*-triangle* $N[-1] \xrightarrow{\delta} L \to M \xrightarrow{\rho} N$, *if* N *is* a_*locally projective then* M *is* a_*locally projective iff so is* L.

(To see this, one may suppose that X is affine, say $X = \mathrm{Spec}(A)$. If N and L are a_locally projective, then one may assume they are sheafifications of bounded projective A-complexes vanishing in all degrees $< a$, so that by the dual versions of [Lp, (2.3.4) and (2.3.8)(v)], δ comes from a $\mathbf{K}(X)$-morphism $\delta_0 \colon N[-1] \to L$; and M is isomorphic to the mapping cone of δ_0, an a_projective complex. Similarly, if N and M are sheafifications of bounded projective A-complexes vanishing in all degrees $< a$, then ρ comes from a $\mathbf{K}(X)$-morphism $\rho_0 \colon M \to N$, and since $L[1]$ is isomorphic to the mapping cone of ρ_0, therefore L is a_locally projective.)

[4] Another term for "q-injective" is "K-injective"—see [Lp, (2.3.2.3), (2.3.5)].

Suppose next that $X = X_1 \cup X_2$ with X_1 and X_2 quasi-compact open subsets for which there exists a $t > 0$ such that Lemma 3.3 holds with this t for all three of X_1, X_2, and $X_1 \cap X_2$. As in the proof of Lemma 3.2, there is a $\mathbf{D}(Y)$-triangle

$$\mathbf{R}f_*E \longrightarrow \mathbf{R}(fi)_*(E|_{X_1}) \oplus \mathbf{R}(fj)_*(E|_{X_2}) \longrightarrow \mathbf{R}(fk)_*(E|_{X_1 \cap X_2}) \longrightarrow \mathbf{R}f_*E[1]$$

in which the two vertices other than $\mathbf{R}f_*E$ are $(a-d-t)_-$projective, whence, by $(*)$, so is $\mathbf{R}f_*E$.

As before, this completes the proof of Lemma 3.3, and so of Theorem 3.1.

4. Approximation by perfect complexes

Terminology remains as in §2.

The main results in this section are the following two theorems.

THEOREM 4.1. *For any scheme X, there exists a positive integer $B = B(X)$ such that for any $E \in \mathbf{D}_{\mathsf{qc}}(X)$ and integer m, if E is $(m-B)$-pseudo-coherent then there exists in $\mathbf{D}_{\mathsf{qc}}(X)$ an m-isomorphism $P \to E$ with P perfect.*

THEOREM 4.2. *Let X be a scheme. Then $\mathbf{D}_{\mathsf{qc}}(X)$ has a perfect generator, i.e., there is a perfect $\mathcal{O}_X$-complex S such that for each $E \neq 0$ in $\mathbf{D}_{\mathsf{qc}}(X)$ there is an $n \in \mathbb{Z}$ and a nonzero $\mathbf{D}_{\mathsf{qc}}(X)$-morphism $S[n] \to E$.*

Moreover, for each such S there is an integer $A = A(S)$ such that for all $E \in \mathbf{D}_{\mathsf{qc}}(X)$ and $j \in \mathbb{Z}$ with $H^j(E) \neq 0$,

$$\operatorname{Hom}\big(S[n], E\big) \neq 0 \quad \text{for some } n \leq A - j.$$

Theorem 4.1 may be compared to [I, p. 173, 2.2.8(b)]). The first statement in Theorem 4.2 comes from [BB, p. 9, Thm. 3.1.1].

Proofs are given in section 5 below.

COROLLARY 4.3.1. *If a map f is either perfect or quasi-perfect, then the functor $f^\times$ is bounded.*

Proof. As mentioned in the Introduction, $f^\times$ commutes with translation of complexes, and $f^\times$ is bounded below. So to show that $f^\times$ is bounded, it is enough to find a j_0 such that for every $m \in \mathbb{Z}$ and $F \in \mathbf{D}_{\mathsf{qc}}(Y)$ with $H^i(F) = 0$ for all $i > m$, it holds that $H^j f^\times F = 0$ for all $j \geq m + j_0$.

Suppose $H^j(f^\times F) \neq 0$. With S and A as in Theorem 4.2, there exists $k \leq A$ and a nonzero $\mathbf{D}(X)$-morphism $S \to f^\times F[j-k]$, the latter corresponding under adjunction to a *nonzero* morphism $\lambda\colon \mathbf{R}f_*S \to F[j-k]$.

For some a, $\mathbf{R}f_*S$ is a_-locally projective—when f is perfect, that results from Lemma 3.3, and when f is quasi-perfect, it's because $\mathbf{R}f_*S$ is perfect. It follows from Lemma 3.2 that there is an integer $s = s(Y)$ such that λ cannot exist if $j \geq m + A - a + s$. With $j_0 := A - a + s$, we must have then that $H^j(f^\times F) = 0$ for all $j \geq m + j_0$; and so $f^\times$ is indeed bounded. □

COROLLARY 4.3.2. *For a map $f\colon X \to Y$, the following are equivalent.*
(i) *f is quasi-proper.*
(ii) *For any perfect $\mathcal{O}_X$-complex P, $\mathbf{R}f_*P$ is pseudo-coherent.*
(iii) *If S is as in Theorem 4.2, then $\mathbf{R}f_*S$ is pseudo-coherent.*

Proof. (i) ⇒ (ii) ⇒ (iii). The first implication is clear (since perfect complexes are pseudo-coherent); and the second is trivial.

(iii) ⇒ (ii). Let R be the smallest triangulated subcategory of $\mathbf{D}_{\mathsf{qc}}(X)$ containing S, and let $\widehat{R}$ be the full subcategory of $\mathbf{D}_{\mathsf{qc}}(X)$ whose objects are all the direct summands of objects in R. The subcategory $\widehat{R} \subset \mathbf{D}_{\mathsf{qc}}(X)$ is triangulated, and closed under formation of direct summands [N2, p. 99, 2.1.39].

The full subcategory R^c of R whose objects are the compact ones in R is triangulated, whence every object in R—and in $\widehat{R}$—is compact. Consequently, [N1, p. 222, Lemma 3.2] shows that the smallest full subcategory of $\mathbf{D}_{\mathsf{qc}}(X)$ which contains $\widehat{R}$ and is closed with respect to coproducts is $\mathbf{D}_{\mathsf{qc}}(X)$ itself. Hence, by [N1, p. 214, Thm. 2.1.3], every perfect complex lies in $\widehat{R}$. (Alternatively, see [N2, p. 285, Prop. 8.4.1 and p. 140, Lemma 4.4.5].)

Since the pseudo-coherent complexes in $\mathbf{D}_{\mathsf{qc}}(Y)$ are the objects of a triangulated subcategory closed under formation of direct summands [I, p. 99, b) and p. 105, 2.12], therefore the complexes $Q \in \mathbf{D}_{\mathsf{qc}}(X)$ such that $\mathbf{R}f_*Q$ is pseudo-coherent are the objects of a triangulated subcategory closed under formation of direct summands. Thus if S is such a Q then every complex in $\widehat{R}$—and so every perfect complex—is such a Q.

(ii) ⇒ (i). Let E be a pseudo-coherent $\mathcal{O}_X$-complex, let $m \in \mathbb{Z}$, and let

$$P \xrightarrow{\ \alpha\ } E \longrightarrow Q \longrightarrow P[1]$$

be a triangle with α an m-isomorphism as in Theorem 4.1. Thus $H^k(Q) = 0$ for all $k \geq m$. As $\mathbf{R}f_*$ is bounded above [Lp, (3.9.2)], there is an integer t depending only on f such that $H^k(\mathbf{R}f_*Q) = 0$ for all $k \geq m + t$, that is, $\mathbf{R}f_*\alpha$ is an $(m+t)$-quasi-isomorphism. So if $\mathbf{R}f_*P$ is pseudo-coherent then $\mathbf{R}f_*E$ is $(m+t)$-pseudo-coherent; and since m is arbitrary, therefore $\mathbf{R}f_*E$ is pseudo-coherent. □

From 4.3.2(ii) we get:

COROLLARY 4.3.3. *Every quasi-perfect map is quasi-proper.*

Next, we deduce "stability" of quasi-properness.

PROPOSITION 4.4. *Let*

$$\begin{array}{ccc} X' & \xrightarrow{\ v\ } & X \\ {\scriptstyle g}\big\downarrow & & \big\downarrow{\scriptstyle f} \\ Y' & \xrightarrow[\ u\ ]{} & Y \end{array}$$

be a tor-independent square. If f is quasi-proper then so is g.

Proof. Since pseudo-coherence is a local property, it suffices to prove the Proposition when Y' is affine and $u(Y')$ is contained in an affine subset of Y. So we can assume that $u = u'u''$ where u' is an open immersion and u'' is an affine map. It follows that it suffices to prove the Proposition (a) when u—hence v—is an open immersion and (b) when u—hence v—is an affine map (see [GD, p. 358, (9.1.16)(iii), (9.1.17)]).

In either of these two cases, it holds that

(∗) *if S is as in Theorem 4.2 then $\mathbf{L}v^*S$ is a generator of $\mathbf{D}_{\mathsf{qc}}(X')$.*

Indeed, in case v is an open immersion and $0 \neq E \in \mathbf{D}_{\mathsf{qc}}(X')$ then $0 \neq \mathbf{R}v_*E \in \mathbf{D}_{\mathsf{qc}}(X)$ (since $E \cong v^*\mathbf{R}v_*E$); and the same holds in case v is affine, by [Lp, (3.10.2.2]. So in either case, for some n,

$$0 \neq \mathrm{Hom}_{\mathbf{D}_{\mathsf{qc}}(X)}(S[n], \mathbf{R}v_*E) \cong \mathrm{Hom}_{\mathbf{D}_{\mathsf{qc}}(X')}(\mathbf{L}v^*S[n], E),$$

proving (∗).

It is easy to see that the complex $\mathbf{L}v^*S$ is perfect. So by Corollary 4.3.2, to prove the Proposition for u as in (∗) it suffices to show that $\mathbf{R}g_*\mathbf{L}v^*S$ is pseudo-coherent. But by [Lp, (3.10.3)], $\mathbf{R}g_*\mathbf{L}v^*S \cong \mathbf{L}u^*\mathbf{R}f_*S$; and since $\mathbf{R}f_*S$ is pseudo-coherent, therefore, by [I, p. 111, 2.16.1], so is $\mathbf{L}u^*\mathbf{R}f_*S$.

5. Proofs of Theorems 4.1 and 4.2

Heavy use will be made of the following technical notion.

DEFINITION 5.1. Let $\mathcal{T}$ be a triangulated category, and let $\mathcal{S} \subset \mathcal{T}$ be a class of objects. Let $m \leq n$ be integers. The full subcategory $\mathcal{S}[m,n] \subset \mathcal{T}$ is the smallest among (= intersection of) all full subcategories $\mathscr{S} \subset \mathcal{T}$ such that:

(i) 0 is contained in $\mathscr{S}$.
(ii) If $E \in \mathcal{S}$, then $E[-\ell] \in \mathscr{S}$ for all integers ℓ in the interval $[m,n]$.
(iii) For any $\mathcal{T}$-triangle

$$E \longrightarrow F \longrightarrow G \longrightarrow E[1],$$

if E and G are in $\mathscr{S}$ then so is F.

REMARK 5.2. One checks that $\mathcal{S}[m,n] = \big(\bigcup_{\ell=m}^{n} \mathcal{S}[-\ell]\big)[0,0]$.

REMARK 5.3. Defn. 5.1 expands to allow $m = -\infty$ or $n = \infty$. For example, $\mathcal{S}[m,\infty) := \bigcup_{n=m}^{\infty} \mathcal{S}[m,n]$. Furthermore, $\mathcal{S}(\infty,\infty) := \bigcup_{m\le n} \mathcal{S}[m,n]$, being closed under translation (see 5.4(i)), is the smallest triangulated subcategory of $\mathcal{T}$ containing $\mathcal{S}$ [N2, p. 60, Defn. 1.5.1].

REMARK 5.4. The following are easy observations.

(i) *If $E \in \mathcal{S}[m,n]$ and $j \in \mathbb{Z}$ then $E[-j] \in \mathcal{S}[m+j, n+j]$.*
Indeed, (i), (ii) and (iii) in 5.1 hold for the full subcategory $\mathcal{S} \subset \mathcal{T}$ whose objects are those $E \in \mathcal{S}[m,n]$ such that $E[-j] \in \mathcal{S}[m+j, n+j]$. One deduces that, with $\mathcal{S}[m,n]_\text{o}$ the class of objects in $\mathcal{S}[m,n]$,

$$\big(\mathcal{S}[m,n]_\text{o}\big)[m',n'] = \mathcal{S}[m+m', n+n'].$$

(ii) *If every object of $\mathcal{S}$ is compact, then so is every object of $\mathcal{S}[m,n]$.*
Indeed, (i), (ii) and (iii) in 5.1 hold for the full subcategory $\mathcal{S} \subset \mathcal{T}$ whose objects are those $E \in \mathcal{S}[m,n]$ which are compact.

(iii) *Let $\mathcal{A}$ be an abelian category, and $H : \mathcal{T} \longrightarrow \mathcal{A}$ a cohomological functor, see* [N2, p. 32, 1.1.9]. *If for every object $F \in \mathcal{S}$ we have $H(F[-i]) = 0$ for all i in some interval $[a,b]$, then for all $E \in \mathcal{S}[m,n]$,*

$$H(E[-j]) = 0 \text{ for all } j \in [a-m, b-n]. \tag{5.4.1}$$

Indeed, (i), (ii) and (iii) in 5.1 hold for the full subcategory $\mathcal{S} \subset \mathcal{T}$ whose objects are those $E \in \mathcal{S}[m,n]$ which satisfy (5.4.1).

(iv) *Let $\phi\colon \mathcal{T} \to \mathcal{T}'$ be a triangle-preserving additive functor* [Lp, §1.5]. *Then*

$$\phi\big(\mathcal{S}[m,n]\big) \subset \{\phi\mathcal{S}\}[m,n].$$

Indeed, (i), (ii) and (iii) in 5.1 hold for the full subcategory $\mathcal{S} \subset \mathcal{T}$ whose objects are those $E \in \mathcal{S}[m,n]$ such that $\phi E \in \{\phi\mathcal{S}\}[m,n]$.

(v) *Let $A \xrightarrow{\alpha} B \xrightarrow{\beta} C$ be morphisms inside $\mathcal{T}$-triangles*

$$E \longrightarrow A \xrightarrow{\ \alpha\ } B \longrightarrow E[1],$$

$$F \longrightarrow A \xrightarrow{\ \beta\alpha\ } C \longrightarrow F[1],$$

$$G \longrightarrow B \xrightarrow{\ \beta\ } C \longrightarrow G[1].$$

If E and G are in $\mathcal{S}[m,n]$ then so is F.

Indeed, the octahedral axiom [N2, p. 60, 1.4.7] produces a triangle

$$E \longrightarrow F \longrightarrow G \longrightarrow E[1].$$

EXAMPLE 5.5. Remark 5.4(iii) will be used thus. Let G be an object of $\mathcal{T}$, and H the cohomological functor $H(-) := \operatorname{Hom}(-,G)$, see [N2, p. 33, 1.1.11]. Then for $a = m$ and $b = n$ the assertion becomes:

If for every object $F \in \mathcal{S}$ we have $\operatorname{Hom}(F[-i],G) = 0$ *for all $i \in [m,n]$, then* $\operatorname{Hom}(E,G) = 0$ *for all $E \in \mathcal{S}[m,n]$.*

A key role in the proofs will be played by *Koszul complexes.*

EXAMPLE 5.6. Let R be a commutative ring, $(f_1, f_2, \dots, f_r)$ a sequence in R, and $(n_1, n_2, \dots, n_r)$ a sequence of positive integers. The associated Koszul complex (see, e.g., [EGA, III, (1.1.1)]) is

$$K_\bullet(f_1^{n_1}, \dots, f_r^{n_r}) := \otimes_{i=1}^r K_\bullet(f_i^{n_i}),$$

where $K_\bullet(f_i^{n_i})$ is $R \xrightarrow{f_i^n} R$ in degrees -1 and 0, and (0) elsewhere.

For $r = 0$, set $K_\bullet(\phi) := R$. For all $r \ge 0$, $K_\bullet(f_1^{n_1}, \dots, f_r^{n_r})$ is a complex with perfect amplitude in $[-r, 0]$, and homology killed by each $f_i^{n_i}$.

For any complex E, and $f \in R$, $K_\bullet(f) \otimes E$ is the mapping cone of the endomorphism "multiplication by f" of E. Thus for $1 \le i < r$, $K_\bullet(f_i^{n_i}, \dots, f_r^{n_r})$ is the mapping cone of the endomorphism "multiplication by $f_i^{n_i}$" of the complex $K_\bullet(f_{i+1}^{n_{i+1}}, f_{i+2}^{n_{t+2}}, \dots, f_r^{n_r})$. It follows that

$$K_\bullet(f_1^{n_1}, f_2^{n_2}, \dots, f_r^{n_r}) \in \{K_\bullet(f_1, f_2, \dots, f_r)\}\,[0, 0].$$

This is shown by a straightforward induction, based on application of 5.4(v) to the following three natural triangles (where $\widehat{\ }$ signifies "omit,"):

$$K_\bullet(f_1^{n_1}, .., f_i^{n_i}, f_{i+1}, .., f_r) \longrightarrow$$
$$K_\bullet(f_1^{n_1}, .., \widehat{f_i^{n_i}}, f_{i+1}, .., f_r)[1] \xrightarrow{f_i^{n_i}} K_\bullet(f_1^{n_1}, .., \widehat{f_i^{n_i}}, f_{i+1}, .., f_r)[1] \xrightarrow{+}$$

$$K_\bullet(f_1^{n_1}, .., f_i^{n_i+1}, f_{i+1}, .., f_r) \longrightarrow$$
$$K_\bullet(f_1^{n_1}, .., \widehat{f_i^{n_i+1}}, f_{i+1}, .., f_r)[1] \xrightarrow{f_i^{n_i+1}} K_\bullet(f_1^{n_1}, .., \widehat{f_i^{n_i+1}}, f_{i+1}, .., f_r)[1] \xrightarrow{+}$$

$$K_\bullet(f_1^{n_1}, .., f_i, f_{i+1}, .., f_r) \longrightarrow$$
$$K_\bullet(f_1^{n_1}, .., \widehat{f_i}, f_{i+1}, .., f_r)[1] \xrightarrow{f_i} K_\bullet(f_1^{n_1}, .., \widehat{f_i}, f_{i+1}, .., f_r)[1] \xrightarrow{+}$$

The proofs of Theorems 4.1 and 4.2 will involve induction on the number of affine open subschemes needed to cover X. One needs to begin with some results on affine schemes.

In the situation of Example 5.6, denote the sequence $(f_1^n, \dots, f_r^n)$ $(n > 0)$ by $\mathbf{f}^n$, omitting the superscript "n" when $n = 1$. Let $C_\bullet(\mathbf{f}^n)$ be the cokernel of that map of complexes $R[-1] \to K_\bullet(\mathbf{f}^n)[-1]$ which is the identity map of R in degree 1. The complex $C_\bullet(\mathbf{f}^n)$ has perfect amplitude in $[1-r, 0]$; and there is a natural homotopy triangle

$$C_\bullet(\mathbf{f}^n) \longrightarrow R \longrightarrow K_\bullet(\mathbf{f}^n) \longrightarrow C_\bullet(\mathbf{f}^n)[1]. \tag{5.6.1}$$

There is a map of complexes $K_\bullet(f^{n+m}) \to K_\bullet(f^n)$ ($f \in R$; $m, n > 0$) depicted by

$$\begin{array}{ccc} R & = & R \\ {\scriptstyle f^{n+m}}\uparrow & & \uparrow{\scriptstyle f^n} \\ R & \xrightarrow[f^m]{} & R \end{array}$$

Tensoring such maps gives a map $K_\bullet(\mathbf{f}^{n+m}) \to K_\bullet(\mathbf{f}^n)$, and hence a map $C_\bullet(\mathbf{f}^{n+m}) \to C_\bullet(\mathbf{f}^n)$. For any R-complex E, we have then the *Čech complex*

$$\check{C}^\bullet(\mathbf{f}, E) := \varinjlim_n \operatorname{Hom}^\bullet_R(C_\bullet(\mathbf{f}^n), E).$$

Let U be the complement of the closed subscheme $\operatorname{Spec}(R/\mathbf{f}R) \subset \operatorname{Spec}(R)$, with inclusion $\iota\colon U \hookrightarrow X$. From [EGA, III, §1.3] it follows readily that, with $E^\sim$ the quasi-coherent complex corresponding to E, *$\mathbf{R}\iota_*\iota^*E^\sim$ is naturally $\mathbf{D}$-isomorphic to the sheafified Čech complex $\check{\mathcal{C}}^\bullet(\mathbf{f}, E) := \check{C}^\bullet(\mathbf{f}, E)^\sim$.*

In particular, if the homology of E is $\mathbf{f}R$-torsion (i.e., for all $i \in \mathbb{Z}$, each element of $H^i(E)$ is annihilated by a power of $\mathbf{f}R$)—or equivalently, if $\iota^*E^\sim$ is exact—then $\check{C}^\bullet(\mathbf{f}, E)$ is exact; and since the complex $C_\bullet(\mathbf{f}^n)$ is bounded and projective, therefore

$$H^0\operatorname{Hom}^\bullet_R(C_\bullet(\mathbf{f}^n), E) \cong H^0\mathbf{R}\operatorname{Hom}^\bullet_R(C_\bullet(\mathbf{f}^n), E) \cong \operatorname{Hom}_{\mathbf{D}(R)}(C_\bullet(\mathbf{f}^n), E),$$

whence

$$\varinjlim_n \operatorname{Hom}_{\mathbf{D}(R)}(C_\bullet(\mathbf{f}^n), E) \cong H^0\check{C}^\bullet(\mathbf{f}, E) = 0.$$

Consequently, the commutative diagrams of the following form, with exact rows coming from (5.6.1), and columns from the maps described above:

$$\begin{array}{ccccc} \operatorname{Hom}_{\mathbf{D}(R)}(K_\bullet(\mathbf{f}), E) & \longrightarrow & \operatorname{Hom}_{\mathbf{D}(R)}(R, E) & \longrightarrow & \operatorname{Hom}_{\mathbf{D}(R)}(C_\bullet(\mathbf{f}), E) \\ \downarrow & & \| & & \downarrow \\ \operatorname{Hom}_{\mathbf{D}(R)}(K_\bullet(\mathbf{f}^n), E) & \longrightarrow & \operatorname{Hom}_{\mathbf{D}(R)}(R, E) & \longrightarrow & \operatorname{Hom}_{\mathbf{D}(R)}(C_\bullet(\mathbf{f}^n), E) \end{array}$$

show that *for any $\lambda \in \operatorname{Hom}_{\mathbf{D}(R)}(R, E)$ there is an $n > 0$ such that λ factors through a $\mathbf{D}(R)$-morphism $K_\bullet(\mathbf{f}^n) \to E$.*

If P is a bounded complex of finitely generated projective R-modules, then the homology of $\operatorname{Hom}^\bullet_R(P, E)$ is still $\mathbf{f}R$-torsion (as one sees, e.g., by induction on the number of nonvanishing components of P); and replacing E in what precedes by $\operatorname{Hom}^\bullet_R(P, E)$, one obtains, via Hom-$\otimes$ adjunction, that *for any $\lambda \in \operatorname{Hom}_{\mathbf{D}(R)}(P, E)$ there is an $n > 0$ such that λ factors through a $\mathbf{D}(R)$-morphism $K_\bullet(\mathbf{f}^n) \otimes P \to E$.*

Lemma 5.7. *Let E be an R-complex such that $H^j(E)$ is $\mathbf{f}R$-torsion for all $j \geq -r$, P an R-complex with perfect amplitude in $[0, b]$ for some $b \geq 0$,* *and* $\lambda \in \mathrm{Hom}_{\mathbf{D}(R)}(P,E)$. *Then there is an integer* $n>0$ *and a homomorphism of* R*-complexes* $\lambda_n\colon K_\bullet(\mathbf{f}^n)\otimes P\to E$ *such that for all* $j\ge -r$, *the homology map* $H^j(\lambda)\colon H^j(P)\to H^j(E)$ *factors as*

$$H^j(P)=H^j(R\otimes P)\xrightarrow{\text{natural}} H^j\big(K_\bullet(\mathbf{f}^n)\otimes P\big)\xrightarrow{H^j(\lambda_n)} H^j(E).$$

Proof. We may assume that P is a complex of finitely generated projective R-modules, vanishing in all degrees outside $[0,b]$, see [I, p. 175, b)]. Let $\tau_{\ge -r}E$ be the usual truncation, and $\pi\colon E\to\tau_{\ge -r}E$ the natural map, which induces homology isomorphisms in all degrees $\ge -r$ (see, e.g., [Lp, §1.10]). By the preceding remarks, $\pi\lambda$ factors in $\mathbf{D}(R)$ as

$$P=R\otimes P\xrightarrow{\text{natural}} K_\bullet(\mathbf{f}^n)\otimes P\xrightarrow{\bar\lambda_n}\tau_{\ge -r}E.$$

Since $K_\bullet(\mathbf{f}^n)\otimes P$ is bounded and projective, we may assume that $\bar\lambda_n$ is a map of R-complexes. Then the R-homomorphism

$$(\bar\lambda_n)^{-r}\colon\big(K_\bullet(\mathbf{f}^n)\otimes P\big)^{-r}=P^0\to(\tau_{\ge -r}E)^{-r}=\mathrm{coker}(E^{-r-1}\to E^{-r})$$

lifts to a map $P^0\to E^{-r}$, giving a map λ_n with the desired properties. □

COROLLARY 5.7.1. *Set* $I:=\mathbf{f}R=(f_1,f_2,\dots,f_r)R$. *Let* $m\in\mathbb{Z}$ *and let* E *be an* R*-complex such that* $H^i(E)$ *is* I*-torsion for all* $i\ge m-r$.

(i) *If* E *is* m*-pseudocoherent, and* $p\ge m$ *is such that* $H^i(E)=0$ *for all* $i>p$, *then there exists in the homotopy category of* R*-complexes an* m*-quasi-isomorphism* $P\to E$ *with* $P\in\{K_\bullet(\mathbf{f})\}[m,p]$.

(ii) *For any* $i\ge m$ *with* $H^i(E)\ne 0$, *there is a nonzero map* $K_\bullet(\mathbf{f})[-i]\to E$.

Proof. (i) By [I, p. 103, 2.10(b)], $H^p(E)$ is a finitely generated R-module. So there is an $\ell>0$ and a surjective homomorphism $R^\ell\twoheadrightarrow H^p(E)$, which lifts to $R^\ell\to\ker(E^p\to E^{p+1})$, and thus there is a homomorphism $R^\ell\to E[p]$, or equivalently, $\lambda\colon R^\ell[m-p]\to E[m]$, giving rise, by Lemma 5.7, to an R-homomorphism

$$\lambda_n[-m]\colon P_1:=\big(K_\bullet(\mathbf{f}^n)\otimes R^\ell[-p]\big)\to E$$

such that $H^p\big(\lambda_n[-m]\big)$ is surjective. By Example 5.6 and Remark 5.4(i), we have $K_\bullet(\mathbf{f}^n)[-p]\in\{K_\bullet(\mathbf{f})\}[p,p]$. So we get a homotopy triangle

$$P_1\longrightarrow E\xrightarrow{\alpha}Q_1\longrightarrow P_1[1]$$

with $P_1\in\{K_\bullet(\mathbf{f})\}[p,p]$ and $H^i(Q_1)=0$ for all $i\ge p$, giving (i) when $p=m$.

In any case, Q_1 is m-pseudocoherent [I, p.100, 2.6]; and since all the homology of P_1 is I-torsion, the exact homology sequence of the preceding triangle shows that $H^i(Q_1)$ is I-torsion for all $i\ge m-r$. If $m<p$ then, using induction on $p-m$, one may assume that there is a homotopy triangle

$$P_2\longrightarrow Q_1\xrightarrow{\beta}Q\longrightarrow P_2[1]$$

with $P_2 \in \{K_\bullet(\mathbf{f})\}[m, p-1] \subset \{K_\bullet(\mathbf{f})\}[m, p]$ and $H^i(Q) = 0$ for all $i \geq m$. There exists then a homotopy triangle

$$P \longrightarrow E \xrightarrow{\beta\alpha} Q \longrightarrow P[1]$$

which, by Remark 5.4(v), is as desired.

(ii) There is, by assumption, a nonzero map $R \to H^i(E)$, which lifts to a map $R \to \ker(E^i \to E^{i+1})$; and so there is a nonzero map $\lambda\colon R \to E[i]$ with $H^0(\lambda) \neq 0$. If $j \geq -r$ then $j + i \geq i - r \geq m - r$, so $H^j(E[i]) = H^{j+i}(E)$ is I-torsion, whence by Lemma 5.7, there is for some $n > 0$ a nonzero map $K_\bullet(\mathbf{f}^n) \to E[i]$. By Example 5.6, $K_\bullet(\mathbf{f}^n) \in K_\bullet(\mathbf{f})[0,0]$; so by Example 5.5, there is a nonzero map $K_\bullet(\mathbf{f}) \to E[i]$, proving (ii). □

For dealing with the nonaffine situation, we need to set up some notation.

NOTATION 5.8. A scheme X can be covered by finitely many open affine subsets, say $X = \bigcup_{k=1}^t U_k$, with $U_k = \operatorname{Spec}(R_k)$. For $1 \leq k \leq t$, set

(i) $V_k := \bigcup_{i=k}^t U_i$.
(ii) $Y_k := X - V_{k+1}$ ($:= X$ when $k = t$).

So we have a filtration by closed subschemes $Y_1 \subset Y_2 \subset \cdots \subset Y_t = X$.

Both U_k and V_{k+1} are quasi-compact open subsets of the (quasi-separated) scheme X, whence so is $U_k \cap V_{k+1}$. So there is a sequence

$$\mathbf{f}_k = \{f_{k1}, f_{k2}, \cdots, f_{kr_k}\}$$

in R_k such that

$$U_k \cap V_{k+1} = \bigcup_{i=1}^{r_k} \operatorname{Spec}(R_k[1/f_{ki}]) .$$

Set

(iii) $I_k := \mathbf{f}_k R_k$, (so that $U_k \cap V_{k+1}$ is the complement of the closed subscheme $\operatorname{Spec}(R_k/I_k) \subset U_k$).
(iv) $C_k := \big(K_\bullet(\mathbf{f}_k) \oplus K_\bullet(\mathbf{f}_k)[1]\big)^\sim = \big(K_\bullet(0, f_{k1}, f_{k2}, \cdots, f_{kr_k})\big)^\sim$ with $K_\bullet(*)$ the Koszul complex over R_k associated to the sequence $(*)$, and $(-)^\sim$ the sheafification functor from R_k-modules to quasi-coherent $\mathcal{O}_{U_k}$-modules—so that C_k is a perfect $\mathcal{O}_{U_k}$-complex.
(The reason for introducing this $\oplus$ will emerge shortly.)

We have the cartesian diagram of (open) inclusion maps

$$\begin{array}{ccc} U_k \cap V_{k+1} & \xrightarrow{\ \nu\ } & V_{k+1} \\ {\scriptstyle\lambda}\big\downarrow & & \big\downarrow{\scriptstyle\xi} \\ U_k & \xrightarrow[\ \mu\ ]{} & V_k = U_k \cup V_{k+1} \end{array}$$

The restriction λ^*C_k is homotopically trivial, whence, in $\mathbf{D}(V_{k+1})$,

$$\xi^*\mathbf{R}\mu_*C_k \cong \mathbf{R}\nu_*\lambda^*C_k = 0.$$

Thus, the restrictions of $\mathbf{R}\mu_*C_k$ to both V_{k+1} and U_k are perfect, and so $\mathbf{R}\mu_*C_k$ is itself perfect.

For any $\mathcal{O}_{V_k}$-complex G, the obvious triangle

$$G \xrightarrow{\;0\;} G \longrightarrow G \oplus G[1] \longrightarrow G[1]$$

shows that the complex $G \oplus G[1]$ vanishes in the Grothendieck group $\mathcal{K}_0(V_k)$. Taking $G := \mathbf{R}\mu_*\big(K_\bullet(\mathbf{f}_k)\big)^{\sim}$, we deduce then from Thomason's localization theorem [TT, p. 338, 5.2.2(a)] that the perfect $\mathcal{O}_{V_k}$-complex $\mathbf{R}\mu_*C_k$ is $\mathbf{D}(V_k)$-isomorphic to the restriction of a perfect $\mathcal{O}_X$-complex.

(v) Let $S_k \in \mathbf{D}_{\mathsf{qc}}(X)$ be a perfect $\mathcal{O}_X$-complex whose restriction to V_k is $\mathbf{D}(V_k)$-isomorphic to $\mathbf{R}\mu_*C_k$.

(vi) Let $\mathcal{S}_k$ be the finite set $\{S_1, S_2, \dots, S_k\}$.

According to Lemma 3.2, there is for each k an integer $N_k > 0$ such that, if $Q \in \mathbf{D}_{\mathsf{qc}}(X)$ satisfies $H^\ell(Q) = 0$ for all $\ell \ge -N_k$ then $\mathrm{Hom}_{\mathbf{D}(X)}(S_k, Q) = 0$. After enlarging N_k if necessary, we have also that $\mathrm{Hom}_{\mathbf{D}(X)}(Q, S_k) = 0$.

Set

(vii) $N := \max\{\, N_1, N_2, \dots, N_t, r_1, r_2, \dots, r_t \,\} + 1.$

Next comes the key statement.

PROPOSITION 5.9. *With the preceding notation, let $m, k \in \mathbb{Z}$, $1 \le k \le t$, let $E \in \mathbf{D}_{\mathsf{qc}}(X)$ be such that $H^j(E)$ is supported in Y_k for all $j \ge m - kN$, and set*

$$a_k = \binom{k+1}{2} N \qquad (1 \le k \le t).$$

(i) *If E is $(m-(k-1)N)$-pseudo-coherent then there is an m-isomorphism $P \to E$ with $P \in \mathcal{S}_k[m - a_k, \infty)$ (so that P is perfect, see* 5.4(ii)*).*

(ii) *If $H^\ell(E) \ne 0$ for some $\ell \ge m$, then for some $i \ge m - a_k$ and some $j \in [1, k]$, there is a nonzero map $S_j[-i] \to E$.*

Before proving this, let us see how to derive Theorems 4.1 and 4.2.

Since $Y_t = X$, Theorem 4.1, with $B := (t-1)N$, is contained in 5.9(i).

Next, 5.9(ii) with $k = t$ shows that if $H^\ell(E) \ne 0$, then there exist integers $i \ge \ell - a_t$ and $j \in [1, t]$, and a non-zero map $S_j[-i] \to E$. This gives Theorem 4.2 for the specific choices

$$S = S_1 \oplus S_2 \oplus \cdots \oplus S_t, \qquad A(S) := a_t = \binom{t+1}{2} N.$$

The rest of Theorem 4.2 results from the following general fact, applied to $\mathcal{H} = \{\, E \in \mathbf{D}_{\mathsf{qc}}(X) \mid H^\ell(E) \ne 0 \,\}$, $\mathcal{T} = \mathbf{D}_{\mathsf{qc}}(X)$ and $A = A(S) - \ell$.

PROPOSITION 5.10. *Let $\mathcal{T}$ be a triangulated category with coproducts. Let $\mathcal{H}$ be a collection of objects of $\mathcal{T}$. Suppose there exists a compact generator $S \in \mathcal{T}$ and an integer A such that*

$$E \in \mathcal{H} \implies \operatorname{Hom}\big(S[n], E\big) \neq 0 \text{ for some } n \leq A.$$

Then every compact generator has a similar property: for each compact generator $S' \in \mathcal{T}$ there is an integer A' such that

$$E \in \mathcal{H} \implies \operatorname{Hom}\big(S'[n], E\big) \neq 0 \text{ for some } n \leq A'.$$

Proof. Let $\widehat{R}$ be the full subcategory of $\mathcal{T}$ whose objects are all the direct summands of objects in $\{S'\}(-\infty, \infty) = \bigcup_{M \geq 0}\{S'\}[-M, M]$ (see Remark 5.3). As in the proof of Corollary 4.3.2, (iii) $\Rightarrow$ (ii), one sees that $S \in \widehat{R}$, i.e., there is an $S^* \in \widehat{R}$ and an $M \geq 0$ such that $S \oplus S^* \in \{S'\}[-M, M]$.

Now if $E \in \mathcal{H}$ then, since $\operatorname{Hom}\big(S[k], E\big) \neq 0$ for some $k \leq A$, and $S[k] \oplus S^*[k] \in \{S'\}[-M-k, M-k]$ (Remark 5.4(i)), therefore Example 5.5 gives $\operatorname{Hom}(S'[n], E) \neq 0$ for some n with $n \leq M + k \leq M + A$. □

It remains to prove Proposition 5.9, which we do now by induction on k.

For $k = 1$, $a_1 = N$, so $H^j(E)$ is supported in $Y_1 = \operatorname{Spec}(R_1/I_1)$ for all $j > m - r_1 - 1$ $(\geq m - N)$. As usual, when considering the restriction $E|_{U_1}$ we may assume it to be a quasi-coherent complex, then relate facts about it to facts about the corresponding complex E of R_1-modules. For example, it holds that $H^i(\mathrm{E})$ is I_1-torsion for all $i \geq m - r_1 - 1$.

Thus, from Corollary 5.7.1(i), applied to $I_1 = (0, f_{11}, f_{22}, \cdots, f_{1r_1})R_1$, it follows via 5.8(iv)) that, if E is m-pseudo-coherent then there exists an m-isomorphism $P \to E|_{U_1}$ with $P \in \{C_1\}[m, \infty)$. Likewise (and more easily), Corollary 5.7.1(ii) gives that if $H^i(E) \neq 0$ for some $i \geq m$—whence, $H^i(E)$ being supported in $Y_1 \subset U_1$, $H^i(E|_{U_1}) \neq 0$—then there is a nonzero map

$$C_1[-i] = \big(K_\bullet(\mathbf{f}_1)[-i]\big)^\sim \oplus \big(K_\bullet(\mathbf{f}_1)[-i+1]\big)^\sim \to E|_{U_1}.$$

Let $\mu\colon U_1 \hookrightarrow X$ be the inclusion. Note that $Y_1 = U_1 \setminus V_2$. Since C_1 is exact outside Y_1, so is $P \in \{C_1\}[m, \infty)$ (argue as in Remark 5.4(i)–(iv)), as is $\mathbf{R}\mu_* P$; and by assumption, $H^i(E)$ vanishes outside Y_1 for all $i \geq m$. With all this in mind, we can extend the preceding statements from U_1 to $X = U_1 \cup V_2$, by applying the following Lemma to $U = U_1$, $V = V_2$, and $C = C_1$ or P.

LEMMA 5.11. *Let U and V be open subsets of a scheme X, and let*

$$\begin{array}{ccc} U \cap V & \xrightarrow{\ \nu\ } & V \\ {\scriptstyle\lambda}\downarrow & & \downarrow{\scriptstyle\xi} \\ U & \xrightarrow[\ \mu\ ]{} & U \cup V \end{array}$$

be the natural diagram of inclusion maps. Let $C \in \mathbf{D}(U)$ *satisfy* $\lambda^*C = 0$. *Let* $E \in \mathbf{D}(U \cup V)$. *Then:*

(i) *Every* $\mathbf{D}(U)$*-morphism* $C \to \mu^*E$ *extends uniquely to a* $\mathbf{D}(U \cup V)$*-morphism* $\mathbf{R}\mu_*C \to E$.

(ii) *If* C *is perfect then so is* $\mathbf{R}\mu_*C$.

(iii) *If* $\mathcal{S} \subset \mathbf{D}(U)$ *and* $m \le n \in \mathbb{Z}$ *then* $\mathbf{R}\mu_*\big(\mathcal{S}[m,n]\big) \subset \big\{\mathbf{R}\mu_*\mathcal{S}\big\}[m,n]$.

Proof. (i) In view of the natural isomorphisms

$$\mathrm{Hom}_{\mathbf{D}(U)}(C, \mu^*E) \cong \mathrm{Hom}_{\mathbf{D}(U)}(\mu^*\mathbf{R}\mu_*C, \mu^*E) \cong \mathrm{Hom}_{\mathbf{D}(U\cup V)}(\mathbf{R}\mu_*C, \mathbf{R}\mu_*\mu^*E)$$

we need only show that the natural map is an isomorphism

$$\mathbf{R}\mathrm{Hom}^\bullet(\mathbf{R}\mu_*C, E) \xrightarrow{\ \sim\ } \mathbf{R}\mathrm{Hom}^\bullet(\mathbf{R}\mu_*C, \mathbf{R}\mu_*\mu^*E)$$

(to which we can apply the homology functor H^0). Thus for any triangle

$$G \longrightarrow E \xrightarrow{\text{natural}} \mathbf{R}\mu_*\mu^*E \longrightarrow G[1]$$

we'd like to see that $\mathbf{R}\mathrm{Hom}^\bullet(\mathbf{R}\mu_*C, G) = 0$. But $\mu^*G = 0 = \lambda^*C$, so that

$$\mu^*\mathbf{R}\mathrm{Hom}^\bullet(\mathbf{R}\mu_*C, G) \cong \mathbf{R}\mathrm{Hom}^\bullet(\mu^*\mathbf{R}\mu_*C, \mu^*G) = 0, \quad \text{and}$$
$$\xi^*\mathbf{R}\mathrm{Hom}^\bullet(\mathbf{R}\mu_*C, G) \cong \mathbf{R}\mathrm{Hom}^\bullet(\xi^*\mathbf{R}\mu_*C, \xi^*G) \cong \mathbf{R}\mathrm{Hom}^\bullet(\mathbf{R}\nu_*\lambda^*C, \xi^*G) = 0,$$

whence the conclusion.

(ii) Since both $\xi^*\mathbf{R}\mu^*C \cong \mathbf{R}\nu_*\lambda^*C = 0$ and $\mu^*\mathbf{R}\mu_*C = C$ are perfect, therefore so is $\mathbf{R}\mu_*C$.

(iii) This is a special case of Remark 5.4(iv). □

LEMMA 5.12. *For* $k > 1$, *suppose Proposition 5.9(i) holds with* $k-1$ *in place of* k. *Then for any* $E \in \mathbf{D}_{\mathsf{qc}}(X)$ *and* $\mathbf{D}(U_k)$*-morphism*

$$\psi\colon F \longrightarrow E|_{U_k} \qquad \big(F \in \{C_k\}[m,\infty)\big),$$

there exists a $\mathbf{D}(X)$*-morphism*

$$\tilde\psi\colon \widetilde{F} \to E \qquad \big(\widetilde{F} \in \mathcal{S}_k[m - N - a_{k-1}\,, \infty)\big)$$

whose restriction $\tilde\psi|_{U_k}$ *is isomorphic to* ψ.

Before proving this Lemma, let us see how it is used to establish the induction step in the proof of Proposition 5.9. With reference to that Proposition, we show, for $k > 1$:

(1) Assertion (i) for $k-1$ implies assertion (i) for k.

(2) Assertions (i) and (ii) for $k-1$, together, imply assertion (ii) for k.

To prove (1), let $E \in \mathbf{D}_{\mathsf{qc}}(X)$ be $(m-(k-1)N)$-pseudocoherent, with $H^j(E)$ supported in Y_k for all $j \ge m-kN$. Since $m-(k-1)N-r_k \ge m-kN$, therefore (after replacement of $K_\bullet(\mathbf{f}_k)^\sim$ by C_k, see above) Corollary 5.7.1 provides a $\mathbf{D}(U_k)$-triangle

$$P_k \longrightarrow E|_{U_k} \longrightarrow Q_k \longrightarrow P_k[1] \tag{5.12.1}$$

with $P_k \in \{C_k\}[m-(k-1)N\,,\,\infty)$ and $H^j(Q_k) = 0$ for all $j \geq m-(k-1)N$. By Lemma 5.12, the map $P_k \longrightarrow E|_{U_k}$ is isomorphic to the restriction of a $\mathbf{D}(X)$-morphism $\psi'\colon P' \to E$, with $P' \in \mathcal{S}_k[m-(k-1)N-N-a_{k-1}\,,\,\infty)$, i.e., since

$$a_{k-1} + kN \quad = \quad \binom{k}{2}N + kN \quad = \quad \binom{k+1}{2}N \quad = \quad a_k\,,$$

with $P' \in \mathcal{S}_k[m-a_k, \infty)$. Any $\mathbf{D}_{\mathsf{qc}}(X)$-triangle

$$P' \xrightarrow{\ \psi'\ } E \xrightarrow{\ \alpha\ } Q' \longrightarrow P'[1]\,,$$

restricts on U_k to one isomorphic to (5.12.1). So when $j \geq m-(k-1)N$, then $H^j(Q')$ vanishes on U_k; furthermore, $H^j(E)$ is supported on Y_k, and since all the members of $\mathcal{S}_k$ are exact outside Y_k therefore so is P' (argue as in Remark 5.4(i)–(iv)); and thus $H^j(Q')$ is supported in $(Y_k \setminus U_k) = Y_{k-1}$.

Moreover, Q' is $(m-(k-2)N)$-pseudocoherent, since both P' and E are [I, p. 100, 2.6]. So now the inductive assumption produces a triangle

$$P'' \longrightarrow Q' \xrightarrow{\ \beta\ } Q \longrightarrow P''[1]$$

with $P'' \in \mathcal{S}_{k-1}[m-a_{k-1}\,, \infty)$, and $H^j(Q) = 0$ whenever $j \geq m$.

There is then a triangle

$$P \xrightarrow{\ \psi'\ } E \xrightarrow{\ \beta\alpha\ } Q \longrightarrow P[1]\,,$$

and the assertion 5.9(i), for the integer k, results from Remark 5.4(v).

As for (2), let E satisfy the hypotheses of 5.9(ii) for k. If $H^i(E|_{U_k}) = 0$ for all $i \geq m-(k-1)N$ then $H^j(E)$ is supported in Y_{k-1} for all $j \geq m-(k-1)N$, $H^\ell(E)$ is non-zero for some $\ell \geq m$, and $m-a_{k-1} \geq m-a_k$; so in this case assertion (ii) for k is already given by assertion (ii) for $k-1$.

If, on the other hand, $H^i(E|_{U_k}) \neq 0$ for some $i \geq m-(k-1)N$, then, since $m-(k-1)N-r_k \geq m-kN$, Corollary 5.7.1 (suitably modified) provides a nonzero map $C_k[-i] \to E|_{U_k}$. By Remark 5.4(i),

$$C_k[-i] \in \{C_k\}[i, \infty) \subset \{C_k\}[m-(k-1)N, \infty),$$

so by Lemma 5.12, there exists a nonzero $\mathbf{D}(X)$-morphism $\widetilde{F} \to E$ with

$$\widetilde{F} \in \mathcal{S}_k[m-(k-1)N-N-a_{k-1}\,, \infty) \;=\; \mathcal{S}_k[m-a_k, \infty)\,.$$

Hence, by Example 5.5, 5.9(ii) holds for k.

We come finally to the *proof of Lemma 5.12.*

Let $\mathcal{S} \subset \mathbf{D}(U_k)$ be the full subcategory with objects those $F \in \{C_k\}[m, \infty)$ for which the Lemma holds. We need to verify the conditions in Definition 5.1, i.e., we need to show:

(a) $C_k[-\ell] \in \mathcal{S}$ for all $\ell \geq m$; and

(b) for any $\mathbf{D}(U_k)$-triangle

$$F' \longrightarrow F \longrightarrow F'' \longrightarrow F'[1],$$

if $F', F'' \in \mathcal{S}$ then $F \in \mathcal{S}$.

For (a), we first use Lemma 5.11 to extend $\psi\colon C_k[-\ell] \to E|_{U_k}$ to a $\mathbf{D}(V_k)$-morphism $\phi\colon S_k[-\ell]\big|_{V_k} \to E|_{V_k}$. By Thomason's localization theorem, as formulated in [N2, p. 214, 2.1.5] (and further elucidated in [*ibid.*, p. 216, proof of Lemma 2.6]),[5] there is then a $\mathbf{D}_{\mathsf{qc}}(X)$-diagram, with top row a triangle of perfect complexes:

$$\begin{array}{ccccccc}
\widetilde{P} & \longrightarrow & \widetilde{F}_1 & \xrightarrow{\ f\ } & S_k[-\ell] & \longrightarrow & \widetilde{P}[1] \\
 & & {\scriptstyle g}\big\downarrow & & & & \\
 & & E & & & &
\end{array}$$

and with $\widetilde{P}$ exact on V_k, so that $f|_{V_k}$ is an isomorphism; and furthermore,

$$\phi = (g|_{V_k}) \circ (f|_{V_k})^{-1}.$$

Since $S_k[-\ell] \in \mathcal{S}_k[\ell, \infty]$ (see Remark 5.4(i)), we need only show that we can choose $\widetilde{P} \in \mathcal{S}_{k-1}[\ell - N - a_{k-1}\,, \infty)$, because then we'll have

$$\widetilde{F}_1 \in \mathcal{S}_k[\ell - N - a_{k-1}\,, \infty) \subset \mathcal{S}_k[m - N - a_{k-1}\,, \infty).$$

The perfect complex $\widetilde{P}$ is exact outside $X - V_k = Y_{k-1}$, and we are assuming that 5.9(i) is true for $k-1$. It follows that there exists a triangle

$$P \longrightarrow \widetilde{P} \longrightarrow Q \longrightarrow P[1]$$

with $P \in \mathcal{S}_{k-1}[\ell - N - a_{k-1}\,, \infty)$ and $H^i(Q) = 0$ for all $i \geq \ell - N$. Since all the members of $\mathcal{S}_{k-1}$ are exact on V_k, the same is true of P (argue as in Remark 5.4(i)–(iv)).

Now [N2, p. 58, 1.4.6] produces an octahedron on $P \to \widetilde{P} \to \widetilde{F}_1$, where the rows and columns are triangles:

$$\begin{array}{ccccccc}
P & = & P & & & & \\
\big\downarrow & & \big\downarrow & & & & \\
\widetilde{P} & \longrightarrow & \widetilde{F}_1 & \xrightarrow{\ f\ } & S_k[-\ell] & \longrightarrow & \widetilde{P}[1] \\
\big\downarrow & & \big\downarrow{\scriptstyle \beta} & & \big\| & & \big\downarrow \\
Q & \longrightarrow & F' & \xrightarrow[\ f'\ ]{} & S_k[-\ell] & \xrightarrow[\ g\ ]{} & Q[1] \\
\big\downarrow & & \big\downarrow{\scriptstyle h} & & & & \\
P[1] & = & P[1] & & & &
\end{array}$$

Since $H^i\big(Q[1+\ell]\big) = 0$ for all $i \geq -N-1$, the definition of N (see Notation 5.8(vii)) forces the map $g\colon S_k[-\ell] \to Q[1]$ to vanish. The exact sequence

[5] where in the absence of separatedness, $j_{\bullet *}$ should become $\mathbf{R}j_{\bullet *}$.

$$\mathrm{Hom}(S_k[-\ell], F') \xrightarrow{\text{via } f'} \mathrm{Hom}(S_k[-\ell], S_k[-\ell]) \xrightarrow{\text{via } g=0} \mathrm{Hom}(S_k[-\ell], Q[1])$$

shows there is a map $\iota\colon S_k[-\ell] \to F'$ with $f'\iota$ the identity map of $S_k[-\ell]$.

This gives rise to yet another octahedron, on $S_k[-\ell] \xrightarrow{\iota} F' \xrightarrow{h} P[1]$:

$$\begin{array}{ccccccc}
P & = & P & & & & \\
\downarrow & & \downarrow & & & & \\
\widetilde{F} & \xrightarrow{\gamma} & \widetilde{F}_1 & \longrightarrow & Q & \longrightarrow & \widetilde{F}[1] \\
{\scriptstyle\alpha}\downarrow & & \downarrow{\scriptstyle\beta} & & \| & & \downarrow \\
S_k[-\ell] & \xrightarrow{\iota} & F' & \longrightarrow & Q & \longrightarrow & S_k[-\ell+1] \\
\downarrow & & \downarrow{\scriptstyle h} & & & & \\
P[1] & = & P[1] & & & &
\end{array}$$

The first column is a triangle, with $\widetilde{F} \in \mathcal{S}_k[\ell - N - a_{k-1}, \infty)$, and $P|_{V_k}$ exact, so that $\alpha|_{V_k}$ is an isomorphism.

Moreover, if $\tilde{\psi}\colon \widetilde{F} \to E$ is the composite $\widetilde{F} \xrightarrow{\gamma} \widetilde{F}_1 \xrightarrow{g} E$, then

$$\alpha = f'\iota\alpha = f'\beta\gamma = f\gamma,$$

so that on V_k,

$$\phi\alpha = \phi f\gamma = g\gamma = \tilde{\psi},$$

proving (a).

Proof of (b).

Let $\psi\colon F \to E|_{U_k}$ be a $\mathbf{D}(U_k)$-morphism. Since $F' \in \mathcal{S}$, there exists a complex $\widetilde{F'} \in \mathcal{S}_k[m - N - a_{k-1}, \infty)$ and a $\mathbf{D}(X)$-morphism $\widetilde{\psi'}\colon \widetilde{F'} \to E$ whose restriction to U_k is isomorphic to the composite $F' \to F \xrightarrow{\psi} E|_{U_k}$. There results a triangle

$$\widetilde{F'} \xrightarrow{\widetilde{\psi'}} E \xrightarrow{\gamma'} E' \longrightarrow \widetilde{F'}[1]\,,$$

and hence a commutative $\mathbf{D}(U_k)$-diagram (part of an octahedron):

$$\begin{array}{ccccccc}
F' & \longrightarrow & F & \longrightarrow & F'' & \longrightarrow & F'[1] \\
\| & & {\scriptstyle\psi}\downarrow & & \downarrow{\scriptstyle g} & & \| \\
F' & \longrightarrow & E|_{U_k} & \xrightarrow{\gamma'|_{U_k}} & E'|_{U_k} & \longrightarrow & F'[1] \\
 & & {\scriptstyle\chi}\downarrow & & \downarrow{\scriptstyle h} & & \\
 & & G & = & G & &
\end{array}$$

Since $F'' \in \mathcal{S}$, there is an $\widetilde{F}'' \in \mathcal{S}_k[m-N-a_{k-1}\,,\,\infty)$ and a $\mathbf{D}(X)$-morphism $\widetilde{\psi''}\colon \widetilde{F}'' \to E'$ whose restriction to U_k is isomorphic to $g : F'' \to E'|_{U_k}$. So there is a triangle

$$\widetilde{F}'' \xrightarrow{\ \widetilde{\psi''}\ } E' \xrightarrow{\ \gamma''\ } E'' \longrightarrow \widetilde{F}''[1]\ ;$$

whose restriction to U_k is isomorphic to

$$F'' \xrightarrow{\ g\ } E'|_{U_k} \xrightarrow{\ h\ } G \longrightarrow F''[1]\ .$$

The restriction to U_k of the composite $E \xrightarrow{\gamma'} E' \xrightarrow{\gamma''} E''$ is isomorphic to the composite $\chi\colon E|_{U_k} \xrightarrow{\gamma'|_{U_k}} E'|_{U_k} \xrightarrow{h} G$. Completing $\gamma''\gamma'$ to a triangle

$$\widetilde{F} \xrightarrow{\ \tilde{\psi}\ } E \xrightarrow{\ \gamma''\gamma'\ } E'' \longrightarrow \widetilde{F}[1]$$

and restricting to U_k, we obtain a triangle isomorphic to

$$F \xrightarrow{\ \psi\ } E|_{U_k} \xrightarrow{\ \chi\ } G \longrightarrow \widetilde{F}[1]\ .$$

That $\widetilde{F} \in \mathcal{S}_k[m - N - a_{k-1}\,,\,\infty)$ follows from Remark 5.4(v). $\square$

References

[BB] A. Bondal and M. van den Bergh, *Generators and representability of functors in commutative and noncommutative geometry*, Moscow Math. J. **3** (2003), 1–36. MR 1996800 (2004h:18009)

[BN] M. Bökstedt and A. Neeman, *Homotopy limits in triangulated categories*, Compositio Math. **86** (1993), 209–234. MR 1214458 (94f:18008)

[EGA] A. Grothendieck and J. Dieudonné, *Éléments de géométrie algébrique. I*, Inst. Hautes Études Sci. Publ. Math. **4** (1960); *II*, ibid. **8** (1961); *III*, ibid. **11** (1961), **17** (1963); *IV*, ibid. **20** (1964), **24** (1965), **28** (1966), **32** (1967).

[GD] ———, *Élements de Géométrie Algébrique* **I**, Springer-Verlag, New York, 1971.

[H] R. Hartshorne, *Residues and duality*, Lecture notes of a seminar on the work of A. Grothendieck, given at Harvard 1963/64. With an appendix by P. Deligne. Lecture Notes in Mathematics, No. 20, Springer-Verlag, Berlin, 1966. MR 0222093 (36 #5145)

[I] L. Illusie, *Généralités sur les conditions de finitude dans les catégories dérivées, etc.*, Théorie des Intersections et Théorème de Riemann-Roch, SGA 6, Lecture Notes in Mathematics, vol. 225, Springer-Verlag, New York, 1971, pp. 78–296.

[Kf] G. R. Kempf, *Some elementary proofs of basic theorems in the cohomology of quasicoherent sheaves*, Rocky Mountain J. Math. **10** (1980), 637–645. MR 590225 (81m:14015)

[Kl] R. Kiehl, *Ein "Descente"-Lemma und Grothendiecks Projektionssatz für nichtnoethersche Schemata*, Math. Ann. **198** (1972), 287–316. MR 0382280 (52 #3165)

[Lp] J. Lipman, *Notes on derived functors and Grothendieck duality*, preprint, available at `http://www.math.purdue.edu/~lipman`.

[N1] A. Neeman, *The Grothendieck duality theorem via Bousfield's techniques and Brown representability*, J. Amer. Math. Soc. **9** (1996), 205–236. MR 1308405 (96c:18006)

[N2] ———, *Triangulated categories*, Annals of Mathematics Studies, vol. 148, Princeton University Press, Princeton, NJ, 2001. MR 1812507 (2001k:18010)

[TT] R. W. Thomason and T. Trobaugh, *Higher algebraic K-theory of schemes and of derived categories*, The Grothendieck Festschrift, Vol. III, Progr. Math., vol. 88, Birkhäuser Boston, Boston, MA, 1990, pp. 247–435. MR 1106918 (92f:19001)

[V] J.-L. Verdier, *Base change for twisted inverse image of coherent sheaves*, Algebraic Geometry (Internat. Colloq., Tata Inst. Fund. Res., Bombay, 1968), Oxford Univ. Press, London, 1969, pp. 393–408. MR 0274464 (43 #227)

Joseph Lipman, Department of Mathematics, Purdue University, W. Lafayette, IN 47907, USA

E-mail address: jlipman@purdue.edu

Amnon Neeman, Centre for Mathematics and its Applications, Mathematical Sciences Institute, John Dednam Building, The Australian National University, Canberra, ACT 0200, Australia

E-mail address: Amnon.Neeman@anu.edu.au

Illinois Journal of Mathematics
Volume 51, Number 1, Spring 2007, Pages 237–254
S 0019-2082

ANNIHILATORS OF LOCAL COHOMOLOGY IN CHARACTERISTIC ZERO

PAUL ROBERTS, ANURAG K. SINGH, AND V. SRINIVAS

To Phil Griffith

ABSTRACT. This paper discusses the problem of whether it is possible to annihilate elements of local cohomology modules by elements of arbitrarily small order under a fixed valuation. We first discuss the general problem and its relationship to the Direct Summand Conjecture, and next present two concrete examples where annihilators with small order are shown to exist. We then prove a more general theorem, where the existence of such annihilators is established in some cases using results on abelian varieties and the Albanese map.

1. Almost vanishing of local cohomology

The concept of almost vanishing that we use here comes out of recent work on *Almost Ring Theory* by Gabber and Ramero [4]. This theory was developed to give a firm foundation to the results of Faltings on *Almost étale extensions* [3], and these ideas have their origins in a classic work of Tate on *p-divisible groups* [21]. The use of the general theory, for our purposes, is comparatively straightforward, but it illustrates the main questions in looking at certain homological conjectures, as discussed later in the section. The approach is heavily influenced by Heitmann's proof of the Direct Summand Conjecture for rings of dimension three [8].

Let A be an integral domain, and let v be a valuation on A with values in the abelian group of rational numbers; more precisely, v is a function from A to $\mathbb{Q} \cup \{\infty\}$ such that

(1) $v(a) = \infty$ if and only if $a = 0$,
(2) $v(ab) = v(a) + v(b)$ for all $a, b \in A$, and
(3) $v(a + b) \geqslant \min\{v(a), v(b)\}$ for all $a, b \in A$.

Received November 6, 2006; received in final form April 4, 2007.
2000 *Mathematics Subject Classification.* Primary 13D22. Secondary 13D45, 14K05.
P.R. and A.K.S. were supported in part by grants from the National Science Foundation. V.S. was supported by a Swarnajayanthi Fellowship of the DST.

We will also assume that $v(a) \geqslant 0$ for all elements $a \in A$.

DEFINITION 1.1. An A-module M is *almost zero* if for every $m \in M$ and every real number $\varepsilon > 0$, there exists an element a in A with $v(a) < \varepsilon$ and $am = 0$. When it is necessary to specify the valuation, we say that M *is almost zero with respect to the valuation* v.

We note some properties of almost zero modules:

(1) For an exact sequence
$$0 \longrightarrow M' \longrightarrow M \longrightarrow M'' \longrightarrow 0,$$
the module M is almost zero if and only if each of M' and M'' is almost zero.

(2) If $\{M_i\}$ is a directed system consisting of almost zero modules, then its direct limit $\varinjlim_i M_i$ is almost zero.

In [4] Gabber and Ramero define a module to be almost zero if it is annihilated by a fixed ideal $\mathfrak{m}$ of A with $\mathfrak{m} = \mathfrak{m}^2$. This set of modules also satisfies conditions (1) and (2), though in many cases their condition is stronger than the one in Definition 1.1.

The *absolute integral closure* R^+ of a domain R is the integral closure of R in an algebraic closure of its fraction field. An important situation for us will be where $(R, \mathfrak{m})$ is a complete local ring. In this case, fix a valuation $v \colon R \longrightarrow \mathbb{Z} \cup \{\infty\}$ which is positive on $\mathfrak{m}$. By Izumi's Theorem [16], two such valuations are bounded by constant multiples of each other. Since R^+ is an integral extension, v extends to a valuation $v \colon R^+ \longrightarrow \mathbb{Q} \cup \{\infty\}$. Let A be a subring of R^+ containing R; we often take A to be R^+. Note that v is positive on the maximal ideal of A. The ring A need not be Noetherian, and by a *system of parameters* for A, we shall mean a system of parameters for some Noetherian subring of A that contains R.

The main question we consider is whether the local cohomology modules $H^i_{\mathfrak{m}}(A)$ are almost zero for $i < \dim A$. Let $x_1, \ldots, x_d$ be a system of parameters for R. Then the local cohomology module $H^i_{\mathfrak{m}}(A)$ is the i-th cohomology module of the Čech complex
$$0 \longrightarrow A \longrightarrow \oplus A_{x_i} \longrightarrow \oplus A_{x_i x_j} \longrightarrow \cdots \longrightarrow A_{x_1 \cdots x_d} \longrightarrow 0.$$
The question whether $H^i_{\mathfrak{m}}(A)$ is almost zero for $i = 0, \ldots, d-1$ is closely related to the question whether the x_i come close to forming a regular sequence in the following sense.

DEFINITION 1.2. A sequence of elements $x_1, \ldots, x_d \in A$ is an *almost regular sequence* if for each $i = 1, \ldots, d$, the module
$$((x_1, \ldots, x_{i-1}) :_A x_i)/(x_1, \ldots, x_{i-1})$$

is almost zero. If every system of parameters for A is an almost regular sequence, we say that A is *almost Cohen-Macaulay*.

The usual inductive argument as in [20, Theorem IV.2.3] shows that if A is almost Cohen-Macaulay, then the modules $H^i_{\mathfrak{m}}(A)$ are almost zero for $i < \dim A$. However, we do not know whether the converse holds in general.

As motivation for the definitions introduced above, we discuss how these are related to the homological conjectures. Let $x_1, \dots, x_d$ be a system of parameters for a local ring R. Hochster's *Monomial Conjecture* states that

$$x_1^t \cdots x_d^t \notin \left(x_1^{t+1}, \dots, x_d^{t+1}\right)R \qquad \text{for all } t \geqslant 0.$$

This is known to be true for local rings containing a field, and Heitmann [8] proved it for local rings of mixed characteristic of dimension up to three. It remains open for mixed characteristic rings of higher dimension, where it is equivalent to several other conjectures such as the Direct Summand Conjecture (which states that regular local rings are direct summands of their module-finite extension rings), the Canonical Element Conjecture, and the Improved New Intersection Conjecture; for some of the related work, we mention [2], [10], [11], [18] and [19].

The connection between the Monomial Conjecture and the almost Cohen-Macaulay property is evident from the following proposition.

PROPOSITION 1.3. *Let R be a local domain with an integral extension which is almost Cohen-Macaulay. Then the Monomial Conjecture holds for R, i.e., for each system of parameters $x_1, \dots, x_d$ of R, we have*

$$x_1^t \cdots x_d^t \notin \left(x_1^{t+1}, \dots, x_d^{t+1}\right)R \qquad \textit{for all } t \geqslant 0\,.$$

Proof. Let A be an integral extension of R which is almost Cohen-Macaulay with respect to a valuation v which is positive on the maximal ideal of R. Then $v(x_i) > 0$ for each $i = 1, \dots, d$; let ε be the minimum of these positive rational numbers. If $x_1^t \cdots x_d^t \in (x_1^{t+1}, \dots, x_d^{t+1})R$ for some integer t, then

$$x_1^t \cdots x_d^t = a_1 x_1^{t+1} + \cdots + a_d x_d^{t+1}$$

for elements a_i of A. (The a_i can be chosen in R, though we will only consider them as elements of A.) Rearranging terms in the above equation, we have

$$x_1^t (x_2^t \cdots x_d^t - a_1 x_1) \in \left(x_2^{t+1}, \dots, x_d^{t+1}\right)A\,.$$

Since A is almost Cohen-Macaulay, the elements $x_1^t, x_2^{t+1}, \dots, x_d^{t+1}$ form an almost regular sequence. Hence there exists $c_1 \in A$ with $v(c_1) < \varepsilon/d$ and

$$c_1 (x_2^t \cdots x_d^t - a_1 x_1) \in \left(x_2^{t+1}, \dots, x_d^{t+1}\right)A\,.$$

This implies that $c_1 x_2^t \cdots x_d^t \in (x_1, x_2^{t+1}, \dots, x_d^{t+1})A$. We now repeat the process for x_2, i.e., we have

$$c_1 x_2^t \cdots x_d^t = b_1 x_1 + b_2 x_2^{t+1} + \cdots + b_d x_d^{t+1}$$

with $b_i \in A$, so

$$x_2^t(c_1 x_3^t \cdots x_d^t - b_2 x_2) \in \left(x_1, x_3^{t+1}, \dots, x_d^{t+1}\right)A\,.$$

By an argument similar to the one above, there is an element $c_2 \in A$ with $v(c_2) < \varepsilon/d$ and

$$c_1 c_2 x_3^t \cdots x_d^t \in \left(x_1, x_2, x_3^{t+1}, \dots, x_d^{t+1}\right)A\,.$$

Repeating this procedure $d-2$ more times, we obtain elements $c_1, c_2, \dots, c_d$ in A with $v(c_i) < \varepsilon/d$ and

$$c_1 c_2 \cdots c_d = u_1 x_1 + \cdots + u_d x_d$$

for some $u_i \in A$. But then

$$v(c_1 \cdots c_d) = v(c_1) + \cdots + v(c_d) < d(\varepsilon/d) = \varepsilon,$$

whereas, since $v(u) \geqslant 0$ for all $u \in A$, we also have

$$v(u_1 x_1 + \cdots + u_d x_d) \geqslant \min\{v(u_i x_i)\} \geqslant \min\{v(u_i) + v(x_i)\} \geqslant \min\{v(x_i)\} = \varepsilon\,,$$

which is a contradiction. □

To put the results of the remainder of the paper in context, we recall the situation in positive characteristic. Let R be a complete local domain containing a field of characteristic $p > 0$. We let R_∞ denote the perfect closure of R, that is, R_∞ is the ring obtained by adjoining to R the p^n-th roots of all its elements.

PROPOSITION 1.4. *Let $(R, \mathfrak{m})$ be a complete local domain containing a field of prime characteristic. Then R_∞ is almost Cohen-Macaulay with respect to any valuation which is positive on $\mathfrak{m}$.*

Proof. Let v be such a valuation, and let $x_1, \dots, x_d$ be a system of parameters for R_∞. Suppose that

$$ax_i = b_1 x_1 + \cdots + b_{i-1} x_{i-1} \tag{1.4.1}$$

for $a, b_j \in R_\infty$. Let R' be the Noetherian subring of R_∞ generated over R by a, $b_1, \dots, b_{i-1}$, and $x_1, \dots, x_d$. By Cohen's structure theorem, R' is a finite extension of a power series ring $S = K[[x_1, \dots, x_d]]$, where K is a coefficient field. Let m be the largest integer such that R' contains a free S-module of rank m. In this case, the cokernel of

$$S^m \subseteq R'$$

is a torsion S-module, so there exists a nonzero element $c \in S$ such that $cR' \subseteq S^m$. Taking p^n-th powers in equation (1.4.1) gives us

$$a^{p^n} x_i^{p^n} \in \left(x_1^{p^n}, \dots, x_{i-1}^{p^n}\right)R' \qquad \text{for all } n \geqslant 0\,.$$

Multiplying the above by c and using $cR' \subseteq S^m$, we get

$$ca^{p^n} x_i^{p^n} \in \left(x_1^{p^n}, \dots, x_{i-1}^{p^n}\right)S^m\,.$$

Since $x_1^{p^n}, \dots, x_i^{p^n}$ is a regular sequence on the free module S^m, it follows that

$$ca^{p^n} \in \left(x_1^{p^n}, \dots, x_{i-1}^{p^n}\right) S^m \subseteq \left(x_1^{p^n}, \dots, x_{i-1}^{p^n}\right) R' .$$

Taking p^n-th roots in an equation for $ca^{p^n} \in (x_1^{p^n}, \dots, x_{i-1}^{p^n}) R'$ gives us

$$c^{1/p^n} a \in \left(x_1, \dots, x_{i-1}\right) R_\infty \qquad \text{for all } n \geqslant 0 .$$

Since the limit of $v(c^{1/p^n})$ is zero as $n \longrightarrow \infty$, it follows that R_∞ is almost Cohen-Macaulay. □

In [14] Hochster and Huneke proved the much deeper fact that for an excellent local domain R of positive characteristic, the ring R^+ is Cohen-Macaulay; see also [15]. We remark that the subring R_∞ may not be Cohen-Macaulay in general: if R is an F-pure ring which is not Cohen-Macaulay, then, since $R \hookrightarrow R_\infty$ is pure, R_∞ is not Cohen-Macaulay as well.

If R is a local domain containing a field of characteristic zero, then R^+ is typically not a big Cohen-Macaulay algebra. For example, let R be a normal ring of characteristic zero which is not Cohen-Macaulay. Then the field trace map shows that R splits from finite integral extensions. Consequently a nontrivial relation on a system of parameters for R remains nontrivial in finite extensions, and hence in R^+. Specifically, for a ring $(R, \mathfrak{m})$ of characteristic zero, the map

$$H^i_\mathfrak{m}(R) \longrightarrow H^i_\mathfrak{m}(R^+)$$

is injective for all i. This leads to the following question.

QUESTION 1.5. Let $(R, \mathfrak{m})$ be a complete local domain. For $i < \dim R$, is the image of natural map

$$H^i_\mathfrak{m}(R) \longrightarrow H^i_\mathfrak{m}(R^+)$$

almost zero?

The answer is affirmative if the ring R contains a field of positive characteristic: this follows from Proposition 1.4, or from either of the stronger statements [14, Theorem 1.1] or [15, Theorem 2.1]. If R is a three-dimensional ring of mixed characteristic p, Heitmann [8] proved that the image of $H^2_\mathfrak{m}(R)$ in $H^2_\mathfrak{m}(R^+)$ is killed by $p^{1/n}$ for all integers $n \geqslant 0$; more recently, he proved the stronger statement [9, Theorem 2.9] that $H^2_\mathfrak{m}(R^+)$ is annihilated by $c^{1/n}$ for all $c \in \mathfrak{m}$ and $n \geqslant 0$. Hence the answer to Question 1.5 is also affirmative for mixed characteristic rings of dimension less than or equal to three.

2. Examples

In this section, we present two nontrivial examples where local cohomology modules of characteristic zero rings are annihilated by elements of arbitrarily small positive order. The examples are $\mathbb{N}$-graded, and in such cases it is natural to use the valuation arising from the grading: $v(r)$ is the least integer n such that the n-th degree component of r is nonzero.

PROPOSITION 2.1. *Let K be a field of characteristic zero, and consider the hypersurface $S = K[x, y, z, w]/(xy - zw)$. For distinct elements α_i of K, let η be a square root of*

$$\prod_{i=1}^{4}(x - \alpha_i z).$$

Then the integral closure of $S[\eta]$ in its field of fractions is the ring

$$R = S\left[\eta, \frac{w}{x}\eta, \frac{w^2}{x^2}\eta\right].$$

Proof. The element $(w^2/x^2)\eta$ is integral over $S[\eta]$ since

$$\left(\frac{w^2}{x^2}\eta\right)^2 = \prod_{i=1}^{4}(w - \alpha_i y).$$

A similar computation shows that $(w/x)\eta$ is integral over $S[\eta]$, and it remains to prove that the integral closure of $S[\eta]$ is generated by these elements. An element of the fraction field of $S[\eta]$ can be written as $a + b\eta$, with a and b from the fraction field of S. Now $a + b\eta$ is integral over S if and only if its trace and norm of are elements of S. Since 2 is a unit in S, this is equivalent to $a \in S$ and $b^2\eta^2 \in S$. Thus the integral closure of $S[\eta]$ is $S \oplus I\eta$, where I is the fractional ideal consisting of elements b with $b^2\eta^2 \in S$.

Since S is a normal domain, $b^2\eta^2$ belongs to S if and only if $v_{\mathfrak{p}}(b^2\eta^2) \geqslant 0$ for all valuations $v_{\mathfrak{p}}$ corresponding to height one prime ideals $\mathfrak{p}$ of S. Note that $v_{\mathfrak{p}}(\eta^2) > 0$ precisely for the primes $\mathfrak{p}_0 = (x, z)$ and $\mathfrak{p}_i = (x - \alpha_i z, w - \alpha_i y)$ for $1 \leqslant i \leqslant 4$. Since $v_{\mathfrak{p}_0}(\eta^2) = 4$ and $v_{\mathfrak{p}_i}(\eta^2) = 1$ for $1 \leqslant i \leqslant 4$, the condition for b to be an element of I is that

$$v_{\mathfrak{p}_0}(b) \geqslant -2 \qquad \text{and} \qquad v_{\mathfrak{p}}(b) \geqslant 0 \text{ for all } \mathfrak{p} \neq \mathfrak{p}_0.$$

This implies that $v_{\mathfrak{p}}(bx^2) \geqslant 0$ for all height one primes $\mathfrak{p}$, i.e., that $bx^2 \in S$. Let $b = s/x^2$. Then $v_{(x,w)}(b) \geqslant 0$ implies that s must be in the ideal $(x, w)^2$. Hence I is generated over S by 1, w/x, and w^2/x^2. □

EXAMPLE 2.2. We continue in the notation of Proposition 2.1, i.e., R is the normalization of $S[\eta]$. The ring R is normal by construction, and has dimension three. It follows that $H^0_{\mathfrak{m}}(R) = 0 = H^1_{\mathfrak{m}}(R)$, where $\mathfrak{m}$ is the homogeneous maximal ideal of R. We show that there are elements of R^+ of arbitrarily small positive order annihilating the image of $H^2_{\mathfrak{m}}(R)$ in $H^2_{\mathfrak{m}}(R^+)$.

Note that x, y, $z + w$ form a homogeneous system of parameters for the hypersurface S, and hence also for R. In the ring R, we have a relation on these elements given by the equation

$$\frac{w}{x}\eta \cdot (z+w) = \eta \cdot y + \frac{w^2}{x^2}\eta \cdot x\,.$$

This is a nontrivial relation in the sense that $(w/x)\eta$ does not belong to the ideal generated by x and y, so the ring R is not Cohen-Macaulay. Consider the element of $H^2_{\mathfrak{m}}(R)$ given by this relation; it turns out that $H^2_{\mathfrak{m}}(R)$ is a one-dimensional K-vector space generated by this element; see Remark 2.3.

Let v be the valuation defined by the grading on R, i.e., v takes value 1 on x, y, z, and w, and $v(\eta) = 2$. We construct elements x_n in finite extensions R_n of R with $v(x_n) = 1/2^n$ and $x_n(w/x)\eta \in (x,y)R_n$; it then follows that each x_n annihilates the image of the map $H^2_{\mathfrak{m}}(R) \longrightarrow H^2_{\mathfrak{m}}(R^+)$.

Let R_1 be the extension ring of R obtained by adjoining $\sqrt{x-\alpha_i z}$ for $1 \leqslant i \leqslant 4$ and normalizing. We claim that the element $x_1 = \sqrt{x-\alpha_1 z}$ multiplies $(w/x)\eta$ into the ideal $(x,y)R_1$. To see this, note that

$$\begin{aligned} x_1\frac{w}{x}\eta = x_1\frac{w}{x}\prod_{i=1}^{4}\sqrt{x-\alpha_i z} &= (x-\alpha_1 z)\frac{w}{x}\prod_{i=2}^{4}\sqrt{x-\alpha_i z} \\ &= x\left(\frac{w}{x}\prod_{i=2}^{4}\sqrt{x-\alpha_i z}\right) - y\left(\alpha_1\prod_{i=2}^{4}\sqrt{x-\alpha_i z}\right). \end{aligned}$$

The element x_1 has $v(x_1) = 1/2$. To find an annihilator x_2 with $v(x_2) = 1/4$, we first write

$$x - \alpha_3 z = \beta(x-\alpha_1 z) - \gamma(x-\alpha_2 z)$$

for suitable $\beta, \gamma \in K$, and then factor as a difference of squares to obtain

$$\begin{aligned} &x - \alpha_3 z \\ &\quad = \left(\sqrt{\beta(x-\alpha_1 z)} + \sqrt{\gamma(x-\alpha_2 z)}\right)\left(\sqrt{\beta(x-\alpha_1 z)} - \sqrt{\gamma(x-\alpha_2 z)}\right). \end{aligned}$$

We let

$$x_2 = \sqrt{\sqrt{\beta(x-\alpha_1 z)} + \sqrt{\gamma(x-\alpha_2 z)}}\,,$$

which is an element with $v(x_2) = 1/4$. Now

$$x_2\sqrt{x-\alpha_3 z} = \lambda\left(\sqrt{\beta(x-\alpha_1 z)} + \sqrt{\gamma(x-\alpha_2 z)}\right),$$

where

$$\lambda = \sqrt{\sqrt{\beta(x-\alpha_1 z)} - \sqrt{\gamma(x-\alpha_2 z)}}\,,$$

and so

$$\begin{aligned} x_2\eta = \lambda(x-\alpha_1 z)&\sqrt{\beta(x-\alpha_2 z)(x-\alpha_4 z)} \\ &+ \lambda(x-\alpha_2 z)\sqrt{\gamma(x-\alpha_1 z)(x-\alpha_4 z)}\,. \end{aligned}$$

Using this, we get

$$\begin{aligned} x_2\frac{w}{x}\eta = x\left(\lambda\frac{w}{x}\sqrt{\beta(x-\alpha_2 z)(x-\alpha_4 z)}\right) - y\left(\lambda\alpha_1\sqrt{\beta(x-\alpha_2 z)(x-\alpha_4 z)}\right) \\ + x\left(\lambda\frac{w}{x}\sqrt{\gamma(x-\alpha_1 z)(x-\alpha_4 z)}\right) - y\left(\lambda\alpha_2\sqrt{\gamma(x-\alpha_1 z)(x-\alpha_4 z)}\right) \end{aligned}$$

and consequently $x_2(w/x)\eta \in (x,y)R_2$, where R_2 is the finite extension of R obtained by adjoining the various roots occurring in the previous equation and normalizing.

We describe briefly the process of constructing x_n for $n \geqslant 3$. The first step is to write $\sqrt{x-\alpha_4 z}$ in terms of $\sqrt{x-\alpha_1 z}$ and $\sqrt{x-\alpha_2 z}$ as we did for $\sqrt{x-\alpha_3 z}$ above. This enables us to write $\sqrt{x-\alpha_3 z}\sqrt{x-\alpha_4 z}$ as a product of four square roots, each of which is a linear combination of $\sqrt{x-\alpha_1 z}$ and $\sqrt{x-\alpha_2 z}$. We can now repeat the process used to construct x_2, essentially replacing x by $\sqrt{x-\alpha_3 z}$ and z by $\sqrt{x-\alpha_4 z}$. Finally, we can repeat this process indefinitely, obtaining elements x_n with $v(x_n) = 1/2^n$ which annihilate the given element of local cohomology.

REMARK 2.3. The ring R in the previous example can be obtained as a Segre products of rings of lower dimension, and we briefly discuss this point of view.

Let A and B be $\mathbb{N}$-graded normal rings which are finitely generated over a field $A_0 = B_0 = K$. Their *Segre product* is the ring

$$R = A\#B = \bigoplus_{n\geqslant 0} A_n \otimes_K B_n\,,$$

which inherits a natural grading where $R_n = A_n \otimes_K B_n$. If K is algebraically closed, then the tensor product $A \otimes_K B$ is a normal ring, and hence so is its direct summand $A\#B$. If M and N are $\mathbb{Z}$-graded modules over A and B respectively, then their Segre product is the R-module

$$M\#N = \bigoplus_{n\in\mathbb{Z}} M_n \otimes_K N_n\,.$$

Using $\mathfrak{m}$ to denote the homogeneous maximal ideal of R, the local cohomology modules $H^k_{\mathfrak{m}}(R)$ can be computed using the Künneth formula due to Goto and Watanabe, [5, Theorem 4.1.5]:

$$\begin{aligned} H^k_{\mathfrak{m}}(R) = \left(A\#H^k_{\mathfrak{m}_B}(B)\right) \oplus \left(H^k_{\mathfrak{m}_A}(A)\#B\right) \\ \oplus \bigoplus_{i+j=k+1} \left(H^i_{\mathfrak{m}_A}(A)\#H^j_{\mathfrak{m}_B}(B)\right)\,. \end{aligned}$$

It follows that if A and B have positive dimension, then

$$\dim(A\#B) = \dim A + \dim B - 1\,.$$

We claim that the ring R in Example 2.2 is isomorphic to the Segre product $A\#B$, where

$$A = K[a,b,c]/\big(c^2 - \prod_{i=1}^{4}(a-\alpha_i b)\big)$$

is a hypersurface with $\deg a = \deg b = 1$ and $\deg c = 2$, and $B = K[s,t]$ is a standard graded polynomial ring. The map

$$x \longmapsto as\,, \qquad y \longmapsto bt\,, \qquad z \longmapsto bs\,, \qquad w \longmapsto at\,,$$

$$\eta \longmapsto cs^2\,, \qquad (w/x)\eta \longmapsto cst\,, \qquad (w/x)^2\eta \longmapsto ct^2$$

extends to a K-algebra homomorphism $\varphi\colon R \longrightarrow A\#B$. This is a surjective homomorphism of integral domains of equal dimension, so it must be an isomorphism. Since A and B are Cohen-Macaulay rings of dimension 2, the Künneth formula for $H^2_{\mathfrak{m}}(R)$ reduces to

$$H^2_{\mathfrak{m}}(R) = \big(A\#H^2_{\mathfrak{m}_B}(B)\big) \ \oplus\ \big(H^2_{\mathfrak{m}_A}(A)\#B\big)\,.$$

The module $H^2_{\mathfrak{m}_B}(B)$ vanishes in nonnegative degrees, which implies that $A\#H^2_{\mathfrak{m}_B}(B) = 0$. The component of $H^2_{\mathfrak{m}_A}(A)$ in nonnegative degree is the one-dimensional vector space spanned by the degree 0 element

$$\left[\frac{c}{ab}\right] \in H^2_{\mathfrak{m}_A}(A)\,.$$

Hence $H^2_{\mathfrak{m}}(R)$ is the one-dimensional vector space spanned by $[c/ab]\otimes 1$. The search for elements $x_n \in R^+$ of small degree annihilating the image of $H^2_{\mathfrak{m}}(R)$ in $H^2_{\mathfrak{m}}(R^+)$ is essentially the search for homogeneous elements of A^+, of small degree, multiplying c into the ideal $(a,b)A^+$.

EXAMPLE 2.4. Let K be an algebraically closed field of characteristic zero, $\theta \in K$ a primitive cube root of unity, and set

$$A = K[x,y,z]/\big(\theta x^3 + \theta^2 y^3 + z^3\big)\,.$$

Let R be the Segre product of A and the polynomial ring $K[s,t]$. Then R is a normal ring of dimension 3, and the elements sx, ty, $sy+tx$ form a homogeneous system of parameters for R. Using the Künneth formula as in Remark 2.3, the local cohomology module $H^2_{\mathfrak{m}}(R)$ is a one-dimensional vector space spanned by an element corresponding to the relation

$$sztz(sy+tx) = (sz)^2ty + (tz)^2sx\,.$$

To annihilate this relation by an element of R^+ of positive degree $\varepsilon \in \mathbb{Q}$, it suffices to find an element $u \in A^+$ of degree ε such that

$$uz^2 \in (x,y)A^+\,;$$

indeed, if $uz^2 = vx + wy$ for homogeneous $v, w \in A^+$ of degree $1+\varepsilon$, then

$$(s^{\varepsilon}u)(sztz) = (s^{\varepsilon}tv)(sx) + (s^{1+\varepsilon}w)(ty)\,,$$

and $s^{\varepsilon}tv$ and $s^{1+\varepsilon}w$ are easily seen to be integral over S.

We have now reduced our problem to working over the hypersurface A, where we are looking for elements $u \in A^+$ of small degree which annihilate

$$\left[\frac{z^2}{xy}\right] \in H^2_{\mathfrak{m}_A}(A^+)\,.$$

Let A_1 be the extension of A obtained by adjoining x_1, y_1, z_1, where

$$x_1^3 = \theta^{1/3}x + \theta^{2/3}y\,, \qquad y_1^3 = \theta^{1/3}x + \theta^{5/3}y\,, \qquad z_1^3 = \theta^{1/3}x + \theta^{8/3}y\,.$$

Note that x and y can be written as K-linear combinations of x_1^3 and y_1^3. Moreover,

$$\begin{aligned}(x_1y_1z_1)^3 &= \big(\theta^{1/3}x + \theta^{2/3}y\big)\big(\theta^{1/3}x + \theta^{5/3}y\big)\big(\theta^{1/3}x + \theta^{8/3}y\big)\\ &= \theta x^3 + \theta^2 y^3 = -z^3\,,\end{aligned}$$

so z belongs to the K-algebra generated by x_1, y_1, and z_1. Now

$$\begin{aligned}\theta x_1^3 + \theta^2 y_1^3 + z_1^3 &= \theta\big(\theta^{1/3}x + \theta^{2/3}y\big) + \theta^2\big(\theta^{1/3}x + \theta^{5/3}y\big) + \big(\theta^{1/3}x + \theta^{8/3}y\big)\\ &= \big(\theta^{4/3} + \theta^{7/3} + \theta^{1/3}\big)x + \big(\theta^{5/3} + \theta^{11/3} + \theta^{8/3}\big)y = 0\,,\end{aligned}$$

which implies that

$$A_1 = K[x_1, y_1, z_1]/\big(\theta x_1^3 + \theta^2 y_1^3 + z_1^3\big)$$

is a ring isomorphic to A. Thus $A \subset A_1$ gives a finite embedding of A into itself under which the generators of degree 1 go to elements of degree 3; or, in terms of the original degree, the new generators of the homogeneous maximal ideal have degree 1/3. Since $[H^2_{\mathfrak{m}_A}(A)]_0$ is annihilated by all elements of positive degree, the image of $[H^2_{\mathfrak{m}_A}(A)]_0$ in $H^2_{\mathfrak{m}_A}(A_1)$ is annihilated by elements of degree 1/3. Iterating this construction, we conclude that there are elements of arbitrarily small positive degree annihilating the image of $[H^2_{\mathfrak{m}_A}(A)]_0$ in $H^2_{\mathfrak{m}_A}(A^+)$. Quite explicitly, we have a tower of extensions

$$A = A_0 \subset A_1 \subset A_2 \subset \cdots\,, \qquad \text{where} \qquad A_n = K[x_n, y_n, z_n]/\big(\theta x_n^3 + \theta^2 y_n^3 + z_n^3\big)\,.$$

The maps $H^2_{\mathfrak{m}}(A_n) \longrightarrow H^2_{\mathfrak{m}}(A_{n+1})$ preserve degrees, so $[H^2_{\mathfrak{m}_A}(A)]_0$ maps to the socle of $H^2_{\mathfrak{m}}(A_n)$ which is killed by all elements of A_n of positive degree, e.g., by the elements x_n, y_n, z_n which have degree $1/3^n$.

REMARK 2.5. In [9, Theorem 2.9] Heitmann proves that if $(R, \mathfrak{m})$ is a mixed characteristic excellent local domain of dimension three, then the image of $H^2_{\mathfrak{m}}(R)$ in $H^2_{\mathfrak{m}}(R^+)$ is annihilated by arbitrarily small powers of every nonunit. The corresponding statement is false for three-dimensional domains of characteristic zero: for the ring R of Example 2.4, we claim that $\sqrt{sz}$ does not annihilate the image of $H^2_{\mathfrak{m}}(R) \longrightarrow H^2_{\mathfrak{m}}(R^+)$. Because of the splitting provided by field trace, it suffices to verify that

$$\sqrt{sz}\,(sztz) \notin (sx, ty)T\,,$$

where T is any normal subring of R^+ containing $R[\sqrt{sz}]$. Take T to be the Segre product of $\widetilde{A} = A[\sqrt{x}, \sqrt{y}, \sqrt{z}]$ and $\widetilde{B} = B[\sqrt{s}, \sqrt{t}]$. Note that $\widetilde{B}$ is a polynomial ring in $\sqrt{s}$ and $\sqrt{t}$, and that $\widetilde{A}$ is the hypersurface

$$K[\sqrt{x}, \sqrt{y}, \sqrt{z}]/\left(\theta(\sqrt{x})^6 + \theta^2(\sqrt{y})^6 + (\sqrt{z})^6\right).$$

It is enough to check that $\sqrt{sz}\,(sztz) \notin (sx, ty)(\widetilde{A} \otimes_K \widetilde{B})$, and after specializing $\sqrt{s} \longmapsto 1$ and $\sqrt{t} \longmapsto 1$ to check that

$$(\sqrt{z})^5 \notin \left((\sqrt{x})^2, (\sqrt{y})^2\right)\widetilde{A},$$

which is immediately seen to be true. The same argument shows that $\sqrt{sx}$, $\sqrt{sy}$, etc. do not annihilate the image of $H^2_{\mathfrak{m}}(R) \longrightarrow H^2_{\mathfrak{m}}(R^+)$. The situation is quite similar with Example 2.2.

3. Annihilators using the Albanese map

For an $\mathbb{N}$-graded domain R which is finitely generated over a field R_0, let $R^{+\,\mathrm{GR}}$ be the $\mathbb{Q}_{\geqslant 0}$-graded ring generated by those elements of R^+ which can be assigned a degree such that they satisfy a homogeneous equation of integral dependence over R. If R_0 is a field of prime characteristic, Hochster and Huneke [14, Theorem 5.15] proved that the induced map

$$H^i_{\mathfrak{m}}(R) \longrightarrow H^i_{\mathfrak{m}}(R^{+\,\mathrm{GR}})$$

is zero for all $i < \dim R$. Translating to projective varieties, one immediately has the vanishing theorem:

THEOREM 3.1 ([14, Theorem 1.2]). *Let X be an irreducible closed subvariety of $\mathbb{P}^n_K$, where K is a field of positive characteristic. Then for all integers i with $0 < i < \dim X$, and all integers t, there exists a projective variety Y over a finite extension of K with a finite surjective morphism $f\colon Y \longrightarrow X$, such that the induced map*

$$H^i(X, \mathcal{O}_X(t)) \longrightarrow H^i(Y, f^*\mathcal{O}_X(t))$$

is zero.

Over fields of characteristic zero, the corresponding statements are false because of the splitting provided by the field trace. However, the following graded analogue of Question 1.5 remains open.

QUESTION 3.2. Let R be an $\mathbb{N}$-graded domain, finitely generated over a field R_0 of characteristic zero. For $i < \dim R$, is every element of the image of

$$H^i_{\mathfrak{m}}(R) \longrightarrow H^i_{\mathfrak{m}}(R^{+\,\mathrm{GR}})$$

killed by elements of $R^{+\,\mathrm{GR}}$ of arbitrarily small positive degree?

This question, when considered for Segre products, leads to the following:

QUESTION 3.3. Let R be an $\mathbb{N}$-graded domain of dimension d, finitely generated over a field R_0 of characteristic zero. Is the image of

$$\left[H^d_{\mathfrak{m}}(R)\right]_{\geqslant 0} \longrightarrow H^d_{\mathfrak{m}}(R^{+\,\mathrm{GR}})$$

killed by elements of $R^{+\,\mathrm{GR}}$ of arbitrarily small positive degree?

In Examples 2.2 and 2.4, we obtained affirmative answers to Question 3.2 by explicitly constructing the annihilators. In this section, we obtain an affirmative answer for the image of $[H^2_{\mathfrak{m}}(R)]_0$ and also settle Question 3.3 for rings of dimension two. We first recall some basic facts about the relationship between graded rings and very ample divisors.

If R is a standard graded ring, the associated projective scheme $X = \operatorname{Proj} R$ has a very ample line bundle $\mathcal{O}_X(1)$ with sections defined by elements of degree one, which generate the line bundle. Conversely, a very ample line bundle defines a standard graded ring and an embedding of X into projective space.

The strategy is to find, for an arbitrarily large positive integer n, a finite surjective map from an integral projective scheme Y to X, together with an ample line bundle $\mathcal{L}$ on Y, such that $\mathcal{L}^{\otimes n}$ is the pullback of $\mathcal{O}_X(1)$ and such that a section of $\mathcal{L}$ annihilates the pullback of the given element of cohomology. This will essentially be accomplished by mapping X to its Albanese variety, and pulling back by the multiplication by N map for large integers N. The precise result that we prove is as follows.

THEOREM 3.4. *Let R be an $\mathbb{N}$-graded domain which is finitely generated over a field R_0 of characteristic 0. Let $X = \operatorname{Proj} R$ and let η be an element of $H^1(X, \mathcal{O}_X)$. Then, for every $\varepsilon > 0$, there exists a finite extension $R \subseteq S$ of $\mathbb{Q}$-graded domains such that the image of η under the induced map*

$$H^1(X, \mathcal{O}_X) \longrightarrow H^1(Y, \mathcal{O}_Y), \qquad \text{where } Y = \operatorname{Proj} S,$$

is annihilated by an element of S of degree less that ε.

We remark that since $H^1(X, \mathcal{O}_X)$ corresponds to the component of $H^2_{\mathfrak{m}}(R)$ of degree zero, this theorem only implies that we can annihilate elements of $H^2_{\mathfrak{m}}(R)$ of degree zero by elements of small degree. If $H^2_{\mathfrak{m}}(R)$ is generated by its degree zero elements—and this happens in several interesting examples—we can deduce the result for all elements of $H^2_{\mathfrak{m}}(R)$.

Proof. Replacing R by its normalization, it suffices to work throughout with normal rings. We also reduce to the case where R is a standard $\mathbb{N}$-graded ring as follows. Using [6, Lemme 2.1.6], R has a Veronese subring $R^{(t)}$ which is generated by elements of equal degree. The local cohomology of $R^{(t)}$ supported at its homogeneous maximal ideal $\mathfrak{m}$ can be obtained by [5,

Theorem 3.1.1], which states that

$$H^i_{\mathfrak{m}}(R^{(t)}) = \bigoplus_{n\in\mathbb{Z}} [H^i_{\mathfrak{m}}(R)]_{nt}\,.$$

In particular, we have

$$[H^2_{\mathfrak{m}}(R^{(t)})]_0 = [H^2_{\mathfrak{m}}(R)]_0 = H^1(X, \mathcal{O}_X)\,.$$

If elements of this cohomology group can be annihilated in graded finite extensions of $R^{(t)}$, then the same can be achieved in extensions of R.

We next treat the special case where $\operatorname{Proj} R$ is itself an abelian variety, which we denote A. For each integer N, let $[N_A]\colon A \longrightarrow A$ be the morphism corresponding to multiplication by N. Assume further that the very ample sheaf $\mathcal{O}_A(1)$ defining the graded ring R satisfies the condition that

$$[(-1)_A]^*(\mathcal{O}_A(1)) = \mathcal{O}_A(1)\,. \tag{3.4.1}$$

Note that, if $\mathcal{L}$ is any very ample line bundle on A, the new very ample line bundle

$$\mathcal{O}_A(1) = \mathcal{L} \otimes [(-1)_A]^*\mathcal{L}$$

satisfies this further assumption.

We recall two facts about abelian varieties from Mumford [17]:

(1) $H^1(A, \mathcal{O}_A(1)) = 0$. By "The Vanishing Theorem" [17, page 150], given a line bundle $\mathcal{L}$, there is a unique integer i such that $H^i(A, \mathcal{L})$ is nonzero. Since $\mathcal{O}_A(1)$ is very ample, this integer must be 0.

(2) $[N_A]^*(\mathcal{O}_A(1)) = \mathcal{O}_A(N^2)$. This follows from [17, Corollary II.6.3] since we are assuming (3.4.1).

The theorem in this case follows from these two properties: the morphism $[N_A]$ induces a map

$$R = \bigoplus_n \Gamma(A, \mathcal{O}_A(n)) \longrightarrow \bigoplus_n \Gamma(A, [N_A]^*\mathcal{O}_A(n))\,,$$

and, by the second property above,

$$\Gamma(A, [N_A]^*\mathcal{O}_A(n)) = \Gamma(A, \mathcal{O}_A(N^2 n))\,.$$

Thus we have a map of graded rings from R to itself that takes an element of degree 1 to an element of degree N^2. Denote the new copy of R by S, and regrade S with a $\mathbb{Q}$-grading such that the map $R \longrightarrow S$ preserves degrees. This implies that S has elements s of degree $1/N^2$ under the new grading. Such an element s must annihilate the image of $\eta \in H^1(A, \mathcal{O}_A)$, since the product $s \cdot \eta$ lies in $H^1(A, \mathcal{O}_A(1)) = 0$. Hence for each positive integer N, we have found a finite extension of R with an element of degree $1/N^2$ that annihilates the image of η.

The remainder of the proof is devoted to reducing to the previous case. Let R be a graded domain such that $X = \operatorname{Proj} R$ is normal. Let A be the *strict Albanese variety* of X as defined in [1]. It is the dual abelian variety to the Picard variety of X (in the sense of Chevalley-Grothendieck) which parametrizes line bundles algebraically equivalent to 0. Let $\varphi\colon X \longrightarrow A$ be the corresponding Albanese morphism. Then (since the ground field has characteristic 0) φ induces an isomorphism

$$H^1(A, \mathcal{O}_A) \cong H^1(X, \mathcal{O}_X)\,;$$

see Chevalley [1].

Since A is an abelian variety, it has a very ample invertible sheaf $\mathcal{O}_A(1)$; see, for example, [17, pp. 60–62]. After replacing $\mathcal{O}_A(1)$ by $\mathcal{O}_A(1) \otimes [(-1)_A]^*\mathcal{O}_A(1)$ if necessary, we may assume that

$$[(-1)_A]^*(\mathcal{O}_A(1)) \cong \mathcal{O}_A(1)\,.$$

We let $\mathcal{O}_X(1)$ denote the very ample invertible sheaf defined by the grading on R. Let $\pi_1\colon Y_1 \longrightarrow X$ be the pullback of multiplication by N on A, and let $\varphi_1\colon Y_1 \longrightarrow A$ be the map induced by φ, so that we have the fiber product diagram below.

$$\begin{CD} Y_1 @>{\varphi_1}>> A \\ @V{\pi_1}VV @VV{[N_A]}V \\ X @>{\varphi}>> A \end{CD}$$

Let $\mathcal{M}_1 = \varphi_1^*(\mathcal{O}_A(1))$. Then

$$\pi_1^*(\varphi^*(\mathcal{O}_A(1))) = \varphi_1^*([N_A]^*(\mathcal{O}_A(1))) = \varphi_1^*(\mathcal{O}_A(N^2)) = \mathcal{M}_1^{\otimes N^2}\,.$$

Now let m be an integer such that $\varphi^*(\mathcal{O}_A(-1)) \otimes \mathcal{O}_X(m)$ is globally generated; such an m exists since $\mathcal{O}_X(1)$ is ample, and we fix one such m.

Since the sheaf $\varphi^*(\mathcal{O}_A(-1)) \otimes \mathcal{O}_X(m)$ is globally generated, there exists a map $\psi\colon X \longrightarrow \mathbb{P}^n$ such that

$$\psi^*(\mathcal{O}_{\mathbb{P}^n}(1)) = \varphi^*(\mathcal{O}_A(-1)) \otimes \mathcal{O}_X(m)\,.$$

Let $\alpha\colon \mathbb{P}^n \longrightarrow \mathbb{P}^n$ be a finite map such that $\alpha^*(\mathcal{O}_{\mathbb{P}^n}(1)) = \mathcal{O}_{\mathbb{P}^n}(N)$; for example, we can take α to be the map defined by the ring homomorphism on a polynomial ring that sends the variables to their N-th powers.

Let Y_2 be the fiber product of ψ and α, which gives us a diagram

$$\begin{CD} Y_2 @>{\varphi_2}>> \mathbb{P}^n \\ @V{\pi_2}VV @VV{\alpha}V \\ X @>{\psi}>> \mathbb{P}^n\,. \end{CD}$$

Let $\mathcal{M}_2 = \varphi_2^*(\mathcal{O}_{\mathbb{P}^n}(1))$. We then have

$$\begin{aligned}\pi_2^*(\varphi^*(\mathcal{O}_A(-1)) \otimes \mathcal{O}_X(m)) \cong \pi_2^*(\psi^*(\mathcal{O}_{\mathbb{P}^n}(1))) &= \varphi_2^*(\alpha^*(\mathcal{O}_{\mathbb{P}^n}(1))) \\ &\cong \varphi_2^*(\mathcal{O}_{\mathbb{P}^n}(N)) \cong \mathcal{M}_2^{\otimes N}.\end{aligned}$$

Note that the above morphisms $\pi_i \colon Y_i \longrightarrow X$, $i = 1, 2$ are *finite and surjective.*

Let Y be a component of the normalization of the reduced fibre product of $\pi_1 \colon Y_1 \longrightarrow X$ and $\pi_2 \colon Y_2 \longrightarrow X$, with the property that $Y \longrightarrow X$ is surjective. Since $Y_1 \times_X Y_2 \longrightarrow X$ is finite and surjective, some irreducible component of this fiber product maps onto X via a finite morphism, and we may take Y to be the normalization of any such component.

We then have an induced finite surjective map $\pi \colon Y \longrightarrow X$, and induced maps $\mu_1 \colon Y \longrightarrow Y_1$ and $\mu_2 \colon Y \longrightarrow Y_2$, giving a commutative diagram

$$\begin{array}{ccc} Y & \xrightarrow{\mu_1} & Y_1 \\ {\scriptstyle\mu_2}\downarrow & \searrow^{\pi} & \downarrow{\scriptstyle\pi_1} \\ Y_2 & \xrightarrow{\pi_2} & X. \end{array}$$

By construction, we have

$$\begin{aligned}\pi^*\mathcal{O}_X(m) &= \pi^*(\varphi^*\mathcal{O}_A(1) \otimes \varphi^*\mathcal{O}_A(-1) \otimes \mathcal{O}_X(m)) \\ &= \mu_1^*\pi_1^*(\varphi^*(\mathcal{O}_A(1))) \otimes \mu_2^*\pi_2^*(\varphi^*(\mathcal{O}_A(-1)) \otimes \mathcal{O}_X(m)) \\ &= \mu_1^*(\mathcal{M}_1^{\otimes N^2}) \otimes \mu_2^*(\mathcal{M}_2^{\otimes N}) = \mathcal{M}^{\otimes N},\end{aligned}$$

where $\mathcal{M} = \mu_1^*\mathcal{M}_1^{\otimes N} \otimes \mu_2^*\mathcal{M}_2$.

Now $\mathcal{M}_1 = \varphi_1^*(\mathcal{O}_A(1))$ is generated by global sections of the form $\varphi_1^*(u)$ with $u \in H^0(A, \mathcal{O}_A(1))$. Choose any such element u such that its image in $H^0(Y, \mu_1^*\mathcal{M}_1)$ is nonzero. Let v be a nonzero element of $H^0(Y, \mu_2^*\mathcal{M}_2)$, and let s be the image of $\mu_1^*\varphi_1^*(u^N) \otimes v$ in $H^0(Y, \mathcal{M})$. Then s is a nonzero section of $\mathcal{M}$, and since $\pi^*(\mathcal{O}_X(m)) = \mathcal{M}^{\otimes N}$, the degree of s in the grading induced from that on R is m/N. We claim that the composition

$$H^1(X, \mathcal{O}_X) \xrightarrow{\ \pi^*\ } H^1(Y, \mathcal{O}_Y) \xrightarrow{\ \cdot s\ } H^1(Y, \mathcal{M})$$

vanishes. For this, it suffices to show that the composition

$$H^1(X, \mathcal{O}_X) \xrightarrow{\ \pi^*\ } H^1(Y, \mathcal{O}_Y) \xrightarrow{\ \mu_1^*\varphi_1^*(u)\ } H^1(Y, \mu_1^*\mathcal{M}_1)$$

vanishes, which in turn reduces to showing that

$$H^1(X, \mathcal{O}_X) \xrightarrow{\ \pi_1^*\ } H^1(Y_1, \mathcal{O}_{Y_1}) \xrightarrow{\ \varphi_1^*(u)\ } H^1(Y_1, \mathcal{M}_1)$$

vanishes. Since $\varphi^* \colon H^1(A, \mathcal{O}_A) \longrightarrow H^1(X, \mathcal{O}_X)$ is an isomorphism, we further reduce to showing that

$$H^1(A, \mathcal{O}_A) \xrightarrow{\ [N_A]^*\ } H^1(A, \mathcal{O}_A) \xrightarrow{\ \cdot u\ } H^1(A, \mathcal{O}_A(1))$$

vanishes. But this is true since $H^1(A, \mathcal{O}_A(1)) = 0$.

Since we can make m/N arbitrarily small by choosing N large, this completes the proof. □

As a corollary, we see that the answer to Question 3.3 is affirmative for rings of dimension two:

COROLLARY 3.5. *Let R be an $\mathbb{N}$-graded domain of dimension 2, which is finitely generated over a field R_0 of characteristic zero. Then the image of*

$$\left[H^2_{\mathfrak{m}}(R)\right]_{\geqslant 0} \longrightarrow H^2_{\mathfrak{m}}(R^{+\,\mathrm{GR}})$$

is killed by elements of $R^{+\,\mathrm{GR}}$ of arbitrarily small positive degree.

Proof. By adjoining roots of elements if necessary, we may assume that R has a system of parameters x, y consisting of linear forms. Theorem 3.4 implies that the image of $[H^2_{\mathfrak{m}}(R)]_0$ is killed by elements of $R^{+\,\mathrm{GR}}$ of arbitrarily small positive degree, so it suffices to prove that $[H^2_{\mathfrak{m}}(R)]_{\geqslant 0}$ is the R-module generated by $[H^2_{\mathfrak{m}}(R)]_0$.

Since x and y are linear forms, we have $[R_{xy}]_{n+1} = [R_{xy}]_n \cdot R_1$ for all integers n. Computing $H^2_{\mathfrak{m}}(R)$ using the Čech complex on x, y, it follows that

$$\left[H^2_{\mathfrak{m}}(R)\right]_{n+1} = \left[H^2_{\mathfrak{m}}(R)\right]_n \cdot R_1 \qquad \text{for all } n \in \mathbb{Z}\,.$$ □

4. Closure operations

The issues discussed here are closely related to closure operations considered by Hochster and Huneke, and by Heitmann. The *plus closure* of an ideal $\mathfrak{a}$ of a domain R is defined as $\mathfrak{a}^+ = \mathfrak{a}R^+ \cap R$. It has desirable properties for rings of prime characteristic; e.g., it bounds colon ideals on systems of parameters: if $x_1, \dots, x_d$ is a system of parameters for an excellent local domain R containing a field of prime characteristic, then

$$(x_1, \dots, x_{i-1}) :_R x_i \subseteq (x_1, \dots, x_{i-1})^+ \qquad \text{for all } i\,.$$

In general, plus closure does not have this colon-capturing property for rings of characteristic zero or of mixed characteristic. Several alternative closure operations are defined by Heitmann in [7] including the *extended plus closure.* Building on these ideas, he settled the Direct Summand Conjecture for mixed characteristic rings of dimension three [8]. In [9, Theorem 1.3] Heitmann proved that extended plus closure has the colon-capturing property for arbitrary sets of three parameters in excellent domains of mixed characteristic.

Let $(R, \mathfrak{m})$ be a complete local domain and fix, as usual, a valuation $v\colon R \longrightarrow \mathbb{Z} \cup \{\infty\}$ which is positive on $\mathfrak{m}$ and extend to $v\colon R^+ \longrightarrow \mathbb{Q} \cup \{\infty\}$. In [13] Hochster and Huneke define the *dagger closure* $\mathfrak{a}^\dagger$ of an ideal $\mathfrak{a}$ as the ideal consisting of all elements $x \in R$ for which there exist elements $u \in R^+$, of arbitrarily small positive order, with $ux \in \mathfrak{a}R^+$. In [13, Theorem 3.1] it is proved that the dagger closure $\mathfrak{a}^\dagger$ agrees with the tight closure $\mathfrak{a}^*$ for ideals of complete local domains of prime characteristic; see also [12, § 6]. While tight closure is defined in characteristic zero by reduction to prime characteristic, the definition of dagger closure is characteristic-free.

QUESTION 4.1 (Hochster-Huneke). Does the dagger closure operation have the colon-capturing property, i.e., if $x_1, \ldots, x_d$ is a system of parameters for a complete local domain R, is it true that

$$(x_1, \ldots, x_{i-1}) :_R x_i \subseteq (x_1, \ldots, x_{i-1})^\dagger\,?$$

According to Hochster and Huneke [13, page 244] "it is important to raise (and answer) this question." If Question 4.1 has an affirmative answer, then so does Question 1.5.

Consider the hypersurface $K[[x, y, z]]/(x^3 + y^3 + z^3)$. If K has prime characteristic, a straightforward calculation—performed in many an introductory lecture on tight closure theory—shows that $z^2 \in (x, y)^*$. If K has characteristic zero, the "reduction modulo p" nature of the definition of tight closure [12, § 3] immediately yields $z^2 \in (x, y)^*$ once again. In contrast, the computation that $z^2 \in (x, y)^\dagger$ is quite delicate and, aside from the linear change of variables, is the computation we performed in Example 2.4. While concrete descriptions of the multipliers of small order are available in this and some other examples, dagger closure remains quite mysterious even in simple examples such as diagonal hypersurfaces:

QUESTION 4.2. Let K be a field of characteristic zero, and let

$$R = K[[x_0, \ldots, x_d]]/(x_0^n + \cdots + x_d^n), \qquad \text{where } n > d.$$

Does x_0^d belong to the dagger closure of the ideal $(x_1, \ldots, x_d)$?

A routine computation of tight closure shows that $x_0^d \in (x_1, \ldots, x_d)^*$. In the case $d = 2$, we have $x_0^2 \in (x_1, x_2)^\dagger$ by Corollary 3.5.

REFERENCES

[1] C. Chevalley, *Sur la théorie de la variété de Picard*, Amer. J. Math. **82** (1960), 435–490. MR 0118723 (22 #9494)

[2] E. G. Evans and P. Griffith, *The syzygy problem*, Ann. of Math. (2) **114** (1981), 323–333. MR 632842 (83i:13006)

[3] G. Faltings, *Almost étale extensions*, Astérisque (2002), 185–270, Cohomologies p-adiques et applications arithmétiques, II. MR 1922831 (2003m:14031)

[4] O. Gabber and L. Ramero, *Almost ring theory*, Lecture Notes in Mathematics, vol. 1800, Springer-Verlag, Berlin, 2003. MR 2004652 (2004k:13027)

[5] S. Goto and K. Watanabe, *On graded rings. I*, J. Math. Soc. Japan **30** (1978), 179–213. MR 494707 (81m:13021)

[6] A. Grothendieck, *Éléments de géométrie algébrique. II. Étude globale élémentaire de quelques classes de morphismes*, Inst. Hautes Études Sci. Publ. Math. **8** (1961). MR 0217084 (36 #177b)

[7] R. C. Heitmann, *Extensions of plus closure*, J. Algebra **238** (2001), 801–826. MR 1823785 (2002a:13005)

[8] ———, *The direct summand conjecture in dimension three*, Ann. of Math. (2) **156** (2002), 695–712. MR 1933722 (2003m:13008)

[9] ———, *Extended plus closure and colon-capturing*, J. Algebra **293** (2005), 407–426. MR 2172347 (2007d:13005)

[10] M. Hochster, *Topics in the homological theory of modules over commutative rings*, CBMS Regional Conference, Lincoln, Neb., 1974, Conference Board of the Mathematical Sciences Regional Conference Series in Mathematics, vol. 24, American Mathematical Society, Providence, R.I., 1975. MR 0371879 (51 #8096)

[11] ———, *Canonical elements in local cohomology modules and the direct summand conjecture*, J. Algebra **84** (1983), 503–553. MR 723406 (85j:13021)

[12] M. Hochster and C. Huneke, *Tight closure, invariant theory, and the Briançon-Skoda theorem*, J. Amer. Math. Soc. **3** (1990), 31–116. MR 1017784 (91g:13010)

[13] ———, *Tight closure and elements of small order in integral extensions*, J. Pure Appl. Algebra **71** (1991), 233–247. MR 1117636 (92i:13002)

[14] ———, *Infinite integral extensions and big Cohen-Macaulay algebras*, Ann. of Math. (2) **135** (1992), 53–89. MR 1147957 (92m:13023)

[15] C. Huneke and G. Lyubeznik, *Absolute integral closure in positive characteristic*, Adv. Math. **210** (2007), 498–504. MR 2303230

[16] S. Izumi, *A measure of integrity for local analytic algebras*, Publ. Res. Inst. Math. Sci. **21** (1985), 719–735. MR 817161 (87i:32014)

[17] D. Mumford, *Abelian varieties*, Tata Institute of Fundamental Research Studies in Mathematics, No. 5, Tata Institute of Fundamental Research, Bombay, 1970. MR 0282985 (44 #219)

[18] C. Peskine and L. Szpiro, *Dimension projective finie et cohomologie locale. Applications à la démonstration de conjectures de M. Auslander, H. Bass et A. Grothendieck*, Inst. Hautes Études Sci. Publ. Math. (1973), 47–119. MR 0374130 (51 #10330)

[19] P. Roberts, *Le théorème d'intersection*, C. R. Acad. Sci. Paris Sér. I Math. **304** (1987), 177–180. MR 880574 (89b:14008)

[20] J.-P. Serre, *Local algebra*, Springer Monographs in Mathematics, Springer-Verlag, Berlin, 2000. MR 1771925 (2001b:13001)

[21] J. T. Tate, *p-divisible groups*, Proc. Conf. Local Fields (Driebergen, 1966), Springer, Berlin, 1967, pp. 158–183. MR 0231827 (38 #155)

Paul Roberts, Department of Mathematics, University of Utah, 155 South 1400 East, Salt Lake City, UT 84112, USA
E-mail address: roberts@math.utah.edu

Anurag K. Singh, Department of Mathematics, University of Utah, 155 South 1400 East, Salt Lake City, UT 84112, USA
E-mail address: singh@math.utah.edu

V. Srinivas, School of Mathematics, Tata Institute of Fundamental Research, Homi Bhabha Road, Mumbai 400005, India
E-mail address: srinivas@math.tifr.res.in

Illinois Journal of Mathematics
Volume 51, Number 1, Spring 2007, Pages 255–285
S 0019-2082

SEMIDUALIZING MODULES AND THE DIVISOR CLASS GROUP

SEAN SATHER-WAGSTAFF

Dedicated to Phillip Griffith on the occasion of his retirement

ABSTRACT. Among the finitely generated modules over a Noetherian ring R, the semidualizing modules have been singled out due to their particularly nice duality properties. When R is a normal domain, we exhibit a natural inclusion of the set of isomorphism classes of semidualizing R-modules into the divisor class group of R. After a description of the basic properties of this inclusion, it is employed to investigate the structure of the set of isomorphism classes of semidualizing R-modules. In particular, this set is described completely for determinantal rings over normal domains.

1. Introduction

Semidualizing modules arise naturally in the investigations of various duality theories in commutative algebra. One instance of this is Grothendieck and Hartshorne's local duality wherein a dualizing module, or more generally a dualizing complex, is employed to study local cohomology [30], [31]. Another instance is Auslander and Bridger's methodical study of duality properties with respect to a rank 1 free module that gives rise to the Gorenstein dimension [3], [4]. A free module of rank 1 and a dualizing module are both examples of semidualizing modules.

Let R be a Noetherian ring. A finitely generated R-module C is *semidualizing* if the natural homothety map $R \to \operatorname{Hom}_R(C, C)$ is an isomorphism and $\operatorname{Ext}^i_R(C, C) = 0$ for each integer $i > 0$. The study of such modules in the abstract was initiated by Foxby [21] and Golod [28], where they were called "suitable" modules, and has been continued recently by others; see, for example, [12], [24], [26], [27].

Received November 17, 2005; received in final form December 20, 2006.
2000 *Mathematics Subject Classification.* Primary 13C05, 13C13, 13C20. Secondary 13C40, 13D05, 13D25.
This research was conducted in part while the author was an NSF Mathematical Sciences Postdoctoral Research Fellow and a visitor at the University of Nebraska-Lincoln.

The semidualizing modules and more generally the semidualizing *complexes* are useful, for example, in identifying local homomorphisms of finite Gorenstein dimension with particularly nice properties as in [7], [22]. This utility, along with our desire to expand upon it, motivates our investigation of the basic properties of such modules and the structure of the entire set of isomorphism classes of semidualizing R-modules, which we denote $\mathfrak{S}_0(R)$. Surprisingly little is known about his set. For instance, researchers in this subject have been grappling with the following open question for several years now; see [13, (1)] for recent progress.

QUESTION 1.1. When R is local, must $\mathfrak{S}_0(R)$ be finite?

This work is part of a research effort focused on determining the overall structure of $\mathfrak{S}_0(R)$. Much of the ground work for this effort, motivated by [12], [26], is found in [23]. Initial evidence of the richness of the structure of this set is found in the fact that it admits an ordering described in terms of a reflexivity relation; see 2.4. Further structure is uncovered in [24], where numerical data from this ordering is used to build a nontrivial metric on $\mathfrak{S}_0(R)$. While the existence of a metric does not itself provide answers to any of the open questions about the structure of $\mathfrak{S}_0(R)$, it represents a new perspective from which to view this set. This perspective has proved particularly useful for identifying questions that we would not have otherwise thought to ask. For instance, our investigation into the nontriviality of the metric led us to the fact that, when R is Cohen-Macaulay and $\mathfrak{S}_0(R)$ is nontrivial, there exist elements of $\mathfrak{S}_0(R)$ that are incomparable under the reflexivity ordering; see [24, (3.5)].

In the current paper, we forward another new perspective from which to investigate the set $\mathfrak{S}_0(R)$. It is motivated by Bruns' work [9], wherein the divisor class group is used to describe the dualizing module of certain Cohen-Macaulay normal domains. Accordingly, when R is a normal domain, we exhibit a natural inclusion $\mathfrak{S}_0(R) \hookrightarrow \mathrm{Cl}(R)$ that behaves well with respect to standard operations. This inclusion allows us to exploit the known behavior of $\mathrm{Cl}(R)$ to gain insight into the structure of $\mathfrak{S}_0(R)$. For instance, if $\mathrm{Cl}(R)$ is finite, then $\mathfrak{S}_0(R)$ is also finite. The basic results from this analysis are presented in Section 3.

Section 4 contains the meat of this investigation and demonstrates the power of this new perspective. It consists of analyses showing how the divisor class group can be used to give a complete description of the set of semidualizing modules for certain classes of rings. We recount here three such situations, focusing our attention on the finiteness and size of $\mathfrak{S}_0(R)$. First, taking our cues from [9], we describe the semidualizing modules over a determinantal ring in Theorem 4.5.

THEOREM 1.2. *Let A be a normal domain and m, n, r nonnegative integers such that $r < \min\{m, n\}$. With $\mathbf{X} = \{X_{ij}\}$ an $m \times n$ matrix of variables, set*

$R = A[\mathbf{X}]/I_{r+1}(\mathbf{X})$, where $I_{r+1}(\mathbf{X})$ is the ideal generated by the minors of $\mathbf{X}$ of size $r+1$. The set $\mathfrak{S}_0(R)$ is finite if and only if $\mathfrak{S}_0(A)$ is so. More specifically, one has the following cases.

(a) *If $r = 0$ or $m = n$, then there is a bijection $\mathfrak{S}_0(R) \approx \mathfrak{S}_0(A)$.*
(b) *If $r > 0$ and $m \neq n$, then there is a bijection $\mathfrak{S}_0(R) \approx \mathfrak{S}_0(A) \times \{0, 1\}$.*

When A is a graded Cohen-Macaulay (super-)normal domain with A_0 local (and complete), this result extends to the localization (and to the completion) of R at its graded maximal ideal; see Corollaries 4.7 and 4.8.

The next two results demonstrate how this technique yields information about rings that are themselves not normal domains; they are contained in Corollary 4.11. Theorem 1.4 is new even when B is a field.

THEOREM 1.3. *Let $A = \coprod_{i \geq 0} A_i$ be a graded super-normal domain with A_0 local and complete. Let $\mathfrak{n}$ be the graded maximal ideal of A and $\widehat{A}$ the $\mathfrak{n}$-adic completion of A. Let $\mathbf{y} = y_1, \ldots, y_q \in \mathfrak{n}A_\mathfrak{n}$ be an $A_\mathfrak{n}$-sequence and fix an integer $m \geq 1$. There are bijections*

$$\mathfrak{S}_0(\widehat{A}/(\mathbf{y})^m) \approx \mathfrak{S}_0(A_\mathfrak{n}/(\mathbf{y})^m) \approx \begin{cases} \mathfrak{S}_0(A_\mathfrak{n}) & \text{if } m = 1 \text{ or } q = 1 \\ \mathfrak{S}_0(A_\mathfrak{n}) \times \{0,1\} & \text{if } m, q > 1. \end{cases}$$

THEOREM 1.4. *With A as in Theorem 1.3, let B denote either $A_\mathfrak{n}$ or $\widehat{A}$, and let t be a positive integer. For $l = 1, \ldots, t$ fix a positive integer q_l and set*

$$S = (B \ltimes B^{q_1}) \otimes_B \cdots \otimes_B (B \ltimes B^{q_t}).$$

If s is the number of indices l with $q_l > 1$, then there is a bijection

$$\mathfrak{S}_0(B) \times \{0,1\}^s \approx \mathfrak{S}_0(S).$$

The statements of our main results are module-theoretic in nature. However, we often employ tools from the derived category. We include a summary of the relevant notions in Section 2 along with basic facts about semidualizing modules and the divisor class group.

2. Background

In this paper, the term "ring" is used for a commutative Noetherian ring with identity, and "module" is used for a unital module. Let R be a ring.

2.1. An *R-complex* is a sequence of R-module homomorphisms

$$X = \cdots \xrightarrow{\partial^X_{i+1}} X_i \xrightarrow{\partial^X_i} X_{i-1} \xrightarrow{\partial^X_{i-1}} \cdots$$

with $\partial^X_i \partial^X_{i+1} = 0$ for each i. We work occasionally in the derived category $\mathsf{D}(R)$ whose objects are the R-complexes; references on the subject include [25], [30], [36], [37]. The category of R-modules $\mathrm{Mod}(R)$ is naturally identified with the full subcategory of $\mathsf{D}(R)$ whose objects are the complexes homologically concentrated in degree 0. For R-complexes X and Y the left derived tensor product complex is denoted $X\otimes^{\mathbf{L}}_R Y$ and the right derived homomorphism complex is $\mathbf{R}\operatorname{Hom}_R(X,Y)$. For an integer n, the nth *shift* or *suspension* of X is denoted $\Sigma^n X$, where $(\Sigma^n X)_i = X_{i-n}$ and $\partial_i^{\Sigma^n X} = (-1)^n \partial_{i-n}^X$. The symbol "$\simeq$" indicates an isomorphism in $\mathsf{D}(R)$, and "$\sim$" is isomorphism up to shift.

A complex X is *homologically finite*, respectively *homologically degreewise finite*, if its total homology module $\mathrm{H}(X)$, respectively each individual homology module $\mathrm{H}_i(X)$, is a finite R-module. The *infimum*, *supremum*, and *amplitude* of X are

$$\begin{aligned}\inf(X) &= \inf\{i \in \mathbb{Z} \mid \mathrm{H}_i(X) \neq 0\},\\ \sup(X) &= \sup\{i \in \mathbb{Z} \mid \mathrm{H}_i(X) \neq 0\},\\ \operatorname{amp}(X) &= \sup(X) - \inf(X),\end{aligned}$$

respectively, with the conventions $\inf\emptyset = \infty$ and $\sup\emptyset = -\infty$. When R is local with residue field k, the *depth* of a homologically finite complex X is

$$\operatorname{depth}_R X = -\sup(\mathbf{R}\operatorname{Hom}_R(k,X)).$$

The *Bass series* of X is the formal Laurent series $I_R^X(t) = \sum_i \mu_R^i(X)t^i$, where

$$\mu_R^i(X) = \operatorname{rank}_k \mathrm{H}_{-i}(\mathbf{R}\operatorname{Hom}_R(k,X))$$

for each integer i. From Foxby [20, (13.11)] the quantity $\operatorname{id}_R X$ is finite if and only if $I_R^X(t)$ is a Laurent polynomial.

2.2. Associated to a complex K is a natural homothety morphism

$$\chi_K^R\colon R \to \mathbf{R}\operatorname{Hom}_R(K,K).$$

When K is homologically finite, it is *semidualizing* if χ_K^R is an isomorphism. A complex D is *dualizing* if it is semidualizing and has finite injective dimension. The set of shift-isomorphism classes of semidualizing R-complexes is denoted $\mathfrak{S}(R)$, and the class of a semidualizing complex K in $\mathfrak{S}(R)$ is denoted $[K]_R$ or simply $[K]$ when there is no danger of confusion. The ring R is $\mathfrak{S}$-*finite* if $\mathfrak{S}(R)$ is a finite set.

A finitely generated R-module C is *semidualizing* if the natural homothety map $R \to \operatorname{Hom}_R(C,C)$ is an isomorphism and $\operatorname{Ext}_R^{\geqslant 1}(C,C) = 0$. The module R is semidualizing. When R is Cohen-Macaulay, a *dualizing module* (or *canonical module*) is a semidualizing module of finite injective dimension. The set of isomorphism classes of semidualizing R-modules is denoted $\mathfrak{S}_0(R)$. The identification of $\operatorname{Mod}(R)$ with a subcategory of $\mathsf{D}(R)$ provides a natural inclusion $\mathfrak{S}_0(R) \hookrightarrow \mathfrak{S}(R)$, and we identify $\mathfrak{S}_0(R)$ with its image in $\mathfrak{S}(R)$. In particular, the class of a semidualizing module C in $\mathfrak{S}_0(R)$ is denoted $[C]_R$ or $[C]$. The ring R is $\mathfrak{S}_0$-*finite* if $\mathfrak{S}_0(R)$ is a finite set.

Some of our favorite ring theoretic properties have characterizations in terms of semidualizing objects. If R is Cohen-Macaulay local, then $\mathfrak{S}(R) = \mathfrak{S}_0(R)$. If R is Gorenstein local, then $\mathfrak{S}(R) = \{[R]\}$. The converses hold when R admits a dualizing complex; see Christensen [12, (3.7),(8.6)].

2.3. Let K be a semidualizing complex. A homologically finite complex X is *K-reflexive* if $\mathbf{R}\mathrm{Hom}_R(X, K)$ is homologically bounded and the natural biduality morphism

$$\delta_X^K \colon X \to \mathbf{R}\mathrm{Hom}_R(\mathbf{R}\mathrm{Hom}_R(X, K), K)$$

is an isomorphism. For instance, the complexes R and K are both K-reflexive. When $\dim(R)$ is finite, the complex K is dualizing if and only if every homologically finite R-complex is K-reflexive. The G_K*-dimension* of X is

$$\mathrm{G}_K\text{-}\dim_R X = \begin{cases} \inf K - \inf \mathbf{R}\mathrm{Hom}_R(X, K) & \text{when } X \text{ is } K\text{-reflexive} \\ \infty & \text{otherwise.} \end{cases}$$

If R is local and X is K-reflexive, then the AB formula [12, (3.14)] reads

$$\mathrm{G}_K\text{-}\dim_R X = \operatorname{depth} R - \operatorname{depth}_R X.$$

When C is a semidualizing module, the G_C-dimension of a finitely generated R-module M can be described in terms of resolutions. We first describe the modules used in the resolutions. A finitely generated R-module G is *totally C-reflexive* if the natural biduality map $G \to \mathrm{Hom}_R(\mathrm{Hom}_R(G, C), C)$ is bijective, and $\mathrm{Ext}_R^{\geqslant 1}(G, C) = 0 = \mathrm{Ext}_R^{\geqslant 1}(\mathrm{Hom}_R(G, C), C)$. A finitely generated R-module M then has finite G_C-dimension if and only if it admits a resolution

$$0 \to G_g \to \cdots \to G_0 \to M \to 0$$

with each G_i totally C-reflexive; the G_C-dimension of M is then the minimum integer g for which M admits such a resolution.

2.4. The above notion of reflexivity gives rise to orderings on the sets $\mathfrak{S}_0(R)$ and $\mathfrak{S}(R)$: write $[C] \trianglelefteq [C']$ whenever C' is C-reflexive. This ordering is trivially reflexive: $[C] \trianglelefteq [C]$. Also, when R is local, the ordering is antisymmetric: if $[C] \trianglelefteq [C']$ and $[C'] \trianglelefteq [C]$, then $[C] = [C']$; see [2, (5.3)]. The question of the transitivity of this ordering has been of interest in this area for some time:

QUESTION 2.5. If $[C] \trianglelefteq [C']$ and $[C'] \trianglelefteq [C'']$, then must one have $[C] \trianglelefteq [C'']$?

2.6. For $1, \dots, n$ let U_i be a set with a relation $\trianglelefteq_i$. The Cartesian product $U_1 \times \cdots \times U_n$ is endowed with the *product relation*: $(u_1, \dots, u_n) \trianglelefteq (u'_1, \dots, u'_n)$ when $u_i \trianglelefteq_i u'_i$ for each $i = 1, \dots, n$. It follows immediately from the definition that $\trianglelefteq$ is transitive if and only if each $\trianglelefteq_i$ is transitive. A map $\alpha \colon U_1 \to U_2$ is *order-respecting* when $u \trianglelefteq_1 u'$ implies $\alpha(u) \trianglelefteq_2 \alpha(u')$, and it is *perfectly order-respecting* when the converse also holds. Observe that, when α is a perfectly order-respecting bijection, the relations $\trianglelefteq_1, \trianglelefteq_2$ are simultaneously transitive. The symbol $\approx$ indicates a perfectly order-respecting bijection.

We pose one final question in this section, motivated by the following well-known equality of Hilbert-Samuel multiplicities: If R is a local Cohen-Macaulay ring with dualizing module ω, then $e(\omega) = e(R)$.

QUESTION 2.7. Let $(R, \mathfrak{m})$ be a local ring and I an $\mathfrak{m}$-primary ideal. If C is a semidualizing R-module, must there be an equality $e(I, C) = e(I, R)$?

Consult Matsumura [34, §14] for the basics of the Hilbert-Samuel multiplicity. We answer Question 2.7 in the affirmative for several classes of rings in Corollaries 3.14 and 4.11. To do so, we need the following lemma, which addresses this question when R is generically Gorenstein.

LEMMA 2.8. *Let $(R, \mathfrak{m})$ be a local ring and C a semidualizing R-module.*

(a) *If $R_\mathfrak{p}$ is Gorenstein for each $\mathfrak{p} \in \operatorname{Spec}(R)$ with $\dim(R/\mathfrak{p}) = \dim(R)$, then $e(J, C) = e(J, R)$ for each $\mathfrak{m}$-primary ideal J.*

(b) *Assume that R is equidimensional and, for each $\mathfrak{p} \in \operatorname{Min}(R)$, the rings $R_\mathfrak{p}$ and $R_\mathfrak{p}/\mathfrak{p}R_\mathfrak{p} \otimes_R \widehat{R}$ are Gorenstein. If $\mathbf{y} \in \mathfrak{m}$ is an R-sequence and $R' = R/\mathbf{y}$, then $e(J, C \otimes R') = e(J, R')$ for each $\mathfrak{m}R'$-primary ideal J.*

Proof. (a) Set $\operatorname{Minh}(R) = \{\mathfrak{p} \in \operatorname{Spec}(R) \mid \dim(R/\mathfrak{p}) = \dim(R)\}$. For each $\mathfrak{p} \in \operatorname{Minh}(R)$, there is an isomorphism $C_\mathfrak{p} \cong R_\mathfrak{p}$ by [12, (8.6)] since $R_\mathfrak{p}$ is Gorenstein. This provides the second equality in the following sequence

$$e(J, C) = \sum_{\mathfrak{p} \in \operatorname{Minh}(R)} \operatorname{length}(C_\mathfrak{p}) e(J, R/\mathfrak{p}) = \sum_{\mathfrak{p} \in \operatorname{Minh}(R)} \operatorname{length}(R_\mathfrak{p}) e(J, R/\mathfrak{p}) = e(J, R),$$

while the others are the additivity formulas for multiplicies [10, (4.7.t)].

(b) Recall the following fact: If $(A, \mathfrak{m}_A) \to (B, \mathfrak{m}_B)$ is a flat local homomorphism such that $\mathfrak{m}_B = \mathfrak{m}_A B$, and if M is a finite A module, then for each $\mathfrak{m}_A$-primary ideal I there is an equality of Hilbert functions

$$\operatorname{length}_A(M/I^n M) = \operatorname{length}_B((M \otimes_A B)/(IB)^n(M \otimes_A B)),$$

which yields an equality of multiplicities; see [32, (2.3)]:

$$e(I, M) = e(IB, M \otimes_A B). \tag{2.8.1}$$

Next, from [29, (0.10.3.1)] one has a flat local homomorphism $\varphi\colon (R, \mathfrak{m}) \to (S, \mathfrak{n})$ such that S has infinite residue field and $\mathfrak{n} = \mathfrak{m}S$. Since S is flat over R, the sequence $\mathbf{y}$ is S-regular. Set $S' = S/(\mathbf{y})S$ and let $\tau\colon S \to S'$ denote the natural surjection. Note that the induced map $R' \to S'$ is flat and local and that the maximal ideal of R' extends to that of S'. It follows that the ideal JS' is $\mathfrak{m}S'$-primary.

For every $\mathfrak{q} \in \operatorname{Minh}(S)$, the local ring $S_{\mathfrak{q}}$ is Gorenstein. To see this, fix $\mathfrak{q} \in \operatorname{Minh}(S)$ and set $\mathfrak{p} = \mathfrak{q} \cap R$. Since φ is faithfully flat, the going-down theorem implies $\operatorname{ht}(\mathfrak{p}) \leq \operatorname{ht}(\mathfrak{q}) = 0$, and so $\mathfrak{p} \in \operatorname{Min}(R)$. By assumption, the rings $R_{\mathfrak{p}}/\mathfrak{p}R_{\mathfrak{p}} \otimes_R \widehat{R}$ and $S/\mathfrak{m}S$ are Gorenstein, and it follows from [6, Main Theorem] that the fibre $R_{\mathfrak{p}}/\mathfrak{p}R_{\mathfrak{p}} \otimes_R S$ is Gorenstein. Thus, the induced map $\varphi_{\mathfrak{q}} \colon R_{\mathfrak{p}} \to S_{\mathfrak{q}}$ is flat and local with Gorenstein source and Gorenstein closed fibre. This implies that $S_{\mathfrak{q}}$ is Gorenstein.

The ring S' has a parameter ideal $\mathbf{x} = (x_1, \ldots, x_r)S' \subseteq JS'$ such that

$$(2.8.2) \quad e(JS', C \otimes_R S') = e(\mathbf{x}, C \otimes_R S') \qquad \text{and} \qquad e(JS', S') = e(\mathbf{x}, S');$$

see [10, (4.6.10)]. Fix $\widetilde{x}_1, \ldots, \widetilde{x}_r \in S$ such that $\tau(\widetilde{x}_i) = x_i$ for each $i = 1, \ldots, r$, and set $\widetilde{J} = (\widetilde{\mathbf{x}}, \mathbf{y})S$. The third equality in the next sequence is from part (a), the first one comes from the isomorphism $(C \otimes_R S) \otimes_S S' \cong C \otimes_R S'$, and the second and fourth hold by [34, (14.11)].

$$e(\mathbf{x}, C \otimes_R S') = e(\mathbf{x}, (C \otimes_R S) \otimes_S S') = e(\widetilde{J}, C \otimes_R S) = e(\widetilde{J}, S) = e(\mathbf{x}, S').$$

This yields the third equality below, while the remaining ones are from equations (2.8.1) and (2.8.2).

$$\begin{aligned} e(J, C \otimes_R R') &= e(JS', C \otimes_R S') = e(\mathbf{x}, C \otimes_R S') \\ &= e(\mathbf{x}, S') = e(JS', S') = e(J, R'). \end{aligned}$$

□

2.9. Let $\varphi \colon R \to S$ be a ring homomorphism. The *flat dimension* of φ is $\operatorname{fd}(\varphi) = \operatorname{fd}_R(S)$. Assume that φ is surjective and $\operatorname{fd}(\varphi) < \infty$. The map φ is *Cohen-Macaulay* of grade d if S is a perfect R-module of grade d. The map is *Gorenstein* of grade d if it is Cohen-Macaulay of grade d and, for each prime ideal $\mathfrak{q} \subset S$, the $S_{\mathfrak{q}}$-module $\operatorname{Ext}^d_R(S, R)_{\mathfrak{q}}$ is cyclic.

2.10. This section concludes with the definition of the divisor class group of a normal domain R with field of fractions Q. Let $(-)^{\star}$ denote the functor $\operatorname{Hom}_R(-, R)$. An R-module M is *reflexive*[1] if it is finitely generated and the natural biduality map $b^S_M \colon M \to M^{\star\star}$ is bijective. The *divisor class group* of R, denoted $\operatorname{Cl}(R)$, is the set of isomorphism classes of reflexive R-modules of rank 1. The isomorphism class of a reflexive module M is denoted $[M]_R$ or $[M]$ when there is no risk of confusion. The set $\operatorname{Cl}(R)$ admits an Abelian group structure: when M, N are rank 1 reflexive modules, then

$$[M] + [N] = [(M \otimes_R N)^{\star\star}], \qquad [M] - [N] = [\operatorname{Hom}_R(N, M)].$$

If $\mathfrak{a}, \mathfrak{b}$ are ideals with $\mathfrak{a} \cong M$ and $\mathfrak{b} \cong N$, then $[M] + [N] = [\mathfrak{a}] + [\mathfrak{b}] = [(\mathfrak{a}\mathfrak{b})^{\star\star}]$.

The fact that the operations described above make $\operatorname{Cl}(R)$ into an Abelian group seems to be part of the folklore of this subject; see [33, Sec. 0]. We sketch a proof of this fact below, which also indicates why this definition is equivalent to other formulations that may be more familiar to some readers.

[1] Not to be confused with "K-reflexive" or "totally C-reflexive".

A *fractionary* ideal of R is a nonzero finitely generated R-submodule of Q. From the proof of [19, (2.2.iv)], one has $\operatorname{Hom}_R(\mathfrak{a},\mathfrak{b}) \cong \mathfrak{a} :_Q \mathfrak{b}$ for every pair of fractionary ideals $\mathfrak{a},\mathfrak{b}$. Let $D(R)$ denote the set of reflexive fractionary ideals of R, and let $P(R)$ denote the set of principal fractionary ideals of R. As R is a normal domain, we learn from [19, (3.4)] that $D(R)$ is an Abelian group via the operations

$$\mathfrak{a}+\mathfrak{b} = R :_Q (R :_Q (\mathfrak{a}\mathfrak{b})), \qquad \mathfrak{a}-\mathfrak{b} = \mathfrak{a} :_Q \mathfrak{b}$$

with identity R. One checks that $P(R)$ is a subgroup of $D(R)$ and that $\overline{\mathfrak{a}} = \overline{\mathfrak{b}}$ in $D(R)/P(R)$ if and only if $\mathfrak{a} = a\mathfrak{b}$ for some $a \in Q^\times$ if and only if $\mathfrak{a} \cong \mathfrak{b}$.

Let $Z(R)$ denote the set of all height 1 prime ideals $\mathfrak{p} \subset R$. For each $\mathfrak{p} \in Z(R)$, the localization $R_\mathfrak{p}$ is a discrete valuation ring because R is a normal domain, and we let $v_\mathfrak{p}\colon Q^\times \to \mathbb{Z}$ denote the associated valuation. For each $\mathfrak{a} \in D(R)$ and $\mathfrak{p} \in Z(R)$, set $v_\mathfrak{p}(\mathfrak{a}) = \inf\{v_\mathfrak{p}(a) \mid a \in \mathfrak{a}\}$. Set $\operatorname{Div}(R) = \oplus_{\mathfrak{p}\in Z(R)}\mathbb{Z}\cdot[R/\mathfrak{p}]$ and consider the function $\operatorname{div}\colon D(R) \to \operatorname{Div}(R)$ given by $\operatorname{div}(\mathfrak{a}) = (v_\mathfrak{p}(\mathfrak{a})[R/\mathfrak{p}])_{\mathfrak{p}\in Z(R)}$. This is an Abelian group isomorphism by [19, (5.9)] and we set $\operatorname{Prin}(R) = \operatorname{div}(P(R)) \subseteq \operatorname{Div}(R)$. It is routine to verify that div induces a group isomorphism $D(R)/P(R) \cong \operatorname{Div}(R)/\operatorname{Prin}(R)$.

Many readers will undoubtedly recognize $\operatorname{Div}(R)/\operatorname{Prin}(R)$ as the definition of the divisor class group from [8]. To see that this is equivalent to the definition formulated above (and that our formulation yields an Abelian group) it suffices to construct a bijection $f\colon D(R)/P(R) \to \operatorname{Cl}(R)$ such that

$$f(\overline{\mathfrak{a}}+\overline{\mathfrak{b}}) = f(\overline{\mathfrak{a}})+f(\overline{\mathfrak{b}}), \qquad f(\overline{\mathfrak{a}}-\overline{\mathfrak{b}}) = f(\overline{\mathfrak{a}})-f(\overline{\mathfrak{b}}), \qquad f(\overline{R}) = [R]$$

for each $\mathfrak{a},\mathfrak{b} \in D(R)$. Since $\overline{\mathfrak{a}} = \overline{\mathfrak{b}}$ if and only if $\mathfrak{a} \cong \mathfrak{b}$, one sees that the assignment $\overline{\mathfrak{a}} \mapsto [\mathfrak{b}]$ describes a well-defined injection $f\colon D(R)/P(R) \to \operatorname{Cl}(R)$. That this map is surjective follows from a standard exercise; see, for instance, [10, (1.4.18)]. For the displayed relations it suffices to show, for each $\mathfrak{a},\mathfrak{b} \in D(R)$,

$$R :_Q (R :_Q \mathfrak{a}\mathfrak{b}) \cong (\mathfrak{a}\otimes_R \mathfrak{b})^{**}, \qquad \mathfrak{a} :_Q \mathfrak{b} \cong \operatorname{Hom}_R(\mathfrak{b},\mathfrak{a}).$$

(The third condition is obvious from the definition of f.) The second of these has already been discussed. For the first, it suffices to show

$$(\mathfrak{a}\mathfrak{b})^{**} \cong (\mathfrak{a}\otimes_R \mathfrak{b})^{**}.$$

Let K be the kernel of the multiplication map $\mu\colon \mathfrak{a}\otimes_R\mathfrak{b} \to \mathfrak{a}\mathfrak{b}$. One checks that the map $\mu\otimes_R K$ is an isomorphism, and so K is torsion. It follows that $K^\star = 0$ and so $(\mathfrak{a}\mathfrak{b})^\star \cong (\mathfrak{a}\otimes_R\mathfrak{b})^\star$ and $(\mathfrak{a}\mathfrak{b})^{**} \cong (\mathfrak{a}\otimes_R\mathfrak{b})^{**}$.

It follows readily from the isomorphisms described above that, if $\mathfrak{p} \in Z(R)$ and $\ell > 0$, then $\ell[\mathfrak{p}] = [\mathfrak{p}^{(\ell)}]$ in $\operatorname{Cl}(R)$.

3. Semidualizing modules as divisor classes

The following proposition compares directly to the "classical" result for the dualizing module which is the prime motivation for our techniques; see, e.g.,

[10, (3.3.18)]. Recall that a finite R-module M has *rank* (respectively, *rank* r) if $M_{\mathfrak{p}}$ is free (respectively, free of rank r) over $R_{\mathfrak{p}}$ for each $\mathfrak{p} \in \operatorname{Ass}(R)$. Of course, condition (i) is satisfied if R is a domain. Also, using $C = R$ one sees that the implication (ii) $\Longrightarrow$ (i) fails in general; see Proposition 3.2(c).

PROPOSITION 3.1. *Let C be a semidualizing R-module, and consider the following conditions.*

(i) *For each $\mathfrak{p} \in \operatorname{Ass}(R)$, the localization $R_{\mathfrak{p}}$ is Gorenstein.*
(ii) *C has rank 1.*
(iii) *C has rank.*
(iv) *C is isomorphic to an ideal of R.*
(v) *C is isomorphic to an ideal $\mathfrak{a}$ of R with torsion quotient $R/\mathfrak{a}$.*

The implications (i) $\Longrightarrow$ (ii) $\Longleftrightarrow$ (iii) $\Longleftrightarrow$ (iv) $\Longleftrightarrow$ (v) *hold.*

Proof. (i) $\Longrightarrow$ (ii). For each associated prime $\mathfrak{p}$, the ring $R_{\mathfrak{p}}$ is Gorenstein and therefore the semidualizing $R_{\mathfrak{p}}$-module $C_{\mathfrak{p}}$ is isomorphic to $R_{\mathfrak{p}}$ by [12, (8.6)].

(ii) $\Longrightarrow$ (iii) is trivial. For the converse, since $C_{\mathfrak{p}}$ is semidualizing for $R_{\mathfrak{p}}$, it is routine to check that, if $C_{\mathfrak{p}}$ is free over $R_{\mathfrak{p}}$, then it is free of rank 1.

(ii) $\Longleftrightarrow$ (iv) $\Longleftrightarrow$ (v). It is straightforward to show that the semidualizing module C is torsion-free; in fact, $\operatorname{Ass}(R) = \operatorname{Ass}_R(C)$. The desired biimplications now follow from a standard exercise; see, for instance, [10, (1.4.18)]. □

A *semidualizing ideal* is an ideal that is semidualizing as an R-module. One consequence of Proposition 3.1 is that, when R is a domain, every semidualizing module is isomorphic to a semidualizing ideal. The next result provides basic properties of such ideals; it compares directly to [10, (3.3.18)]. We restrict our attention to proper ideals as the case $\mathfrak{a} = R$ is tedious. Since a principal ideal generated by a non-zerodivisor is semidualizing, but is dualizing if and only if R is Gorenstein, the implication (iii) $\Longrightarrow$ (ii) fails in general.

PROPOSITION 3.2. *Let R be a Cohen-Macaulay ring of dimension d and $\mathfrak{a}$ a proper semidualizing ideal of R.*

(a) $\operatorname{ht}(\mathfrak{a}) = 1$ *and $R/\mathfrak{a}$ is Cohen-Macaulay of dimension $d-1$.*
(b) *The quotient $R/\mathfrak{a}$ has $G_{\mathfrak{a}}$-dimension 1 and there are isomorphisms*

$$\operatorname{Ext}^i_R(R/\mathfrak{a}, \mathfrak{a}) \cong \begin{cases} R/\mathfrak{a} & \text{if } i = 1 \\ 0 & \text{otherwise.} \end{cases}$$

(c) *Consider the following conditions:*
   (i) *The quotient $R/\mathfrak{a}$ is a Gorenstein ring.*
   (ii) *The ideal $\mathfrak{a}$ is dualizing for R.*
   (iii) *R is generically Gorenstein.*

   The implications (i) $\Longleftrightarrow$ (ii) $\Longrightarrow$ (iii) hold.

Proof. The proof of (a) is nearly identical to that of [10, (3.3.18.b)], so we omit it here. For part (b), use the exact sequence

$$0 \to \mathfrak{a} \to R \to R/\mathfrak{a} \to 0 \tag{3.2.1}$$

with the fact that $\mathfrak{a}$ and R are both totally $\mathfrak{a}$-reflexive to conclude that $R/\mathfrak{a}$ has $\mathrm{G}_{\mathfrak{a}}$-dimension at most 1. In particular, $\mathrm{Ext}^i_R(R/\mathfrak{a},\mathfrak{a}) = 0$ for $i > 1$. Furthermore, since $\mathfrak{a}$ has rank, it contains an element that is both R-regular and $\mathfrak{a}$-regular, and thus $\mathrm{Hom}_R(R/\mathfrak{a},\mathfrak{a}) = 0$. Applying $\mathrm{Hom}_R(-,\mathfrak{a})$ to (3.2.1) supplies the exact sequence

$$0 \to \underbrace{\mathrm{Hom}_R(R,\mathfrak{a})}_{\mathfrak{a}} \to \underbrace{\mathrm{Hom}_R(\mathfrak{a},\mathfrak{a})}_{R} \to \mathrm{Ext}^1_R(R/\mathfrak{a},\mathfrak{a}) \to 0,$$

which yields an isomorphism $\mathrm{Ext}^1_R(R/\mathfrak{a},\mathfrak{a}) \cong R/\mathfrak{a}$ and the desired conclusions.

For part (c) we may assume that R is local. In the following sequence of formal equalities of series, the first is by [12, (1.6.7)] and the third is standard, while the second is a consequence of part (b).

$$I^{\mathfrak{a}}_R(t) = I^{\mathbf{R}\mathrm{Hom}_R(R/\mathfrak{a},\mathfrak{a})}_{R/\mathfrak{a}}(t) = I^{\Sigma^{-1}R/\mathfrak{a}}_{R/\mathfrak{a}}(t) = t \cdot I^{R/\mathfrak{a}}_{R/\mathfrak{a}}(t).$$

It follows that $\mathrm{id}_R(\mathfrak{a})$ and $\mathrm{id}_{R/\mathfrak{a}}(R/\mathfrak{a})$ are simultaneously finite. This gives the equivalence of (i) and (ii), and the implication (ii) $\Longrightarrow$ (iii) is part of [10, (3.3.18)]. □

The next result simplifies the computation of $[\mathfrak{a}] + [\mathfrak{b}]$ in $\mathrm{Cl}(R)$ for certain semidualizing modules $\mathfrak{a}, \mathfrak{b}$ and is a key tool for the proof of Theorem 4.2.

PROPOSITION 3.3. *Let $\mathfrak{a}$ and $\mathfrak{b}$ be semidualizing ideals such that $\mathfrak{a} \otimes_R \mathfrak{b}$ is semidualizing. The natural multiplication map $\mathfrak{a} \otimes_R \mathfrak{b} \to \mathfrak{a}\mathfrak{b}$ is an isomorphism.*

Proof. The map $\mathfrak{a} \otimes_R \mathfrak{b} \to \mathfrak{a}\mathfrak{b}$ is always surjective, so it remains to verify injectivity. Let U denote the complement in R of the union of the associated primes of R. Since $\mathfrak{a}$ and $\mathfrak{b}$ have rank, the same is true of $\mathfrak{a} \otimes_R \mathfrak{b}$. Furthermore, the fact that $\mathfrak{a} \otimes_R \mathfrak{b}$ is semidualizing implies that $\mathfrak{a} \otimes_R \mathfrak{b}$ is torsion-free. This yields the injectivity of the localization map $\mathfrak{a} \otimes_R \mathfrak{b} \to U^{-1}(\mathfrak{a} \otimes_R \mathfrak{b})$ in the following commuting diagram, where the maps (1) and (2) are given by the appropriate multiplication and the others are the natural ones.

$$\begin{array}{ccccccc}
\mathfrak{a} \otimes_R \mathfrak{b} & \xrightarrow{(1)} & \mathfrak{a}\mathfrak{b} & \hookrightarrow & R & \hookrightarrow & U^{-1}R \\
\downarrow & & & & & & \downarrow = \\
U^{-1}(\mathfrak{a} \otimes_R \mathfrak{b}) & \xrightarrow{\cong} & U^{-1}\mathfrak{a} \otimes_{U^{-1}R} U^{-1}\mathfrak{b} & \xhookrightarrow{(2)} & (U^{-1}\mathfrak{a})(U^{-1}\mathfrak{b}) & \hookrightarrow & U^{-1}R
\end{array}$$

The map (2) is injective, since $U^{-1}\mathfrak{a}$ and $U^{-1}\mathfrak{b}$ are $U^{-1}R$-free of rank 1. It follows that the map (1) must be injective, as desired. □

The next result supplies the main tool for this investigation. Note that Theorem 4.2 shows that $\mathfrak{S}_0(R)$ cannot be given a group structure making the inclusion into a group homomorphism.

PROPOSITION 3.4. *Let R be a normal domain. Each semidualizing R-module C is a rank 1 reflexive module, so there is a natural inclusion $\mathfrak{S}_0(R) \subseteq \mathrm{Cl}(R)$.*

Proof. It suffices to verify the first statement. Proposition 3.1 shows that C has rank 1. For each prime ideal $\mathfrak{p}$ of height 1, the ring $R_\mathfrak{p}$ is regular as R is (R_1), and so $C_\mathfrak{p} \cong R_\mathfrak{p}$. Since R is (S_2) and $\mathrm{depth}_{R_\mathfrak{p}}(C_\mathfrak{p}) = \mathrm{depth}(R_\mathfrak{p})$ for each prime ideal $\mathfrak{p}$, the reflexivity of C follows from [10, (1.4.1)]. □

We record an immediate corollary.

COROLLARY 3.5. *Every normal domain with finite divisor class group is $\mathfrak{S}_0$-finite, and every Cohen-Macaulay normal domain with finite divisor class group is $\mathfrak{S}$-finite.* □

Since a Cohen-Macaulay normal domain R with $\mathrm{Cl}(R) = 0$ is Gorenstein, we note that there are non-Gorenstein rings that satisfy the hypotheses of the corollary. For instance, if k is a field and $\mathbf{X}$ a symmetric $n \times n$ matrix of variables and r an integer such that $0 < r < n$, then the ring $R = k[\mathbf{X}]/I_{r+1}(\mathbf{X})$ is a Cohen-Macaulay normal domain with $\mathrm{Cl}(R) \cong \mathbb{Z}/(2)$ and is non-Gorenstein if and only if $r \equiv n \pmod 2$. Here $I_{r+1}(X)$ is the ideal generated by the minors of $\mathbf{X}$ of size $r + 1$; see [10, (7.3.7.c)]. Determinantal rings will be of particular interest in Section 4.

Proposition 3.4 points toward a plethora of examples of nonlocal rings that are neither $\mathfrak{S}$-finite nor $\mathfrak{S}_0$-finite. Hence the local hypothesis in Question 1.1.

3.6. The *Picard group* of a normal domain R, denoted $\mathrm{Pic}(R)$, is the set of isomorphism classes of finitely generated locally free (i.e., projective) R-modules of rank 1 with operation given by tensor product. The inverse of an element $[P] \in \mathrm{Pic}(R)$ is the class $[P]^{-1} = [\mathrm{Hom}_R(P, R)]$; see [19, p. 105]. It is straightforward to show that there are natural inclusions $\mathrm{Pic}(R) \subseteq \mathfrak{S}_0(R) \subseteq \mathrm{Cl}(R)$ for any normal domain R. Using [23, (3.2)] one sees that the first inclusion is an equality when R is Gorenstein. Each inclusion is an equality when R is a Dedekind domain by Fossum [19, (18.5)].

A result of Claborn [19, (14.10)] states that any Abelian group G can be realized as the divisor class group of a Dedekind domain. In particular, for any Abelian group G, regardless of the cardinality, there is a Dedekind domain R such that the sets $\mathfrak{S}(R)$ and $\mathfrak{S}_0(R)$ are in bijection with G.

We observe that [23, (3.1.b)] implies that the hypothesis of the next result is satisfied when C is C''-reflexive and $C' = \mathrm{Hom}_R(C, C'')$. Compare to Proposition 3.3.

PROPOSITION 3.7. *Let R be a normal domain and C, C' semidualizing R-modules. If $C' \otimes_R C$ is R-semidualizing, then $[C' \otimes_R C] = [C'] + [C]$ in* $\mathrm{Cl}(R)$.

Proof. Since $C' \otimes_R C$ is semidualizing, it is reflexive, so $(C' \otimes_R C)^{\star\star} \cong C' \otimes_R C$, and so $[C' \otimes_R C] = (C' \otimes_R C)^{\star\star} = [C'] + [C]$; see 2.10. □

The inclusion $\mathfrak{S}_0(R) \subseteq \mathrm{Cl}(R)$ is well-behaved with respect to certain operations that are defined on both sets. The remainder of this section is devoted to describing some of this behavior. We begin by describing base-change maps for Picard groups and sets of semidualizing objects.

3.8. Let $\varphi\colon R \to S$ be a ring homomorphism.

(a) The assignment $L \mapsto L \otimes^{\mathbf{L}}_R S$ yields a well-defined group homomorphism $\mathrm{Pic}(\varphi)\colon \mathrm{Pic}(R) \to \mathrm{Pic}(S)$; see [19, discussion after (18.3)].
(b) When $\mathrm{fd}(\varphi)$ is finite and K is a semidualizing complex, the S-complex $K \otimes^{\mathbf{L}}_R S$ is semidualizing by [23, (4.5)], and the assignment $K \mapsto K \otimes^{\mathbf{L}}_R S$ gives rise to a well-defined order-respecting map $\mathfrak{S}(\varphi)\colon \mathfrak{S}(R) \to \mathfrak{S}(S)$ by [23, (4.7)].
(c) When $\mathrm{fd}(\varphi)$ is finite and C is a semidualizing R-module, the S-module $C \otimes_R S$ is semidualizing and $\mathrm{Tor}^R_{\geqslant 1}(C, S) = 0$ by [23, (4.5)]; the assignment $C \mapsto C \otimes_R S$ induces a well-defined order-respecting map $\mathfrak{S}_0(\varphi)\colon \mathfrak{S}_0(R) \to \mathfrak{S}_0(S)$ by [23, (4.7)].

Next we consider the divisor class group. Part (d) in the following lemma is well-known, but we include it here for completeness; see [19, Sec. 6].

LEMMA 3.9. *Let $\varphi\colon R \to S$ be a homomorphism of finite flat dimension between normal domains.*

(a) *Fix $\mathfrak{q} \in \mathrm{Spec}(S)$ and set $\mathfrak{p} = \varphi^{-1}(\mathfrak{q})$. If $\mathrm{ht}(\mathfrak{q}) \leq 1$, then $R_{\mathfrak{p}}$ is regular.*
(b) *If M is a reflexive R-module of rank 1, then* $\mathrm{rank}_S(M \otimes_R S) = 1$.
(c) *If φ is module-finite, the assignment $M \mapsto \mathrm{Hom}_S(\mathrm{Hom}_S(M \otimes_R S, S), S)$ yields a well-defined group homomorphism $\mathrm{Cl}(\varphi)\colon \mathrm{Cl}(R) \to \mathrm{Cl}(S)$.*
(d) *If φ is flat, then the assignment $M \mapsto M \otimes_R S$ yields a well-defined group homomorphism $\mathrm{Cl}(\varphi)\colon \mathrm{Cl}(R) \to \mathrm{Cl}(S)$.*

Proof. (a) The induced map $\varphi_{\mathfrak{q}}\colon R_{\mathfrak{p}} \to S_{\mathfrak{q}}$ has finite flat dimension. Since S is normal, it satisfies Serre's condition (R_1) and so the local ring $S_{\mathfrak{q}}$ is regular. It follows from [1, Theorem R] that $R_{\mathfrak{p}}$ is also regular.

(b) To show that $M \otimes_R S$ has rank 1, it suffices to set $\mathfrak{q} = (0)S$ and exhibit an isomorphism $(M \otimes_R S)_{\mathfrak{q}} \cong S_{\mathfrak{q}}$. With $\mathfrak{p} = \mathrm{Ker}(\varphi)$, part (a) implies $\mathrm{Cl}(R_{\mathfrak{p}}) = 0$ and so $M_{\mathfrak{p}} \cong R_{\mathfrak{p}}$. The next isomorphisms now follow readily

$$(M \otimes_R S)_{\mathfrak{q}} \cong M_{\mathfrak{p}} \otimes_{R_{\mathfrak{p}}} S_{\mathfrak{q}} \cong R_{\mathfrak{p}} \otimes_{R_{\mathfrak{p}}} S_{\mathfrak{q}} \cong S_{\mathfrak{q}}$$

and provide the desired conclusion.

(c) Using part (a), this follows from [35, (1.2.1)].

(d) For a finitely generated R-module U, one has a natural S-linear map

$$\begin{aligned} f_U\colon \operatorname{Hom}_R(U,R)\otimes_R S &\to \operatorname{Hom}_S(U\otimes_R S, S)\\ \psi\otimes s &\mapsto [u\otimes s' \mapsto \varphi(\psi(u))ss'],\end{aligned}$$

which is readily seen to be an isomorphism because φ is flat.

Let M and N be rank 1 reflexive R-modules. The S-module $M\otimes_R S$ has rank 1 by part (b). The flatness of φ provides the following commutative diagram from which one concludes that $M\otimes_R S$ is a reflexive S-module.

$$\begin{array}{ccc} M\otimes_R S & \xrightarrow[\cong]{b^R_M\otimes_R S} & \operatorname{Hom}_R(\operatorname{Hom}_R(M,R),R)\otimes_R S \\ {\scriptstyle b^S_{M\otimes_R S}}\downarrow & & \downarrow{\scriptstyle f_{\operatorname{Hom}_R(M,R)}}\ \cong \\ \operatorname{Hom}_S(\operatorname{Hom}_S(M\otimes_R S,S),S) & \xrightarrow[\cong]{\operatorname{Hom}_S(f_M,S)} & \operatorname{Hom}_S(\operatorname{Hom}_R(M,R)\otimes_R S,S) \end{array}$$

Hence, the map $\operatorname{Cl}(\varphi)$ is well-defined. The fact that $\operatorname{Cl}(\varphi)$ is a group homomorphism follows from the next sequence of isomorphisms

$$\begin{aligned}\operatorname{Hom}_R(\operatorname{Hom}_R(M\otimes_R N,R),R)\otimes_R S &\overset{(1)}{\cong} \operatorname{Hom}_S(\operatorname{Hom}_R(M\otimes_R N,R)\otimes_R S,S)\\ &\overset{(2)}{\cong} \operatorname{Hom}_S(\operatorname{Hom}_S((M\otimes_R N)\otimes_R S,S),S)\\ &\overset{(3)}{\cong} \operatorname{Hom}_S(\operatorname{Hom}_S((M\otimes_R S)\otimes_S (N\otimes_R S),S),S),\end{aligned}$$

where (1) is $f_{\operatorname{Hom}_R(M\otimes_R N,R)}$, (2) is $\operatorname{Hom}_S(f_{M\otimes_R N},S)$, and (3) is standard. □

LEMMA 3.10. *Let $\varphi\colon R\to S$ be a homomorphism of finite flat dimension.*

(a) *If R and S are normal domains and φ is module-finite or flat, then the following diagram commutes.*

$$\begin{array}{ccc} \mathfrak{S}_0(R) & \hookrightarrow & \operatorname{Cl}(R) \\ {\scriptstyle \mathfrak{S}_0(\varphi)}\downarrow & & \downarrow{\scriptstyle \operatorname{Cl}(\varphi)} \\ \mathfrak{S}_0(S) & \hookrightarrow & \operatorname{Cl}(S) \end{array}$$

In particular, if $\operatorname{Cl}(\varphi)$ is injective, then so is $\mathfrak{S}_0(\varphi)$.

(b) *Assume that the image of the map $\operatorname{Spec}(\varphi)\colon \operatorname{Spec}(S)\to\operatorname{Spec}(R)$ contains $\operatorname{m-Spec}(R)$. If the map $\operatorname{Pic}(\varphi)\colon \operatorname{Pic}(R)\to\operatorname{Pic}(S)$ is injective, e.g., if φ is surjective or local, then $\mathfrak{S}_0(\varphi)$ and $\mathfrak{S}(\varphi)$ are also injective.*

(c) *Assume that R,S are normal domains and φ is faithfully flat. If $\operatorname{Cl}(\varphi)$ is surjective, then so is $\mathfrak{S}_0(\varphi)$.*

(d) *Assume that R,S are normal domains and φ is faithfully flat. If $\operatorname{Cl}(\varphi)$ is bijective, then $\mathfrak{S}_0(\varphi)$ is a perfectly order-respecting bijection.*

Proof. When φ is flat, the commutativity of the diagram in (a) follows readily from the definitions. When φ is module finite and C is a semidualizing R-module, the fact that $C \otimes_R S$ is semidualizing for S implies that it is reflexive, and the commutativity of the diagram follows easily. Part (b) is contained in [23, (4.9),(4.11)], and (c) is in [23, (4.5)]. When $\mathrm{Cl}(\varphi)$ is bijective, the map $\mathrm{Pic}(\varphi)$ is injective, so (d) follows from parts (b) and (c) with [23, (4.8)]. □

When $\varphi\colon R \to S$ is a local homomorphism of finite flat dimension, it is a straightforward exercise to show that, if $\mathfrak{S}(\varphi)$ is a perfectly order-respecting bijection, then it is also an isometry with respect to the metric structure defined in [24]. For instance, this holds under the hypotheses of Lemma 3.10(d) when S is Cohen-Macaulay. Some particular instances of this are provided in the next corollary. Others are given in Corollary 3.13 and Proposition 3.15.

COROLLARY 3.11. *Let R be a normal domain and $\mathbf{X} = X_1, \ldots, X_n$ a sequence of variables. For the following flat R-algebras S, the map $\mathfrak{S}_0(R) \to \mathfrak{S}_0(S)$ is a perfectly order-respecting bijection:*

(a) $S = R[\mathbf{X}]$;
(b) $S = R[\mathbf{X}][f_1^{-1}, \ldots, f_i^{-1}]$, *where $f_1, \ldots, f_i$ are prime elements of $R[\mathbf{X}]$ and the ring homomorphism $R \to R[\mathbf{X}][f_1^{-1}, \ldots, f_i^{-1}]$ is faithfully flat;*
(c) $S = R[\mathbf{X}]_{\mathfrak{m}R[\mathbf{X}]}$ *when R is local with maximal ideal $\mathfrak{m}$;*
(d) $S = R[\![\mathbf{X}]\!]_{\mathfrak{m}R[\![\mathbf{X}]\!]}$ *when R is local with maximal ideal $\mathfrak{m}$ and the $\mathfrak{m}$-adic completion of R is normal.*

Proof. By Lemma 3.10(d), it suffices to note that the maps $\mathrm{Cl}(R) \to \mathrm{Cl}(S)$ are bijective; see [19, (7.3),(8.1),(8.9),(19.15)]. □

The following is an important case when localization induces a bijection on the set of semidualizing modules.

PROPOSITION 3.12. *Let $R = \coprod_{i \geq 0} R_i$ be a graded normal domain such that $(R_0, \mathfrak{m}_0)$ is local. Setting $\mathfrak{m} = \mathfrak{m}_0 + \coprod_{i \geq 1} R_i$, the natural map $\mathfrak{S}_0(R) \to \mathfrak{S}_0(R_{\mathfrak{m}})$ is a perfectly order-respecting bijection.*

Proof. Let $\varphi\colon R \to R_{\mathfrak{m}}$ be the localization map. Using [23, (2.14)], the argument of [19, (10.3)] shows that $\mathrm{Cl}(\varphi)$ is bijective. From Lemma 3.10(a) it follows that $\mathfrak{S}_0(\varphi)$ is injective. To show surjectivity, fix a semidualizing $R_{\mathfrak{m}}$-module L. Use the surjectivity of $\mathrm{Cl}(\varphi)$ and [19, (10.2)] to obtain a homogeneous reflexive ideal $\mathfrak{a}$ of R such that $\mathfrak{a}_{\mathfrak{m}} \cong L$. Since L is $R_{\mathfrak{m}}$-semidualizing, the R-module $\mathfrak{a}$ is R-semidualizing by [23, (2.15.a)], and it follows that $\mathfrak{S}_0(\varphi)$ is bijective. The fact that $\mathfrak{S}_0(\varphi)$ is perfectly order-respecting then follows from [23, (2.15.b)]. □

If R is a local ring with completion map $\varphi\colon R \to \widehat{R}$, then the map $\mathfrak{S}_0(\varphi)$ is not usually surjective. Indeed, there exists a Cohen-Macaulay local ring R that does not admit a dualizing module; the complete local ring $\widehat{R}$ does admit a dualizing module ω, and it is straightforward to show that $[\omega] \in \mathfrak{S}_0(\widehat{R})$ cannot be in the image of $\mathfrak{S}_0(\varphi)$. However, a result of Flenner [18, (1.4)] can be applied in certain cases to provide bijectivity. See Corollary 3.14(a) for a generalization, and also [14, Thm. D]. Recall that a ring is *super-normal* if it satisfies Serre's conditions (S_3) and (R_2). Further, note that a ring R satisfying the hypotheses of the next result is excellent because it is finitely generated over the complete local ring R_0. In particular, the complete ring $\widehat{R}$ is also a super-normal domain.

COROLLARY 3.13. *Let $R = \coprod_{i\geq 0} R_i$ be a graded super-normal domain with $(R_0, \mathfrak{m}_0)$ local and complete, and set $\mathfrak{m} = \mathfrak{m}_0 + \coprod_{i\geq 1} R_i$ and $\widehat{R} = \prod_{i\geq 0} R_i$. The induced maps $\mathfrak{S}_0(R) \to \mathfrak{S}_0(\widehat{R})$ and $\mathfrak{S}_0(R_\mathfrak{m}) \to \mathfrak{S}_0(\widehat{R})$ are perfectly order-respecting bijections.*

Proof. The ring $\widehat{R}$ is the $\mathfrak{m}$-adic completion of $R_\mathfrak{m}$, and since R is excellent and super-normal, the same is true of $R_\mathfrak{m}$ and $\widehat{R}$. Let $\varphi\colon R \to R_\mathfrak{m}$ be the localization map and $\psi\colon R_\mathfrak{m} \to \widehat{R}$ the completion map. By Proposition 3.12, the map $\mathfrak{S}_0(\varphi)$ is a perfectly order-respecting bijection, so the equality $\mathfrak{S}_0(\psi\varphi) = \mathfrak{S}_0(\psi)\mathfrak{S}_0(\varphi)$ shows that we need only verify the same for $\mathfrak{S}_0(\psi)$. Since ψ is flat and local, Lemma 3.10(b) supplies the injectivity of $\mathfrak{S}_0(\psi)$. The surjectivity is a consequence of Lemma 3.10 (c), as [18, (1.4)] guarantees that $\mathrm{Cl}(\psi\varphi) = \mathrm{Cl}(\psi)\,\mathrm{Cl}(\varphi)$ is surjective, and therefore that $\mathrm{Cl}(\psi)$ is surjective. □

Here is the first indication that our methods have applications outside the normal domain arena. See Corollary 4.11 for a more general statement.

COROLLARY 3.14. *With $R, \mathfrak{m}, \widehat{R}$ as in Corollary 3.13, fix an $R_\mathfrak{m}$-regular sequence $\mathbf{y} = y_1, \ldots, y_q \in \mathfrak{m}R_\mathfrak{m}$.*

(a) *The natural homomorphisms $R_\mathfrak{m} \to R_\mathfrak{m}/(\mathbf{y}) \to \widehat{R}/(\mathbf{y})$ induce perfectly order-respecting bijections*

$$\mathfrak{S}_0(R_\mathfrak{m}) \xrightarrow{\approx} \mathfrak{S}_0(R_\mathfrak{m}/(\mathbf{y})) \xrightarrow{\approx} \mathfrak{S}_0(\widehat{R}/(\mathbf{y})).$$

(b) *Let R' denote either $R_\mathfrak{m}/(\mathbf{y})$ or $\widehat{R}/(\mathbf{y})$ and fix an $\mathfrak{m}R'$-primary ideal $J \subset R'$. If C' is a semidualizing R'-module, then $e(J, C) = e(J, R')$.*

(c) *If $\mathbf{y}$ is an R-regular sequence in $\mathfrak{m}$, then the composition of induced maps*

$$\mathfrak{S}_0(R) \to \mathfrak{S}_0(R/(\mathbf{y})) \to \mathfrak{S}_0(R_\mathfrak{m}/(\mathbf{y}))$$

is a perfectly order-respecting bijection; thus, the first map is a perfectly order-respecting injection and the second is surjective.

Proof. (a) The rings under consideration fit into the commutative diagram of local ring homomorphisms on the left

$$\begin{array}{ccc} R_{\mathfrak{m}} & \xrightarrow{\beta} & \widehat{R} \\ {\scriptstyle \alpha_{\mathfrak{m}}}\downarrow & & \downarrow{\scriptstyle \widehat{\alpha}} \\ R_{\mathfrak{m}}/(\mathbf{y}) & \xrightarrow{\beta'} & \widehat{R}/(\mathbf{y}) \end{array} \qquad \begin{array}{ccc} \mathfrak{S}_0(R_{\mathfrak{m}}) & \xrightarrow[\approx]{\mathfrak{S}_0(\beta)} & \mathfrak{S}_0(\widehat{R}) \\ {\scriptstyle \mathfrak{S}_0(\alpha_{\mathfrak{m}})}\downarrow & & \downarrow{\scriptstyle \mathfrak{S}_0(\widehat{\alpha})}\ \approx \\ \mathfrak{S}_0(R_{\mathfrak{m}}/(\mathbf{y})) & \xrightarrow{\mathfrak{S}_0(\beta')} & \mathfrak{S}_0(\widehat{R}/(\mathbf{y})) \end{array}$$

and the second commutative diagram arises by applying $\mathfrak{S}_0(-)$ to the first. The maps $\mathfrak{S}_0(\beta)$ and $\mathfrak{S}_0(\widehat{\alpha})$ are perfectly order-respecting bijections by Corollary 3.13 and [24, (4.2)], respectively; see also Gerko [27, (3)]. From the diagram, it follows that $\mathfrak{S}_0(\beta')$ is surjective, and so is bijective by Lemma 3.10(b). That it is perfectly order-respecting is then a consequence of [23, (4.8)]. Thus, $\mathfrak{S}_0(\alpha_{\mathfrak{m}})$ is a perfectly order-respecting bijection, as well.

(b) Set $\widetilde{R} = R_{\mathfrak{m}}$ or $\widehat{R}$, according to whether $R' = R_{\mathfrak{m}}/(\mathbf{y})$ or $\widehat{R}/(\mathbf{y})$. Fix a semidualizing $\widetilde{R}$-module C such that $C' \cong C \otimes_{\widetilde{R}} R'$. Since $\widetilde{R}$ is an excellent local domain, one applies Lemma 2.8(b) to obtain the desired conclusion.

(c) When $\mathbf{y}$ is an R-regular sequence in $\mathfrak{m}$, there is a commutative diagram

$$\begin{array}{ccc} \mathfrak{S}_0(R) & \xrightarrow[\approx]{\mathfrak{S}_0(\gamma)} & \mathfrak{S}_0(R_{\mathfrak{m}}) \\ {\scriptstyle \mathfrak{S}_0(\alpha)}\downarrow & & \downarrow{\scriptstyle \mathfrak{S}_0(\alpha_{\mathfrak{m}})}\ \approx \\ \mathfrak{S}_0(R/(\mathbf{y})) & \xrightarrow{\mathfrak{S}_0(\gamma')} & \mathfrak{S}_0(R_{\mathfrak{m}}/(\mathbf{y})) \end{array}$$

where $\mathfrak{S}_0(\gamma)$ and $\mathfrak{S}_0(\alpha_{\mathfrak{m}})$ are perfectly order-respecting bijections by Proposition 3.12 and part (a). The injectivity of $\mathfrak{S}_0(\alpha)$ and surjectivity of $\mathfrak{S}_0(\gamma')$ follow immediately. To see that $\mathfrak{S}_0(\alpha)$ is perfectly order-respecting, fix $[C], [C'] \in \mathfrak{S}_0(R)$ such that $\mathfrak{S}_0(\alpha)([C]) \trianglelefteq \mathfrak{S}_0(\alpha)([C'])$. It follows that

$$\begin{aligned} \mathfrak{S}_0(\alpha_{\mathfrak{m}})(\mathfrak{S}_0(\gamma)([C])) &= \mathfrak{S}_0(\gamma)(\mathfrak{S}_0(\alpha)([C])) \\ &\trianglelefteq \mathfrak{S}_0(\gamma)(\mathfrak{S}_0(\alpha)([C'])) \\ &= \mathfrak{S}_0(\alpha_{\mathfrak{m}})(\mathfrak{S}_0(\gamma)([C'])), \end{aligned}$$

and so $[C] \trianglelefteq [C']$ since $\mathfrak{S}_0(\alpha_{\mathfrak{m}})$ and $\mathfrak{S}_0(\gamma)$ are perfectly order-respecting. □

The surjectivity of the natural map $\mathfrak{S}_0(R_{\mathfrak{m}}) \to \mathfrak{S}_0(R_{\mathfrak{m}}/(\mathbf{y}))$ in the corollary does not hold for more general local rings. Indeed, let $(A, \mathfrak{n})$ be a local Cohen-Macaulay ring that does not admit a dualizing module. If $\mathbf{y} \in \mathfrak{n}$ is a system of parameters of A, then the map $\mathfrak{S}_0(A) \to \mathfrak{S}_0(A/(\mathbf{y}))$ is not surjective, as $A/(\mathbf{y})$ is Artinian and therefore admits a dualizing module. See [14, (5.5),(5.6)] for further discussion.

When the homomorphism $\varphi\colon R \to S$ is part of a retract pair, the next result sometimes allows one to conclude that $\mathfrak{S}_0(\varphi)$ is bijective. Examples of such retract pairs can be found in power series and localized polynomial extensions:

(a) The natural maps $R \to R[\![\mathbf{X}]\!]$ and $R[\![\mathbf{X}]\!] \to R$.
(b) The natural maps $R \to R[\mathbf{X}]_{\mathfrak{n}}$ and $R[\mathbf{X}]_{\mathfrak{n}} \to R$, when $(R, \mathfrak{m})$ is local and $\mathfrak{n} = (\mathfrak{m}, X_1 - a_1, \ldots, X_n - a_n)R[\mathbf{X}]$ for a sequence $a_1, \ldots, a_n \in R$.

Note that the rings involved are not assumed to be normal domains, so one cannot use the divisor class group directly. However, the method of proof is taken directly from the corresponding results for divisor class groups.

PROPOSITION 3.15. *Let $\varphi\colon R \to S$ and $\psi\colon S \to R$ be homomorphisms of finite flat dimension such that the composition $\psi\varphi$ is the identity on R. If $\ker(\psi)$ is contained in the Jacobson radical of S, then the induced maps $\mathfrak{S}_0(\varphi), \mathfrak{S}_0(\psi), \mathfrak{S}(\varphi), \mathfrak{S}(\psi)$ are perfectly order-respecting bijections.*

Proof. (a) Since the composition $\psi\varphi\colon R \to R$ is the identity, the same is true of the composition $\mathfrak{S}_0(\psi)\mathfrak{S}_0(\varphi)$. In particular, $\mathfrak{S}_0(\psi)$ is surjective. Since $\ker(\psi)$ is in the Jacobson radical of S, Lemma 3.10(b) guarantees that this map is bijective, and therefore so is $\mathfrak{S}_0(\varphi)$. The same argument works for $\mathfrak{S}(\psi)$ and $\mathfrak{S}(\varphi)$. □

This is a surprising departure from the parallels we have seen between the behavior of $\mathrm{Cl}(-)$ and $\mathfrak{S}_0(-)$, as it is known that, when R is a normal domain, the map $\mathrm{Cl}(R) \to \mathrm{Cl}(R[\![\mathbf{X}]\!])$ need not be bijective; see Danilov [15], [16], [17].

The proof of Proposition 3.15 can be translated easily to show that the natural maps $R \to R[\mathbf{X}] \to R$ induce injections and surjections respectively on sets of semidualizing objects. However, we can only say more about these maps when we assume that R is a normal domain, by using Corollary 3.11.

QUESTION 3.16. Must the induced maps $\mathfrak{S}_0(R) \to \mathfrak{S}_0(R[X]) \to \mathfrak{S}_0(R)$ and $\mathfrak{S}(R) \to \mathfrak{S}(R[X]) \to \mathfrak{S}(R)$ all be bijective?

4. Analysis of special cases

We begin with some notation and facts on determinantal rings from [11].

4.1. Let A be a Noetherian ring and m, n, r nonnegative integers satisfying $r < \min\{m, n\}$. If $\mathbf{X} = \{X_{ij}\}$ is an $m \times n$ matrix of variables, then set

$$R = R_{r+1}(A; m, n) = A[\mathbf{X}]/I_{r+1}(\mathbf{X}),$$

where $I_{r+1}(\mathbf{X})$ denotes the ideal generated by the minors of $\mathbf{X}$ of size $r+1$. If A is a normal domain (respectively, is Cohen-Macaulay or is (S_3) or is (R_2)), then so is R; see [11, (6.3),(5.17),(5.16),(6.12)]. The ring R is Gorenstein if and only if A is Gorenstein and either $m = n$ or $r = 0$ by [11, (8.9)].

Assume that A is a normal domain and $r > 0$. Let $\mathfrak{p}$ be the ideal of R generated by the r-minors of the first r rows of the residue matrix $\mathbf{x}$. The ideal $\mathfrak{p}$ is prime, and there is an isomorphism $\mathrm{Cl}(R) \cong \mathrm{Cl}(B) \oplus \mathbb{Z}$, where the summand $\mathbb{Z}$ is generated by $[\mathfrak{p}]$; see [11, (8.4)]. For $\ell > 0$ one has $-[\mathfrak{p}] = [\mathrm{Hom}_R(\mathfrak{p}, R)] = [\mathfrak{q}]$, where $\mathfrak{q}$ is the prime ideal of R generated by the r-minors of the first r columns of $\mathbf{x}$. For $\ell \geq 0$ one has $\ell[\mathfrak{p}] = [\mathfrak{p}^{(\ell)}] = [\mathfrak{p}^\ell]$ and $-\ell[\mathfrak{p}] = [\mathfrak{q}^{(\ell)}] = [\mathfrak{q}^\ell]$ by [11, (7.10)], and we write $\mathfrak{p}^{-\ell}$ in place of $\mathfrak{q}^\ell$. If A is also Gorenstein local and $m \geq n$, then R admits a unique (up to isomorphism) dualizing module $\omega \cong \mathfrak{p}^{(m-n)} = \mathfrak{p}^{m-n}$; see [11, (7.10),(8.8)].

Assume that A is a field and $m \geq n \geq r > 1$. Let $\mathbf{Y} = \{Y_{pq}\}$ be an $(m-1) \times (n-1)$ matrix of variables and set $R' = R_r(A; m-1, n-1)$. Let $\mathfrak{p}'$ be the ideal of R' generated by the $(r-1)$-minors of the first $r-1$ rows of the residue matrix $\mathbf{y}$. The discussion above implies that R' has a unique dualizing module, namely, the ideal $(\mathfrak{p}')^{(m-1)-(n-1)} = (\mathfrak{p}')^{m-n}$. We consider three homomorphisms of normal domains

$$R \xrightarrow{\varphi} R_{x_{11}} \underset{\cong}{\xleftarrow{\rho}} R'[X_{11}, \ldots, X_{m1}, X_{12}, \ldots, X_{1n}]_{X_{11}} \xleftarrow{\psi} R',$$

where ρ is given by $y_{ij} \mapsto x_{i+1j+1} - x_{1j+1}x_{i+1,1}x_{11}^{-1}$ and φ and ψ are the natural flat maps. By [10, (7.3.3)] the map ρ is an isomorphism. Further, the induced maps are all isomorphisms between groups isomorphic to $\mathbb{Z}$,

$$\mathrm{Cl}(R) \xrightarrow{\cong} \mathrm{Cl}(R_{x_{11}}) \xleftarrow{\cong} \mathrm{Cl}(R'[X_{11}, \ldots, X_{m1}, X_{12}, \ldots, X_{1n}]_{X_{11}}) \xleftarrow{\cong} \mathrm{Cl}(R'),$$

and $\mathrm{Cl}(\varphi)([\mathfrak{p}]) = \mathrm{Cl}(\rho\psi)([\mathfrak{p}'])$; see Lemma 3.9(d) and [10, proof of (7.3.6)]. For each integer $\ell \geq 0$, the additivity of $\mathrm{Cl}(\varphi)$ and $\mathrm{Cl}(\rho\psi)$ provides the second equality in the next sequence while the others are from the previous paragraph.

$$\mathrm{Cl}(\varphi)([\mathfrak{p}^\ell]) = \mathrm{Cl}(\varphi)(\ell[\mathfrak{p}]) = \mathrm{Cl}(\rho\psi)(\ell[\mathfrak{p}']) = \mathrm{Cl}(\rho\psi)([(\mathfrak{p}')^\ell]).$$

Lemma 3.10(d) implies that $\mathfrak{S}_0(\psi)$ is bijective. Let C be a semidualizing R-module and let c be the unique integer with $[C] = c[\mathfrak{p}] = [\mathfrak{p}^c]$ in $\mathrm{Cl}(R)$. The $R_{x_{11}}$-module $C \otimes_R R_{x_{11}}$ is semidualizing by 3.8(c), and we compute its class in $\mathfrak{S}_0(R_{x_{11}})$ in the next sequence, where the equalities follow from Lemma 3.10(a), the choice of c, and the previous displayed sequence.

$$\mathfrak{S}_0(\varphi)([C]) = \mathrm{Cl}(\varphi)([C]) = \mathrm{Cl}(\varphi)([\mathfrak{p}^c]) = \mathrm{Cl}(\rho\psi)([(\mathfrak{p}')^c]).$$

As $\rho\psi$ is flat, this provides an isomorphism of $R_{x_{11}}$-modules

$$(\mathfrak{p}')^c \otimes_{R'} R_{x_{11}} \cong C \otimes_R R_{x_{11}}.$$

Since C is R-semidualizing, the module $C \otimes_R R_{x_{11}}$ is $R_{x_{11}}$-semidualizing, and the last isomorphism implies that $(\mathfrak{p}')^c$ is R'-semidualizing by [23, (4.5)].

THEOREM 4.2. *Let k be a field and m, n, r nonnegative integers such that $r < \min\{m, n\}$. The ring $R = R_{r+1}(k; m, n)$ satisfies $\mathfrak{S}_0(R) = \{[R], [\omega]\}$, where ω is a dualizing module for R. In particular, the cardinality of $\mathfrak{S}_0(R)$*

is

$$\operatorname{card}\mathfrak{S}_0(R) = \begin{cases} 1 & \textit{when } m = n \textit{ or } r = 0 \\ 2 & \textit{when } m \neq n \textit{ and } r \neq 0. \end{cases}$$

Proof. If $r = 0$ or $m = n$, then R is Gorenstein and the result follows from 3.6. Assume for the remainder of the proof that $r > 0$ and $m \neq n$. We may also assume that $n \leq m$, as replacing $\mathbf{X}$ with its transpose yields an isomorphism $R_{r+1}(k; m, n) \cong R_{r+1}(k; n, m)$. Let x_{ij} denote the residue of X_{ij} in R.

We have the containment $\mathfrak{S}_0(R) \supseteq \{[R], [\mathfrak{p}^{m-n}]\}$ from 4.1, so it remains to verify the containment $\mathfrak{S}_0(R) \subseteq \{[R], [\mathfrak{p}^{m-n}]\}$. Let C be a semidualizing R-module and let c be the unique integer with $[C] = c[\mathfrak{p}]$ in $\operatorname{Cl}(R) \cong \mathbb{Z}[\mathfrak{p}]$.

Following the proof of [10, (7.3.6)], we use induction on r to reduce to the case $r = 1$. Suppose that $r > 1$ and employ the notation of the last paragraph of 4.1. The final conclusion of 4.1 says that $(\mathfrak{p}')^c$ is R'-semidualizing, and the induction hypothesis states that $\mathfrak{S}_0(R') = \{[R'], [(\mathfrak{p}')^{m-n}]\}$. Hence, either $c = 0$ or $c = m - n$. By our choice of c, this implies either $[C] = [\mathfrak{p}^0] = [R]$ or $[C] = [\mathfrak{p}^{m-n}]$, as desired.

Assume $r = 1$. As above, it suffices to show that $c = 0$ or $c = m - n$. The ring R is a standard graded ring over a field and $\mathfrak{p}$ is a homogeneous prime ideal. For each $v \geq 0$ the power $\mathfrak{p}^v$ is homogeneous, and so we may speak of its minimal number of generators, denoted $\beta_0(\mathfrak{p}^v)$. As is noted in [11, (9.20)], the homogeneous minimal generators of $\mathfrak{p}^v$ are in bijection with the monomials of degree v in the ring $k[Z_1, \ldots, Z_n]$. Since $n \geq 2$, a routine argument shows

$$\beta_0(\mathfrak{p}^u)\beta_0(\mathfrak{p}^v) > \beta_0(\mathfrak{p}^{u+v}) > \beta_0(\mathfrak{p}^v). \tag{4.2.1}$$

Also, when $u \leq v$, the following sequence implies $\mathfrak{p}^{v-u} \cong \operatorname{Hom}_R(\mathfrak{p}^u, \mathfrak{p}^v)$:

$$[\mathfrak{p}^{v-u}] = (v-u)[\mathfrak{p}] = v[\mathfrak{p}] - u[\mathfrak{p}] = [\mathfrak{p}^v] - [\mathfrak{p}^u] = [\operatorname{Hom}_R(\mathfrak{p}^u, \mathfrak{p}^v)].$$

Suppose that $0 < c < m - n$. Then $C \cong \mathfrak{p}^c$, and $\mathfrak{p}^{m-n}$ is a dualizing module for R. It follows from [23, (2.14.a),(3.1.a)] and [12, (3.4.a)] that $\operatorname{Hom}_R(\mathfrak{p}^c, \mathfrak{p}^{m-n}) \cong \mathfrak{p}^{m-n-c}$ is semidualizing. Furthermore, Proposition 3.3 yields an isomorphism

$$\mathfrak{p}^c \otimes_R \mathfrak{p}^{m-n-c} \xrightarrow{\cong} \mathfrak{p}^{m-n}$$

and thus the equality $\beta_0(\mathfrak{p}^{m-n}) = \beta_0(\mathfrak{p}^c)\beta_0(\mathfrak{p}^{m-n-c})$, contradicting (4.2.1).

Next, suppose that $c > m - n$. As above, we have

$$\beta_0(\mathfrak{p}^c) > \beta_0(\mathfrak{p}^{m-n}) = \beta_0(\mathfrak{p}^c)\beta_0(\operatorname{Hom}_R(\mathfrak{p}^c, \mathfrak{p}^{m-n})) > \beta_0(\mathfrak{p}^c),$$

again yielding a contradiction.

Finally, suppose that $c < 0$. Then $\operatorname{Hom}_R(C, \mathfrak{p}^{m-n}) \cong \mathfrak{p}^{m-n-c}$ is semidualizing. However, $c < 0$ implies that $m - n - c > m - n$, contradicting the previous case. □

Next, we present the local analogue of Theorem 4.2.

COROLLARY 4.3. *With notation as in Theorem 4.2, let $\mathfrak{m}$ denote the maximal ideal of R generated by the residues of the variables X_{ij}. If R' is either the localization $R_{\mathfrak{m}}$ or its $\mathfrak{m}$-adic completion $\widehat{R}$, then $\mathfrak{S}_0(R') = \{[R'], [\omega']\}$, where ω' is a dualizing module for R'. In particular, the cardinality of $\mathfrak{S}_0(R')$ is*

$$\operatorname{card} \mathfrak{S}_0(R') = \begin{cases} 1 & \text{when } m = n \text{ or } r = 0 \\ 2 & \text{when } m \neq n \text{ and } r \neq 0. \end{cases}$$

Proof. The ring R satisfies the hypotheses of Corollary 3.13 as it is Cohen-Macaulay and (R_2) by [11, (6.12)]. □

The next result is a considerable generalization of Theorem 4.2 that encompasses Theorem 1.2 from the introduction. Its proof requires more notation.

4.4. Let A be a normal domain and m, n, r nonnegative integers such that $r < \min\{m, n\}$. Set $R = R_{r+1}(A; m, n)$ and consider the commutative diagram of natural ring homomorphisms

(4.4.1)

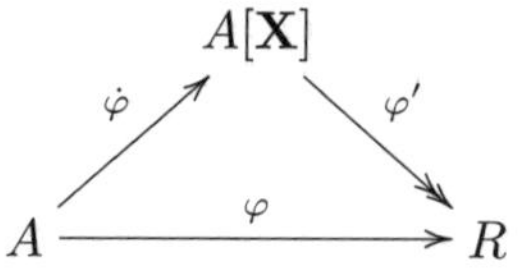

wherein φ and $\dot{\varphi}$ are faithfully flat, and φ' is surjective and Cohen-Macaulay of grade $d = mn - r(m + n - r)$ by [11, (5.18)]; see 2.9 for terminology. If C is a semidualizing A-module, then the following R-module is semidualizing by [23, (6.1)]:

$$C(\varphi) = \operatorname{Ext}^d_{A[\mathbf{X}]}(R, C \otimes_A A[\mathbf{X}]).$$

THEOREM 4.5. *Let A be a normal domain and m, n, r nonnegative integers such that $r < \min\{m, n\}$. The ring $R = R_{r+1}(A; m, n)$ is $\mathfrak{S}_0$-finite if and only if A is so, and the ordering on $\mathfrak{S}_0(R)$ is transitive if and only if the ordering on $\mathfrak{S}_0(A)$ is so. More specifically, one has the following cases.*

(a) *If $r = 0$ or $m = n$, then $\mathfrak{S}_0(\varphi)$ is a perfectly order-respecting bijection*

$$\mathfrak{S}_0(\varphi) \colon \mathfrak{S}_0(A) \xrightarrow{\approx} \mathfrak{S}_0(R).$$

(b) *If $r > 0$ and $m \neq n$, then the assignment*

$$([C]_A, 0) \mapsto [C(\varphi)]_R, \qquad ([C]_A, 1) \mapsto [C \otimes_A R]_R$$

describes a perfectly order-respecting bijection

$$h \colon \mathfrak{S}_0(A) \times \{0, 1\} \xrightarrow{\approx} \mathfrak{S}_0(R).$$

The proof of this result is rather long, so it is presented at the end of the section in 4.14. For now we focus on some consequences of the theorem.

4.6. Continue with the notation of 4.12. Let $\mathfrak{n}$ be a prime ideal of A and consider the prime ideals $\mathfrak{N} = (\mathfrak{n}, \mathbf{X})A[\mathbf{X}]$ and $\mathfrak{m} = (\mathfrak{n}, \mathbf{x})R$. Localizing and completing the diagram (4.4.1) yield similar commutative diagrams

$$\begin{array}{ccc} & A[\mathbf{X}]_{\mathfrak{N}} & \\ {\scriptstyle \dot\varphi_{\mathfrak{N}}}\nearrow & & \searrow{\scriptstyle \varphi'_{\mathfrak{m}}} \\ A_{\mathfrak{n}} & \xrightarrow{\varphi_{\mathfrak{m}}} & R_{\mathfrak{m}} \end{array} \qquad \begin{array}{ccc} & \widehat{A[\mathbf{X}]_{\mathfrak{N}}} & \\ {\scriptstyle \widehat{\dot\varphi_{\mathfrak{N}}}}\nearrow & & \searrow{\scriptstyle \widehat{\varphi'_{\mathfrak{m}}}} \\ \widehat{A} = \widehat{A_{\mathfrak{n}}} & \xrightarrow{\widehat{\varphi}=\widehat{\varphi_{\mathfrak{m}}}} & \widehat{R_{\mathfrak{m}}} = \widehat{R} \end{array}$$

For semidualizing $A_{\mathfrak{n}}$- and $\widehat{A}$-modules C_0 and C_1, respectively, we set

$$C_0(\varphi_{\mathfrak{m}}) = \operatorname{Ext}^d_{A[\mathbf{X}]_{\mathfrak{N}}}(R_{\mathfrak{m}}, C_0 \otimes_{A_{\mathfrak{n}}} A[\mathbf{X}]_{\mathfrak{N}}) \qquad \text{(semidualizing for } R_{\mathfrak{m}}\text{)},$$
$$C_1(\widehat{\varphi}) = \operatorname{Ext}^d_{\widehat{A[\mathbf{X}]_{\mathfrak{N}}}}(\widehat{R}, C_1 \otimes_{\widehat{A}} \widehat{A[\mathbf{X}]_{\mathfrak{N}}}) \qquad \text{(semidualizing for } \widehat{R}\text{)}.$$

These local constructions are discussed extensively in [23, Section 6].

What follows is the localized version of Theorem 4.5.

Corollary 4.7. *Let $A = \coprod_{i\geq 0} A_i$ be a graded normal domain with $(A_0, \mathfrak{n}_0)$ local, and set $\mathfrak{n} = \mathfrak{n}_0 + \coprod_{i\geq 1} A_i$.*

(a) *If $r = 0$ or $m = n$, then $\mathfrak{S}_0(\varphi_{\mathfrak{m}})$ is a perfectly order-respecting bijection*

$$\mathfrak{S}_0(\varphi_{\mathfrak{m}})\colon \mathfrak{S}_0(A_{\mathfrak{n}}) \xrightarrow{\approx} \mathfrak{S}_0(R_{\mathfrak{m}}).$$

(b) *If $r > 0$ and $m \neq n$, then the assignment*

$$([C]_{A_{\mathfrak{n}}}, 0) \mapsto [C(\varphi_{\mathfrak{n}})]_{R_{\mathfrak{m}}}, \qquad ([C]_{A_{\mathfrak{n}}}, 1) \mapsto [C \otimes_{A_{\mathfrak{n}}} R_{\mathfrak{m}}]_{R_{\mathfrak{m}}}$$

describes a perfectly order-respecting bijection

$$\mathfrak{S}_0(A_{\mathfrak{n}}) \times \{0,1\} \xrightarrow{\approx} \mathfrak{S}_0(R_{\mathfrak{m}}).$$

Proof. The following diagrams (one for each of our cases) commute.

$$\begin{array}{ccc} \mathfrak{S}_0(A) & \xrightarrow{\approx} & \mathfrak{S}_0(A_{\mathfrak{n}}) \\ {\scriptstyle\approx}\Big\downarrow{\scriptstyle \mathfrak{S}_0(\varphi)} & & \Big\downarrow{\scriptstyle \mathfrak{S}_0(\varphi_{\mathfrak{m}})} \\ \mathfrak{S}_0(R) & \xrightarrow{\approx} & \mathfrak{S}_0(R_{\mathfrak{m}}) \end{array} \qquad \begin{array}{ccc} \mathfrak{S}_0(A) \times \{0,1\} & \xrightarrow{\approx} & \mathfrak{S}_0(A_{\mathfrak{n}}) \times \{0,1\} \\ {\scriptstyle\approx}\Big\downarrow{\scriptstyle h} & & \Big\downarrow{\scriptstyle h_{\mathfrak{m}}} \\ \mathfrak{S}_0(R) & \xrightarrow{\approx} & \mathfrak{S}_0(R_{\mathfrak{m}}) \end{array}$$

The four horizontal maps are perfectly order-respecting bijections by Proposition 3.12, as are two of the vertical ones by Theorem 4.5. Thus, the two remaining maps are so as well. □

Corollary 4.8. *Let $A = \coprod_{i\geq 0} A_i$ be a graded super-normal domain with $(A_0, \mathfrak{n}_0)$ local and complete. Set $\mathfrak{n} = \mathfrak{n}_0 + \coprod_{i\geq 1} A_i$ and let $\mathfrak{m}, \widehat{A}, \widehat{R}$ be as in 4.6.*

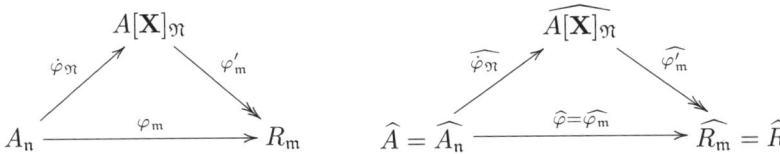

(a) *If $r = 0$ or $m = n$, then $\mathfrak{S}_0(\widehat{\varphi})$ is a perfectly order-respecting bijection*

$$\mathfrak{S}_0(\widehat{\varphi})\colon \mathfrak{S}_0(\widehat{A}) \xrightarrow{\approx} \mathfrak{S}_0(\widehat{R}).$$

(b) *If $r > 0$ and $m \neq n$, then the assignment*

$$([C]_{\widehat{A}}, 0) \mapsto [C(\widehat{\varphi})]_{\widehat{R}}, \qquad ([C]_{\widehat{A}}, 1) \mapsto [C \otimes_{\widehat{A}} \widehat{R}]_{\widehat{R}}$$

describes a perfectly order-respecting bijection

$$\mathfrak{S}_0(\widehat{A}) \times \{0,1\} \xrightarrow{\approx} \mathfrak{S}_0(\widehat{R}).$$

Proof. The proof is almost identical to the previous one, using Corollary 3.13 in place of Proposition 3.12. One needs only note that, since A is super-normal, the same is true of R by [11, (5.17),(6.12)]. □

The next step is to iterate the previous three results.

COROLLARY 4.9. *Let A be a normal domain and t a positive integer. For $l = 1, \dots, t$ fix integers r_l, m_l, n_l such that $0 \leq r_l < \min\{m_l, n_l\}$ and let $\mathbf{X}_{l**} = \{X_{lij}\}$ be an $m_l \times n_l$ matrix of variables. Let $\mathbf{X}$ denote the entire list of variables $X_{111}, \dots, X_{tm_tn_t}$ and set*

$$R = A[\mathbf{X}]/\textstyle\sum_{l=1}^{t} I_{r_l+1}(\mathbf{X}_{l**})$$

with $\mathbf{x}$ the image in R of the sequence $\mathbf{X}$. Let s be the number of indices l such that $r_l > 0$ and $m_l \neq n_l$.

(a) *There is a perfectly order-respecting bijection*

$$\mathfrak{S}_0(A) \times \{0,1\}^s \xrightarrow{\approx} \mathfrak{S}_0(R).$$

(b) *With $A, \mathfrak{n}$ as in Corollary 4.7 and $\mathfrak{m} = (\mathfrak{n}, \mathbf{x})R$, there is a perfectly order-respecting bijection*

$$\mathfrak{S}_0(A_\mathfrak{n}) \times \{0,1\}^s \xrightarrow{\approx} \mathfrak{S}_0(R_\mathfrak{m}).$$

(c) *With $A, \mathfrak{n}$ as in Corollary 4.8 and $\mathfrak{m} = (\mathfrak{n}, x)R$, let $\widehat{A}$ and $\widehat{R}$ denote the $\mathfrak{n}$-adic and $\mathfrak{m}$-adic completions of A and R, respectively. There is a perfectly order-respecting bijection*

$$\mathfrak{S}_0(\widehat{A}) \times \{0,1\}^s \xrightarrow{\approx} \mathfrak{S}_0(\widehat{R}).$$

Proof. Write $R_0 = A$ and for $l = 1, \dots, t$ set $R_l = R_{r_l+1}(R_{l-1}; m_l, n_l)$. Then $R_t \cong R$ and part (a) is proved by induction on t using Theorem 4.5. Parts (b) and (c) now follow from Proposition 3.12 and Corollary 3.13, respectively. □

Before continuing, we present some notation.

4.10. Let A be a ring and fix an integer $m \geq 1$ and an A-regular sequence $\mathbf{y} = y_1, \dots, y_q \in A$. Set $n = m + q - 1$ and let $\mathbf{X}$ be an $m \times n$ matrix of variables. The discussion before and after [11, (2.14)] exhibits a surjection

$$A[\mathbf{X}]/I_m(\mathbf{X}) \twoheadrightarrow A/(\mathbf{y})^m$$

whose kernel is generated by an $A[\mathbf{X}]/I_m(\mathbf{X})$-regular sequence.

For $l = 1, \dots, t$ fix integers $m_l, q_l \geq 1$ and a sequence $\mathbf{y}_{l*} = y_{l1}, \dots, y_{lq_l} \in A$. Assume that the sequence $\mathbf{y}_{**}$ is A-regular and set

$$B(A, \mathbf{y}, \mathbf{m}, \mathbf{q}) = A/\textstyle\sum_{l=1}^{t} (\mathbf{y}_{l*})^{m_l}.$$

With R as in Corollary 4.9, tensoring copies of the surjection from the previous paragraph provides a surjection

$$\tau \colon R \twoheadrightarrow B(A, \mathbf{y}, \mathbf{m}, \mathbf{q}) \tag{4.10.1}$$

whose kernel is generated by an R-regular sequence.

Let B be a ring and u a positive integer. For $l = 1, \dots, u$ fix a positive integer p_l and variables $\mathbf{Z}_{l*} = Z_{l1}, \dots, Z_{lp_l}$. We consider the ring

$$S = S(B, \mathbf{p}) = B[\mathbf{Z}_{1*}]/(\mathbf{Z}_{1*})^2 \otimes_B \cdots \otimes_B B[\mathbf{Z}_{u*}]/(\mathbf{Z}_{u*})^2,$$

which can be thought of in several different ways. Each ring $B[\mathbf{Z}_{l*}]/(\mathbf{Z}_{l*})^2$ is isomorphic to the trivial extension $B \ltimes B^{q_l}$, so there is an isomorphism

$$S \cong (B \ltimes B^{q_1}) \otimes_B \cdots \otimes_B (B \ltimes B^{q_u}).$$

Next, set $S_0 = B$ and take successive trivial extensions $S_l = S_{l-1} \ltimes (S_{l-1})^{q_l}$. From the previous description, there is an isomorphism $S \cong S_u$. Finally, let $\mathbf{Z}$ denote the full list of variables $\mathbf{Z} = Z_{11}, \dots, Z_{up_u}$ and let $\mathbf{z}$ denote the image in S of the sequence $\mathbf{Z}$. From the definition of S, one obtains the isomorphism

$$S \cong B[\mathbf{Z}]/\textstyle\sum_{l=1}^{u} (\mathbf{Z}_{l*})^2.$$

If B is (complete) local with maximal ideal $\mathfrak{r}$, then S is (complete) local with maximal ideal $(\mathfrak{r}, \mathbf{z})S$.

The final result of this paper generalizes Corollary 3.14 and contains Theorems 1.3 and 1.4 from the introduction.

COROLLARY 4.11. *With $A, \mathfrak{n}$ as in Corollary 4.8, let t, u be nonnegative integers. For $l = 1, \dots, t$ fix positive integers m_l, q_l and a sequence $\mathbf{y}_{l*} = y_{l1}, \dots, y_{lq_l} \in \mathfrak{n}\widehat{A}$, and let s be the number of indices l such that $m_l, q_l > 1$. For $l = 1, \dots, u$ fix a positive integer p_l, and let r denote the number of indices l such that $p_l > 1$.*

(a) *Set $\widehat{B} = B(\widehat{A}, \mathbf{y}, \mathbf{m}, \mathbf{q})$ and $\widehat{S} = S(\widehat{B}, \mathbf{p})$. If $\mathbf{y}$ is $\widehat{A}$-regular, then there is a perfectly order-respecting bijection*

$$\mathfrak{S}_0(\widehat{A}) \times \{0,1\}^{r+s} \xrightarrow{\approx} \mathfrak{S}_0(\widehat{S}).$$

Furthermore, if $J \subset \widehat{S}$ is an ideal primary to the maximal ideal of $\widehat{S}$ and C is a semidualizing $\widehat{S}$-module, then $e(J,C) = e(J,\widehat{S})$.

(b) *Assume that $\mathbf{y}$ is an $A_{\mathfrak{n}}$-sequence in $\mathfrak{n}A_{\mathfrak{n}}$, and set $B' = B(A_{\mathfrak{n}}, \mathbf{y}, \mathbf{m}, \mathbf{q})$ and $S' = S(B', \mathbf{p})$. There are perfectly order-respecting bijections*

$$\mathfrak{S}_0(A_{\mathfrak{n}}) \times \{0,1\}^{r+s} \xrightarrow{\approx} \mathfrak{S}_0(S') \xrightarrow{\approx} \mathfrak{S}_0(\widehat{S}).$$

Furthermore, if $J \subset S'$ is an ideal primary to the maximal ideal of S' and C is a semidualizing S'-module, then $e(J,C) = e(J,S')$.

(c) *Assume that $\mathbf{y}$ is an A-regular sequence in $\mathfrak{n}$, and set $B = B(A, \mathbf{y}, \mathbf{m}, \mathbf{q})$ and $S = S(B, \mathbf{p})$. There is a perfectly order-respecting injection*

$$\mathfrak{S}_0(A) \times \{0,1\}^{r+s} \hookrightarrow \mathfrak{S}_0(S).$$

Proof. We prove part (a); argue similarly for the other parts. Let $\mathbf{Z}$ be as in 4.10. Then there are isomorphisms

$$\widehat{S} \cong \widehat{B}[\mathbf{Z}]/\textstyle\sum_{l=1}^{u}(\mathbf{Z}_{l*})^2 \cong \widehat{A}[\mathbf{Z}]/(\sum_{l=1}^{t}(\mathbf{y}_{l*})^{k_l} + \sum_{l=1}^{u}(\mathbf{Z}_{l*})^2).$$

By Corollary 3.11(a), the natural map $\mathfrak{S}_0(\widehat{A}) \to \mathfrak{S}_0(\widehat{A}[\mathbf{Z}])$ is a perfectly order-respecting bijection. Pass to the ring $\widehat{A}[\mathbf{Z}]$ and use the fact that $\mathbf{Z}$ is $\widehat{A}[\mathbf{Z}]/(\mathbf{y})$-regular, to reduce to the case $u = 0 = r$, that is, $\widehat{S} = \widehat{B}$.

For $l = 1, \ldots, t$ set $r_l = m_l - 1$ and $n_l = m_l + q_l - 1$, and let $R, \mathfrak{m}, \widehat{R}$ be as in Corollary 4.9. The surjection (4.10.1) completes to a surjection $\widehat{\tau}\colon \widehat{R} \to \widehat{S}$ whose kernel is generated by a $\widehat{R}$-regular sequence. The perfectly order-respecting bijections in the next sequence are in Corollary 4.9(c) and [24, (4.2)], respectively:

$$\mathfrak{S}_0(\widehat{A}) \times \{0,1\}^{r+s} \xrightarrow{\approx} \mathfrak{S}_0(\widehat{R}) \xrightarrow{\approx} \mathfrak{S}_0(\widehat{S}).$$

The equality of multiplicities follows from Corollary 3.14(b). □

4.12. To keep things tangible, we give an explicit description of the injection

$$\mathfrak{S}_0(A) \times \{0,1\}^{r+s} \hookrightarrow \mathfrak{S}_0(S)$$

from the previous corollary. (The two bijections are described analogously.) For $l = t+1, \ldots, t+u$ set $m_l = 2$ and $q_l = p_{l-t}$. Set $S_0 = A$. For $l = 1, \ldots, t$ take quotients $S_l = S_{l-1}/(\mathbf{y}_{l*})^{m_l}$, and for $l = t+1, \ldots, t+u$ take trivial extensions $S_l = S_{l-1} \ltimes (S_{l-1})^{p_l}$, so that $S \cong S_{t+u}$. Each homomorphism $\varphi_{l-1}\colon S_{l-1} \to S_l$ induces an injective map:

(a) If $q_l = 1$ or $m_l = 1$, then set $f_{l-1} = \mathfrak{S}_0(\varphi_{l-1})\colon \mathfrak{S}_0(S_{l-1}) \to \mathfrak{S}_0(S_l)$.

(b) If $m_l, q_l > 1$, then let $f_{l-1}\colon \mathfrak{S}_0(S_{l-1}) \times \{0,1\} \to \mathfrak{S}_0(S_l)$ be given by

$$([C]_{S_{l-1}}, 0) \mapsto \begin{cases} [\operatorname{Ext}^{q_l}_{S_{l-1}}(S_l, C)]_{S_l} & \text{if } l \leq t \\ [\operatorname{Hom}_{S_{l-1}}(S_l, C)]_{S_l} & \text{if } l > t \end{cases}$$

$$([C]_{S_{l-1}}, 1) \mapsto [C \otimes_{S_{l-1}} S_l]_{S_l}.$$

The desired inclusion is exactly the composition $f_{t+u-1}\cdots f_0$.

The calculations of this section motivate a refinement of Question 1.1.

QUESTION 4.13. If R is a local ring, must the cardinalities of the sets $\mathfrak{S}_0(R)$ and $\mathfrak{S}(R)$ be powers of 2?

Paragraph 3.6 explains the need for the "local" hypothesis. Beyond the results of this section, evidence justifying this question can be found in [24, (3.4)]: If R is a non-Gorenstein ring admitting a dualizing complex and $\mathfrak{S}(R)$ is a finite set, then $\mathfrak{S}(R)$ has even cardinality.

We conclude this section with the proof of Theorem 4.5.

4.14. *(Proof of Theorem 4.5.)* Let x_{ij} denote the residue of X_{ij} in R and set

$$\Delta=\det\begin{pmatrix}X_{11} & \dots & X_{1r}\\ \vdots & & \vdots\\ X_{r1} & \dots & X_{rr}\end{pmatrix}\in A[\mathbf{X}],\qquad \delta=\det\begin{pmatrix}x_{11} & \dots & x_{1r}\\ \vdots & & \vdots\\ x_{r1} & \dots & x_{rr}\end{pmatrix}\in R,$$

noting $\delta=\varphi'(\Delta)$. Also, set $e=\dim R-\dim A$ and note that [11, (5.18)] implies $e=(m+n-r)r$. By [11, (6.4)] there is a prime element $\zeta\in A[T_1,\dots,T_e]$ and an isomorphism ϵ as in the next display; the isomorphism τ is clear.

$$R\otimes_{A[\mathbf{X}]}A[\mathbf{X}]_\Delta \xrightarrow[\tau]{\cong} R_\delta \xleftarrow[\epsilon]{\cong} A[T_1,\dots,T_e]_\zeta\,.$$

Furthermore, the natural map $\alpha\colon A\to A[T_1,\dots,T_e]_\zeta$ is faithfully flat, so $\mathfrak{S}_0(\alpha)$ is bijective by Corollary 3.11(b).

Set $U=A\smallsetminus(0)$ and $F=U^{-1}A$. Using Lemma 3.10(a), the natural maps $\beta\colon R\to R_\delta$ and $\gamma\colon R\to U^{-1}R$ along with $\epsilon\alpha$ yield a commutative diagram

$$(4.14.1)\qquad \begin{array}{ccccc}\mathfrak{S}_0(R) & \xrightarrow{f} & \mathfrak{S}_0(R_\delta)\times\mathfrak{S}_0(U^{-1}R) & \xleftarrow{g} & \mathfrak{S}_0(A)\times\mathfrak{S}_0(U^{-1}R)\\ \big\downarrow & & \big\downarrow & & \big\downarrow\\ \mathrm{Cl}(R) & \xrightarrow[f']{\cong} & \mathrm{Cl}(R_\delta)\times\mathrm{Cl}(U^{-1}R) & \xleftarrow[-g']{\cong} & \mathrm{Cl}(A)\times\mathrm{Cl}(U^{-1}R)\end{array}$$

where the horizontal maps are given by

$$(4.14.2)\qquad \begin{array}{c}[C]\overset{f}{\longmapsto}([C\otimes_R R_\delta],[C\otimes_R U^{-1}R])\\ ([C'\otimes_A R_\delta],[C''])\overset{g}{\longleftarrow\!\!\!\shortmid}([C'],[C''])\end{array}$$

and the vertical arrows are induced by the respective inclusions. The maps f' and g' are bijective by [11, (8.3)] and [19, (7.3),(8.1)], respectively. In particular, the maps f,g are injective, and g is bijective by Corollary 3.11(b).

(a) Assuming that $r = 0$ or $m = n$, Theorem 4.2 implies that $\mathfrak{S}_0(U^{-1}R)$ is trivial since $U^{-1}R \cong R_{r+1}(F; m, n)$. Thus, the top row of (4.14.1) reduces to

$$\mathfrak{S}_0(R) \xhookrightarrow{\mathfrak{S}_0(\beta)} \mathfrak{S}_0(R_\delta) \xleftarrow[\approx]{\mathfrak{S}_0(\epsilon\alpha)} \mathfrak{S}_0(A). \tag{4.14.3}$$

The functoriality of $\mathfrak{S}_0(-)$ and the following commutative diagram of ring homomorphisms

$$\begin{array}{ccc} A & \xrightarrow{\alpha} & A[T_1, \ldots, T_e]_\zeta \\ \downarrow{\scriptstyle\varphi} & & \downarrow{\scriptstyle\epsilon} \\ R & \xrightarrow{\beta} & R_\delta \end{array}$$

yield the equality $\mathfrak{S}_0(\beta)\mathfrak{S}_0(\varphi) = \mathfrak{S}_0(\epsilon\alpha)$. Since (4.14.3) shows that $\mathfrak{S}_0(\epsilon\alpha)$ is bijective, it follows that $\mathfrak{S}_0(\beta)$ is surjective. As noted above, $\mathfrak{S}_0(\beta)$ is also injective, so it follows that $\mathfrak{S}_0(\varphi)$ is bijective. That it is a perfectly order-respecting bijection follows from [23, (4.8)] since φ is faithfully flat.

(b) Assume now that $r > 0$ and $m \neq n$. The isomorphism $U^{-1}R \cong R_{r+1}(F; m, n)$ in conjunction with Theorem 4.2 yields a bijection $i: \{0, 1\} \xrightarrow{\approx} \mathfrak{S}_0(U^{-1}R)$ given by $i(0) = [\omega_{U^{-1}R}]$ and $i(1) = [U^{-1}R]$, where $\omega_{U^{-1}R}$ is a dualizing module for $U^{-1}R$. Let $i': \mathfrak{S}_0(A) \times \{0, 1\} \to \mathfrak{S}_0(A) \times \mathfrak{S}_0(U^{-1}R)$ be the induced bijection.

Recall that $h: \mathfrak{S}_0(A) \times \{0, 1\} \to \mathfrak{S}_0(R)$ is defined as $h([C]_A, 0) = [C(\varphi)]_R$ and $h([C]_A, 1) = [C \otimes_A R]_R$. Below we construct a bijection

$$j: \mathfrak{S}_0(R_\delta) \times \mathfrak{S}_0(U^{-1}R) \to \mathfrak{S}_0(R_\delta) \times \mathfrak{S}_0(U^{-1}R)$$

such that the following diagram commutes.

$$\begin{array}{ccccc} \mathfrak{S}_0(A) \times \{0, 1\} & \xrightarrow{h} & \mathfrak{S}_0(R) & \xhookrightarrow{f} & \mathfrak{S}_0(R_\delta) \times \mathfrak{S}_0(U^{-1}R) \\ {\scriptstyle i'}\downarrow{\scriptstyle\approx} & & & & {\scriptstyle j}\downarrow{\scriptstyle\approx} \\ \mathfrak{S}_0(A) \times \mathfrak{S}_0(U^{-1}R) & & \xrightarrow[\approx]{g} & & \mathfrak{S}_0(R_\delta) \times \mathfrak{S}_0(U^{-1}R) \end{array}$$

Once this is done, a simple diagram chase provides the bijectivity of h.

Localize the surjection $\varphi': A[\mathbf{X}] \to R$ by inverting Δ to obtain a surjection $\rho: A[\mathbf{X}]_\Delta \to R_\delta$. We claim that ρ is Gorenstein; see 2.9 for terminology. In 4.4 it is observed that φ' is Cohen-Macaulay of grade d. Hence, the same is true of ρ. The diagram (4.4.1) fits in the next commutative diagram of ring homomorphisms.

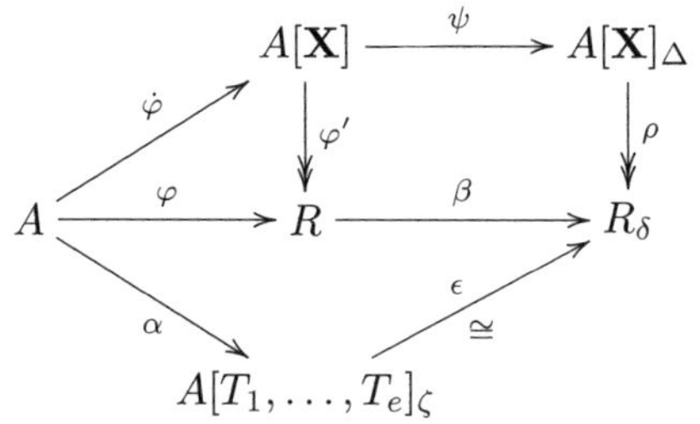

The map α is faithfully flat. Furthermore, for each prime ideal $\mathfrak{p}$ of A, the fibre $\kappa(\mathfrak{p}) \otimes_A A[T_1, \ldots, T_e]_\zeta \cong \kappa(\mathfrak{p})[T_1, \ldots, T_e]_\zeta$ is Gorenstein. That ρ is Gorenstein now follows from Avramov and Foxby [5, (6.2),(6.3)].

Set $\omega_\rho = \operatorname{Ext}^d_{A[\mathbf{X}]_\Delta}(R_\delta, A[\mathbf{X}]_\Delta)$, which is R_δ-semidualizing by [23, (5.6.a)]. Moreover, it is locally free of rank 1 by [23, (5.6.b)]. Setting

$$\omega_\rho^{-1} = \operatorname{Hom}_{R_\delta}(\omega_\rho, R_\delta),$$

the discussion in 3.6 yields an isomorphism

$$\omega_\rho \otimes_{R_\delta} \omega_\rho^{-1} \cong R_\delta. \tag{4.14.4}$$

We now define the aforementioned map j and demonstrate that it has the desired properties. For each semidualizing R_δ-module C, set

$$j([C], [U^{-1}R]) = ([C], [U^{-1}R]), \qquad j([C], [\omega_{U^{-1}R}]) = ([\omega_\rho^{-1} \otimes_{R_\delta} C], [\omega_{U^{-1}R}]).$$

It follows from the isomorphism (4.14.4) that the assignment

$$([C], [U^{-1}R]) \mapsto ([C], [U^{-1}R]), \qquad ([C], [\omega_{U^{-1}R}]) \mapsto ([\omega_\rho \otimes_{R_\delta} C], [\omega_{U^{-1}R}])$$

describes an inverse of j, so that j is bijective. It remains only to show that $gi' = jfh$, so fix a semidualizing A-module C. First, there are isomorphisms

$$C \otimes_A U^{-1}R \cong (C \otimes_A U^{-1}A) \otimes_{U^{-1}A} U^{-1}R \cong U^{-1}A \otimes_{U^{-1}A} U^{-1}R \cong U^{-1}R,$$

the first and third of which are standard, and the second of which is due to the fact that $U^{-1}A$ is a field. This yields equality (1) in the following sequence

$$\begin{aligned} jfh([C], 1) &= jf([C \otimes_A R]) \\ &= j([C \otimes_A R \otimes_R R_\delta)], [C \otimes_A R \otimes_R U^{-1}R]) \\ &\overset{(1)}{=} j([C \otimes_A R_\delta], [U^{-1}R]) \\ &= ([C \otimes_A R_\delta], [U^{-1}R]) \\ &= g([C], [U^{-1}R]) \\ &= gi'([C], 1), \end{aligned}$$

where each of the unmarked equalities follows either from a definition (e.g., (4.14.2)) or by a standard isomorphism.

To compute $jfh([C],0)$, we first describe some isomorphisms:

$$\begin{aligned}
C(\varphi)\otimes_R R_\delta &= \operatorname{Ext}^d_{A[\mathbf{X}]}(R, C\otimes_A A[\mathbf{X}])\otimes_R R_\delta \\
&\overset{(2)}{\cong} \operatorname{Ext}^d_{A[\mathbf{X}]}(R, C\otimes_A A[\mathbf{X}])\otimes_R (R\otimes_{A[\mathbf{X}]} A[\mathbf{X}]_\Delta) \\
&\cong \operatorname{Ext}^d_{A[\mathbf{X}]}(R, C\otimes_A A[\mathbf{X}])\otimes_{A[\mathbf{X}]} A[\mathbf{X}]_\Delta \\
&\overset{(3)}{\cong} \operatorname{Ext}^d_{A[\mathbf{X}]_\Delta}(R\otimes_{A[\mathbf{X}]} A[\mathbf{X}]_\Delta, C\otimes_A A[\mathbf{X}]\otimes_{A[\mathbf{X}]} A[\mathbf{X}]_\Delta) \\
&\overset{(4)}{\cong} \operatorname{Ext}^d_{A[\mathbf{X}]_\Delta}(R_\delta, C\otimes_A A[\mathbf{X}]_\Delta) \\
&\overset{(5)}{\cong} \omega_\rho\otimes_{R_\delta} (C\otimes_A A[\mathbf{X}]_\Delta\otimes_{A[\mathbf{X}]_\Delta} R_\delta) \\
&\cong \omega_\rho\otimes_{R_\delta} (C\otimes_A R_\delta).
\end{aligned}$$

Each of the unmarked isomorphisms is either by definition or standard. Isomorphisms (2) and (4) are via the isomorphism τ, whereas (3) is from the flatness of ψ. For (5) use the equality $\operatorname{pd}_{A[\mathbf{X}]_\Delta}(R_\delta)=d$ to apply [23, (1.7.b)] and the definition of ω_ρ. Similar explanations yield all but one of the following isomorphisms.

$$\begin{aligned}
C(\varphi)\otimes_R U^{-1}R &= \operatorname{Ext}^d_{A[\mathbf{X}]}(R, C\otimes_A A[\mathbf{X}])\otimes_R U^{-1}R \\
&\cong \operatorname{Ext}^d_{A[\mathbf{X}]}(R, C\otimes_A A[\mathbf{X}])\otimes_R (R\otimes_{A[\mathbf{X}]} U^{-1}A[\mathbf{X}]) \\
&\cong \operatorname{Ext}^d_{A[\mathbf{X}]}(R, C\otimes_A A[\mathbf{X}])\otimes_{A[\mathbf{X}]} U^{-1}A[\mathbf{X}] \\
&\cong \operatorname{Ext}^d_{U^{-1}A[\mathbf{X}]}(R\otimes_{A[\mathbf{X}]} U^{-1}A[\mathbf{X}], C\otimes_A A[\mathbf{X}]\otimes_{A[\mathbf{X}]} U^{-1}A[\mathbf{X}]) \\
&\cong \operatorname{Ext}^d_{U^{-1}A[\mathbf{X}]}(U^{-1}R, C\otimes_A U^{-1}A[\mathbf{X}]) \\
&\cong \operatorname{Ext}^d_{U^{-1}A[\mathbf{X}]}(U^{-1}R, (C\otimes_A U^{-1}A)\otimes_{U^{-1}A} U^{-1}A[\mathbf{X}]) \\
&\cong \operatorname{Ext}^d_{U^{-1}A[\mathbf{X}]}(U^{-1}R, U^{-1}A\otimes_{U^{-1}A} U^{-1}A[\mathbf{X}]) \\
&\cong \operatorname{Ext}^d_{U^{-1}A[\mathbf{X}]}(U^{-1}R, U^{-1}A[\mathbf{X}]) \\
&\overset{(6)}{\cong} \omega_{U^{-1}R}.
\end{aligned}$$

For isomorphism (6), the ring $U^{-1}A[\mathbf{X}]$ is regular and surjects onto $U^{-1}R$ so that $\operatorname{Ext}^d_{U^{-1}A[\mathbf{X}]}(U^{-1}R, U^{-1}A[\mathbf{X}])$ is a dualizing module for $U^{-1}R$, and is therefore isomorphic to $\omega_{U^{-1}R}$ since the dualizing module of $U^{-1}R$ is unique up to isomorphism.

The preceding isomorphisms yield equality (7) in the next computation, while (8) is by (4.14.4), and the others are by definition; see, e.g., (4.14.2).

$$\begin{aligned}
jfh([C],0) &= jf([C(\varphi)]) \\
&= j([C(\varphi)\otimes_R R_\delta], [C(\varphi)\otimes_R U^{-1}R]) \\
&\overset{(7)}{=} j([\omega_\rho\otimes_{R_\delta}(C\otimes_A R_\delta)], [\omega_{U^{-1}R}])
\end{aligned}$$

$$= ([\omega_\rho^{-1} \otimes_{R_\delta} \omega_\rho \otimes_{R_\delta} (C \otimes_A R_\delta)], [\omega_{U^{-1}R}])$$
$$\overset{(8)}{=} ([C \otimes_A R_\delta], [\omega_{U^{-1}R}])$$
$$= g([C], [\omega_{U^{-1}R}])$$
$$= gi'([C], 0).$$

To complete the proof, we verify the behavior of the orderings. One implication follows from [23, (4.7),(5.7),(5.12)]: If $[C]_A \trianglelefteq [C']_A$ and $i \leq i'$, then $h([C]_A, i) \trianglelefteq h([C']_A, i')$. For the converse, assume that $h([C]_A, i) \trianglelefteq h([C']_A, i')$. By way of contradiction, suppose that $i > i'$, that is, $i = 1$ and $i' = 0$. Then our assumption is $[C \otimes_A R]_R \trianglelefteq [C'(\varphi)]_R$. The computations above provide isomorphisms

$$C \otimes_A R \otimes_R U^{-1}R \cong U^{-1}R \qquad \text{and} \qquad C'(\varphi) \otimes_R U^{-1}R \cong \omega_{U^{-1}R},$$

so that the order-respecting map $\mathfrak{S}_0(\gamma)$ yields $[U^{-1}R]_{U^{-1}R} \trianglelefteq [\omega_{U^{-1}R}]_{U^{-1}R}$, a contradiction since $U^{-1}R$ is not Gorenstein. Thus, we have $i \leq i'$. The final conclusion $[C]_A \trianglelefteq [C']_A$ follows from [23, (4.8),(5.8),(5.13)]. □

Acknowledgments. I am grateful to Luchezar Avramov, Lars Christensen, Neil Epstein, Anders Frankild, Srikanth Iyengar, Graham Leuschke, Paul Roberts, and Roger Wiegand for stimulating conversations about this research. Many thanks also to the referee for his/her very thorough comments. I am especially grateful to Phillip Griffith for his tireless support and motivation. His work on the divisor class group and my many conversations with him inspired the research contained in this paper.

References

[1] D. Apassov, *Almost finite modules*, Comm. Algebra **27** (1999), 919–931. MR 1672015 (2000a:13037)

[2] T. Araya, R. Takahashi, and Y. Yoshino, *Homological invariants associated to semi-dualizing bimodules*, J. Math. Kyoto Univ. **45** (2005), 287–306. MR 2161693 (2006e:16014)

[3] M. Auslander, *Anneaux de Gorenstein, et torsion en algèbre commutative*, Séminaire d'Algèbre Commutative dirigé par Pierre Samuel, 1966/67, École Normale Supérieure de Jeunes Filles, Secrétariat mathématique, Paris, 1967. MR 0225844 (37 #1435)

[4] M. Auslander and M. Bridger, *Stable module theory*, Memoirs of the American Mathematical Society, No. 94, American Mathematical Society, Providence, R.I., 1969. MR 0269685 (42 #4580)

[5] L. L. Avramov and H.-B. Foxby, *Locally Gorenstein homomorphisms*, Amer. J. Math. **114** (1992), 1007–1047. MR 1183530 (93i:13019)

[6] ———, *Grothendieck's localization problem*, Commutative algebra: syzygies, multiplicities, and birational algebra (South Hadley, MA, 1992), Contemp. Math., vol. 159, Amer. Math. Soc., Providence, RI, 1994, pp. 1–13. MR 1266174 (94m:13011)

[7] ———, *Ring homomorphisms and finite Gorenstein dimension*, Proc. London Math. Soc. (3) **75** (1997), 241–270. MR 1455856 (98d:13014)

[8] N. Bourbaki, *Éléments de mathématique. Fasc. XXXI. Algèbre commutative. Chapitre 7: Diviseurs*, Actualités Scientifiques et Industrielles, No. 1314, Hermann, Paris, 1965. MR 0260715 (41 #5339)

[9] W. Bruns, *The canonical module of a determinantal ring*, Commutative algebra: Durham 1981 (Durham, 1981), London Math. Soc. Lecture Note Ser., vol. 72, Cambridge Univ. Press, Cambridge, 1982, pp. 109–120. MR 693630 (84i:13016)

[10] W. Bruns and J. Herzog, *Cohen-Macaulay rings*, Cambridge Studies in Advanced Mathematics, vol. 39, Cambridge University Press, Cambridge, 1993. MR 1251956 (95h:13020)

[11] W. Bruns and U. Vetter, *Determinantal rings*, Lecture Notes in Mathematics, vol. 1327, Springer-Verlag, Berlin, 1988. MR 953963 (89i:13001)

[12] L. W. Christensen, *Semi-dualizing complexes and their Auslander categories*, Trans. Amer. Math. Soc. **353** (2001), 1839–1883 (electronic). MR 1813596 (2002a:13017)

[13] L. W. Christensen and S. Sather-Wagstaff, *A Cohen-Macaulay algebra has only finitely many semidualizing modules*, Math. Proc. Cambridge Philos. Soc., to appear, `arXiv:0704.2734v2`.

[14] ———, *Descent via Koszul extensions*, preprint (2007), `arXiv:math/0612311`.

[15] V. I. Danilov, *Rings with a discrete group of divisor classes*, Mat. Sb. (N.S.) **83 (125)** (1970), 372–389. MR 0282980 (44 #214)

[16] ———, *Samuel's conjecture*, Mat. Sb. (N.S.) **81 (123)** (1970), 132–144. MR 0252374 (40 #5595)

[17] ———, *Rings with a discrete group of divisor classes*, Mat. Sb. (N.S.) **88 (130)** (1972), 229–237. MR 0306184 (46 #5311)

[18] H. Flenner, *Divisorenklassengruppen quasihomogener Singularitäten*, J. Reine Angew. Math. **328** (1981), 128–160. MR 636200 (83a:13009)

[19] R. M. Fossum, *The divisor class group of a Krull domain*, Springer-Verlag, New York, 1973, Ergebnisse der Mathematik und ihrer Grenzgebiete, vol. 74. MR 0382254 (52 #3139)

[20] H.-B. Foxby, *Hyperhomological algebra & commutative rings*, in preparation.

[21] ———, *Gorenstein modules and related modules*, Math. Scand. **31** (1972), 267–284 (1973). MR 0327752 (48 #6094)

[22] A. Frankild, *Quasi Cohen-Macaulay properties of local homomorphisms*, J. Algebra **235** (2001), 214–242. MR 1807663 (2001j:13023)

[23] A. Frankild and S. Sather-Wagstaff, *Reflexivity and ring homomorphisms of finite flat dimension*, Comm. Algebra **35** (2007), 461–500. MR 2294611

[24] ———, *The set of semidualizing complexes is a nontrivial metric space*, J. Algebra **308** (2007), 124–143. MR 2290914

[25] S. I. Gelfand and Y. I. Manin, *Methods of homological algebra*, Springer-Verlag, Berlin, 1996. MR 1438306 (97j:18001)

[26] A. Gerko, *On the structure of the set of semidualizing complexes*, Illinois J. Math. **48** (2004), 965–976. MR 2114263 (2005h:13026)

[27] ——— *On suitable modules and G-perfect ideals*, Uspekhi Mat. Nauk **56** (2001), 141–142. MR 1861448

[28] E. S. Golod, *G-dimension and generalized perfect ideals*, Trudy Mat. Inst. Steklov. **165** (1984), 62–66, Algebraic geometry and its applications. MR 752933 (85m:13011)

[29] A. Grothendieck, *Éléments de géométrie algébrique. III. Étude cohomologique des faisceaux cohérents. I*, Inst. Hautes Études Sci. Publ. Math. **11** (1961). MR 0163910 (29 #1209)

[30] R. Hartshorne, *Residues and duality*, Lecture Notes in Mathematics, vol. 20, Springer-Verlag, Berlin, 1966. MR 0222093 (36 #5145)

[31] ———, *Local cohomology*, A seminar given by A. Grothendieck, Harvard University, Fall, vol. 1961, Springer-Verlag, Berlin, 1967. MR 0224620 (37 #219)
[32] B. Herzog, *On the macaulayfication of local rings*, J. Algebra **67** (1980), no. 2, 305–317. MR 0602065 (82c:13029)
[33] J. Lipman, *Rings with discrete divisor class group: theorem of Danilov-Samuel*, Amer. J. Math. **101** (1979), no. 1, 203–211. MR 527832 (80g:13002)
[34] H. Matsumura, *Commutative ring theory*, second ed., Studies in Advanced Mathematics, vol. 8, Cambridge University Press, Cambridge, 1989. MR 1011461 (90i:13001)
[35] S. Spiroff, *The behavior on the restriction of divisor classes to sequences of hypersurfaces*, Ph.D. Thesis, University of Illinois at Urbana-Champaign, 2003.
[36] J.-L. Verdier, *Catégories dérivées*, SGA $4\frac{1}{2}$, Lecture Notes in Mathematics, vol. 569, Springer-Verlag, Berlin, 1977, pp. 262–311. MR 0463174 (57 #3132)
[37] ———, *Des catégories dérivées des catégories abéliennes*, Astérisque **239** (1996). MR 1453167 (98c:18007)

SEAN SATHER-WAGSTAFF, DEPARTMENT OF MATHEMATICS, CALIFORNIA STATE UNIVERSITY, DOMINGUEZ HILLS, 1000 E. VICTORIA ST., CARSON, CA 90747, USA

Current address: Department of Mathematics, North Dakota State University, 300 Minard Hall, Fargo, ND 58105-5075, USA

E-mail address: Sean.Sather-Wagstaff@ndsu.edu

URL: http://math.ndsu.nodak.edu/faculty/ssatherw/

Illinois Journal of Mathematics
Volume 51, Number 1, Spring 2007, Pages 287–298
S 0019-2082

LOCAL COHOMOLOGY AND PURE MORPHISMS

ANURAG K. SINGH AND ULI WALTHER

Dedicated to Professor Phil Griffith

ABSTRACT. We study a question raised by Eisenbud, Mustaţă, and Stillman regarding the injectivity of natural maps from Ext modules to local cohomology modules. We obtain some positive answers to this question which extend earlier results of Lyubeznik. In the process, we also prove a vanishing theorem for local cohomology modules which connects theorems previously known in the case of positive characteristic and in the case of monomial ideals.

1. Introduction

Throughout this paper, the rings we consider are commutative, Noetherian, and contain an identity element. For an ideal $\mathfrak{a}$ of a ring R, the local cohomology modules $H^i_{\mathfrak{a}}(R)$ may be obtained as

$$H^i_{\mathfrak{a}}(R) = \varinjlim_t \operatorname{Ext}^i_R(R/\mathfrak{a}_t, R) \qquad \text{for } i \geqslant 0,$$

where $\{\mathfrak{a}_t\}_{t\geqslant 0}$ is a decreasing chain of ideals cofinal with the chain $\{\mathfrak{a}^t\}_{t\geqslant 0}$, and the maps in the directed system are those induced by the natural surjections

$$R/\mathfrak{a}_{t+1} \longrightarrow R/\mathfrak{a}_t .$$

Any chain of ideals which is cofinal with the chain $\{\mathfrak{a}^t\}_{t\geqslant 0}$ yields the same direct limit. In this context, Eisenbud, Mustaţă, and Stillman have raised the following questions:

QUESTION 1.1 ([EMS, Question 6.1]). Let R be a polynomial ring over a field. For which ideals $\mathfrak{a}$ of R does there exist a chain of ideals $\{\mathfrak{a}_t\}_{t\geqslant 0}$ as above, such that for all $i \geqslant 0$ and all $t \geqslant 0$, the natural map

$$\operatorname{Ext}^i_R(R/\mathfrak{a}_t, R) \longrightarrow H^i_{\mathfrak{a}}(R)$$

Received August 7, 2006; received in final form January 18, 2007.
2000 *Mathematics Subject Classification.* Primary 13D45. Secondary 13A35, 13H05.
A.K.S. was supported by NSF grants DMS 0300600 and DMS 0600819. U.W. was supported by NSF grant DMS 0555319 and by NSA grant H98230-06-1-0012.

is injective?

QUESTION 1.2 ([EMS, Question 6.2]). Given a polynomial ring R over a field, for which ideals $\mathfrak{a}$ is the natural map $\operatorname{Ext}^i_R(R/\mathfrak{a}, R) \longrightarrow H^i_{\mathfrak{a}}(R)$ an inclusion?

Question 1.1 is motivated by the fact that the R-modules $H^i_{\mathfrak{a}}(R)$ are typically not finitely generated, whereas modules of the form $\operatorname{Ext}^i_R(R/\mathfrak{b}, R)$ are finitely generated. Consequently, a chain of ideals as in Question 1.1 yields a filtration of $H^i_{\mathfrak{a}}(R)$ by a natural family of finitely generated submodules.

Let R be a polynomial ring over a field. For an ideal $\mathfrak{a}$ generated by square-free monomials $m_1, \dots, m_r$, set $\mathfrak{a}^{[t]} = (m_1^t, \dots, m_r^t)$ for integers $t \geqslant 0$. Lyubeznik [Ly1, Theorem 1 (i)] proved that the natural maps

$$\operatorname{Ext}^i_R(R/\mathfrak{a}^{[t]}, R) \longrightarrow H^i_{\mathfrak{a}}(R)$$

are injective for all $i \geqslant 0$ and $t \geqslant 0$; see also Mustaţă [Mu, Theorem 1.1]. If R has positive characteristic, an ideal $\mathfrak{a}$ generated by square-free monomials has the property that $R/\mathfrak{a}$ is *F-pure*; see § 2. Our main result, Theorem 2.8, recovers Lyubeznik's result and also provides a positive answer to Question 1.1 for ideals defining F-pure rings, a case we single out for mention here:

THEOREM 1.3. *Let R be a regular ring containing a field of characteristic $p > 0$, and $\mathfrak{a}$ an ideal such that $R/\mathfrak{a}$ is F-pure. Then the natural maps*

$$\operatorname{Ext}^i_R(R/\mathfrak{a}^{[p^t]}, R) \longrightarrow H^i_{\mathfrak{a}}(R)$$

are injective for all $i \geqslant 0$ and all $t \geqslant 0$.

REMARK 1.4. If $d = \operatorname{depth}_R(\mathfrak{a}, R)$, then the natural map

$$\operatorname{Ext}^d_R(R/\mathfrak{a}, R) \longrightarrow H^d_{\mathfrak{a}}(R)$$

is injective. To see this, let $E^\bullet$ be a minimal injective resolution of R. Then $H^\bullet_{\mathfrak{a}}(R)$ is the cohomology of the complex $\Gamma_{\mathfrak{a}}(E^\bullet)$ and $\operatorname{Ext}^\bullet_R(R/\mathfrak{a}, R)$ is the cohomology of its subcomplex $\operatorname{Hom}_R(R/\mathfrak{a}, E^\bullet) = (0 :_{E^\bullet} \mathfrak{a})$. Since d is the least integer i such that $\Gamma_{\mathfrak{a}}(E^i)$ is nonzero, we are considering the cohomology of the rows of the diagram

$$\begin{array}{ccccccccc}
\cdots & \longrightarrow & 0 & \longrightarrow & \Gamma_{\mathfrak{a}}(E^d) & \longrightarrow & \Gamma_{\mathfrak{a}}(E^{d+1}) & \longrightarrow & \cdots \\
 & & \uparrow & & \uparrow & & \uparrow & & \\
\cdots & \longrightarrow & 0 & \longrightarrow & (0 :_{E^d} \mathfrak{a}) & \longrightarrow & (0 :_{E^{d+1}})\mathfrak{a} & \longrightarrow & \cdots,
\end{array}$$

and the desired inclusion follows.

REMARK 1.5. It is easy to see that Question 1.1 has a positive answer if $\mathfrak{a}$ is a set-theoretic complete intersection: if $f_1, \dots, f_n$ is a regular sequence generating $\mathfrak{a}$ up to radical, then the ideals $\mathfrak{a}_t = (f_1^t, \dots, f_n^t)$ form a descending chain with $\mathrm{Ext}^i_R(R/\mathfrak{a}_t, R) \hookrightarrow H^i_{\mathfrak{a}}(R)$ for all $i \geqslant 0$ and $t \geqslant 0$; for $i = n$, this follows from Remark 1.4, whereas if $i \neq n$, then $\mathrm{Ext}^i_R(R/\mathfrak{a}_t, R) = 0 = H^i_{\mathfrak{a}}(R)$.

We thank David Eisenbud, Srikanth Iyengar, and Oana Veliche for useful discussions. Our work also owes a great intellectual debt to Gennady Lyubeznik's paper [Ly2].

2. Pure homomorphisms and F-pure rings

DEFINITION 2.1. A ring homomorphism $\varphi\colon R \longrightarrow S$ is *pure* if the map

$$\varphi \otimes 1\colon R \otimes_R M \longrightarrow S \otimes_R M$$

is injective for each R-module M. If R contains a field of characteristic $p > 0$, then R is *F-pure* if the Frobenius homomorphism $r \mapsto r^p$ is pure.

Evidently, pure homomorphisms are injective. Let R be a subring of S. If the inclusion $R \hookrightarrow S$ splits as a maps of R-modules, then it is pure. The converse is also true for module-finite extensions; see [HR2, Corollary 5.3].

EXAMPLE 2.2. Let $R = \mathbb{K}[x_1, \dots, x_d]$ be a polynomial ring over a field $\mathbb{K}$, and let t be a positive integer. Then there is a $\mathbb{K}$-linear endomorphism φ of R with $\varphi(x_i) = x_i^t$ for $1 \leqslant i \leqslant d$. The inclusion $\varphi(R) \subseteq R$ splits since R is a free module over $\varphi(R)$ with basis $x_1^{e_1} \cdots x_d^{e_d}$, where $0 \leqslant e_i \leqslant t-1$. It follows that $\varphi\colon R \longrightarrow R$ is pure.

Let $\mathfrak{a}$ be an ideal of R generated by square-free monomials. Then $\varphi(\mathfrak{a}) \subseteq \mathfrak{a}$, so φ induces an endomorphism $\overline{\varphi}$ of $R/\mathfrak{a}$. The image of $\overline{\varphi}$ is spanned, as a $\mathbb{K}$-vector space, by those monomials in $x_1^t, \dots, x_d^t$ which are not in $\mathfrak{a}$. Using the map which is the identity on these monomials, and kills the rest, we obtain a splitting of $\overline{\varphi}$. It follows that the endomorphism $\overline{\varphi}\colon R/\mathfrak{a} \longrightarrow R/\mathfrak{a}$ is pure.

REMARK 2.3. The notion of F-pure rings was introduced by Hochster and Roberts in the course of their study of rings of invariants [HR1], [HR2]. Examples of F-pure rings include regular rings, determinantal rings, Plücker embeddings of Grassmannians, polynomial rings modulo square-free monomial ideals, normal affine semigroup rings, homogeneous coordinate rings of ordinary elliptic curves, and, more generally, homogeneous coordinate rings of ordinary Abelian varieties. Moreover, pure subrings of F-pure rings are F-pure, and if R and S are F-pure algebras over a perfect field $\mathbb{K}$, then their tensor product $R \otimes_{\mathbb{K}} S$ is also F-pure.

REMARK 2.4. Let $\varphi\colon R \longrightarrow S$ be a ring homomorphism. If $f \in R$, then φ localizes to give a map $R_f \longrightarrow S_{\varphi(f)}$. Similarly, if $\boldsymbol{f}$ is a sequence of elements of R, then φ induces a map of Čech complexes

$$\check{C}^\bullet_{\boldsymbol{f}}(R) \longrightarrow \check{C}^\bullet_{\varphi(\boldsymbol{f})}(S)\,.$$

Setting $\mathfrak{a} = (\boldsymbol{f})$, we have an induced map of local cohomology groups

$$\varphi_* : H^i_{\mathfrak{a}}(R) \longrightarrow H^i_{\mathfrak{a}S}(S) \qquad \text{for all } i \geqslant 0\,.$$

Note that for $r \in R$ and $\eta \in H^i_{\mathfrak{a}}(R)$, we have $\varphi(r)\varphi_*(\eta) = \varphi_*(r\eta)$.

Now suppose φ is an endomorphism of R with $\operatorname{rad}\mathfrak{a} = \operatorname{rad}\varphi(\mathfrak{a})R$. Then one obtains an induced *action*

$$\varphi_* : H^i_{\mathfrak{a}}(R) \longrightarrow H^i_{\varphi(\mathfrak{a})R}(R) = H^i_{\mathfrak{a}}(R)\,,$$

which is an endomorphism of the underlying Abelian group.

The archetypal example is the one where φ is the Frobenius endomorphism of a ring R of prime characteristic; in this case, for all ideals $\mathfrak{a}$ of R and integers $i \geqslant 0$, there is an induced action φ_* on $H^i_{\mathfrak{a}}(R)$ known as the *Frobenius action.*

If $\varphi\colon R \longrightarrow S$ is pure, then for all ideals $\mathfrak{a}$ of R and all integers $i \geqslant 0$, the induced map $\varphi_* : H^i_{\mathfrak{a}}(R) \longrightarrow H^i_{\mathfrak{a}S}(S)$ is injective; see [HR1, Corollary 6.8] or [HR2, Lemma 2.1]. In another direction, we have the following lemma, which will be a key ingredient in the proof of Theorem 2.8.

LEMMA 2.5. *Let $(R, \mathfrak{m})$ be a local ring with a pure endomorphism φ such that $\varphi(\mathfrak{m})R$ is $\mathfrak{m}$-primary. Then, for all $i \geqslant 0$, the induced action*

$$\varphi_* : H^i_{\mathfrak{m}}(R) \longrightarrow H^i_{\mathfrak{m}}(R)$$

is surjective up to R-span, i.e., $\varphi_*(H^i_{\mathfrak{m}}(R))$ generates $H^i_{\mathfrak{m}}(R)$ as an R-module.

Proof. Consider an element $\eta \in H^i_{\mathfrak{m}}(R)$; we need to show that it belongs to the R-module spanned by $\varphi_*(H^i_{\mathfrak{m}}(R))$. The descending chain of R-modules

$$\langle \eta, \varphi_*(\eta), \varphi_*^2(\eta), \dots\rangle \supseteq \langle \varphi_*(\eta), \varphi_*^2(\eta), \dots\rangle \supseteq \langle \varphi_*^2(\eta), \varphi_*^3(\eta), \dots\rangle$$

stabilizes since $H^i_{\mathfrak{m}}(R)$ is Artinian. Hence there exists $e \geqslant 0$ such that

$$\varphi_*^e(\eta) \in \langle \varphi_*^{e+1}(\eta), \varphi_*^{e+2}(\eta), \dots\rangle\,. \tag{2.5.1}$$

Let e be the least such integer. If $e = 0$ we are done, whereas if $e \geqslant 1$ then the R-module

$$M = \frac{\langle \varphi_*^{e-1}(\eta), \varphi_*^{e}(\eta), \varphi_*^{e+1}(\eta), \dots\rangle}{\langle \varphi_*^{e}(\eta), \varphi_*^{e+1}(\eta), \dots\rangle}$$

is nonzero. But then, by the purity of φ, so is its image under

$$\varphi \otimes 1 \colon R \otimes_R M \longrightarrow R \otimes_R M\,,$$

which contradicts (2.5.1). □

REMARK 2.6. Let R be a regular ring with a flat endomorphism φ. We use R^φ to denote the R-bimodule which has R as its underlying Abelian group, the usual action of R on the left, and the right R-action with $r'r = \varphi(r)r'$ for $r \in R$ and $r' \in R^\varphi$. Let Φ be the functor on the category of R-modules with

$$\Phi(M) = R^\varphi \otimes_R M\,,$$

where $\Phi(M)$ is viewed as an R-module via the left R-module structure of R^{φ}. The iteration Φ^t is the functor with

$$\Phi^t(M) = R^{\varphi} \otimes_R \Phi^{t-1}(M) \qquad \text{for } t \geqslant 1,$$

where Φ^0 is interpreted as the identity functor. It is easily seen that

$$\Phi^t(M) = R^{\varphi^t} \otimes_R M.$$

(1) There is an isomorphism $\Phi(R) \cong R$ given by $r' \otimes r \mapsto r'\varphi(r)$. It follows that if M is a free R-module, then $\Phi(M) \cong M$. For a map α of free modules given by a matrix (α_{ij}), the map $\Phi(\alpha)$ is given by the matrix $(\varphi(\alpha_{ij}))$. Since φ is flat, Φ is an exact functor, and so it takes finite free resolutions to finite free resolutions. If M and N are R-modules, then there are natural isomorphisms

$$\Phi(\operatorname{Ext}^i_R(M,N)) \cong \operatorname{Ext}^i_R(\Phi(M),\Phi(N)) \qquad \text{for all } i \geqslant 0. \tag{2.6.1}$$

In particular, if $\mathfrak{a}$ is an ideal of R, then (2.6.1) implies that

$$\Phi(\operatorname{Ext}^i_R(R/\mathfrak{a},R)) \cong \operatorname{Ext}^i_R(R/\varphi(\mathfrak{a})R,R).$$

(2) Suppose that the ideals $\{\varphi^t(\mathfrak{a})R\}_{t\geqslant 0}$ form a descending chain cofinal with the chain $\{\mathfrak{a}^t\}_{t\geqslant 0}$. Then, for each $i \geqslant 0$, the above isomorphism and its iterations fit into a commutative diagram

$$\begin{array}{ccccccc}
\cdots \longrightarrow & \operatorname{Ext}^i_R(R/\varphi^t(\mathfrak{a})R,R) & \longrightarrow & \operatorname{Ext}^i_R(R/\varphi^{t+1}(\mathfrak{a})R,R) & \longrightarrow \cdots \\
& \big\downarrow & & \big\downarrow & \\
\cdots \longrightarrow & \Phi^t(\operatorname{Ext}^i_R(R/\mathfrak{a},R)) & \longrightarrow & \Phi^{t+1}(\operatorname{Ext}^i_R(R/\mathfrak{a},R)) & \longrightarrow \cdots
\end{array}$$

where the maps in the top row are those induced by the natural surjections $R/\varphi^{t+1}(\mathfrak{a})R \longrightarrow R/\varphi^t(\mathfrak{a})R$, and the vertical maps are isomorphisms. Hence the bottom row has direct limit $H^i_{\mathfrak{a}}(R)$. It follows that $H^i_{\mathfrak{m}}(R) \cong \Phi(H^i_{\mathfrak{m}}(R))$.

(3) Assume in addition that $(R,\mathfrak{m})$ is a regular local ring of dimension d, and that φ is a flat local endomorphism. In this case, the dimension formula

$$\dim R + \dim R/\varphi(\mathfrak{m})R = \dim R$$

implies that $\varphi(\mathfrak{m})R$ is $\mathfrak{m}$-primary. Let E denote the injective hull of $R/\mathfrak{m}$ as an R-module, and set $(-)^{\vee} = \operatorname{Hom}_R(-,E)$. Since R is Gorenstein, we have

$$E \cong H^d_{\mathfrak{m}}(R) \cong \Phi(H^d_{\mathfrak{m}}(R)) \cong \Phi(E).$$

Hence (2.6.1) implies that $\Phi(M^{\vee}) \cong (\Phi(M))^{\vee}$ for each R-module M. Setting $M = \operatorname{Ext}^i_R(R/\mathfrak{a},R)$ and using local duality, we get

$$(\Phi(\operatorname{Ext}^i_R(R/\mathfrak{a},R)))^{\vee} \cong \Phi(H^{d-i}_{\mathfrak{m}}(R/\mathfrak{a})).$$

Since $\Phi^t(-) = R^{\varphi^t} \otimes_R (-)$, we immediately obtain the isomorphisms

$$(\Phi^t(\operatorname{Ext}^i_R(R/\mathfrak{a},R)))^{\vee} \cong \Phi^t(H^{d-i}_{\mathfrak{m}}(R/\mathfrak{a})) \qquad \text{for all } t \geqslant 0.$$

Applying $(-)^\vee$ to the diagram in (2), we get the commutative diagram

$$\begin{array}{ccccccc} \cdots & \longleftarrow & H^{d-i}_{\mathfrak{m}}(R/\varphi^t(\mathfrak{a})R) & \longleftarrow & H^{d-i}_{\mathfrak{m}}(R/\varphi^{t+1}(\mathfrak{a})R) & \longleftarrow & \cdots \\ & & \uparrow & & \uparrow & & \\ \cdots & \longleftarrow & \Phi^t(H^{d-i}_{\mathfrak{m}}(R/\mathfrak{a})) & \longleftarrow & \Phi^{t+1}(H^{d-i}_{\mathfrak{m}}(R/\mathfrak{a})) & \longleftarrow & \cdots \end{array}$$

where the vertical maps are isomorphisms, and the maps in the first row are those induced by the natural surjections $R/\varphi^{t+1}(\mathfrak{a})R \longrightarrow R/\varphi^t(\mathfrak{a})R$.

In the archetypal example, R is a regular ring containing a field of positive characteristic, and φ is the Frobenius endomorphism. In this case, φ is flat by Kunz's theorem [Ku, Theorem 2.1]. The functor Φ is the Peskine-Szpiro functor of [PS], and the commutative diagram in Remark 2.6 (2) is precisely that obtained by Lyubeznik in [Ly2, Lemma 2.1]. The following is a mild generalization of [Ly2, Lemma 2.2].

LEMMA 2.7. *Let $(R, \mathfrak{m})$ be a regular local ring with a flat local endomorphism φ, and let $\mathfrak{a}$ be an ideal such that $\varphi(\mathfrak{a}) \subseteq \mathfrak{a}$. Then φ induces an endomorphism $\overline{\varphi}$ of $R/\mathfrak{a}$, and hence an action $\overline{\varphi}_*\colon H^i_{\mathfrak{m}}(R/\mathfrak{a}) \longrightarrow H^i_{\mathfrak{m}}(R/\mathfrak{a})$. The composition*

$$R^\varphi \otimes_R H^i_{\mathfrak{m}}(R/\mathfrak{a}) \xrightarrow{\ \cong\ } H^i_{\mathfrak{m}}(R/\varphi(\mathfrak{a})R) \xrightarrow{\ \pi\ } H^i_{\mathfrak{m}}(R/\mathfrak{a})$$

is the map with $r' \otimes \eta \longmapsto r' \cdot \overline{\varphi}_*(\eta)$, where π is the map induced by the natural surjection $R/\varphi(\mathfrak{a})R \longrightarrow R/\mathfrak{a}$.

Proof. Since $\varphi(\mathfrak{m})R$ is $\mathfrak{m}$-primary, if $\boldsymbol{x}$ is a system of parameters for R, then so is its image $\varphi(\boldsymbol{x})$. The displayed isomorphism is a consequence of the flatness of φ as we saw in Remark 2.6. To analyze this isomorphism, let $\widetilde{\eta}$ be a lift of $\eta \in H^i_{\mathfrak{m}}(R/\mathfrak{a})$ to the module $\check{C}^i_{\boldsymbol{x}}(R/\mathfrak{a})$ of the Čech complex $\check{C}^\bullet_{\boldsymbol{x}}(R/\mathfrak{a})$. Then

$$\varphi(\widetilde{\eta}) \in \check{C}^i_{\varphi(\boldsymbol{x})}(R/\varphi(\mathfrak{a})R)$$

and the image of $r' \otimes \eta$ under the isomorphism is the image of $r' \cdot \varphi(\widetilde{\eta})$ in $H^i_{\mathfrak{m}}(R/\varphi(\mathfrak{a})R)$. Lastly, π maps this to $r' \cdot \overline{\varphi}_*(\eta) \in H^i_{\mathfrak{m}}(R/\mathfrak{a})$. □

We are now ready to prove the main result:

THEOREM 2.8. *Let R be a regular ring and $\mathfrak{a}$ an ideal of R. Suppose R has a flat endomorphism φ such that $\{\varphi^t(\mathfrak{a})R\}_{t \geqslant 0}$ is a decreasing chain of ideals cofinal with $\{\mathfrak{a}^t\}_{t \geqslant 0}$, and the induced endomorphism $\overline{\varphi}\colon R/\mathfrak{a} \longrightarrow R/\mathfrak{a}$ is pure. Then, for all $i \geqslant 0$ and $t \geqslant 0$, the natural map*

$$\operatorname{Ext}^i_R(R/\varphi^t(\mathfrak{a})R, R) \longrightarrow \operatorname{Ext}^i_R(R/\varphi^{t+1}(\mathfrak{a})R, R)$$

is injective.

Proof. It suffices to verify the injectivity after localizing at maximal ideals, so we assume that $(R, \mathfrak{m})$ is a regular local ring. Let $d = \dim R$, and let E be the injective hull of $R/\mathfrak{m}$ as an R-module. Using $(-)^\vee = \operatorname{Hom}_R(-, E)$, local duality gives an isomorphism

$$\operatorname{Ext}^i_R(R/\varphi^t(\mathfrak{a})R, R)^\vee \cong H^{d-i}_\mathfrak{m}(R/\varphi^t(\mathfrak{a})R),$$

and it suffices to show that the map

$$H^{d-i}_\mathfrak{m}(R/\varphi^{t+1}(\mathfrak{a})R) \longrightarrow H^{d-i}_\mathfrak{m}(R/\varphi^t(\mathfrak{a})R) \tag{2.8.1}$$

induced by the natural surjection

$$R/\varphi^{t+1}(\mathfrak{a})R \longrightarrow R/\varphi^t(\mathfrak{a})R$$

is surjective for each $t \geqslant 0$. In view of the isomorphisms

$$R^\varphi \otimes_R H^{d-i}_\mathfrak{m}(R/\varphi^t(\mathfrak{a})R) \cong H^{d-i}_\mathfrak{m}(R/\varphi^{t+1}(\mathfrak{a})R)$$

and the right exactness of tensor, it suffices to verify the surjectivity of (2.8.1) in the case $t = 0$. By Lemma 2.7, this reduces to checking that the $\overline{\varphi}$-action

$$\overline{\varphi}_* \colon H^{d-i}_\mathfrak{m}(R/\mathfrak{a}) \longrightarrow H^{d-i}_\mathfrak{m}(R/\mathfrak{a})$$

is surjective up to taking the R-span. This follows from Lemma 2.5. □

Theorem 1.3 follows immediately from Theorem 2.8 by taking φ to be the Frobenius endomorphism. To recover the result for square-free monomial ideals [Ly1, Theorem 1 (i)], take φ as in Example 2.2.

3. Examples

We first construct an example of a module M over a regular local ring $(R, \mathfrak{m})$ such that $H^i_\mathfrak{m}(M) = 0$ but $\operatorname{Ext}^i_R(R/\mathfrak{a}, M)$ is nonzero for every $\mathfrak{m}$-primary ideal $\mathfrak{a}$ of R. It then follows that $H^i_\mathfrak{m}(M)$ cannot be realized as a union of appropriate Ext-modules. We use the following lemma:

LEMMA 3.1. *Let $(R, \mathfrak{m})$ be a regular local ring of dimension d, and let $\mathfrak{a}$ be an $\mathfrak{m}$-primary ideal. Then, for each R-module M, there is an isomorphism*

$$\operatorname{Ext}^i_R(R/\mathfrak{a}, M) \cong \operatorname{Tor}^R_{d-i}(\operatorname{Ext}^d_R(R/\mathfrak{a}, R), M) \qquad \textit{for all } 0 \leqslant i \leqslant d.$$

Proof. Let $P_\bullet$ be a minimal free resolution of $R/\mathfrak{a}$. The complex $\operatorname{Hom}_R(P_\bullet, R)$ has homology $\operatorname{Ext}^\bullet_R(R/\mathfrak{a}, R)$. Since $\mathfrak{a}$ is an $\mathfrak{m}$-primary ideal of a regular ring R, we have $\operatorname{depth}_\mathfrak{a} R = d$, and so $\operatorname{Ext}^j_R(R/\mathfrak{a}, R)$ is nonzero only for $j = d$. It follows that, with a change of index, $\operatorname{Hom}_R(P_\bullet, R)$ is an acyclic complex of free modules resolving the module $\operatorname{Ext}^d_R(R/\mathfrak{a}, R)$. Hence

$$\begin{aligned}\operatorname{Ext}^i_R(R/\mathfrak{a}, M) = H^i(\operatorname{Hom}(P_\bullet, M)) &\cong H^i(\operatorname{Hom}(P_\bullet, R) \otimes_R M) \\ &\cong \operatorname{Tor}^R_{d-i}(\operatorname{Ext}^d_R(R/\mathfrak{a}, R), M).\end{aligned}$$ □

EXAMPLE 3.2. Let $(R,\mathfrak{m})$ be a regular local ring of dimension $d > 0$, and x a nonzero element of $\mathfrak{m}$. Then $R/(x)$ has dimension $d-1$, so $H^d_\mathfrak{m}(R/(x)) = 0$. However, if $\mathfrak{a}$ is an $\mathfrak{m}$-primary ideal, then Lemma 3.1 implies that

$$\operatorname{Ext}^d_R(R/\mathfrak{a}, R/(x)) \cong \operatorname{Ext}^d_R(R/\mathfrak{a}, R) \otimes_R R/(x)\,,$$

which is nonzero. In particular, if $\{\mathfrak{a}_t\}_{t\geqslant 0}$ is a decreasing family of ideals cofinal with $\{\mathfrak{m}^t\}_{t\geqslant 0}$, then the modules $\operatorname{Ext}^d_R(R/\mathfrak{a}_t, R/(x))$ are nonzero for each t, and so the maps $\operatorname{Ext}^d_R(R/\mathfrak{a}_t, R/(x)) \longrightarrow H^d_\mathfrak{m}(R/(x))$ are not injective.

Example 3.4 below is due to Eisenbud: given positive integers $a \leqslant b-2$, there exists a polynomial ring R and a finitely generated graded R-module M, such that the natural map $\operatorname{Ext}^i_R(R/\mathfrak{a}, M) \longrightarrow H^i_\mathfrak{m}(M)$ is not injective for all $a < i < b$ and all $\mathfrak{m}$-primary ideals $\mathfrak{a}$. This is based on a construction of Evans and Griffith [EG, Theorem A].

THEOREM 3.3 (Evans-Griffith). *Let $\mathbb{K}$ be an infinite field and take a sequence of positive integers, $n_0 < n_1 < \cdots < n_s$. Then there exists a polynomial ring R over $\mathbb{K}$, with a homogeneous prime ideal $\mathfrak{p}$, such that the local cohomology module $H^i_\mathfrak{m}(R/\mathfrak{p})$ is nonzero if and only if $i \in \{n_0, n_1, \ldots, n_s\}$. Moreover, if $n_0 \geqslant 2$, then $R/\mathfrak{p}$ may be chosen to be a normal domain.* □

EXAMPLE 3.4 (Eisenbud). Let $a \leqslant b-2$ be positive integers. By Theorem 3.3, there exists a polynomial ring R with a homogeneous prime $\mathfrak{p}$, such that $\operatorname{depth} R/\mathfrak{p} = a$, $\dim R/\mathfrak{p} = b$ and $H^j_\mathfrak{m}(R/\mathfrak{p}) = 0$ for all $a < j < b$. Let $\mathfrak{a}$ be an $\mathfrak{m}$-primary ideal. Then $\operatorname{Ext}^a_R(R/\mathfrak{a}, R/\mathfrak{p})$ is nonzero so, by Lemma 3.1,

$$\operatorname{Tor}^R_{d-a}(\operatorname{Ext}^d_R(R/\mathfrak{a}, R), R/\mathfrak{p}) \neq 0 \qquad \text{where } d = \dim R\,.$$

By the rigidity of Tor over regular local rings, [Li], it follows that

$$\operatorname{Tor}^R_j(\operatorname{Ext}^d_R(R/\mathfrak{a}, R), R/\mathfrak{p}) \neq 0 \qquad \text{for all } 0 \leqslant j \leqslant d-a\,.$$

By another application of Lemma 3.1, the module $\operatorname{Ext}^i_R(R/\mathfrak{a}, R/\mathfrak{p})$ is nonzero if $a \leqslant i \leqslant d$. Now if $\{\mathfrak{a}_t\}_{t\geqslant 0}$ is any decreasing family of ideals cofinal with $\{\mathfrak{m}^t\}_{t\geqslant 0}$, it follows that the maps

$$\operatorname{Ext}^i_R(R/\mathfrak{a}_t, R/\mathfrak{p}) \longrightarrow H^i_\mathfrak{m}(R/\mathfrak{p})$$

are not injective for each $a < i < b$ and each $t \geqslant 0$.

EXAMPLE 3.5. Let $\mathbb{K}$ be a field and consider the $\mathbb{K}$-linear ring homomorphism

$$\alpha\colon R = \mathbb{K}[w,x,y,z] \longrightarrow \mathbb{K}[s^4, s^3t, st^3, t^4]$$

where α sends w, x, y, z to the elements s^4, s^3t, st^3, t^4 respectively. Let $\mathfrak{a}$ be the kernel of α. Using vanishing theorems such as [HL, Theorem 2.9], it may be verified that $H^i_\mathfrak{a}(R) = 0$ for $i \geqslant 3$.

If $\mathbb{K}$ has characteristic $p > 0$, Hartshorne [Ha] showed that $\mathfrak{a}$ is a set-theoretic complete intersection, i.e., that there exist elements f, g in R such that $\mathfrak{a} = \operatorname{rad}(f, g)$. In this case, the ideals $\mathfrak{a}_t = (f^t, g^t)$ form a descending chain cofinal with $\{\mathfrak{a}^t\}$ for which the maps $\operatorname{Ext}^i_R(R/\mathfrak{a}_t, R) \longrightarrow H^i_{\mathfrak{a}}(R)$ are injective for all $i \geqslant 0$ and $t \geqslant 0$; see Remark 1.5.

Next, suppose that $\mathbb{K}$ has characteristic 0. If $\mathfrak{b}$ is an ideal with $\operatorname{rad}\mathfrak{b} = \mathfrak{a}$ such that $\operatorname{Ext}^i_R(R/\mathfrak{b}, R) \longrightarrow H^i_{\mathfrak{a}}(R)$ is injective for all $i \geqslant 0$, then

$$\operatorname{Ext}^3_R(R/\mathfrak{b}, R) = 0 = \operatorname{Ext}^4_R(R/\mathfrak{b}, R)$$

and so $R/\mathfrak{b}$ is Cohen-Macaulay. This leads to the following question:

QUESTION 3.6. Let $\mathbb{K}$ be a field of characteristic 0 and, as in Example 3.5, let $\mathfrak{a} \subset R = \mathbb{K}[w, x, y, z]$ be an ideal with $R/\mathfrak{a} \cong \mathbb{K}[s^4, s^3t, st^3, t^4]$. Is the ideal $\mathfrak{a}$ set-theoretically Cohen-Macaulay, i.e., does there exist an ideal $\mathfrak{b} \subset R$ with $\operatorname{rad}\mathfrak{b} = \mathfrak{a}$, such that the ring $R/\mathfrak{b}$ is Cohen-Macaulay?

While the requirement of F-purity in Theorem 1.3 is certainly a strong hypothesis, it appears to be a crucial ingredient. In the following example, we have regular rings $R_p = R/pR$ of prime characteristic p and ideals $\mathfrak{a}_p = \mathfrak{a}R_p$ such that the maps

$$\operatorname{Ext}^4_{R_p}(R_p/\mathfrak{a}_p^{[p^t]}, R_p) \longrightarrow H^4_{\mathfrak{a}_p}(R_p)$$

are injective if and only if $R_p/\mathfrak{a}_p$ is F-pure; the set of primes for which this is the case is infinite, as is its complement.

EXAMPLE 3.7. Let $E \subset \mathbb{P}^2_{\mathbb{Q}}$ be an elliptic curve, and consider the Segre embedding of $E \times \mathbb{P}^1_{\mathbb{Q}}$ in $\mathbb{P}^5_{\mathbb{Q}}$. Clearing denominators in a set of generators for the defining ideal of the homogeneous coordinate ring, we obtain an ideal $\mathfrak{a}$ of $R = \mathbb{Z}[u, v, w, x, y, z]$ such that $R/\mathfrak{a} \otimes_{\mathbb{Z}} \mathbb{Q}$ is the coordinate ring of $E \times \mathbb{P}^1_{\mathbb{Q}}$. For prime integers p, let $R_p = R/pR$ and $\mathfrak{a}_p = \mathfrak{a}R_p$. For all but finitely many primes p, the reduction mod p of E is a smooth elliptic curve E_p and $R_p/\mathfrak{a}_p$ is a homogeneous coordinate ring for $E_p \times \mathbb{P}^1_{\mathbb{Z}/p}$. We restrict our attention to such primes. Since $\operatorname{depth} R_p/\mathfrak{a}_p = 2$, the Auslander-Buchsbaum formula implies that $\operatorname{pd}_{R_p} R_p/\mathfrak{a}_p = 4$. Using the flatness of Frobenius, we see that

$$\operatorname{pd}_{R_p} R_p/\mathfrak{a}_p^{[p^t]} = 4,$$

and hence that

$$\operatorname{Ext}^4_{R_p}(R_p/\mathfrak{a}_p^{[p^t]}, R_p) \neq 0 \qquad \text{for all } t \geqslant 0.$$

On the other hand, $H^4_{\mathfrak{a}_p}(R_p)$ is zero if E_p is supersingular and nonzero if E_p is ordinary; see [HS, Example 3, page 75] or [Ly2, page 219]. By well-know results on elliptic curves, there are infinitely primes p for which E_p

is supersingular, and infinitely many for which it is ordinary. Consider the natural map

$$\operatorname{Ext}^i_{R_p}(R_p/\mathfrak{a}_p^{[p^t]}, R_p) \longrightarrow H^i_{\mathfrak{a}_p}(R_p)\,. \tag{3.7.1}$$

Ordinary primes. If E_p is ordinary, then its coordinate ring is F-pure, and it follows that $R_p/\mathfrak{a}_p$ is F-pure as well. In this case, Theorem 1.3 implies that the map (3.7.1) is injective for all $i \geqslant 0$ and $t \geqslant 0$.

Supersingular primes. If p is a prime such that E_p is supersingular, then $H^4_{\mathfrak{a}_p}(R_p) = 0$, so the map (3.7.1) is not injective for $i = 4$. We do not know whether there exists an $\mathfrak{a}_p$-primary ideal $\mathfrak{b}$ for which the maps

$$\operatorname{Ext}^i_{R_p}(R_p/\mathfrak{b}, R_p) \longrightarrow H^i_{\mathfrak{a}_p}(R_p)$$

are injective for all $i \geqslant 0$. Since $H^i_{\mathfrak{a}_p}(R_p) = 0$ for $i \geqslant 4$ in the supersingular case, the existence of such an ideal would imply that $\mathfrak{a}_p$ is set-theoretically Cohen-Macaulay; see also [SW, § 3].

4. A vanishing criterion

The observations from § 2 yield the following vanishing theorem, which links Lyubeznik's positive characteristic result [Ly2, Theorem 1.1] to a theorem for monomial ideals recorded below as Corollary 4.2.

THEOREM 4.1. *Let $(R, \mathfrak{m})$ be a regular local ring, $\mathfrak{a}$ an ideal, and φ a flat local endomorphism such that $\{\varphi^t(\mathfrak{a})R\}_{t\geqslant 0}$ is a decreasing chain of ideals cofinal with the chain $\{\mathfrak{a}^t\}_{t\geqslant 0}$. Then $H^i_{\mathfrak{a}}(R) = 0$ if and only if some iteration of the induced action*

$$\overline{\varphi}_* : H^{\dim R - i}_{\mathfrak{m}}(R/\mathfrak{a}) \longrightarrow H^{\dim R - i}_{\mathfrak{m}}(R/\mathfrak{a})$$

is zero.

Proof. Let $d = \dim R$. The direct limit

$$H^i_{\mathfrak{a}}(R) = \varinjlim_t \operatorname{Ext}^i_R(R/\varphi^t(\mathfrak{a})R, R)$$

vanishes if and only if for each $t \in \mathbb{N}$, there exists $k \in \mathbb{N}$ such that the map

$$\operatorname{Ext}^i_R(R/\varphi^t(\mathfrak{a})R, R) \longrightarrow \operatorname{Ext}^i_R(R/\varphi^{t+k}(\mathfrak{a})R, R) \tag{4.1.1}$$

induced by the surjection $R/\varphi^{t+k}(\mathfrak{a})R \longrightarrow R/\varphi^t(\mathfrak{a})R$ is zero. By local duality, the map (4.1.1) is zero if and only if

$$H^{d-i}_{\mathfrak{m}}(R/\varphi^{t+k}(\mathfrak{a})R) \longrightarrow H^{d-i}_{\mathfrak{m}}(R/\varphi^t(\mathfrak{a})R)$$

is the zero map. By Remark 2.6 (3) and the flatness of $R^\varphi \otimes_R -$, this is equivalent to the map

$$H^{d-i}_{\mathfrak{m}}(R/\varphi^k(\mathfrak{a})R) \longrightarrow H^{d-i}_{\mathfrak{m}}(R/\mathfrak{a})$$

being zero. By Lemma 2.7, this last condition is equivalent to the k-th iterate of the action $\overline{\varphi}_*\colon H^{d-i}_{\mathfrak{m}}(R/\mathfrak{a}) \longrightarrow H^{d-i}_{\mathfrak{m}}(R/\mathfrak{a})$ being zero. □

Using Theorem 4.1 we next recover a vanishing theorem for monomial ideals, [Ly1, Theorem 1 (iii)]. In [Mi, Corollary 6.7] Miller proves a stronger statement connecting $H^i_{\mathfrak{a}}(S)$ and $H^{\dim S - i}_{\mathfrak{m}}(S/\mathfrak{a})$ via Alexander duality.

COROLLARY 4.2. *Let S be a polynomial ring over a field, and let $\mathfrak{a}$ be an ideal generated by square-free monomials. Then $H^i_{\mathfrak{a}}(S) = 0$ if and only if $H^{\dim S - i}_{\mathfrak{m}}(S/\mathfrak{a}) = 0$.*

Proof. Let $S = \mathbb{K}[x_1, \dots, x_d]$, and let φ be the $\mathbb{K}$-linear endomorphism with $\varphi(x_i) = x_i^2$ for $1 \leqslant i \leqslant d$. Then φ is flat, and induces a pure endomorphism of $S/\mathfrak{a}$; see Example 2.2.

Each of the modules in question is graded, so the issue of vanishing is unchanged under localization at the homogeneous maximal ideal of S. We can therefore work over the regular local ring $(R, \mathfrak{m})$, where we need to show that $H^i_{\mathfrak{a}}(R) = 0$ if and only if $H^{d-i}_{\mathfrak{m}}(R/\mathfrak{a}R) = 0$. The endomorphism φ localizes to give a flat endomorphism of R. Moreover, since purity localizes, φ induces a pure endomorphism $\overline{\varphi}$ of $R/\mathfrak{a}R$. By Theorem 4.1, $H^i_{\mathfrak{a}}(R) = 0$ if and only if some iterate of the action

$$\overline{\varphi}_*\colon H^{d-i}_{\mathfrak{m}}(R/\mathfrak{a}R) \longrightarrow H^{d-i}_{\mathfrak{m}}(R/\mathfrak{a}R)$$

is zero. But $\overline{\varphi}_*$ is injective since $\overline{\varphi}$ is pure, so an iterate of $\overline{\varphi}_*$ is zero precisely if $H^{d-i}_{\mathfrak{m}}(R/\mathfrak{a}R) = 0$. □

REFERENCES

[EMS] D. Eisenbud, M. Mustaţă, and M. Stillman, *Cohomology on toric varieties and local cohomology with monomial supports*, J. Symbolic Comput. **29** (2000), 583–600. MR 1769656 (2002a:14059)

[EG] E. G. Evans, Jr. and P. A. Griffith, *Local cohomology modules for normal domains*, J. London Math. Soc. (2) **19** (1979), 277–284. MR 533326 (81b:14002)

[Ha] R. Hartshorne, *Complete intersections in characteristic $p > 0$*, Amer. J. Math. **101** (1979), 380–383. MR 527998 (80d:14028)

[HS] R. Hartshorne and R. Speiser, *Local cohomological dimension in characteristic p*, Ann. of Math. (2) **105** (1977), 45–79. MR 0441962 (56 #353)

[HR1] M. Hochster and J. L. Roberts, *Rings of invariants of reductive groups acting on regular rings are Cohen-Macaulay*, Advances in Math. **13** (1974), 115–175. MR 0347810 (50 #311)

[HR2] ———, *The purity of the Frobenius and local cohomology*, Advances in Math. **21** (1976), 117–172. MR 0417172 (54 #5230)

[HL] C. Huneke and G. Lyubeznik, *On the vanishing of local cohomology modules*, Invent. Math. **102** (1990), 73–93. MR 1069240 (91i:13020)

[Ku] E. Kunz, *Characterizations of regular local rings for characteristic p*, Amer. J. Math. **91** (1969), 772–784. MR 0252389 (40 #5609)

[Li] S. Lichtenbaum, *On the vanishing of* Tor *in regular local rings*, Illinois J. Math. **10** (1966), 220–226. MR 0188249 (32 #5688)

[Ly1] G. Lyubeznik, *On the local cohomology modules $H^i_{\mathfrak{a}}(R)$ for ideals $\mathfrak{a}$ generated by monomials in an R-sequence*, Complete intersections (Acireale, 1983), Lecture Notes in Math., vol. 1092, Springer, Berlin, 1984, pp. 214–220. MR 775884 (86f:14002)

[Ly2] ———, *On the vanishing of local cohomology in characteristic $p > 0$*, Compos. Math. **142** (2006), 207–221. MR 2197409 (2007b:13029)

[Mi] E. Miller, *The Alexander duality functors and local duality with monomial support*, J. Algebra **231** (2000), 180–234. MR 1779598 (2001k:13028)

[Mu] M. Mustaţă, *Local cohomology at monomial ideals*, J. Symbolic Comput. **29** (2000), 709–720. MR 1769662 (2001i:13020)

[PS] C. Peskine and L. Szpiro, *Dimension projective finie et cohomologie locale. Applications à la démonstration de conjectures de M. Auslander, H. Bass et A. Grothendieck*, Inst. Hautes Études Sci. Publ. Math. (1973), 47–119. MR 0374130 (51 #10330)

[SW] A. K. Singh and U. Walther, *On the arithmetic rank of certain Segre products*, Commutative algebra and algebraic geometry, Contemp. Math., vol. 390, Amer. Math. Soc., Providence, RI, 2005, pp. 147–155. MR 2187332 (2006h:14059)

Anurag K. Singh, Department of Mathematics, University of Utah, 155 South 1400 East, Salt Lake City, UT 84112, USA

E-mail address: singh@math.utah.edu

Uli Walther, Department of Mathematics, Purdue University, 150 N. University Street, West Lafayette, IN 47907, USA

E-mail address: walther@math.purdue.edu

Illinois Journal of Mathematics
Volume 51, Number 1, Spring 2007, Pages 299–311
S 0019-2082

FIELD DEGREES AND MULTIPLICITIES FOR NON-INTEGRAL EXTENSIONS

BERND ULRICH AND CLARENCE W. WILKERSON

Dedicated to Phil Griffith, for his numerous contributions to algebra

ABSTRACT. Let R be a graded subalgebra of a polynomial ring S over a field so that S is algebraic over R. The goal of this paper is to relate the generator degrees of R to the degree $[S : R]$ of the underlying quotient field extension, and to provide a numerical criterion for S to be integral over R that is based on this relationship. As an application we obtain a condition guaranteeing that a ring of invariants of a finite group is a polynomial ring.

1. Introduction

Let k be a field and $S = k[t_1, \ldots, t_d]$ a polynomial ring with variables t_i of degree one. Consider a k-subalgebra R generated by m homogeneous elements $\{x_1, \ldots, x_m\}$. In general, if x is a homogeneous element in a graded object, we denote its degree by $|x|$.

PROBLEM. *If S is algebraic over R, calculate $[S : R]$ from the $\{|x_i|\}$.*

First, one has a form of Bezout's Theorem:

THEOREM 1.1. *If S is integral over R, the following hold:*
(a) *$[S : R]$ divides $\prod |x_i|$.*
(b) *If $m = d$, then $[S : R] = \prod |x_i|$.*

In this paper we obtain a converse to part (b) above:

THEOREM 1.2. *If S is algebraic over R, the following hold:*
(a) *$[S : R] \leq \prod |x_i|$.*

Received November 6, 2006; received in final form March 19, 2007.
2000 *Mathematics Subject Classification.* Primary 13A50, 13B21. Secondary 13A30, 13D40, 13H15.
Bernd Ulrich was supported by NSF grant DMS-0501011. Clarence Wilkerson was supported by an NSF FRG grant, DMS-0354787.

(b) *If* $[S : R] \geq \prod |x_i|$, *then* S *is integral over* R *(equivalently,* S *is finitely generated as an* R*-module) and* $R = k[y_1, \ldots, y_d]$ *is a polynomial ring with variables* $\{y_1, \ldots, y_d\} \subset \{x_1, \ldots, x_m\}$.

We also note that if S is not integral over R, then $[S : R]$ need not divide $\prod |x_i|$ even for $m = d$.

Our proofs rely on reduction to the case of standard graded k-algebras. By a *standard graded* k-algebra we mean a positively graded k-algebra that is Noetherian and generated by its homogeneous elements of degree one (equivalently, is generated by finitely many homogeneous elements of degree one). For such an algebra A the Hilbert function $H_A(n) = \dim_k A_n$ is eventually polynomial,

$$H_A(n) = e(A)\, n^{d-1}/(d-1)! + \text{lower order terms}.$$

Here d is the Krull dimension, $\dim A$, of A, whereas the positive integer $e(A)$ is defined to be the *multiplicity* of A. More generally, if M is a finitely generated graded A-module, one has

$$H_M(n) = e(M)\, n^{d-1}/(d-1)! + \text{ lower order terms}$$

for $n \gg 0$, where $d = \dim M$ is the Krull dimension of M as an A-module and $e(M)$ denotes its multiplicity; see, e.g., [4, 4.1.3].

We will deduce Theorem 1.2 from the next result that provides a criterion for integrality in terms of multiplicities:

THEOREM 1.3. *Let* $A \subset B$ *be an inclusion of standard graded* k*-algebras which are domains and for which* B *is algebraic over* A. *One has:*

(a) $e(B) \geq [B : A]\; e(A)$.

(b) $e(B) = [B : A]\; e(A)$ *if and only if* $A \subset B$ *is integral.*

An interesting application of Theorem 1.2(b) is in the study of rings of invariants of finite groups acting on a polynomial ring:

THEOREM 1.4. *Let* V *be a* d*-dimensional vector space over the field* k, V^* *its* k*-dual, and* $S = k[V^*] = k[t_1, \ldots, t_d]$ *the algebra of polynomial functions on* V. *Let* $W \subset GL(V)$ *be a finite group and consider the induced action on* S. *Then* $R = S^W$ *is a polynomial algebra over* k *if and only if there exist homogeneous elements* $\{x_1, \ldots, x_d\}$ *of* R *such that*

(a) S *is algebraic over* $k[x_1, \ldots, x_d]$, *and*

(b) $|W| \geq \prod |x_i|$.

Notice that we do not assume $\{x_1, \ldots, x_d\}$ to form a system of parameters of S, as was done in [14, 5.5.5].

In the last section, we give examples of rings of invariants for which Theorem 1.4 is useful in providing a proof of polynomial structure.

We note that Theorem 1.3 is a special case of results by Simis-Ulrich-Vasconcelos [13, 6.1]. However, the stronger hypotheses here make a streamlined proof possible. We have also included more details of the graded algebra computations in order to make the paper accessible to the wider audiences of invariant theorists and algebraic topologists.

2. Proof of Theorem 1.1

We borrow the proof from Adams-Wilkerson [2].

Proof of Theorem 1.1. First notice that $\dim R = d$, since S is integral over R and S is a polynomial ring in d variables. Pick a homogeneous generating set $\{s_i | 1 \leq i \leq M\}$ for the finitely generated graded R-module S. Choose a basis for the quotient field L of S over the quotient field K of R consisting of homogeneous elements $\{u_j | 1 \leq j \leq N\}$ from S. Here $N = [S : R]$. Let U be the graded R-submodule of S generated by the $\{u_j\}$. Then U is a free graded R-module. Note that for each i, there exist homogeneous elements $\{a_{ij}\}$ and $\{b_{ij}\}$ in R so that $b_{ij} \neq 0$ and $s_i = \sum (a_{ij}/b_{ij}) u_j$. Then, taking $\Delta = \prod_{i,j} b_{ij}$ one obtains $\Delta S \subset U \subset S$.

We now record some Hilbert-Poincaré series:

(a) $P_S(T) = (1-T)^{-d}$.
(b) $P_{\Delta S}(T) = T^{|\Delta|} P_S(T)$.
(c) $P_U(T) = g(T) P_R(T)$, where g is a polynomial with non-negative integer coefficients and $g(1) = N = [S : R]$.
(d) $P_R(T) = h(T)(1-T)^{m-d} \prod (1 - T^{|x_k|})^{-1}$, for h a polynomial with integer coefficients; indeed, the pole order of this rational function is at most $d = \dim R$ as can be seen from a graded Noether normalization of R.

From the inclusions

$$\Delta S \subset U \subset S$$

one sees that

$$P_{\Delta S}(T) \leq P_U(T) \leq P_S(T).$$

These inequalities should first be interpreted as holding for the non-negative integer coefficients of the powers of T in the respective formal power series. But each of the series also represents a real analytic function for T real and $|T| < 1$. In terms of these functions, we can restate the inequalities as

$$P_{\Delta S}(T) \leq P_U(T) \leq P_S(T) \quad \text{for } 0 \leq T < 1 .$$

After multiplying by $(1-T)^d > 0$, one obtains

$$T^{|\Delta|} \leq g(T) h(T) \prod ((1-T)/(1 - T^{|x_i|})) \leq 1 \quad \text{for } 0 \leq T < 1 .$$

These inequalities have meaning for the functions, although not necessarily for the series. Thus in the limit as $T \to 1$, one has

$$g(1)h(1)\prod |x_i|^{-1} = 1, \quad \text{or} \quad g(1)h(1) = |x_1| \cdot \ldots \cdot |x_m| \,.$$

Since $g(1) = N = [S : R]$, it follows that $[S : R]$ indeed divides $|x_1| \cdot \ldots \cdot |x_m|$. In the special case that $m = d$, R is a polynomial algebra and $h(T) = 1$. Thus (a) and (b) are both established. □

3. Reduction of the proof of Theorem 1.2 to Theorem 1.3

Step One: We first reduce to the case $m = d$. In fact, we may assume that the first d homogeneous generators $\{x_1, \ldots, x_d\}$ of R are algebraically independent over k. Set $R' = k[x_1, \ldots, x_d]$. Notice that $[S : R'] \geq [S : R]$ and $|x_1| \cdot \ldots \cdot |x_d| \leq |x_1| \cdot \ldots \cdot |x_m|$. Thus if (a) holds for $R' \subset S$, it also holds for $R \subset S$. Therefore we may replace R by R' in (a). Furthermore, the assumption of (b) for $R \subset S$ implies

$$[S : R'] \geq [S : R] \geq |x_1| \cdot \ldots \cdot |x_m| \geq |x_1| \cdot \ldots \cdot |x_d| \geq [S : R'] \,.$$

Thus the assumption of (b) is satisfied for $R' \subset S$, and $[S : R'] = [S : R]$. In particular R' and R have the same quotient field. Once we have shown that S is integral over R', then S is integral over R and R is an integral extension of the polynomial ring R'. As R' is integrally closed and R is contained in the quotient field of R', it follows that $R = R'$. Thus indeed S is integral over R, and $R = R'$ is a polynomial ring in the variables $\{x_1, \ldots, x_d\} \subset \{x_1, \ldots, x_m\}$. Therefore in part (b) too, we may replace R by R'.

Step Two: In $R = k[x_1, \ldots, x_d]$ consider the subalgebra $R' = k[x_1^{k_1}, \ldots, x_d^{k_d}]$, where the positive integers $\{k_i\}$ are chosen so that for every i, $|x_i^{k_i}| = N$, the least common multiple of the $|x_i|$. Furthermore let $S^{(N)}$ denote the *Veronese subring* of S generated by all homogeneous elements of S whose degree is an integer multiple of N. Notice that both extensions $R' \subset R$ and $S^{(N)} \subset S$ are integral.

Step Three: We have an inclusion $R' \subset S^{(N)}$ in which all elements of each algebra have degree a multiple of N. Regrade the algebras by declaring the new grading to be the old grading divided by N. Then $R' \subset S^{(N)}$ can be regarded as an inclusion of standard graded domains, so Theorem 1.3 applies:

$$e(S^{(N)}) \geq [S^{(N)} : R']\, e(R')$$

and

$$e(S^{(N)}) = [S^{(N)} : R']\, e(R')$$

if and only if $R' \subset S^{(N)}$ is integral.

We now need some small calculations.

LEMMA 3.1. $e(R') = 1$.

Proof. This follows since R' is a polynomial algebra on degree one generators. □

LEMMA 3.2. $e(S^{(N)}) = N^{d-1}$.

Proof. Since $S^{(N)}$ is a Veronese subring, we have $H_{S^{(N)}}(n) = H_S(Nn)$. Therefore

$$e(S^{(N)})\, n^{d-1}/(d-1)! + \cdots = e(S)\,(Nn)^{d-1}/(d-1)! + \cdots$$

and we obtain $e(S^{(N)}) = N^{d-1}e(S) = N^{d-1}$. □

LEMMA 3.3. $[S : S^{(N)}] = N$.

Proof. Notice that $\{t_1^i \,|\, 0 \le i \le N-1\}$ is a vector space basis for the quotient field of S over that of $S^{(N)}$. □

LEMMA 3.4. *If $m = d$, then $[R : R'] = \prod k_i$.*

Proof. This can be easily seen by considering a basis of R as an R'-module. □

LEMMA 3.5. *If $m = d$, then $[S : R] \prod k_i = [S : R'] = [S^{(N)} : R']\, N$.*

Proof. One has the chains of inclusions $R' \subset R \subset S$ and $R' \subset S^{(N)} \subset S$, and likewise on the quotient field level. Now use Lemmas 3.3 and 3.4. □

LEMMA 3.6. *Assume that $m = d$. Then*

$$\frac{[S : R]}{\prod |x_i|} = \frac{[S^{(N)} : R']\; e(R')}{e(S^{(N)})} \,.$$

Proof. Notice that $|x_i| = N/k_i$. Now divide both sides of the equality in Lemma 3.5 by N^d, and use Lemmas 3.1 and 3.2. □

LEMMA 3.7. *S is integral over R if and only if $S^{(N)}$ is integral over R'.*

Proof. If S is integral over R, then S is integral over R', and hence $S^{(N)}$ is integral over R'. On the other hand, if $S^{(N)}$ is integral over R', then S is integral over R', and hence over R. □

To prove Theorem 1.2(a) we apply Theorem 1.3(a) to the inclusion $R' \subset S^{(N)}$ and use Lemma 3.6, which shows that the inequalities of Theorems 1.2(a) and 1.3(a) are equivalent. Similarly, Theorem 1.2(b) is a consequence of Theorem 1.3(b) and Lemmas 3.6 and 3.7.

4. Proof of Theorem 1.3

To prove Theorem 1.3 it will be convenient to consider the more general class of *quasi-standard graded* algebras over a field k. By this we mean positively graded k-algebras A so that A is Noetherian, $A_0 = k$, and A is integral over the k-subalgebra generated by the homogeneous elements of degree one. As such algebras are finitely generated graded modules over standard graded k-algebras, one can define the concepts of Hilbert functions and multiplicities as in the standard graded case.

In this section, $A \subset B$ is an inclusion of quasi-standard graded k-domains for which B is algebraic over A. The aim is to prove that under suitable restrictions on the multiplicities of A and B, the ring B must be a finitely generated A-module. More specifically, we have the following generalization of Theorem 1.3:

THEOREM 4.1. *Let $A \subset B$ be an inclusion of quasi-standard graded k-algebras which are domains and for which B is algebraic over A. One has:*
(a) $e(B) \geq [B : A]\, e(A)$.
(b) $e(B) = [B : A]\, e(A)$ *if and only if $A \subset B$ is integral.*

Notice that B contains a rank $[B : A]$ graded free module over A. Now part (a) of Theorem 4.1 follows by comparing the two Hilbert functions. One direction of the implication in (b) is standard:

PROPOSITION 4.2. *Let $A \subset B$ be an inclusion of quasi-standard graded k-algebras which are domains. If B is integral over A, then $e(B) = [B : A]\, e(A)$.*

This can be deduced from a more general fact:

PROPOSITION 4.3. *Let A be a quasi-standard graded k-algebra which is a domain, with quotient field K. If M is a finitely generated graded A-module with* $\dim M = \dim A$, *then* $e(M) = \dim_K(M \otimes_A K)\, e(A)$.

The proof of Proposition 4.3 is an easy adaptation of [10, 14.8].

The rest of this section is devoted to proving the other implication in Theorem 4.1(b). We first show that we can reduce to the case where $e(A) = e(B)$ and A and B are standard graded.

Given $A \subset B$ algebraic, choose homogeneous elements $\{c_i \mid 1 \leq i \leq N\}$ in B that form a basis for the quotient field L of B over the quotient field K of A. For each such c_i, there exists a nonzero homogeneous $a_i \in A$ such that $b_i = a_i c_i$ is integral over A. Define A' to be the A-subalgebra of B generated by the $\{b_i\}$. Then A' is integral over A and $[A' : A] = [B : A]$. Notice that A' is still quasi-standard graded, but that it may fail to be standard graded even if A and B are—hence the need to consider the wider class of quasi-standard graded algebras.

Thus by Proposition 4.2, since $A \subset A'$ is an integral extension of quasi-standard graded k-algebras,

$$e(A') = [A' : A]\, e(A) .$$

Hence if $e(B) = [B : A]\, e(A)$, one has $e(A') = e(B)$. Therefore the proof of Theorem 4.1(b) can be reduced to showing the following proposition:

PROPOSITION 4.4. *Let $A \subset B$ be an inclusion of quasi-standard graded k-algebras which are domains and for which B is algebraic over A. If $e(B) = e(A)$, then B is a finitely generated A-module.*

Finally, the reduction to the standard graded case follows from some easy facts:

LEMMA 4.5. *Let A be a quasi-standard graded k-algebra. There exists a positive integer N so that for each positive integer r, the Veronese subring $A^{(rN)}$ is generated by the elements of A_{rN} as a k-algebra. That is, after regrading, $A^{(rN)}$ is a standard graded k-algebra.*

Proof. One can take N to be the maximal degree occurring in a homogeneous minimal generating set of A, considered as a finitely generated module over a standard graded k-subalgebra. □

LEMMA 4.6. *Let $A \subset B$ be an inclusion of quasi-standard graded k-domains such that B is algebraic over A. Let N be a positive integer. If $e(B) = e(A)$, then $e(B^{(N)}) = e(A^{(N)})$.*

Proof. All algebras involved have the same Krull dimension according to [8, Theorem A, p. 286], for instance. Now the lemma follows by comparing Hilbert functions. □

LEMMA 4.7. *Let $A \subset B$ be an inclusion of quasi-standard graded k-domains. Let N be a positive integer. Then B is integral over A if and only if $B^{(N)}$ is integral over $A^{(N)}$.*

In light of Lemmas 4.5, 4.6 and 4.7 it will suffice to prove Proposition 4.4 in the standard graded case.

As in [13] the idea of the proof then is to consider the graded A-module C defined by the short exact sequence

$$0 \to A \to B \to C \to 0 .$$

The module C has no obvious finiteness properties as an A-module, but it does have a Hilbert function, namely

$$H_C(n) = H_B(n) - H_A(n) .$$

Since B is algebraic over A, the two algebras have the same Krull dimension, say d; see, for instance, [8, Theorem A, p. 286]. As furthermore $e(A) = e(B)$, it follows that the leading term of $H_C(n)$ occurs in degree $d-2$ or less.

We need to associate more structure to C in order to utilize this information about $H_C(n)$. Let $I = A_1B$ be the homogeneous B-ideal generated by A_1, the homogeneous elements of A having degree 1. Let G be the graded algebra associated to the filtration of B by powers of I. That is,

$$G = \bigoplus_{i=0}^{\infty} I^i/I^{i+1} \ .$$

Then G is a positively graded Noetherian ring, although in general it is not a domain and G_0 is not a field. It has Krull dimension $d = \dim B$. In fact, one has:

PROPOSITION 4.8. *The associated graded ring G is equidimensional of dimension d. That is, for each minimal prime ideal $\mathfrak{p} \subset G$, $\dim G/\mathfrak{p} = \dim G = d$.*

Proof. The ring G can also be thought of as the quotient of the domain $\mathcal{R} = B[It, t^{-1}]$ (the extended Rees algebra) with respect to the principal ideal generated by t^{-1}. The Krull dimension of $\mathcal{R}$ is $d+1$; see [8, Theorem A, p. 286] or [10, 15.7]. The minimal prime ideals of G correspond to the minimal prime ideals of the principal ideal in $\mathcal{R}$ generated by t^{-1}. Let $\mathfrak{q}$ be such a prime ideal in $\mathcal{R}$. By the Krull Principal Ideal Theorem (see, e.g., [10, 13.5]), the height of $\mathfrak{q}$ is 1. Since $\mathcal{R}$ is an affine domain, we have $\dim \mathcal{R}/\mathfrak{q} = \dim \mathcal{R} - \operatorname{ht} \mathfrak{q}$; see, e.g., [8, 13.4]. Therefore

$$\dim \mathcal{R}/\mathfrak{q} = \dim \mathcal{R} - \operatorname{ht} \mathfrak{q} = (d+1) - 1 = d \ .$$

Hence if $\mathfrak{p} \subset G$ is the corresponding minimal prime ideal in G, then $\dim G/\mathfrak{p} = \dim \mathcal{R}/\mathfrak{q} = d$. That is, G is equidimensional of dimension d. □

PROPOSITION 4.9. *In addition to the assumptions of Proposition 4.4 suppose that A and B are standard graded. Then the homogeneous ideal $B_1G \subset G$ is nilpotent.*

Proof of Proposition 4.4 using Proposition 4.9. In light of Lemmas 4.5, 4.6 and 4.7 we may assume that A and B are standard graded. Since B_1G is nilpotent according to Proposition 4.9, its filtration degree 0 component, B_1B/A_1B, is also nilpotent in G. Hence, back in B, $B_1B \subset \sqrt{A_1B}$. So there exists a positive integer N such that $B_1^N \subset A_1B_{N-1}$. As B is standard graded we deduce that $B_1^N = A_1B_1^{N-1}$ and then $B_n = A_{n-N+1}B_{N-1}$ for every $n \geq N$. Thus a generating set for B as an A-module can be obtained from a k-basis of $\bigoplus_{i=0}^{N-1} B_i$. That is, B is a finitely generated A-module. □

Proof of Proposition 4.9. The algebra G inherits an internal degree from B and a filtration degree from the I-adic filtration. We write $G_{(m,i)}$, where m is the internal degree and i is the filtration degree. For the total degree, we use the sum of the two degrees and set $G_n = \bigoplus_{m+i=n} G_{(m,i)}$. Note that since A and B are standard graded, $A_1^n = A_n$ and $B_1 B_{n-1} = B_n$ for every positive n. Thus

$$G_n = B_n/A_1B_{n-1} \oplus A_1B_{n-1}/A_2B_{n-2} \oplus \quad \cdots \quad \oplus A_{n-1}B_1/A_n \oplus A_n \ ,$$

and

$$(B_1G)_n = B_n/A_1B_{n-1} \oplus A_1B_{n-1}/A_2B_{n-2} \oplus \quad \cdots \quad \oplus A_{n-1}B_1/A_n \ .$$

This last expression gives us the "raison d'être" for G and B_1G in our strategy. When we take lengths, consecutive terms cancel. That is, $\dim_k(B_1G)_n = \dim_k B_n/A_n = \dim_k C_n$. Thus $H_{B_1G}(n) = H_C(n)$. Notice that with respect to the total degree, G is a standard graded (finitely generated) k-algebra and B_1G is a homogeneous G-ideal. Hence the Hilbert function and polynomial for B_1G detect the Krull dimension of B_1G as a module over G:

LEMMA 4.10. $\dim B_1G \le d-1$.

Proof. For $n \gg 0$, $H_{B_1G}(n) = H_C(n) = H_B(n) - H_A(n)$ is a polynomial function of degree $\le d-2$, from the hypothesis that $e(A) = e(B)$. Since B_1G is a finitely generated graded module over the standard graded k-algebra G, its Krull dimension equals the degree of the Hilbert polynomial plus one; see, e.g., [4, 4.1.3]. □

On the other hand, one has $\dim B_1G = \dim G/\operatorname{ann}(B_1G)$, for $\operatorname{ann}(B_1G)$ the annihilator ideal of B_1G in G.

LEMMA 4.11. *Let $\mathfrak{p} \subset G$ be a minimal prime ideal of G. Then $B_1G \subset \mathfrak{p}$.*

Proof. Proposition 4.8 shows that $\dim G/\mathfrak{p} = \dim G = d$ for every such $\mathfrak{p}$. On the other hand, $\dim G/\operatorname{ann}(B_1G) = \dim B_1G \le d-1$ according to Lemma 4.10. Hence $\dim G/\operatorname{ann}(B_1G) < \dim G/\mathfrak{p}$, which implies that $\operatorname{ann}(B_1G) \not\subset \mathfrak{p}$. Since $\operatorname{ann}(B_1G) \cdot B_1G = 0 \subset \mathfrak{p}$ and $\mathfrak{p}$ is prime, it follows that $B_1G \subset \mathfrak{p}$. □

Now Lemma 4.11 immediately gives Proposition 4.9, because the nilradical of G is the intersection of its minimal prime ideals. □

5. Proof of Theorem 1.4 and a counterexample

We begin by recording a proof of Theorem 1.4.

Proof of Theorem 1.4. The forward implication is a consequence of Theorem 1.1(b). To show the converse, write $R' = k[x_1, \dots, x_d]$ and notice that

$$[S : R'] \ge [S : R] = |W| \ge |x_1| \cdot \dots \cdot |x_d| \ .$$

Applying Theorem 1.2 to the extension $R' \subset S$ we conclude that S is integral over R' (by part (b)) and that $[S : R'] = [S : R]$ (by part (a)). Thus R is integral over R' and the two rings have the same quotient field. But R' is a polynomial ring and hence integrally closed. It follows that $R' = R$. □

We finish this section with an example related to Theorem 1.2(a). It shows that the degree of the field extension need not divide the product of the degrees of the algebra generators.

EXAMPLE 5.1. Let $S = k[x, y]$, where $|x| = |y| = 1$. Let R be the k-subalgebra of S generated by the monomials x^3 and xy^2. Let K be the quotient field of R and K' the extension given by adjoining the element x. Then $[K' : K] = 3$ and K' contains y^2. Hence the quotient field L of S is obtained from K' by adjoining y. Thus $[L : K'] = 2$, and $[L : K] = 6$, which does not divide $3^2 = 9$. There are of course no examples involving only one variable.

6. Applications to rings of invariants

A problem of interest in topology and invariant theory for the last thirty-five years has been the determination of which representations of finite groups have polynomial algebras as rings of invariants. Work of Shephard-Todd [11], Chevalley [5], Serre [12], Clark-Ewing [6], and others largely solved this problem in the case that the order of the group is a unit in the ground field. The work of Adams-Wilkerson [1] emphasized the connection of the problem to topology, even in the case where this condition fails.

Wilkerson, in [15, Section III], observed that often polynomial rings of invariants can be verified using this general strategy:

(a) pick d homogeneous elements $\{x_1, \ldots, x_d\}$ of S^W,
(b) verify that S is integral over R', the subalgebra generated by the $\{x_i\}$, and
(c) check that $[S : R'] = |W|$.

According to Theorem 1.1(b), for instance, in the presence of (b) item (c) is equivalent to

(c') check that $\prod |x_i| = |W|$.

He lists several types of finite linear groups in characteristic p for which this strategy works, for example, general linear groups (Dickson invariants), special linear groups and variations on the upper triangular groups. In these cases the integrality condition above is evident from the description of the invariants. Theorem 1.4 tells us that the integrality condition can be replaced by the weaker statement that the $\{x_i\}$ are algebraically independent.

Thus, computationally, given a choice of elements $\{x_i\}$ satisfying the above conditions (a) and (c'), one has two options:

(a) show that $\dim_k(S/I) < \infty$, for I the S-ideal generated by the $\{x_i\}$, or

(b) show that the Jacobian $|\partial x_i/\partial t_j|$, $1 \le i, j \le d$, is nonzero.

Computer algebra programs typically implement option (a) using Gröbner basis algorithms, and the time and memory requirements increase dramatically with the dimension d and the complexity of W. On the other hand, the Jacobian can be computed at points in $V \otimes_k \bar{k}$ by a combination of symbolic and numerical techniques that may be less demanding computationally.

In 1994, C. Xu [16], [17] studied three examples from the [11], [6] list of complex and p-adic reflection groups. These groups are labeled as W_{29}, W_{31} and W_{34} in characteristic 5, 5 and 7, and dimensions 4, 4, and 6 respectively. In each case he obtained a collection of invariant polynomial forms that allow Theorem 1.4 to be applied:

THEOREM 6.1.

(a) *For the group* W_{29} *over* $\mathbb{F}_5$, *there are forty linear forms* $\{L_i\}$ *in* $\mathbb{F}_5[t_1, \ldots, t_4]$ *for which* W_{29} *permutes the powers* $\{L_i^4\}$. *The first, second, third, and fifth elementary symmetric polynomials* $\{x_4, x_8, x_{12}, x_{20}\}$ *in the* $\{L_i^4\}$ *are algebraically independent over* $\mathbb{F}_5$ *and the product of the degrees is* $4 \cdot 8 \cdot 12 \cdot 20 = 7680 = |W_{29}|$.

(b) *For the group* W_{31} *over* $\mathbb{F}_5$, *there is a degree* 4 *polynomial* Y_4 *in* $\mathbb{F}_5[t_1, \ldots, t_4]$ *for which the* W_{31}*-orbit contains only six distinct polynomials,* $\{Y_{4,i}\}$. *The second, third, fifth, and sixth elementary symmetric polynomials* $\{y_8, y_{12}, y_{20}, y_{24}\}$ *in the* $\{Y_{4,i}\}$ *are algebraically independent. The product of the degrees is* $8 \cdot 12 \cdot 20 \cdot 24 = 46080 = |W_{31}|$.

(c) *For the group* W_{34} *over* $\mathbb{F}_7$, *there are* 126 *linear forms* $\{L_i\}$ *in* $\mathbb{F}_7[t_1, \ldots, t_6]$ *for which the sixth powers* $\{L_i^6\}$ *are permuted by the action. The first, second, third, fourth, fifth, and seventh elementary symmetric polynomials* $\{z_6, z_{12}, z_{18}, z_{24}, z_{30}, z_{42}\}$ *in the* $\{L_i^6\}$ *are algebraically independent and the product of the degrees is* $6 \cdot 12 \cdot 18 \cdot 24 \cdot 30 \cdot 42 = 39191040 = |W_{34}|$.

The example W_{34} appears in his thesis [17] only, so we review the strategy briefly here. We begin with a method for finding some invariants. This works well for both W_{29} and W_{34}.

Let W be a subgroup of $GL(V)$, where V is a finite dimensional vector space over $k = \mathbb{F}_p$. Let $\mathcal{S}$ be the set of all reflections in W of order prime to p. These reflections are diagonalizable over k. Denote a typical such reflection by s. Let H_s denote the fixed hyperplane of s and f_s a linear form on V defining H_s.

LEMMA 6.2. *For each* $w \in W$, wH_s *is fixed by* wsw^{-1}. *That is* $wH_s = H_{wsw^{-1}}$.

Proof. Let $v \in H_s$ and $w \in W$. Then $wsw^{-1}(wv) = ws(v) = wv$. □

LEMMA 6.3. *W acts on the set of all H_s, for $s \in \mathcal{S}$.*

Now if f is a linear form on V to k, i.e., $f \in V^*$, define the action of W on V^* by $(wf)(v) = f(w^{-1}v)$, for $v \in V$.

LEMMA 6.4. *Let $s \in \mathcal{S}$ and $w \in W$. Then $wf_s = \theta_{s,w} f_{wsw^{-1}}$, where $\theta_{s,w} \neq 0$ is an element of k. That is, up to units, the action of W permutes the linear forms defining the H_s.*

Proof. By Lemma 6.2, wf_s is zero on the hyperplane $wH_s = H_{wsw^{-1}}$. □

PROPOSITION 6.5. *For V and W as above, the polynomials ${f_s}^{p-1}$ are permuted by the action of W.*

Proof. The multiplicative order of each $\theta_{s,w}$ is a divisor of $p - 1$. □

The group W_{34} has only one conjugacy class of reflections. This class has 126 elements and each has order 2. Thus the preceding theory can be applied to the sixth powers of the linear forms representing the hyperplanes for these reflections.

Xu was able to use the Gröbner basis routines of Macaulay to verify the specified generators for W_{29} and W_{31} (and, unstated, that S is integral over R'). In 1994, however, this failed for W_{34}. He therefore resorted to the Jacobian method. The Jacobian is not expanded as a degree 126 polynomial. Rather, each entry is evaluated at a point of the vector space and the resulting 6×6 determinant computed. In fact, it vanishes at each point of $(\mathbb{F}_7)^6$. However, Xu finds a point in $(\mathbb{F}_{49})^6$ where it is not zero. Thus the proposed generators are algebraically independent and by Theorem 1.4 the rings of invariants are as claimed by Xu:

COROLLARY 6.6.

(a) $\mathbb{F}_5[t_1, \ldots, t_4]^{W_{29}} = \mathbb{F}_5[x_4, x_8, x_{12}, x_{20}]$.
(b) $\mathbb{F}_5[t_1, \ldots, t_4]^{W_{31}} = \mathbb{F}_5[y_8, y_{12}, y_{20}, y_{24}]$.
(c) $\mathbb{F}_7[t_1, \ldots, t_6]^{W_{34}} = \mathbb{F}_7[z_6, z_{12}, z_{18}, z_{24}, z_{30}, z_{42}]$.

Here the degree of x_k, y_k, or z_k is k.

These examples can also be found in Aguadé [3], without proof of the polynomial nature of the rings of invariants. They are also included in [9].

REFERENCES

[1] J. F. Adams and C. W. Wilkerson, *Finite H-spaces and algebras over the Steenrod algebra*, Ann. of Math. (2) **111** (1980), 95–143. MR 558398 (81h:55006)

[2] ———, *A correction: "Finite H-spaces and algebras over the Steenrod algebra"*, Ann. of Math. (2) **113** (1981), 621–622. MR 621021 (82i:55010)

[3] J. Aguadé, *Constructing modular classifying spaces*, Israel J. Math. **66** (1989), 23–40. MR 1017153 (90m:55016)

[4] W. Bruns and J. Herzog, *Cohen-Macaulay rings*, Cambridge Studies in Advanced Mathematics, vol. 39, Cambridge University Press, Cambridge, 1993. MR 1251956 (95h:13020)

[5] C. Chevalley, *Invariants of finite groups generated by reflections*, Amer. J. Math. **77** (1955), 778–782. MR 0072877 (17,345d)

[6] A. Clark and J. Ewing, *The realization of polynomial algebras as cohomology rings*, Pacific J. Math. **50** (1974), 425–434. MR 0367979 (51 #4221)

[7] W. G. Dwyer and C. W. Wilkerson, *The elementary geometric structure of compact Lie groups*, Bull. London Math. Soc. **30** (1998), 337–364. MR 1620888 (99g:22007)

[8] D. Eisenbud, *Commutative algebra with a view toward algebraic geometry*, Graduate Texts in Mathematics, vol. 150, Springer-Verlag, New York, 1995. MR 1322960 (97a:13001)

[9] G. Kemper and G. Malle, *The finite irreducible linear groups with polynomial ring of invariants*, Transform. Groups **2** (1997), 57–89. MR 1439246 (98a:13012)

[10] H. Matsumura, *Commutative ring theory*, Cambridge Studies in Advanced Mathematics, vol. 8, Cambridge University Press, Cambridge, 1986. MR 879273 (88h:13001)

[11] G. C. Shephard and J. A. Todd, *Finite unitary reflection groups*, Canadian J. Math. **6** (1954), 274–304. MR 0059914 (15,600b)

[12] J.-P. Serre, *Groupes finis d'automorphismes d'anneaux locaux réguliers*, Colloque d'Algèbre (Paris, 1967), Exp. 8, Secrétariat mathématique, Paris, 1968, 11 pp. MR 0234953 (38 #3267)

[13] A. Simis, B. Ulrich, and W. V. Vasconcelos, *Codimension, multiplicity and integral extensions*, Math. Proc. Cambridge Philos. Soc. **130** (2001), 237–257. MR 1806775 (2002c:13017)

[14] L. Smith, *Polynomial invariants of finite groups*, Research Notes in Mathematics, vol. 6, A K Peters Ltd., Wellesley, MA, 1995. MR 1328644 (96f:13008)

[15] C. Wilkerson, *A primer on the Dickson invariants*, Proceedings of the Northwestern Homotopy Theory Conference (Evanston, Ill., 1982), Contemp. Math., vol. 19, Amer. Math. Soc., Providence, RI, 1983, pp. 421–434. MR 711066 (85c:55017)

[16] C. Xu, *Computing invariant polynomials of p-adic reflection groups*, Mathematics of Computation 1943–1993: a half-century of computational mathematics (Vancouver, BC, 1993), Proc. Sympos. Appl. Math., vol. 48, Amer. Math. Soc., Providence, RI, 1994, pp. 599–602. MR 1314898

[17] C. Xu, *The existence and uniqueness of simply connected p-compact groups with Weyl groups W such that the order of W is not divisible by the square of p*, Thesis, Purdue University, 1994.

Bernd Ulrich, Department of Mathematics, Purdue University, West Lafayette, IN 47907, USA
E-mail address: `ulrich@math.purdue.edu`

Clarence W. Wilkerson, Department of Mathematics, Purdue University, West Lafayette, IN 47907, USA
E-mail address: `cwilkers@purdue.edu`

Illinois Journal of Mathematics
Volume 51, Number 1, Spring 2007, Pages 313–323
S 0019-2082

STABILITY OF THE DIVISOR CLASS GROUP UPON COMPLETION

DANA WESTON

In honor of Phil Griffith's contributions to mathematics

ABSTRACT. If A is a local, normal approximation domain with finite divisor class group, then $Cl(A) \cong Cl(\widehat{A})$.

1. Introduction

Let $(A, \mathfrak{M})$ be a local, analytically normal domain and $\widehat{A}$ its completion with respect to the maximal ideal $\mathfrak{M}$. We shall use $\mathrm{cl}(\mathfrak{a})$ to denote the equivalence class of a divisorial A-ideal $\mathfrak{a}$ in the divisor class group, $\mathrm{Cl}(A)$, of A. It is well known that the canonical group homomorphism $i : \mathrm{Cl}(A) \to \mathrm{Cl}(\widehat{A})$, defined by $i(\mathrm{cl}(\mathfrak{a})) := \mathrm{cl}(\widehat{\mathfrak{a}})$, is one-to-one, but, in general, far from being onto, whence the question arises for which rings might one expect the above map to be an isomorphism.

Danilov investigated this question as part of a cycle of related problems in a series of papers [4], [5], [6], [7]. Bingener [1], and Bingener and Storch [2], also gave some thought to the vanishing of $\mathrm{Coker}(i)$. In both cases, the authors used geometric techniques (and, in particular, required resolution of singularities and an algebraically closed residue field) to guarantee the surjectivity of i for approximation rings with finitely generated divisor class group.

On the other hand, Griffith and Weston [9, 4.1] leaned on algebraic methods to establish the surjectivity of the map $i| : \mathrm{Cl}(A)_{\mathrm{tor}} \to \mathrm{Cl}(\widehat{A})_{\mathrm{tor}}$ for a local $\mathbb{Q}$-algebra A, at once a normal approximation domain. (Here, $i|$ stands for the restriction of i to the torsion subgroups of the relevant divisor class groups. We mention that the condition in 4.1 on primitive roots of unity can be eliminated.) The gist of the proof involved the manipulation and descent of a Galois extension of $\widehat{A}$, module-isomorphic to $\widehat{A} \oplus \mathfrak{b} \oplus \mathfrak{b}^{(2)} \oplus \cdots \oplus \mathfrak{b}^{(e-1)}$ for a divisorial torsion ideal $\mathfrak{b}$ of $\widehat{A}$ with e the order of $\mathrm{cl}(\mathfrak{b})$ in $\mathrm{Cl}(\widehat{A})$.

Received July 24, 2006; received in final form January 28, 2007.
2000 *Mathematics Subject Classification.* Primary 13C20, 13J10.

Rotthaus' paper [15] represented yet another algebraic approach to the same problem. The author introduced the technique of complete induction which she applied to, among others, the issue of the bijectivity of $i : \mathrm{Cl}(A) \to \mathrm{Cl}(\widehat{A})$ and $i| : \mathrm{Cl}(A)_{\mathrm{tor}} \to \mathrm{Cl}(\widehat{A})_{\mathrm{tor}}$. The setting was that of a local, normal domain A with the complete approximation property (e.g., if A is both a $\mathbb{Q}$-algebra and an approximation ring) satisfying the R_2-regularity condition and with $\mathrm{Cl}(A)$ and $\mathrm{Cl}(A)_{\mathrm{tor}}$ a finite group, respectively.

In the following, we eliminate the R_2-regularity condition, and ask merely for the approximation (rather than the complete approximation) property to conclude that i is an isomorphism when $\mathrm{Cl}(A)$ is finite. The present line of argumentation was motivated by a desire to generalize Hochster's proof (oral communication) of the fact that a local approximation unique factorial domain remains a factorial domain upon completion. It is tempting to generalize Danilov's result [7, Proof of Theorem 1] with the same techniques and claim that i is an isomorphism when A is a local, normal approximation domain with finitely generated divisor class group. However, the methods below, while natural for torsion elements, do not adapt with ease to the torsion free case.

2. Terminology and facts

We present some notation that will be used without repeated mention:

The set of natural numbers $\{0, 1, 2, \dots\}$ will be denoted by $\mathbb{N}$, and the set of positive integers $\{1, 2, \dots\}$ by $\mathbb{P}$. For a ring A, $\mathcal{U}(A)$ will denote the set of invertible elements and $X^1(A)$ the set of all height one prime ideals of A. If A is a domain, K will stand for the field of quotients $A_{(0)}$. When A is a normal domain and $\mathfrak{a}$ a fractional ideal of A, then $\mathfrak{a}^{**}$ is a divisorial ideal (here, $\mathfrak{a}^* := A : \mathfrak{a}$) and $\mathfrak{a}^{**} = \bigcap_{\mathfrak{p} \in X^1(A)} \mathfrak{a}_{\mathfrak{p}}$.

What follows is a pedestrian overview of facts used often in the proof, and provided here for the convenience of the reader. These can be gleaned from standard texts on the topic, such as [3], [8], [11], [12].

THEOREM 2.1 (Approximation Theorem for Krull Domains [8, 5.8]). *Let A be a Krull domain with $\mathfrak{p}_1, \dots, \mathfrak{p}_r$ distinct height one prime ideals of A and $n_1, \dots, n_r$ integers. Then there is a non-zero element x in K such that $v_{\mathfrak{p}_i}(x) = n_i$ and $v_{\mathfrak{p}}(x) \geq 0$ for $\mathfrak{p} \in X^1(A) \backslash \{\mathfrak{p}_1, \dots, \mathfrak{p}_r\}$.*

(Here, $v_{\mathfrak{p}}$ is the valuation associated to the principal valuation ring $A_{\mathfrak{p}}$ and $v_{\mathfrak{p}}(\mathfrak{a}) := \inf\{v_{\mathfrak{p}}(x) | x \in \mathfrak{a}\}$ for a non-zero fractional ideal $\mathfrak{a}$ of A.)

THEOREM 2.2 (Nagata's Theorem [8, 7.2]). *Let S be a multiplicatively closed subset of a Krull domain A. Then $1 \to \langle \mathrm{cl}(\mathfrak{p}) | \mathfrak{p} \in X^1(A)$ and $S \bigcap \mathfrak{p} \neq \emptyset\rangle \to \mathrm{Cl}(A) \to \mathrm{Cl}(S^{-1}A) \to 1$ is a short exact sequence of abelian groups.*

COROLLARY 1. *Let A be a normal domain with* $\mathrm{Cl}(A)$ *finitely generated. Then there is an element $a \in A\backslash\{0\}$ such that A_a is a unique factorization domain.*

Proof. Let $\mathfrak{a}_1, \dots, \mathfrak{a}_k$ be divisorial ideals of A with $\mathrm{Cl}(A) = \langle \mathrm{cl}(\mathfrak{a}_1), \dots, \mathrm{cl}(\mathfrak{a}_k)\rangle$. Then let $\{\mathfrak{p}_1, \dots, \mathfrak{p}_r\} = \{\mathfrak{p} \in X^1(A) \mid v_{\mathfrak{p}}(\mathfrak{a}_i) \neq 0$ for some $i = 1, \dots, k\}$.

By Theorem 1, there are $\mathfrak{p}_{r+1}, \dots, \mathfrak{p}_s \in X^1(A)\backslash\{\mathfrak{p}, \dots, \mathfrak{p}_r\}$ and $n_{r+1}, \dots, n_s \in \mathbb{P}$ such that $(\bigcap_{i=1}^r \mathfrak{p}_i) \bigcap (\bigcap_{j=r+1}^s \mathfrak{p}_j^{(n_j)}) = aA$ for some $a \in A$.

By Theorem 2, $\mathrm{Cl}(A_a) = 0$ since $\mathrm{Cl}(A) = \langle \mathrm{cl}(\mathfrak{p}_1), \dots, \mathrm{cl}(\mathfrak{p}_r)\rangle$. □

We refer the reader to [2] or [8] for the following result.

FACT 1. *Let $\mathfrak{a}$ and $\mathfrak{b}$ be fractional ideals of A.*
(i) *Then,* $\mathrm{div}(\mathfrak{a} \cdot \mathfrak{b}) = \mathrm{div}(\mathfrak{a}) + \mathrm{div}(\mathfrak{b})$.
(ii) *If $v_{\mathfrak{p}}(\mathfrak{a}) \cdot v_{\mathfrak{p}}(\mathfrak{b}) = 0$ for all $\mathfrak{p} \in X^1(A)$, then $(\mathfrak{a} \cdot \mathfrak{b})^{**} = \mathfrak{a}^{**} \bigcap \mathfrak{b}^{**}$.*
(iii) *For $c \in K\backslash\{0\}$, we have $(c\mathfrak{a})^{**} = c(\mathfrak{a}^{**})$.*

Consequently, we have:

FACT 2. *Let $P_1, \dots, P_m \in X^1(A)$ be distinct prime ideals such that $\mathrm{cl}(P_1) = \dots = \mathrm{cl}(P_m)$. (Say $P_i = \alpha_i\mathfrak{p}$ for $i = 1, \dots, m$ with $\alpha_i \in K\backslash\{0\}$.) Let $k_1, \dots, k_m \in \mathbb{P}$.*
(i) *Then $\bigcap_{i=1}^m P_i^{(k_i)} = (\Pi_{i=1}^m P_i^{k_i})^{**} = (\Pi_{i=1}^m \alpha_i^{k_i}\mathfrak{p}^{k_i})^{**} = \alpha\mathfrak{p}^{(k)}$, where $\alpha := \Pi_{i=1}^m \alpha_i^{k_i}$ and $k := \sum_{i=1}^m k_i$.*
(ii) *Suppose* $\mathrm{Cl}(A)$ *is a finite group of order $t \leq m$. Let $b \in A$ be such that $\mathfrak{p}^{(t)} = bA$. Then $\bigcap_{i=1}^t P_i = \gamma A$ with $\gamma = b \cdot \alpha_1 \cdot \dots \cdot \alpha_t$. Since $\bigcap_{i=1}^t P_i \subset A$, then $\gamma \in A\backslash\mathcal{U}(A)$.*
(iii) *Suppose that $t \lneqq m$. Define $\mathfrak{a} := (\bigcap_{i=1}^t P_i^{(k_i-1)}) \bigcap (\bigcap_{i=t+1}^m P_i^{(k_i)})$ and note that $\mathfrak{a}$ is a divisorial ideal, properly contained in A. Then $\bigcap_{i=1}^m P_i^{(k_i)} = \gamma\mathfrak{a}$ for γ defined as in* (ii).

Let
(i) $\underline{X}$ stand for a sequence of variables $X_1, \dots, X_s$;
(ii) $\underline{f}$ stand for a sequence of polynomial functions $f_1(\underline{X}), \dots, f_t(\underline{X}) \in A[\underline{X}] = A[X_1, \dots, X_s]$;
(iii) $\underline{x}$ stand for a sequence of elements $x_1, \dots, x_s$ in A;
(iv) $\underline{\tilde{x}}$ stand for a sequence of elements $\underline{\tilde{x}}_1, \dots, \underline{\tilde{x}}_s$ in $\widehat{A}$;
(v) $\underline{x} \equiv \underline{\tilde{x}} \bmod \widehat{\mathfrak{M}}^n$ stand for a sequence of congruences $x_1 \equiv \tilde{x}_1 \bmod \widehat{\mathfrak{M}}^n$, $\dots$, $x_s \equiv \tilde{x}_s \bmod \widehat{\mathfrak{M}}^n$.

This notation will facilitate the statement of the following definition.

DEFINITION. Let $(A, \mathfrak{M})$ be a local, Noetherian ring.

(i) Suppose that given any $\underline{f} \in A[\underline{X}]$ having a solution $\tilde{\underline{x}} \in \widehat{A}^s$, and given any $n \in \mathbb{P}$, there is a solution $\underline{x} \in A^s$ for $\underline{f}$ such that $\underline{x} \equiv \tilde{\underline{x}} \bmod \widehat{\mathfrak{M}}^n$. Then A is called an *approximation ring.*

(ii) Suppose that given any $\underline{f} \in A[\underline{X}]$, there is an increasing function $\theta : \mathbb{N} \to \mathbb{N}$ with $\theta(n) \geq n$ such that whenever $\underline{x} \in A^s$ is a solution of $\underline{f} \bmod \mathfrak{M}^{\theta(n)}$, then there is a solution $\tilde{\underline{x}} \in A^s$ of $\underline{f}$ with $\underline{x} \equiv \tilde{\underline{x}} \bmod \mathfrak{M}^n$. Then A is called a *strong approximation ring*, and θ is called a *strong approximation function with respect to $\underline{f}$*. (Note that $\tilde{\underline{x}}$ depends on n, in general.)

Pfister and Popescu [13] and van der Put [16], in a more limited setting, have shown that the above two properties yield the same class of rings. Clearly, then, all complete, local rings are strong approximation rings.

Often it will be convenient to make "index shifts" as in the following fact.

FACT 3. *Let A be a strong approximation ring, $\underline{f} = (f_1, \ldots, f_t) \in A[\underline{X}] = A[X_1, \ldots, X_s], \theta : \mathbb{N} \to \mathbb{N}$ be a strong approximation function with respect to $\underline{f}$, and suppose that $\underline{x}(n) := (x_{1n}, \ldots, x_{sn}) \in A^s$ is a solution of $\underline{f} \bmod \mathfrak{M}^n$ for $n \in \mathbb{N}$.*

(i) *Then we may assume that, for $n \gg 0$, there is a solution $\underline{x} \in A^s$ of $\underline{f}$ such that $\underline{x}(n) \equiv \underline{x} \bmod \mathfrak{M}^{\theta^{-1}(n)}$ for $n \in \mathrm{Im}(\theta)$.*

(ii) *Furthermore, we may assume that, for $n \gg 0$, there is a solution $\underline{x} \in A^s$ of $\underline{f}$ such that $\underline{x}(n) \equiv \underline{x} \bmod \mathfrak{M}^n$.*

Proof. (i) Observe that $\theta(1) \geq \theta(2) \geq \cdots$ and $\lim_{n\to\infty} \theta(n) = +\infty$. Let $l_0 := \theta(1)$. Then the statement holds for $n \geq l_0$ since we can replace $\underline{x}(n)$ with $\underline{x}(l)$, where $l = \min\{k \in \mathbb{N} \mid n \leq k \text{ and } k \in \mathrm{Im}(\theta)\}$.

(ii) Let $\underline{x}(n) \equiv \underline{x} \bmod \mathfrak{M}^{\theta^{-1}(n)}$ for $n \in \mathrm{Im}(\theta)$. We can replace $\underline{x}(n)$ with $\underline{x}(\theta(n))$. Then there is a solution $\underline{x} \in A^s$ of $\underline{f}$ such that $\underline{x}(n) \equiv \underline{x} \bmod \mathfrak{M}^n$. □

We may make use of the above artifice without explicit mention.

Most of our manoeuvres will involve the following fact:

FACT 4. *Let $(A, \mathfrak{M})$ be a local ring with $b, b_n \in A$ for $n \in \mathbb{N}$ such that $b_n \equiv b \bmod \mathfrak{M}^n$ and such that b_n enjoys a property $\mathcal{P}$ for infinitely many n. Then we may assume that b_n enjoys the property $\mathcal{P}$ for all n and that $b_n \equiv b \bmod \mathfrak{M}^n$.*

Proof. If b_n does not satisfy $\mathcal{P}$, then replace b_n with b_l, where $l = \min\{k \in \mathbb{N} \mid n \leq k \text{ and } b_k \text{ satisfies } \mathcal{P}\}$. □

To avoid cumbersome extra notation, we shall frequently avail ourselves of the following fact:

FACT 5. *Let $\mathfrak{p} \in X^1(A)$, $w \in A$, and $s \in \mathbb{P}$. Then the statement $w \in \mathfrak{p}^{(s)}$ is synonymous with an equation. (Namely, $w = \sum_{i=1}^k c_i r_i$ for some $r_i \in A$, if $\mathfrak{p}^{(s)} = (c_1, \dots, c_k)A$.)*

Our main proof will hinge on two additional facts.

FACT 6. *Let $(A, \mathfrak{M})$ be a local Noetherian ring. Let $b, b_n \in A$ be ruled by $b_n \equiv b \bmod \mathfrak{M}^n$ for $n \in \mathbb{N}$.*

(i) *Then for each $\mathfrak{p} \in X^1(A)$ there are positive integers k and m for which $b \in \mathfrak{p}^{(m)} \backslash \mathfrak{p}^{(m+1)}$ and $b_n \notin \mathfrak{p}^{(m+1)}$ for all $n \geq k$. (Obviously, k and m depend on $\mathfrak{p} \in X^1(A)$.)*

(ii) *If $b_n \in \mathfrak{p}^{(t)}$ for some $t \in \mathbb{N}$, for $n \gg 0$, then $b \in \mathfrak{p}^{(t)}$.*

Proof. Let $\mathfrak{I} \subseteq A$ be an ideal of A with $b \notin \mathfrak{I}$. Then there is a $k \in \mathbb{N}$ such that $b \notin \mathfrak{I} + \mathfrak{M}^k$. Thus $b_n \notin \mathfrak{I}$ for $n \geq k$.

(i) Since $\bigcap_{n=0}^{\infty} \mathfrak{p}^{(n)} = (0)$, there is an $m \in \mathbb{N}$ with $b \in \mathfrak{p}^{(m)} \backslash \mathfrak{p}^{(m+1)}$. Let $\mathfrak{I} = \mathfrak{p}^{(m+1)}$.

(ii) Let $\mathfrak{I} = \mathfrak{p}^{(t)}$. □

FACT 7. *Suppose that $(A, \mathfrak{M})$ is a local, normal approximation domain with $a \in \mathfrak{M}$. Let $x \in \widehat{A}$ and $x_n \in A$ be ruled by $x_n \equiv x \bmod \widehat{\mathfrak{M}}^n$ for all $n \in \mathbb{N}$.*

(i) *If x is irreducible in $\widehat{A}$, then x_n is irreducible in A for all $n \gg 0$.*

(ii) *If $x \notin \mathcal{U}(\widehat{A}_a)$, then $x_n \notin \mathcal{U}(A_a)$ for $n \gg 0$.*

Proof. (i) By way of contradiction, let $x_n = c_n d_n$ with $c_n, d_n \in A \backslash \mathcal{U}(A)$ for infinitely many n. We can then assume that this state of affairs holds for all n.

Thus $x \equiv c_n d_n \bmod \widehat{\mathfrak{M}}^n$ for all n. Since $\widehat{A}$ is a strong approximation ring, then there are $c, d \in \widehat{A}$ with $x = cd$ and $(c_n, d_n) \equiv (c, d) \bmod \widehat{\mathfrak{M}}^n$ for $n \gg 0$. This duet of congruences together with the assumption that $c_n, d_n \in \mathfrak{M}$ (for all n) force the conclusion that $c, d \in \mathfrak{M}$, a contradiction to the irreducibility of x in $\widehat{A}$. Hence, x_n is irreducible in A for $n \gg 0$.

(ii) Let $aA = \bigcap_{i=1}^s \mathfrak{p}_i^{(l_i)}$, where $\mathfrak{p}_i \in X^1(A)$ and $l_i \in \mathbb{P}$ for $i = 1, \dots, s$. By way of contradiction, assume that $x_n \in \mathcal{U}(A_a)$ for infinitely many (and, hence, for all) $n \in \mathbb{N}$. Then $x_n A = \bigcap_{i=1}^s \mathfrak{p}_i^{(m_{in})}$, where $m_{in} \in \mathbb{N}$. For each $i \in \{1, \dots, s\}$, there is an $m_i \in \mathbb{N}$ such that $m_{in} = m_i$ for $n \gg 0$ by Fact 6. So there are $x_0, u_n, v_n \in A$ for $n \gg 0$ such that $x_0 A = \bigcap_{i=1}^s \mathfrak{p}_i^{(m_i)}$, $x_n = x_0 u_n$ and $1 = u_n v_n$. Then $(x, 1) \equiv (x_0 u_n, u_n v_n) \bmod \widehat{\mathfrak{M}}^n$. Since $\widehat{A}$ is a strong approximation ring, then there are $u, v \in \widehat{A}$ such that $x = x_0 u$ and $uv = 1$ with $(u_n, v_n) \equiv (u, v) \bmod \widehat{\mathfrak{M}}^n$ for $n \gg 0$. But $x_0 \in \mathcal{U}(\widehat{A}_a)$ and $u \in \mathcal{U}(\widehat{A})$, contradicting that $x \notin \mathcal{U}(\widehat{A}_a)$. So $x_n \notin \mathcal{U}(A_a)$ for $n \gg 0$. □

Finally we turn to a paper by Rotthaus [14] in which the author showed that approximation rings are both excellent and Henselian. This result carries with it some crucial consequences:

COROLLARY 2. *A local, normal approximation domain is analytically normal.*

COROLLARY 3. *Let* $(A, \mathfrak{M})$ *be an approximation ring and* $\mathfrak{p} \in \operatorname{Spec}(A)$. *Then* $\widehat{\mathfrak{p}} \in \operatorname{Spec}(\widehat{A})$.

Proof. Note that $A/\mathfrak{p}$ is local, reduced, and excellent. Let $(A/\mathfrak{p})'$ stand for the integral closure of $A/\mathfrak{p}$ in its total field of fractions. By [10, 7.8.3.1 (vii)], it is known that:

(i) $\widehat{A}/\widehat{\mathfrak{p}}$ is reduced,
(ii) $(A/\mathfrak{p})'$ has the same number of maximal ideals as $\widehat{A}/\widehat{\mathfrak{p}}$ has minimal prime ideals, and
(iii) $(A/\mathfrak{p})'$ is a finitely generated $A/\mathfrak{p}$-module.

Since $A/\mathfrak{p}$ is Henselian, then $(A/\mathfrak{p})'$ is a product of a finite number of local rings by (iii).

On the other hand, $(A/\mathfrak{p})'$ is a domain, and hence local. By (ii), $\widehat{A}/\widehat{\mathfrak{p}}$ has only one minimal prime. By (i), $\widehat{\mathfrak{p}} \in \operatorname{Spec}(\widehat{A})$. □

3. Descent of $\operatorname{Cl}(\widehat{A})$

At this point, we are in the position to plunge into the proof.

THEOREM 3.1. *Let* $(A, \mathfrak{M})$ *be a local, normal, approximation domain with finite divisor class group. Then the canonical group monomorphism* $i : \operatorname{Cl}(A) \longrightarrow \operatorname{Cl}(\widehat{A})$, *ruled by* $i(\operatorname{cl}(\mathfrak{a})) = \operatorname{cl}(\widehat{\mathfrak{a}})$, *is an isomorphism.*

Proof. By Corollary 1, there is an element $a \in A$ for which A_a is an unique factorization domain. Let $\mathfrak{p}_1, \ldots, \mathfrak{p}_s \in X^1(A)$ and $l_1, \ldots, l_s$ be positive integers such that $aA = \bigcap_{i=1}^{s} \mathfrak{p}_i^{(l_i)}$. It follows from Theorem 2 that $\operatorname{Cl}(A) = \langle \operatorname{cl}(\mathfrak{p}_i) \mid i = 1, \ldots, s \rangle$.

Also Corollary 3 implies that $a\widehat{A} = \bigcap_{i=1}^{s} \widehat{\mathfrak{p}}_i^{(l_i)}$.

If $\widehat{A}_a$ is an unique factorization domain, then $\operatorname{Cl}(\widehat{A}) = \langle \operatorname{cl}(\widehat{\mathfrak{p}}_i) \mid i = 1, \ldots, s \rangle$, and hence $i : \operatorname{Cl}(A) \longrightarrow \operatorname{Cl}(\widehat{A})$ is an isomorphism. Therefore, our goal is to establish that $\widehat{A}_a$ is an unique factorization domain. Since $\widehat{A}_a$ is a Noetherian domain, it suffices to show that any irreducible element x in $\widehat{A}_a$ is a prime element in $\widehat{A}_a$. Furthermore, one can reduce the problem to proving that x divides either u or v in $\widehat{A}_a$ whenever $x \in \widehat{A}$ is an irreducible element of both $\widehat{A}$ and $\widehat{A}_a$ and $u, v \in \widehat{A}$ are such that $uv = xy$ for some $y \in \widehat{A}$.

Let $x \in \widehat{\mathfrak{p}}_i^{(m_i)} \backslash \widehat{\mathfrak{p}}_i^{(m_i+1)}$ and $u \in \widehat{\mathfrak{p}}_i^{(o_i)} \backslash \widehat{\mathfrak{p}}_i^{(o_i+1)}$, where $\mathfrak{p}_1, \ldots, \mathfrak{p}_s \in X^1(A)$ with $aA = \bigcap_{i=1}^{s} \mathfrak{p}_i^{(l_i)}$. As A is an approximation ring, there are elements u_n, v_n, x_n, y_n of A such that $u_n v_n = x_n y_n$, $x_n \in \mathfrak{p}_i^{(m_i)} \backslash \mathfrak{p}_i^{(m_i+1)}$ and $u_n \in \mathfrak{p}_i^{(o_i)} \backslash \mathfrak{p}_i^{(o_i+1)}$, and $(u_n, v_n, x_n, y_n) \equiv (u, v, x, y) \bmod \widehat{\mathfrak{M}}^n$ (see Facts 5 and 6). Also, x_n is irreducible in A and $x_n \notin \mathcal{U}(A_a)$ for $n \gg 0$ by Fact 7.

At this point we turn to the primary decomposition of $x_n A$. For each n, there are positive integers $m_n, k_{1n}, k_{2n}, \ldots, k_{m_n n}$ and distinct prime ideals $P_{1n}, \ldots, P_{m_n n} \in X^1(A)$ such that $x_n A = \bigcap_{j=1}^{m_n} P_{jn}^{(k_{jn})}$. Our strategy will consist of "stabilizing" m_n and k_{jn}, namely proving that there are positive integers $m, k_1, \ldots, k_m$ such that $x_n A = \bigcap_{j=1}^{m} P_{jn}^{(k_j)}$ for $n \gg 0$.

Suppose that $|\operatorname{Cl}(A)| = t$.

We shall first bound m_n:

Let $\mathcal{S}_n = \{P_{jn} \mid j = 1, \ldots, m_n\}$ with $|\mathcal{S}_n| = m_n$. We partition $\mathcal{S}_n$ into a disjoint union $\mathcal{S}_n = \dot{\bigcup}_{g \in \operatorname{Cl}(A)} \mathcal{S}_{gn}$, where $\mathcal{S}_{gn} := \{P_{jn} \in \mathcal{S}_n \mid \operatorname{cl}(P_{jn}) = g\}$ and $m_{gn} := |\mathcal{S}_{gn}|$. (Here we are allowing $\mathcal{S}_{gn}$ to be the empty set for some $g \in \operatorname{Cl}(A)$, as would be the case if g is the identity in $\operatorname{Cl}(A)$, for instance.)

Let $\mathcal{H}$ be the subset $\{g \in \operatorname{Cl}(A) \mid \mathcal{S}_{gn} \neq \emptyset\}$ of $\operatorname{Cl}(A)$. From now on, any mention of g will presume that $g \in \mathcal{H}$, as opposed to the larger $\operatorname{Cl}(A)$. Since $|\operatorname{Cl}(A)| = t$, then $|\mathcal{H}| < t$.

We fix $\mathfrak{p}_{gn} \in \mathcal{S}_{gn}$ so that $\operatorname{cl}(\mathfrak{p}_{gn}) = g$. For purposes of inventory, I_{gn} will denote the set $\{j \mid 1 \leq j \leq m_n \text{ and } P_{jn} \in \mathcal{S}_{gn}\}$.

If $t \lneqq m_{gn}$, then let I'_{gn} be a fixed subset of I_{gn} of order t, and $I''_{gn} := I_{gn} \backslash I'_{gn}$.

Evocative of the notation in Fact 2, we shall let

(i) $\alpha_{jn} \in K$ be such that $P_{jn} = \alpha_{jn} \mathfrak{p}_{gn}$ for $j \in I_{gn}$,

(ii) $b_{gn} \in A$ be such that $\mathfrak{p}_{gn}^{(t)} = b_{gn} A$,

(iii) $\mathfrak{a}_{gn} := (\bigcap_{j \in I'_g} P_{jn}^{(k_{jn}-1)}) \bigcap (\bigcap_{j \in I''_g} P_{jn}^{(k_{jn})}) \subsetneqq A$ if $t \lneqq m_{gn}$, and
$\mathfrak{a}_{gn} := \bigcap_{j \in I_g} P_{jn}^{(k_{jn})} \subsetneqq A$ if $t \geq m_{gn}$,

(iv) $\gamma_{gn} := (\prod_{j \in I'_g} \alpha_{jn}) b_{gn} \in \bigcap_{j \in I'_g} P_{jn} \subset A \backslash \mathcal{U}(A)$ if $t \lneqq m_{gn}$, and
$\gamma_{gn} := 1$ if $t \geq m_{gn}$.

Then by Fact 2, $x_n A = (\prod_{g \in \mathcal{H}} \gamma_{gn}) \cdot (\bigcap_{g \in \mathcal{H}} \mathfrak{a}_{gn})$.

Since $\mathfrak{a}_{gn} \subseteq A$ for all $g \in \mathcal{H}$, then $\bigcap_{g \in \mathcal{H}} \mathfrak{a}_{gn} = \delta_n A$ for some $\delta_n \in A$. So $x_n = (\prod_{g \in \mathcal{H}} \gamma_{gn}) \cdot \delta_n \cdot u_n$ for some $u_n \in \mathcal{U}(A)$.

If $t \lneqq m_{gn}$ for at least one $g \in \mathcal{H}$, then $\delta_n \in A \backslash \mathcal{U}(A)$ and $\gamma_{gn} \in A \backslash \mathcal{U}(A)$, contradicting that x_n is irreducible in A for $n \gg 0$. Thus, $m_{gn} \leq t$ for all $g \in \mathcal{H}$, for $n \gg 0$.

But $m_n = \sum_{g \in \mathcal{H}} m_{gn} \leq t \cdot (t-1)$ since $|\mathcal{H}| < t$ for $n \gg 0$.

As $t^2 - t < \infty$, then there is an $m \in \{1, 2, \ldots, t^2 - t\}$ such that $m_n = m$ for $n \gg 0$ as per Fact 4.

So $x_nA = \bigcap_{j=1}^{m} P_{jn}^{(k_{jn})}$ for $n \gg 0$.

We now bound k_{jn}:

As before, we can find $b_{jn} \in A$ with $P_{jn}^{(t)} = b_{jn}A$. Let $q_{jn}, r_{jn} \in \mathbb{N}$ be such that $k_{jn} = t \cdot q_{jn} + r_{jn}$ with $0 \leq r_{jn} \lneqq t$. Then $x_n = b_n \cdot c_n$, where $b_n := \Pi_{j=1}^{m} b_{jn}^{q_{jn}}$ and $c_nA = \bigcap_{j=1}^{m} P_{jn}^{(r_{jn})}$. Given that x_n is irreducible for $n \gg 0$, it would seem that either $q_{jn} = 0$ for $j = 1, 2, \ldots, m$; or that $m = 1$ and k_{1n} equals the order of $\mathrm{cl}(P_{1n})$ in $\mathrm{Cl}(A)$. In either case, $k_{jn} < t$ for $j = 1, \ldots, m$. Since $t < \infty$, then there is a $k_j \in \{1, 2 \ldots, t\}$ such that $k_{jn} = k_j$ for $n \gg 0$ as per Fact 4.

So $x_nA = \bigcap_{j=1}^{m} P_{jn}^{(k_j)}$ for $n \gg 0$.

Without loss of generality, we may assume that $P_{jn} \notin \{\mathfrak{p}_1, \ldots, \mathfrak{p}_s\}$ for $j = 1, \ldots, r_n$ and that $P_{jn} \in \{\mathfrak{p}_1, \ldots, \mathfrak{p}_s\}$ for $j = r_n + 1, \ldots, m$. (We observe that $1 \leq r_n \leq m$ as $x_n \notin \mathcal{U}(A_a)$.) Further, since $m < \infty$, there is an $r \in \{1, \ldots, m\}$ such that $r_n = r$ for $n \gg 0$ as per Fact 4.

Then there exist $\kappa_i \in \{0, 1, \ldots, t-1\}$ such that

$$(\dagger) \qquad x_nA = \bigcap_{j=1}^{r} P_{jn}^{(k_j)} \bigcap \left(\bigcap_{i=1}^{s} \mathfrak{p}_i^{(\kappa_i)} \right) \quad \text{for } n \gg 0.$$

Since A_a is an unique factorization domain, there are $z_{jn} \in A$ with $z_{jn}A_a = P_{jn}A_a$ for $j = 1, \ldots, r$. It follows that $z_{jn}A = P_{jn} \bigcap (\bigcap_{i=1}^{s} \mathfrak{p}_i^{(m_{ijn})})$ with $m_{ijn} \in \mathbb{N}$ for $j = 1, \ldots, r$. We may assume that z_{jn} is irreducible in A for $j = 1, \ldots, r$, thereby forcing (by the same line of reasoning as immediately above) that $m_{ijn} < t$ for all $i = 1, \ldots, s$ and all $j = 1, \ldots, m$. Since $t < \infty$, then there are $m_{ij} \in \{1, \ldots, t-1\}$ such that $m_{ijn} = m_{ij}$ for $n \gg 0$ as per Fact 4.

So $z_{jn}A = P_{jn} \bigcap (\bigcap_{i=1}^{s} \mathfrak{p}_i^{(m_{ij})})$ for $n \gg 0$ with $m_{ij} \in \{0, 1, \ldots, t-1\}$ for all i and j.

Let $e \in A \bigcap \mathcal{U}(A_a)$ be such that $eA := \bigcap_{i=1}^{s} \mathfrak{p}_i^{(\kappa_i')}$ with $\kappa_i' \geq \kappa_i$. Then

$$\begin{aligned}
e \cdot \left(\prod_{j=1}^{r} z_{jn}^{k_j} \right) A &= \left(\bigcap_{i=1}^{s} \mathfrak{p}_i^{(\kappa_i')} \right) \cdot \prod_{j=1}^{r} \left(P_{jn}^{(k_j)} \bigcap \left(\bigcap_{i=1}^{s} \mathfrak{p}_i^{(k_j \cdot m_{ij})} \right) \right) \\
&= \left(\left(\left(\prod_{j=1}^{r} P_{jn}^{(k_j)} \right) \cdot \left(\prod_{i=1}^{s} \mathfrak{p}_i^{(\kappa_i)} \right) \right)^{**} \cdot \left(\prod_{i=1}^{s} \mathfrak{p}_i^{\left(\kappa_i' - \kappa_i + \sum_{j=1}^{r} k_j \cdot m_{ij}\right)} \right)^{**} \right)^{**} \\
&= \left(\left(\left(\bigcap_{j=1}^{r} P_{jn}^{(k_j)} \right) \bigcap \left(\bigcap_{i=1}^{s} \mathfrak{p}_i^{(\kappa_i)} \right) \right) \cdot \left(\bigcap_{i=1}^{s} \mathfrak{p}_i^{(\kappa_i' - \kappa_i + \sum_{j=1}^{r} k_j \cdot m_{ij})} \right) \right)^{**} \\
&= x_n \cdot \left(\bigcap_{i=1}^{s} \mathfrak{p}_i^{\left(\kappa_i' - \kappa_i + \sum_{j=1}^{r} k_j \cdot m_{ij}\right)} \right)
\end{aligned}$$

by Fact 1.

Let $t_i := \kappa_i' - \kappa_i + \sum_{j=1}^{r} k_j \cdot m_{ij}$. Then $e \cdot (\prod_{j=1}^{r} z_{jn}^{k_j})A = x_n \cdot (\bigcap_{i=1}^{s} \mathfrak{p}_i^{(t_i)})$ implies that there are $d \in A \bigcap \mathcal{U}(A_a)$ and μ_n , $\eta_n \in A$ such that for $n \gg 0$:

$$dA = \bigcap_{i=1}^{s} \mathfrak{p}_i^{(t_i)},$$

$$(*) \qquad e \cdot \left(\prod_{j=1}^{r} z_{jn}^{k_j} \right) = x_n \cdot \mu_n \cdot d,$$

$$\mu_n \cdot \eta_n = 1.$$

Then $e \cdot (\prod_{j=1}^{r} z_{jn}^{k_j}) \equiv x \cdot \mu_n \cdot d \bmod \widehat{\mathfrak{M}}^n$. Since $z_{nj} \in \mathfrak{p}_i^{(m_{ij})}$, by Fact 5, we also have $z_{nj} \in \widehat{\mathfrak{p}_i}^{(m_{ij})} \bmod \widehat{\mathfrak{M}}^n$. Using the strong approximation property of $\widehat{A}$, it follows that there are elements $\tilde{z}_{1n}, \ldots, \tilde{z}_{rn}, \tilde{\mu}_n, \tilde{\eta}_n \in \widehat{A}$ such that:

$$(**) \qquad e \cdot \left(\prod_{j=1}^{r} \tilde{z}_{jn}^{k_j} \right) = x \cdot \tilde{\mu}_n \cdot d,$$

$$\tilde{\mu}_n \cdot \tilde{\eta}_n = 1,$$

$$(***) \qquad \tilde{z}_{jn} \in \widehat{\mathfrak{p}}_i^{(m_{ij})} \quad \text{for } j = 1, \ldots, r,$$

$$(\tilde{z}_{1n}, \ldots, \tilde{z}_{rn}, \tilde{\mu}_n, \tilde{\eta}_n) \equiv (z_{1n}, \ldots, z_{rn}, \mu_n, \eta_n) \bmod \widehat{\mathfrak{M}}^{\theta^{-1}(n)}$$

for $n \gg 0$. (Here, $\theta : \mathbb{N} \to \mathbb{N}$ is the appropriate strong approximation function.)

Since $\tilde{\mu}_n \cdot d \in \mathcal{U}(\widehat{A}_a)$ and x is irreducible in $\widehat{A}_a$, then all but one of $\tilde{z}_{1n}, \ldots, \tilde{z}_{rn}$ are in $\mathcal{U}(\widehat{A}_a)$. Since $r < \infty$, then there exists a $j \in \{1, \ldots, r\}$ such that $\tilde{z}_{jn} \notin \mathcal{U}(\widehat{A}_a)$, for $n \gg 0$ as per Fact 4. Without loss of generality, let $j = 1$. By the irreducibility of x in $\widehat{A}_a$, $k_1 = 1$.

Since x and d are fixed and $\tilde{\mu}_n \in \mathcal{U}(\widehat{A})$, then there exist $\alpha_i \in \mathbb{N}$ such that $x \cdot \tilde{\mu}_n \cdot d \in \widehat{\mathfrak{p}}_i^{(\alpha_i)} \backslash \widehat{\mathfrak{p}}_i^{(\alpha_i+1)}$ for $i = 1, \ldots, s$. By equation $(**)$, it follows that $\tilde{z}_{jn} \notin \widehat{\mathfrak{p}}_i^{(\alpha_i+1)}$ for $i = 1, \ldots, s$, and $j = 1, \ldots, r$, for $n \gg 0$. Since $\alpha_i < \infty$ for all i, then there are $\beta_{ij} \in \{0, \ldots, \alpha_i\}$ for all $j = 1, \ldots, r$ such that $\tilde{z}_{jn} \in \widehat{\mathfrak{p}}_i^{(\beta_{ij})} \backslash \widehat{\mathfrak{p}}_i^{(\beta_{ij}+1)}$ for $n \gg 0$ as per Fact 4.

By $(***)$, $\beta_{ij} \geq m_{ij}$ for all i and j. We observe that $v_{\mathfrak{p}_i}(x_n \cdot \mu_n \cdot d) = v_{\widehat{\mathfrak{p}}_i}(x \cdot \tilde{\mu}_n \cdot d)$, since $\mu_n \in \mathcal{U}(A)$, $\tilde{\mu}_n \in \mathcal{U}(\widehat{A})$, and $x_n \in \mathfrak{p}_i^{(m_i)} \backslash \mathfrak{p}_i^{(m_i+1)}$, $x \in \widehat{\mathfrak{p}}_i^{(m_i)} \backslash \widehat{\mathfrak{p}}_i^{(m_i+1)}$. So

$$\sum_{j=1}^{r} k_j m_{ij} = v_{\mathfrak{p}_i} \left(\prod_{j=1}^{r} z_{jn}^{k_j} \right) = v_{\mathfrak{p}_i}(x_n \cdot \mu_n \cdot d) - v_{\mathfrak{p}_i}(e)$$

$$= v_{\widehat{\mathfrak{p}}_i}(x \cdot \tilde{\mu}_n \cdot d) - v_{\widehat{\mathfrak{p}}_i}(e) = v_{\widehat{\mathfrak{p}}_i} \left(\prod_{j=1}^{r} \tilde{z}_{jn}^{k_j} \right) = \sum_{j=1}^{r} k_j \beta_{ij},$$

and thus $m_{ij} = \beta_{ij}$ for all i and j.

Since $\tilde{z}_{jn} \in \mathcal{U}(\widehat{A}_a)$ for $j \in \{2, \dots, r\}$, we can find $z_j \in A$,and $\gamma_{jn}, \delta_{jn} \in \widehat{A}$ for $j \in \{2, \dots, r\}$ such that $z_j A = \bigcap_{i=1}^{s} \mathfrak{p}_i^{(m_{ij})}$, $\tilde{z}_{jn} = z_j \gamma_{jn}$ and $\gamma_{jn} \cdot \delta_{jn} = 1$. Then $z_{jn} \equiv z_j \gamma_{jn} \bmod \widehat{\mathfrak{M}}^{\theta^{-1}(n)}$ for $n \gg 0$ and $j \in \{2, \dots, r\}$.

Hence, upon a suitable shift of the index n up, as per Fact 3, $z_{jn} \equiv z_j \gamma_{jn} \bmod \widehat{\mathfrak{M}}^n$ and $\gamma_{jn}\delta_{jn} = 1$ for $n \gg 0$.

So for $j \in \{2, \dots, r\}$, we have $z_{jn} A = z_j \cdot P_{jn}$. This equality implies that P_{jn} is principal, say $P_{jn} = p_{jn} A$ for some $p_{jn} \in P_{jn}$.

By way of contradiction we assume that $r \geq 2$. Equation (†) and the fact that $k_1 = 1$ produce the new equation

$$\begin{aligned} x_n A &= P_{1n} \bigcap \left(\bigcap_{j=2}^{r} P_{jn}^{(k_j)} \right) \bigcap \left(\bigcap_{i=1}^{s} \mathfrak{p}_i^{(\kappa_i)} \right) \\ &= \left(P_{1n} \bigcap \left(\bigcap_{i=1}^{s} \mathfrak{p}_i^{(\kappa_i)} \right) \right) \cdot \left(\prod_{j=2}^{r} p_{jn}^{k_j} \right). \end{aligned}$$

Thus $P_{1n} \bigcap (\bigcap_{i=1}^{s} \mathfrak{p}_i^{(\kappa_i)})$ is a principal ideal generated over A by some $d_n \in A$ such that $x_n = d_n \cdot (\prod_{j=2}^{r} p_{jn}^{k_j})$. Note that $d_n \in A \backslash \mathcal{U}(A)$ since $d_n \in P_{1n}$ and that $p_{jn} \in A \backslash \mathcal{U}(A)$ for $j = 2, \dots, r$ since $p_{jn} \in P_{jn}$, contradicting the irreducibility of x_n in A.

Therefore, $r = 1$. As $x_n \notin \mathcal{U}(A_a)$, then $P_{1n} \notin \{\mathfrak{p}_1, \dots, \mathfrak{p}_s\}$. Also since $x_n \in \mathfrak{p}_i^{(m_i)} \backslash \mathfrak{p}_i^{(m_i+1)}$, then $\kappa_i = m_i$ for all i, and so $x_n A = P_{1n} \bigcap (\bigcap_{i=1}^{s} \mathfrak{p}_i^{(m_i)})$. Thus x_n is a prime element in A_a for $n \gg 0$. Recalling the equations $x_n y_n = u_n v_n$, we conclude then that x_n divides u_n or v_n in A_a. We may assume that x_n divides u_n in A_a for infinitely many (and, hence, for all) $n \gg 0$. So $x_n c_n = u_n a^{s_n}$ for some $c_n \in A$ not divisible by a, for some $s_n \in \mathbb{N}$ and $n \gg 0$.

Recall that $u_n \in \mathfrak{p}_i^{(o_i)} \backslash \mathfrak{p}_i^{(o_i+1)}$ for $n \gg 0$, and $a \in \mathfrak{p}_i^{(l_i)} \backslash \mathfrak{p}_i^{(l_i+1)}$ with $l_i \geq 1$. Then $v_{\mathfrak{p}_i}(c_n) = s_n l_i + o_i - m_i \geq 0$ for all i since $c_n \in A$. On the other hand, since a does not divide c_n, there must be an $i_0 \in \{1, \dots, s\}$ such that $s_n l_{i_0} + o_{i_0} - m_{i_0} \lneqq l_{i_0}$.

Let

$$\begin{aligned} \mathcal{S} &:= \{\sigma \in \mathbb{N} \mid \sigma l_i + o_i - m_i \geq 0 \quad \text{for all } i = 1, \dots, s\}, \\ \mathcal{T} &:= \{\sigma \in \mathbb{N} \mid \sigma l_i + o_i - m_i \lneqq l_i \quad \text{for some } i = 1, \dots, s\}, \end{aligned}$$

Then $s_n \in \mathcal{S} \bigcap \mathcal{T}$. If we let $s_0 := \min \mathcal{S}$, it can be shown that either $s_n = 0$ for $n \gg 0$, or that $\mathcal{S} \bigcap \mathcal{T} = \{s_o\}$. Let $s = 0$ in the former case, and let $s = s_0$ in the latter. We now turn to the new equations: $x_n c_n = u_n a^s$ to conclude that $x c_n \equiv u a^s \bmod \widehat{\mathfrak{M}}^n$. By the strong approximation property of $\widehat{A}$ we can find $c \in \widehat{A}$ such that $xc = ua^s$ and $c_n \equiv c \bmod \widehat{\mathfrak{M}}^{\theta^{-1}(n)}$ for $n \gg 0$. Thus x divides u in $\widehat{A}_a$, which is what we set out to prove. □

References

[1] J. Bingener, *Divisorenklassengruppen der Komplettierungen analytischer Algebren*, Math. Ann. **217** (1975), 113–120. MR 0379908 (52 #812)

[2] J. Bingener and U. Storch, *Zur Berechnung der Divisorenklassengruppen kompletter lokaler Ringe*, Nova Acta Leopoldina (N.F.) **52** (1981), 7–63. MR 642696 (83m:13017)

[3] N. Bourbaki, *Commutative Algebra*, Hermann, Paris, 1969.

[4] V.I. Danilov, *The group of ideal classes of a completed ring*, Mat. Sb. **77 (119)** (1968), 533–541. MR 0237483 (38 #5765)

[5] ———, *On a conjecture of Samuel*, Mat. Sb. **81 (123)** (1970), 132–144. MR 0282980 (44 #214)

[6] ———, *Rings with a discrete group of divisor classes*, Mat. Sb. (N.S.) **83 (125)** (1970), 372–389. MR 0282980 (44 #214)

[7] ———, *On rings with a discrete divisor class group*, Mat. Sb. **88 (130)** (1972), 229-237. MR 0306184 (46 #5311)

[8] R. M. Fossum, *The divisor class group of a Krull domain*, Ergebnisse der Mathematik und ihrer Grenzgebiete, vol. 74, Springer-Verlag, New York, 1973. MR 0382254 (52 #3139)

[9] P. Griffith and D. Weston, *Restrictions of torsion divisor classes to hypersurfaces*, J. Algebra **167** (1994), 473–487. MR 1283298 (95c:13008)

[10] A. Grothendieck, *Éléments de géométrie algébrique. IV. Étude locale des schémas et des morphismes de schémas. II*, Inst. Hautes Études Sci. Publ. Math. **24** (1965). MR 0199181 (33 #7330)

[11] H. Matsumura, *Commutative algebra*, W. A. Benjamin, Inc., New York, 1970. MR 0266911 (42 #1813)

[12] M. Nagata, *Local rings*, Interscience Tracts in Pure and Applied Mathematics, No. 13, John Wiley & Sons, New York-London, 1962. MR 0155856 (27 #5790)

[13] G. Pfister and D. Popescu, *Die strenge Approximationseigenschaft lokaler Ringe*, Invent. Math. **30** (1975), 145–174. MR 0379490 (52 #395)

[14] C. Rotthaus, *Rings with approximation property*, Math. Ann. **287** (1990), 455–466. MR 1060686 (91f:13025)

[15] ———, *Divisorial ascent in rings with approximation property*, J. Algebra **178** (1995), 541–560. MR 1359902 (97b:13026)

[16] M. van der Put, *A problem on coefficient fields and equations over local rings*, Compositio Math. **30** (1975), 235–258. MR 0384796 (52 #5668)

Dana Weston, Department of Mathematics, University of Missouri, Columbia, Missouri 65211, USA

E-mail address: weston@math.missouri.edu